Decontamination and Decommissioning of Nuclear Facilities

Decontamination and Decommissioning of Nuclear Facilities

Edited by
Marilyn M. Osterhout
Argonne National Laboratory
Idaho Falls, Idaho

PLENUM PRESS • NEW YORK AND LONDON

Library of Congress Cataloging in Publication Data

Main entry under title:

Decontamination and decommissioning of nuclear facilities.

"Proceedings of the American Nuclear Society topical meeting in Sun Valley, Idaho, September 16–20, 1979."

Sponsored by the American Nuclear Society's Reactor Operations Division and the American Nuclear Society's Eastern Idaho Section.

1. Nuclear facilities–Decommissioning–Congresses. 2. Radioactive decontamination–Congresses. I. Osterhout, Marilyn M. II. American Nuclear Society. Reactor Operations Division. III. American Nuclear Society. Eastern Idaho Section.

TK9151.4.D42 621.48'35 80-10223

ISBN 0-306-40429-X

Proceedings of the American Nuclear Society Topical Meeting in Sun Valley, Idaho, September 16–20, 1979.

A Division of Plenum Publishing Corporation
227 West 17th Street, New York, N.Y. 10011

Printed in the United States of America

Preface

This volume contains the invited and contributed papers presented at the American Nuclear Society (ANS) meeting on Decontamination and Decommissioning (D & D) of Nuclear Facilities, held September 16-20, 1979, in Sun Valley, Idaho. This was the first U. S. meeting of the ANS which addressed both of these important and related subjects. The meeting was attended by more than 400 engineers, scientists, laymen, and representatives of federal, state, and local governments, including participants from eleven foreign countries.

The technical sessions included several sessions concentrating on ongoing D & D programs in the U. S. and abroad. In addition, "new ground" was broken in such areas as decommissioning costs and cost recovery, advanced programs on reactor coolant filtration, and other areas of continuing and increasing importance to the nuclear industry and to consumers.

The dual sponsorship of the meeting (The ANS Reactor Operations Division and the Eastern Idaho Section of the ANS) helped spur a high quality program, a pleasant location, and a high degree of success in technical interchange between the attendees. As guest speaker, we were honored to have Mr. Vince Boyer of Philadelphia Electric Company. Mr. Boyer is both a past chairman of the ANS Reactor Operations Division and a past president of the American Nuclear Society. His views on the nuclear industry and of its current status were informative and interesting. We were also pleased to have four representatives of several public utilities in a special plenary session during which the recovery programs at Three Mile Island Unit 2 were discussed.

There are many elements which contribute to the overall success of meetings like this one. Companies dedicate large amounts of employee time in the planning and execution phases, many events are sponsored, and - last but not least - attendees and speakers from across the globe come together to share experiences and to learn. For the general and program committees, the work begins 24 months before the meeting and continues until about four months after. We would, therefore, like to take this opportunity to show our appreciation to the committee members for a job well done - in all respects - and to thank the companies which provided the support of their activities.

General Program:

Chairman	Glen A. Mortensen Intermountain Technologies, Inc.
Assistant Chairman	Willis R. Young DOE Idaho
Arrangements	Earl E. Burdick EG&G Idaho
Registration	Clyde R. Toole EG&G Idaho
Finance	Joseph W. Henscheid EG&G Idaho
Transportation and Tours	Ormand L. Cordes Exxon Nuclear Idaho Co., Inc.
Publications	Marilyn M. Osterhout Argonne National Laboratory - West
Public Information	Carl F. Obenchain EG&G Idaho
Guest Program	Patricia L. Mortensen
I.S.U. Liaison	V. P. Hary Charyulu Idaho State University
Meeting Assistance	Helen Brown Intermountain Technologies, Inc.

Technical Program:

Cochairman	R. Jon Stouky QUADREX/NSC
	Philip M. Garrett Technology for Energy Corporation
Assistant Chairman	Richard H. Meservey EG&G Idaho

Committee Members

Harry E. Hootman
Savannah River Lab.

Norm Ziomek
Pacific Gas & Electric Co.

Tom Drolet
Ontario Hydro

Bob Bernero
U. S. NRC

Richard I. Smith
Battelle Pacific Northwest Lab.

Robert Shaw
Electric Power Research Institute

Dewy Lange
Oak Ridge National Lab.

Bill Manion
Nuclear Engineering Services Inc.

George Wagner
Commonwealth Edison Company

Paul Pettitt
Department of Energy

Henry L. Bermanis
United Engineers & Constructors Inc.

Jerome P. Kane
Catalytic Inc.

Charles M. Rice
Energy Incorporated

Paul F. McTigue
Consolidated Edison Company of N. Y., Inc.

Peter F. Santoro
Northeast Utilities Service Co.

Ken Rohde
Exxon Nuclear Idaho Co., Inc.

We also gratefully acknowledge those companies who generously contributed to the sponsorship of various events and functions at the meeting:

EG&G Idaho
NUS Corp.
Allied Chemical
Energy Incorporated
Nuclear Engineering Services

It may not be obvious that one individual performs a significant part of the work in preparing the manuscripts for the trip to the printer. However, in our case, a very high degree of journalistic, coercive, and organizational skill was continually demonstrated by Marilyn Osterhout, while maintaining her sense of humor. We are very grateful for her able assistance.

We are hopeful that this book will be a valuable reference document for you, and we look forward to seeing you at the next D & D meeting.

Glen A. Mortensen
General Chairman

R. Jon Stouky
Technical Program Cochairman

Philip M. Garrett
Technical Program Cochairman

Contents

POWER PLANT DECONTAMINATION (PART II)

POWER REACTOR DECOMMISSIONING (PART I)

POWER REACTOR DECOMMISSIONING (PART II)

DECONTAMINATION AND DECOMMISSIONING OF FUEL CYCLE AND RESEARCH FACILITIES (PART II)

RADIATION FIELD AND EXPOSURE CONTROL

ECONOMIC ASPECTS OF POWER REACTOR DECOMMISSIONING

DECONTAMINATION

THE UTILITY VIEWPOINT

W. P. Worden

Commonwealth Edison

Chicago, Illinois 60690

Decontamination, or chemical cleaning, has always tended to be a controversial topic. In the early years of the industry, decontamination was taken by some vendors to be a fact of life and was considered by them in reactor design. The active literature in the 50's and early 60's reflects this. However, during the latter part of the 60's and much of the 70's, it became a non-issue in this country largely because of the emphasis on rapid expansion in the industry. As operating experience grew and problems in plant operation, maintenance and testing developed, interest in decontamination was reborn.

Probably the most important causes of this rebirth in interest were the concerns of some utilities over the increased man-rem of radiation exposure being experienced and the potential impact of the inservice inspection that was being required by Section XI of the ASME Boiler and Pressure Vessel Code.

The first serious decontamination program since the early 60's in the United States was begun at Dresden Unit 1 in 1973. This project was undertaken mainly to reduce radiation levels in the primary system so as to allow for plant modification and to provide for inservice inspection. Since the start of this project, others have been considered and at least two have been funded and have made significant progress. One of these is the Commonwealth Edison -- Department of Energy, Low Level, BWR primary system decontamination and the other is the Consolidated Edison PWR Steam Generator Chemical Cleaning project at the Indian Point Plant.

There is as yet no clear consensus as to what the future holds in this technology. However, some of us are convinced that decontamination can and will be found to be safe and cost effective and will follow the experience of chemical cleaning of fossil boilers. The early years of chemical cleaning of fossil plants was also controversial but is now a routine and cost effective part of the business.

A REVIEW OF RECENT D & D PROGRAM ACTIVITIES IN THE INTERNATIONAL COMMUNITY

W. J. Manion, Division President

Nuclear Energy Services, Inc.
Danbury, Connecticut 06810

I will report on the first international symposium on decommissioning of nuclear facilities; the work accomplished by the International Atomic Energy Agency on decommissioning; and on current programmatic activities on the subject in the international community.

A symposium sponsored by the International Atomic Energy Agency and the Nuclear Energy Agency was held in Vienna, Austria from November 13-17 of 1978. There were 225 participants from 26 countries, including the Vatican, who listened to 41 individual presentations. The scope of the symposium covered both progress and problems associated with decommissioning. Eight individual sessions were held, covering: national policies and their international significance; policy and standards development; engineering considerations to decommissioning; radiological release considerations and waste classification; decommissioning experience; decontamination and remote operations. The symposium concluded with a panel discussion.

National policy, relative to decommissioning, is under study in many countries. The Federal Republic of Germany has already reached certain conclusions, namely: (1) that the licensing of nuclear power plants must include proof that the plants can be decommissioned and removed at the end of their operating lives; (2) that funds for decommissioning must be accumulated during the life of the plant; (3) that so-called combination modes, which place the facility in a temporary condition of protective storage for a period of thirty to fifty years, are acceptable before final dismantling.

Many decommissioning studies have been performed by the Member States of the IAEA. The studies include deferred dismantling for periods of up to 100 years. They have covered various reactor types, including the French Super Phenix 1, reprocessing plants, mixed oxide fuel fabrication plants, and uranium mill sites. These studies have characterized the nature and volume of the wastes resulting from decommissioning. The impact on the environment of these wastes has been evaluated, and various methodologies examined, to determine the maximum radioactivity level that would qualify for unrestricted use of the material. The methodologies include both pathway analyses for the residual radioactivity and absolute levels, independent of waste location or future use.

A number of papers were devoted to decommissioning experience. The largest decommissioning activity undertaken to date in the United Kingdom has been the partial decommissioning and modification of the fast reactor irradiated fuel reprocessing plant at Dounreay. This program required approximately four years to complete and created about 700 cubic meters of waste. It demonstrated the feasibility of decommissioning an alpha contaminated facility in a high gamma field. A major project nearing completion in the United States is the dismantling of the Sodium Reactor Experiment (SRE). This removal program has extended the state-of-the-art of plasma arc cutting beyond that developed for the Elk River Reactor dismantling and has introduced underwater explosive cutting of difficult piping arrays. The French representatives reported that a total of eight reactors have been decommissioned in France with three of them being dismantled. The 35 MW_{th} Pagase reactor is the largest of this category. Representatives of Canada, Belgium, Switzerland, Italy, and India also reported on decommissioning experiences in their countries.

The French reported on a family of telemanipulators that have been used in decommissioning programs. Their applications have included television viewing; inspection and measurement; and decommissioning functions such as placement and control of plasma torches; and removal of radioactive material. The newest telemanipulators incorporate feedback mechanisms to impart a degree of "learning" to the unit.

The symposium ended with a panel discussion involving representatives of the United States, the Federal Republic of Germany, France, the United Kingdom and Belgium. The conclusions of this panel of experts were not unexpected. They concluded that decommissioning, including complete removal of a facility, is feasible and practical in all aspects. In addition, they chided regulatory agencies for procrastinating on the establishment of unrestricted release criteria, particularly as applied to facility decommissioning. They raised questions as to the financing of decommissioning programs, including: who should pay; who should be responsible for the accumulating fund; and what is a satisfactory financial resolution to a plant prematurely shutdown.

Finally, they pointed out that while much has been accomplished in the broad area of decommissioning, much remains to be done in the areas of demonstration of technical feasibility; development of reasonable regulations; disposal of radioactive waste material; plant design to accommodate dismantling; and recycling of waste material.

I would like to bring to your attention a paper presented at the symposium by Madame Anne Marie Chapuis. This paper was authored by Monsieur Jacquoemin and Madame Chapuis, and is entitled, "Criteria for Admissible Residual Activity." This was a particularly provocative paper in that it deals with one possible approach to unrestricted release criteria for radioactive materials. Simply stated, the paper presents a rationale for drawing the line between controlled and unrestricted use of radioactive materials, potentially resulting in much higher individual doses to the public than presently considered allowable by many countries. The examples presented demonstrated that an individual in the public could receive a dose as high as 100 millirem in any given year from recycled radioactive waste. The logic of this higher allowable dose is based upon a cost-benefit analysis in which a value to the public good is placed on the reclaimed radioactive material, for such uses as structural metal in a ship. In other words, this reclaimed material would be providing a benefit which exceeds the negative impact to the public of an increased cancer incidence. Obviously, different values placed on the reclaimed material would lead to varying allowable levels of exposure. Implementation of such position on a unilateral basis would result in chaos at the international level. This points out two important needs: first, that this type of criteria be developed with the cooperation and agreement of the Member States of the IAEA; and secondly, that further delay in establishing unrestricted release criteria for decommissioning will only cause further confusion through unilateral action. I must say that the paper given by Madame Chapuis was extremely well prepared and well presented and I encourage you to read it. I would be happy to arrange for you to receive copies if you wish.

IAEA TECHNICAL REPORT ON DECOMMISSIONING

The International Atomic Energy Agency convened a meeting in 1973 to examine the decommissioning of nuclear facilities. The Agency was advised by the participants that decommissioning activities should be introduced into the IAEA program and that the Agency should promote the formulation of guides, recommendations and standards for this purpose. A technical committee, comprised of decommissioning experts from the Member States of the IAEA, was organized and meetings were held in 1975, 1977 and 1978 for the purpose of developing and formulating a code and guide concerning the decommissioning of land-based nuclear reactors. The 1975 and 1977 meetings have been documented in IAEA technical documents 179 and 205. The 1978 meeting resulted in a draft code and guide for the decommissioning of land-based nuclear reactors. However, during the fall of 1978, some IAEA Member States strongly disagreed with issuing such a code and guide at the time. Apparently, these Member States felt that completion of the programs being pursued in their countries, was a necessary prerequisite to their acceptance of such a code and guide. As a result, the IAEA responded to their wishes and has decided to issue the work done by the technical committee in the form of a report in the IAEA Safety Series. Specifically, the report will be issued as an IAEA Safety Series Report, Category 4, Procedures and Data. A publication in this category contains information on procedures, techniques and criteria pertaining to safety matters; however, the report's contents are nonbinding for the Member States. Comments on the draft report were gathered by IAEA in June of this year. It is expected that the report will be issued by the end of calendar 1979. Some of the more interesting recommendations of the technical committee are discussed below.

DECOMMISSIONING PLANS

Two decommissioning plans are recommended for preparation. A preliminary plan should be prepared during the reactor plant design period so as to be of guidance to the designers and to provide necessary information to the appropriate regulatory body. The preliminary plan is intended to establish feasible decommissioning schemes which could be accomplished for that specific plant, without undue risk to the health and safety of the public and site personnel; without adverse affects on the environment; and within the requirments of and for the approval of the appropriate regulatory body. This preliminary plan, while not a detailed document, would serve to ensure that decommissioning and ultimate disposal of a nuclear reactor facility is considered during the initial design and licensing of the plant. It is further recommended that a detailed plan for decommissioning be prepared prior to initiation of the actual decommissioning program. This plan would be similar to those that have already been prepared for actual decommissioning programs in this country. The plan would define specific work activities and include safety evaluations of both the method of decommissioning and the end product resulting from the decommissioning program. The document would provide sufficient information to the appropriate regulatory body to achieve necessary approvals to proceed.

RELEASE CRITERIA

In deference to the current "state-of-the-art" of unrestricted release criteria, the technical committee defined their recommendations in purely qualitative terms that, simply stated, required local or individual regulatory body approval.

Although the report, when issued, will not contain any startling new information or suggestions, it does represent an important milestone in that there was complete agreement among the Member States represented on the technical committee as to technical feasibility, regulatory interface, and the importance of protection of public health and safety associated with the decommissioning of nuclear reactors. Certainly from that base, the future efforts aimed at quantifying certain standards and criteria on an international plane should be achievable.

I am not aware of any further actions planned by the IAEA concerning decommissioning of nuclear reactors beyond the issuance of the above-described report in their safety series. However, the Nuclear Energy Agency of the Organization for Economic Cooperation and Development, headquartered in Paris, France, has initiated a program which will followup the topic. The first step was a meeting sponsored by NEA in March of this year which was attended by decommissioning experts from thirteen member countries. This group defined a number of areas in the field of decommissioning which require further study and definition. These include facility design, decommissioning methods, decontamination, and unrestricted release criteria. NEA is actively following up these recommendations with three specific actions. There will be an NEA-sponsored meeting in the spring of 1980 covering the topic of decommissioning requirements in the design of nuclear facilities. The NEA is also planning to issue, in the second half of 1980, two state-of-the-art reports covering decontamination methods for decommissioning and remote handling and cutting techniques. I believe Mr. Maestas of OECD will be giving us more information on the NEA plans in his presentation tomorrow morning.

CURRENT DECOMMISSIONING PROGRAMATIC ACTIVITIES IN THE INTERNATIONAL COMMUNITY

I solicited information concerning the status of current decommissioning programs and related development work from experts within the international community, and I will report on the information I have received. Unfortunately, I did not receive responses from all of those contacted. First, I would call your attention to a paper being given tomorrow by Mr. Essmann who will discuss the status of decommissioning in the Federal Republic of Germany. Therefore, I will not steal any of his thunder here.

Ben Pettersson tells me that at present Sweden has not embarked on any large decommissioning project. What has been done has been restricted to protective storage of research and prototype reactors. However, they are considering preparation of a decommissioning plan covering protective storage of a larger facility and to pursue the development of that plan on an internationally cooperative basis. They are also developing decontamination procedures, but not necessarily associated with decommissioning. The unrestricted release of radioactive material is being considered but not in an absolute sense; that is, case studies are being analyzed which consider the potential use of scrap radioactive material in certain recycled conditions, and the resultant radiation doses due to the end product use via a specific recycling process.

I received an extensive summary of the decommissioning status and anticipated programs in France from my good friend Andre Cregut. The pursuit of the French nuclear power program will result in the final shutdown of four major installations per year by the year 2000. Before that date, 80 installations currently in service would also be decommissioned. These figures reinforce the magnitude of the problem in a country committed to nuclear power development. Work accomplished to date in France concerning decommissioning has been quite extensive. The French have identified the activities associated with decommissioning that are different from normal maintenance work as the following: the presence of very heavy pieces, for example, 200 to 400 tons for a pressure vessel; a large quantity of activated and contaminated material, for example 5,000 tons of graphite, 1,500 tons of steel, and 23,000 cubic yards of concrete for a HTGR; and large effluent volumes from decontamination that could be as high as 800,000 gallons of liquid requiring treatment and/or disposal. They are pursuing programs in decontamination, cutting, remote disassembly, remote handling, preparation for shipping and storage, shipping, radwaste treatment, special storage, and personnel protection. The cutting techniques they are studying include arc saw, plasma arc, hollow-charged explosives, thermal lance, laser and mechanical saw, shear, cutter, and nibbler. They are also pursuing volume reduction programs and special waste disposal processes such as graphite incineration. They are proceeding to incorporate decommissioning considerations into the initial design of their reactor facilities. Some of the objectives of these considerations are to minimize occupational exposure and assure operator safety; reduce decommissioning costs; and reduce radwaste material volumes. Their studies have concluded that achievement of these objectives does not necessarily result in higher investments.

Bill Lunning of the United Kingdom Atomic Energy Agency was kind enough to give me an extensive writeup on the UK decommissioning programs. In summary, the UKAEA has selected the Windscale Advanced Gas-Cooled Reactor (WAGR) for detailed study. The 32 MWe WAGR is to be withdrawn from service in the early 1980's. Combination modes and complete removal options are being considered. The decision will be based on occupational exposure, cost, environmental considerations, and experience value. Complete removal is estimated to take five years after completion of defueling.

NRC'S VIEW OF DECONTAMINATION AND DECOMMISSIONING

Victor Stello, Jr., Director

Office of Inspection and Enforcement
U. S. Nuclear Regulatory Commission

The nuclear regulatory commission is deeply interested in the two basic subjects of this conference, decontamination and decommissioning. We have already licensed about 70 commercial power reactors to operate in this country and see about another 130 in the review and construction pipeline. There are dozens of fuel cycle facilities to support these reactors, and counting both State and NRC licenses, there are more than 20,000 nuclear material activities in the country. All of these nuclear facilities, from the largest reactor to the smallest material facility, will have radioactive residues--they will have to be decommissioned after their useful life is over and these residues disposed of while they are operating, many of these facilities will require some form of decontamination ranging from the cleanup of minor spills to major efforts such as the Dresden-1 Reactor Coolant System Decontamination or the cleanup of Three Mile Island-2. As time passes, NRC is devoting more and more attention to decontamination and decommissioning activities. We are, therefore, more conscious now of our existing policies, regulations, and standards on these subjects.

The NRC is now in the midst of a major re-evaluation of our policy and regulations on decommissioning. As many of you know, our present regulations and standards on the subject are not very specific. Dealing with reactors, for example, part 50.33 imposes financial responsibility and part 50.82 requires the licensee to submit the decommissioning plan for NRC approval before executing it. But we have no firm requirements on when to decommission or how to do so. We published regulatory guide 1.86 about five years ago, which identifies four acceptable options for decommissioning; mothballing, dismantling, entombment, or conversion.

All of these options are being reconsidered. Our basic approach for reconsideration of the policy is to develop an environmental impact statement which weighs the advantages and disadvantages of the alternatives for NRC's decommissioning policy and rules. A lot of data exists on decommissioning now; this conference is witness to that. We are looking to all the information we can from this conference and other sources. In addition, NRC has made a series of contracts with Battelle's Pacific Northwest Laboratories to generate a large body of data on decommissioning alternatives for all sorts of nuclear facilities. I believe Battelle people will be presenting some papers on this work later in this conference.

We are not waiting until we have all our considerations sorted out before we go over these matters with others. NRC is especially conscious of the importance of decommissioning matters to the State Governments. Therefore, we decided at the outset to share our decommissioning deliberations with the states. Just a year ago, we held three regional meetings with State Representatives on Decommissioning Policy Deliberations. Last week in Columbia, South Carolina, and next week in Seattle, Washington, we are holding sequels to those meetings of a year ago. We are dealing with the State Government there at many levels; the Governor's Offices, the Legislatures, the Radiation Control Program Offices and the Public Utility Commissions. These people are interested, and they are letting their views be known. Their comments are having a significant impact on our deliberations.

We have tried to get the views of the Nuclear Industry into this work too. All of our reports have been distributed widely with requests for review and comment. The nuclear industry, after all, should be the principal source of technical comment on this work. I am sorry to say that we have received only a limited number of comments on these reports. We welcome comments and advice from all interested parties in this matter.

A substantial amount of new analysis has been completed in our program. Our PWR decommissioning studies are now essentially complete and the BWR studies are nearing completion. Studies on other facilities are also well along. These analyses have shown so far that the technology already exists for complete decommissioning of large power reactors and the many nuclear material facilities. The dollar and occupational exposure costs for the various alternatives have been calculated. Decommissioning is a labor intensive activity; direct labor cost is estimated to be at least one-fourth of the total decommissioning cost for a large reactor. As a result, we are cautious about the calculated dollar and exposure costs; they could easily be higher.

I'd like to pass on some of the results of our analyses so far and discuss some of the policy implications of these results. I will limit my remarks to power reactors.

The occupational exposure costs of decommissioning are not excessive, even for immediate total dismantling of a large reactor, on the order of 1500 Man-Rem. We do see, though, the possibility of further reducing that exposure by a factor of three or four, by waiting 20 to 30 years before dismantling; that is letting the Cobalt-60 decay off. This raises the possibility of NRC policy permitting, encouraging or even requiring a delay of several decades before final decommissioning. I doubt that the NRC will impose a blanket restraint on decommissioning time. I think we will permit or encourage some delay for radioactive decay, but we will not require it. At the same time, I don't think we will simply permit indefinite delay in decommissioning a facility. I believe that the decommissioning schedule will be established on a site specific basis.

The estimated dollar cost of complete dismantling is relatively low, compared to the cost of construction. Our estimates show costs on the order of $30-40 million. These costs are not an enormous burden for a company or group which operates a reactor, but they are large enough to warrant orderly planning for their expense. This raises the possibility of NRC requiring of licensees some method of setting aside funds for decommissioning. Many of the State Public Utility Commissions have already established such requirements. Many of them would prefer to control such fund set-asides at the local level. At the federal level, the NRC and the FERC might deal with those who are not subject to the Local Public Utility Commission. I don't have to emphasize how deeply anything we or the States do in this depends on the federal tax treatment of such funds. I recommend for your attention a discussion of these matters by Bob Wood of the NRC Staff in NUREG-0584, which we are using as a base of discussion with the state people. I think NRC will require some means of fund set-aside.

One area of our decommissioning policy which requires a lot of thought is the question of cleanup criteria. In regulatory guide 1.86 we published a long used table of criteria for the cleanup of contamination on surfaces of equipment or structures. These criteria have been in use for almost 20 years; they are pretty good, as far as they go. But these are not generally recognized and accepted cleanup criteria, and they do not cover all aspects of contamination cleanup encountered in decommissioning. We need solid, accepted criteria for residual activity in soils, on surfaces, in rubble, and in activated or internally contaminated metal scrap. The technologies exist to measure these forms of contamination and to deal with them as waste.

What is needed are criteria by which we can determine how clean is clean enough. If we are too conservative here, we can set standards which would have us cleaning up nature. We have been considering cleanup criteria based on the maximum annual radiation dose one might receive from exposure to the tolerated residue, perhaps a maximum of about 5 mRem per year from such residue based on realistic pathway analysis. The problem is easier to address for residues at a site than for recycle of scrap materials because of the many pathways or scenarios for exposure which can be postulated for recycled materials. Whatever we do here, it is very important that there be technical and political consensus. Otherwise, we will have chaos with conflicting criteria for cleanup. I believe we will have a standard based on 5 mRem/year and that EPA and most of the state people will support it.

One of the greatest occasions for controversy in NRC Policy will be our action on the entombment alternative. By entombment I do not mean hardened safe storage for 50-100 years, but a deliberate act of sealing radioactive residue in some encasement so durable that the radioactivity will decay to innocuous levels before the encasement fails. Some of our calculations show that modest savings in dollar and occupational exposure cost can be achieved by using such permanent entombment rather than dismantling. We take these apparent savings with a grain of salt, though, because we are hard put to weigh the real cost of the land dedicated to such entombment and the acceptability of waste disposal in so many scattered tombstones under so many jursidictions. I doubt that we will allow permanent entombment routinely.

All of these decommissioning questions are coming to a head now. We need conferences such as this one and vigorous debate in the public rulemaking process. Through both we hope to set a coherent national policy for decommissioning without undue preemption of local authority. While we are now concentrating on decommissioning, NRC is also interested in decontamination since it may provide the resolution of the conflict between the continuing need for timely inservice inspection and maintenance and the growing occupational exposures entailed. Recent experience convinces us of the need to continue requiring rigorous inservice inspection and maintenance. But, as the existing plants operate, the deposits of radioactive corrosion products are building up, and with that buildup the f occupational exposures are increasing. Occupational exposures suffered during plant operation far exceed those foreseen from end-of-life decommissioning. NRC has a clear concern to reduce these exposures, and decontamination may be the answer.

DEPARTMENT OF ENERGY PROGRAMS FOR DECONTAMINATION AND DECOMMISSIONING

Robert W. Ramsey, Chief

Nuclear Technologies Branch
Division of Environmental Control Technology
Office of Environmental Compliance and Overview
Office of Assistant Secretary for Environment
Department of Energy
Washington, D.C. 20545

The purpose of this paper is to provide a brief description of the decontamination and decommissioning (D&D) programs of the Department of Energy (DOE). These fall into two basic categories: Management of surplus contaminated DOE-owned sites and facilities, and remedial actions at locations that have been found to be contaminated from some past government sponsored or conducted operation even though they had been cleared for unrestricted use. Two other papers are included in this symposium which provide details of these programs. One is the paper, "Management of the DOE Inventory of Excess Radioactively Contaminated Facilities," by Dr. A. F. Kluk of DOE and the other is my paper titled "ECT Remedial Action Programs," which are both in session IVB on D&D of Fuel Cycle and Research Facilities.

The management of surplus facilities represents the DOE's effort to reduce the inventory of sites that, because of their contamination with radioactive materials, must be managed to the extent necessary to eliminate potential hazard and must be restricted from excess or use until decontaminated. The program includes about 500 individual locations that have previously been identified and declared surplus. The locations receive appropriate surveillance to prevent inadvertent exposure or possible willful removal and spread of radioactive material. A plan has been developed to allow all of these locations to be stabilized or decontaminated to eliminate this hazard potential and allow alternative use of the land. A program of about 80 projects is planned that could be implemented between now and the end of FY 2000. Implementation of this plan could eliminate the current inventory, and hopefully, with incorporation of projects for the D&D of currently active facilities, could result in no backlog of surplus undisposed nuclear facilities at that time.

The remedial actions, as the name implies, have as their objective the elimination of potentially hazardous or environmentally unacceptable radioactive contamination which has been found at former plants and laboratories, and at certain uranium mill tailings sites. The contamination could cause unacceptable exposure to the public or restrict the use of the land. In the case of Grand Junction structures, the inadvertent misuse of uncontrolled uranium mill tailings materials has caused contamination to exist in private dwellings and cause excessive exposure to occupants. About 800 structures are estimated to qualify for some type of remedial action under the joint State-Federal program of remedial action authorized by Public Law 93-314 as amended. The program is almost half completed.

So far about 30 formerly utilized sites that were regarded as released for unrestricted use have been found to require remedial action, and 22 uranium mill tailings piles are similarly in need of some measures to further control the potential radiological impact they represent.

The authority to undertake remedial action in some cases derives from the Atomic Energy Act of 1954 as amended and from terminal provisions of contracts made between private companies and the Manhattan Engineer District (MED) or the Atomic Energy Commission (AEC) for the formerly utilized sites. Special legislative acts have also been passed, specifically Public Law 92-314 as amended which authorized the cleanup of Grand Junction structures, and Public Law 95-604 passed November 8, 1979, which authorized cleanup of the 22 designated inactive uranium mill trailings sites that produced uranium for sale to the AEC and were never licensed for private commercial processing of uranium ores.

Some known contaminated sites still lack coverage of authorization. These include such sites as the radium processing location recently rediscovered in Denver, Colorado, certain of the formerly utilized sites where the terminating provisions of contracts release of Federal government from responsibility or where the contracts are silent, or because of the destruction of obsolete records, no record exists to allow a proper interpretation of responsibility. In most of these cases there will have to be special legislation to obtain authority to do the remedial action.

The programs of D&D in DOE have been assigned to several organizational divisions over the past 8 years and have recently been reassigned to secure better definition of responsibility and eliminate potential conflicts of interest.

The D&D program for DOE surplus sites originated in the AEC Waste Management Program under the Assistant General Manager for Operations and was later assigned to an Assistant General Manager for Environment. When AEC became the Energy Research and Development Administration (ERDA), D&D was assigned to the Assistant Administrator for Environment, who at the same time was administering programs to identify formerly utilized sites and uranium mill trailings piles and to conduct remedial action at Grand Junction, all of which were forerunners of the remedial action program.

Under the DOE, the D&D and Remedial Action Programs were located in the Office of the Assistant Secretary for Environment (ASEV) where they were consolidated in the Division of Environmental Control Technology (ECT). The need to maintain an independent environmentally sensitive overview of these activities separate from the major project management functions to implement the D&D and remedial action programs have led to a recent shift of responsibilities. The ASEV functioning through the ECT Division and other components of the Office of Environmental Compliance and Overview will conduct the investigations and radiological surveys necessary to determine the need for remedial actions and their priority; will specify the conditions to allow unrestricted use of sites, and, following remedial action, will independently verify the radiological conditions necessary to certify the site as unrestricted for the DOE.

All aspects of the implementation of remedial actions from the time the site is declared in need of remedial action and the specifications and relative priority are set will be conducted by a newly established Remedial Actions Project Office of the Office of Nuclear Waste Management under the Assistant Secretary for Energy Technology.

The program will be decentralized from Headquarters to program site offices established at three of DOE's Operations Offices. The D&D of DOE owned surplus facilities will, as a practical matter, be a responsibility of the field office that owns the site of the surplus facility but all aspects of the program will be coordinated and managed from a site office established at the Richland Operations Office.

The remedial actions at formerly utilized sites (FUSRAP) will be managed from Oak Ridge Operations Office. The Uranium Mill Tailings Remedial Action Program will similarly be managed from a special site office being established within the Albuquerque Operations Office. The Grand Junction remedial action will continue to operate from the Grand Junction Office in close collaboration with the Albuquerque Project Office for Uranium Mill Tailings. By using the talents, proximity and existing support of DOE field offices, we expect to be able to effectively implement these programs.

Finally the matter of final disposition of the wastes from remedial actions is receiving attention. The task of finding suitable policies and technical approaches to the problem of disposal presented by large quantities of very low-level contaminated material, mostly soils, is a great challenge and will require the thoughtful consideration of many members of the societal fabric.

In conclusion, I believe that the DOE has embarked upon a responsive and timely program to assure that the potential burden of radioactive contamination is not passed undiminished to future generation. The program that had its beginnings as a stewardship of the environment is now receiving the resources and management attention it requires as another aspect of the critical area of waste management.

DECOMMISSIONING STANDARDS

W. N. CROFFORD

Office of Radiation Programs

U.S. Environmental Protection Agency

Introductions and Conclusions

EPA has agreed to establish a series of environmental standards for the safe disposal of radioactive waste through participation in the Interagency Review Group on Nuclear Wast Management (IRG). One of the standards required under the I is the standard for decommissioning of radioactive contamina sites, facilities, and materials. This standard is to be proposed by December 1980 and promulgated by December 1981.

Several considerations are important in establishin standards. This paper includes discussions of some of t considerations and attempts to evaluate their relative importance. Items covered include: the form of the sta timing for decommissioning, occupational radiation pro costs and financial provisions.

Other areas which are not addressed in this pap which may become important in considerations of dec standards include: differences between existing fa future facilities regarding decommissioning, concer to dedicated sites and facilities, issues of the r and benefits of various methods such as removal/d fixation in place, and the time period which impa must cover. All of these topics must be examine development of decommissioning standards but th influence on the standard itself is currently r

Decommissioning appears more closely related to radiation protection than to waste management, although it is often carried under waste management programs. Basically, decommissioning is the removal of radioactive contamination from facilities, sites, and materials so that they can be returned to unrestricted use or other actions designed to minimize radiation exposure of the public, such as entombment.

REVIEW OF EXISTING STANDARDS, REGULATION AND CRITERIA

The most difficult problem in establishing radiation protection standards for decommissioning is the choice of the rationale or basis of the standard. Once the rationale is chosen, the approach to setting the standard (or the form of the standard) is defined and the standard effort can proceed in a straightforward fashion. For example, if the existing FRC guidance is to provide the basis for decommissioning standards, the principal task becomes the ALAP approach, the familiar cost-effectiveness assessment, and the standard is established at a level where further expenditures provide little additional health protection. Other rationale are available, however, and must be examined in light of identifying which basis would potentially provide the greatest assurance of maximum public health and environmental protection.

As an initial step in identifying decommissioning rationale, a thorough review is planned of existing standards, regulations and criteria. The objective of the review is to determine how these standards were arrived at. This identification will require a thorough search of the history of the applicable standards and will include such items as:

- o the reason the standard was established;
- o the contribution and opinion of expert advisory groups to the development of the standard;
- o the use or interpretation made of existing authoritative guidance, such as that of the Federal Radiation Council;
- o input by the public and governmental agencies during the public comment period, both written and oral.

This information can be useful in determining the basis decommissioning standards in that it will make available logic previously followed by those who addressed the mmissioning issue, or one closely related. Too often the king performed in the past is neglected because it is ved new or unique issues are being confronted. While is no great assurance that past thinking will provide a rationale for the current effort, it is believed it will sufficient insight to be helpful.

Previous efforts that have been chosen for examination thus far include:

- 10 CFR Part 20, Paragraph 105, U.S. NRC;
- EPA's proposed Federal Radiation Protection Guidance on Dose Limits to Persons Exposed to Transuranium Elements in the Environment;
- the Surgeon General's Recommendations to the State of Colorado for Radiation Protection for Individuals Living in Residence Contaminated with Uranium Mill Tailings;
- Regulatory Guide 1.86, Termination of Operating Licenses for Nuclear Reactors, U.S. NRC;
- ANSI Standard N328, Control of Radioactive Surface Contamination on Materials, Equipment and Facilities to be Released for Uncontrolled Use.

TIMING

At the end of its operating life a nuclear power plant or other nuclear facility will be shut down and retired from operation. These facilities and sites must be restored to a radioactivity level that is acceptable for their unrestricted use in the future. This raises the important question of at what time after shutdown must facilities and sites meet this residual radioactive level requirement. This is especially true of power reactors because at the time of retirement radiation levels will be quite high, primarily due to neutron activation products (Sc). Since most of the activation products have relatively short lifetime, less than 5 1/2 years, there are potentially significant benefits to be gained by delaying decommissioning until there is appreciable decay of the activation products. But the delay of decommissioning is undesirable from the view of both the possibility of institutional failures and the responsibilities of the generation which received the benefits to reduce the impact (or cost) to future generations.

An important consideration here is the ultimate use to which the site or facility is to be put. Obviously, a contaminated laboratory or fuel fabrication plant that has shut down must be decommissioned (or decontaminated) to a level that will protect workers during any future use (Sc). However, the case is not as clear-cut for a nuclear power plant. A nuclear power plant site represents a large investment in equipment and

residual radioactivity at a reactor will be due to activation products which are predominantly short-lived materials. The situation at nuclear facilities other than reactors is not as clear because the contaminating materials may have long half-lives.

Reactor decommissioning has normally followes a scenario which included removal and offsite shipment of all spent fuel, decontamination of internal piping and the site, packaging and offsite shipment of the low-level radioactive waste, and mothballing the reactor for a period of time. This time period is largely controlled by the decay rate of cobalt-60 a significant activation product in terms of external exposure with a half life about 5 years. Longer-lived activation products are also present, such as Ni-59 and Nb-94 but to a lesser degree. Since cobalt-60 is the predominant radiation source of the shorter-lived nuclides, occupational exposure reductions to be gained by delaying decommissioning can be directly related to the half life of cobalt-60. For example, a 20 year delay would produce a reduction of 16, and a 50 year delay in reduction results in a factor of 1,000 reduction. Therefore, it appears likely that a maximum limit of 50 years for reactors could provide an optimum point for occupational exposures balanced with economic and institutional considerations.

Facilities providing the fuel for and managing the waste from reactors, along with most of the defense and research oriented facilities, pose different problems in decommissioning since the contamination is a result of longer-lived materials. First, the front end of the fuel cycle is dominated by uranium and its daughters thorium-230 and radium-226 which lead to the radon-222 decay chain. Because of the long-lived nature of these products, there is nothing to be gained through a delay, and decommissioning activities should be conducted immediately upon plant closings. As for the tail end of the fuel cycle, regardless of decisions concerning recycle, facilities most likely will be highly contaminated with longer-lived materials and little will be gained through delays in decommissioning such facilities. A similar situation can be expected at most of the defense related and research oriented facilities.

DECOMMISSIONING COSTS AND FINANCIAL PROVISIONS

Current estimates indicate decommissioning costs may be large and, what is worse yet, the current estimates have not been verified by actual experience. Thus, no evidence exists

land, such as: transmission lines, switch yards and site improvements, especially reactor cooling water provisions. This results in a strong incentive for continued use of the site for centralized power, nuclear or not. Thus, central power sites may eventually become mixtures of nuclear and fossil fuel power generation. In addition, utilities may choose to recycle certain structures or other parts of a site, such as containment buildings or cooling water structures. These considerations serve to greatly complicate the timing question for decommissioning. However, timing requirements must be established to assure public health and environmental protection for future generations. Thus, such uncertainties should be considered to the extent possible at this time with the concept that future activities may require a reevaluation of the timing requirements.

As an intitial estimate of what the timing for decommissioning should be, it is suggested that EPA's proposed Criteria for Radioactive Wastes be used (EP). Criterion No. 2 of this proposal states, "Controls which are based on institutional functions should not be relied upon for longer than 100 years to assure continued isolation from the biosphere." While this criterion addresses radioactive waste disposal rather than decommissioning, these two issues share a common theme in the timing area, that of the acceptance of continuing reliance on institutions for health and environmental protection. Applied here the result is the scenario:

Nuclear power plant licensing and construction	10 years
Nuclear power plant operations	40 years
Maximum decommissioning time	50 years
Total	100 years

This scenario provides a maximum 50 year period for decommissioning while meeting the intent and spirit of the criterion. A 50 year period following reactor shutdown also provides a most substantial reduction in radiation exposure levels from activation products.

OCCUPATIONAL RADIATION PROTECTION

The major health related issue in determining timing requirements for decommissioning is occupational radiation exposure. In the case of reactors, occupational dose can be reduced during decommissioning by postponing decommissioning operations. This is based on the assumption that most of the

to prove that the costs may not be significantly greater than current estimates. There is no way to avoid some costs since prudent public health policies dictate that nuclear facilities and sites be returned to a condition of insignificant radiation threat to the public. A logical policy regarding such costs would be that the beneficiaries of the activity be responsible for the decommissioning of the activity-related sites and facilities. In addition since the decommissioning costs appear large, it seems logical that financial provisions should be made before the operation is initiated, or at the very least during the early stages of the operation, to assure the existence of sufficient funding when decommissioning takes place.

The diverse nature of nuclear facilities will lead to numerous solutions to the funding problem. For instance, nuclear power plants, as regulated utilities, have a relatively secure economic future, especially when compared to such facilities as uranium mills. The situation for reactors appears relatively simple, because the demand for electric power is not expected to disappear in the foreseeable future, the electric utility is expected to have a continuing role. If the utility could demonstrate that the decommissioning cost was a small fraction of the power production cost, say less than 5 percent to 10 percent, this cost could be included in the operating costs and regulated by the controlling Public Utility Commission (PUC). Perhaps PUCs could even drop the decommissioning cost from the operating costs, if sufficient reliance could be placed on a relatively small decommissioning cost would be borne by the utility. The major obstacle to this approach is the lack of firm costs data for decommissioning. It should be recognized that any funding requirements would require approval of the regulating PUC.

The situation is not as simple for other nuclear facilities, however. First the question of financial responsibility and a continuing role most be addressed since most of these facilities are owned by companies in a competitive market with no oversight provided for the common good, in contrast to utility regulations by PUCs. A case in point here is the company which owns the NFS facility in West Valley, New York. According to a recent study the company can terminate its lease for the site at the end of 1980 with the question as to their financial responsibility unclear, but apparently with little likelihood that the company will be liable. The cost for decommissioning this site is estimated at 536 million (DE), or much greater than the initial cost of the plant, constructed during the early and middle 60's. This illustrates the second point to be made in decommissioning non-reactor sites and facilities -- the cost of decommissioning may be far greater than the initial capital investment. In addition to the West Valley situation, the abandoned uranium mill tailings sites in the western part of the U.S. further amplify this financial problem.

The only logical conclusion that can be drawn here is that arrangements must be made before licensing such activities to assure both responsibility for decommissioning and adequate funding.

References

(Sm) R. I. Smith, et. al., "Technology, Safety and Cost of Decommissioning a Reference Pressurized Water Reactor Power Station" NUREG/CR-0130, June, 1978.

(Sc) C. E. Jenkins, E. S. Murphy, and K. J. Schneider "Technology, Safety and Costs of Decommissioning a Reference Small Mixed Oxide Fuel Fabrication Plant," NUREG/CR-0129, Feb. 1979.

(De) U.S. Department of Energy, "Western New York Nuclear Service Center Study - Final Report for Public," TID-28905-1, Nov. 1978

(EP) U.S. Environmental Protection Agency, "Environmental Protection Criteria for Radioactive waste," 43 F.R. 53262, Nov. 1978.

A STATE'S PERSPECTIVE

David M. Boonin

Pennsylvania Public Utility Commission

Harrisburg, Pennsylvania

The Pennsylvania Public Utility Commission has been confronting the issue of decommissioning nuclear generating plants for several years. Our objective has been to place the cost of decommissioning upon the consumers who benefit from the electricity generated by a nuclear plant, while recognizing the related safety issues. In addition to our usual concerns about "safe and reliable service at just and reasonable rates," the Pennsylvania Public Utility Commission had the enormous task of deciding the state regulatory issues surrounding the nuclear accident at Three Mile Island. The accident has made all of us more keenly aware of the issues surrounding nuclear energy, including the topics of this symposium, decommissioning and decontamination.

I will try to shed some light on three areas this morning:

1. The Commission's treatment of decommissioning expenses in a rate case.

2. The Commission's actions in response to the accident at Three Mile Island.

3. A review of the Commission's order of June 15, 1979 on Three Mile Island.

Rate Case Treatment of Decommissioning

The Commission was first faced with decommissioning in 1975 during a Metropolitan Edison Company rate case. In this case the

Commission denied the company's request for an annual decommissioning expense. Although the Commission agreed conceptually that "current ratepayers should contribute toward the cost of the eventual decommissioning of Three Mile Island's Unit No. 1," it was felt that the uncertainty of the method and the cost of the eventual decommissioning and the principles delineated in a Pennsylvania Superior Court decision (Penn-Sheraton Hotel v. PA PUC 198 Pa. Superior Court 618, 623-627, 1962) constrained the Commission from approving the inclusion of the prospective expense for ratemaking purposes. The Penn-Sheraton decision directed that prospective negative salvage be capitalized and amortized for ratemaking purposes if and when experienced.

The policy was followed until February 1978, when in a rate case concerning the Pennsylvania Electric Company, the Commission allowed an annual decommissioning expense. In its decision the Commission stated that "in reaching this conclusion we are motivated by our concern for the future health and safety of the citizens of the Commonwealth. This concern requires that the company make adequate annual financial provision for a known event in the future; an event that has a substantial cost. We must take the initial step now to protect future ratepayers from bearing a significant revenue burden associated with the decommissioning of a plant from which they will receive no service.

"By providing a mechanism for carrying out the fiscal aspects of control of these hazardous nuclear components, we insure that this will be done. Those who now enjoy the benefits of this technology can do no less than to assure that it will not impose an unreasonable financial burden on future ratepayers.

"Although the total amount determined to be needed may not be accurate, the rejection of the claim would ignore the vital issues of health and safety which are reasonably foreseeable. Changes in the estimates of decommissioning costs may be dealt with through periodic review and adjustment of the total estimate (and its annual provision) within each rate case, or at any time upon the initiative of the Commission when it feels that such review is necessary."

It was the overriding health and safety issues attached to decommissioning which the Commission used to distinguish it from the Penn-Sheraton case findings. Therefore, we only allowed the prospective decommissioning expense for the nuclear structures and disallowed the costs associated with the dismantling of the turbine buildings, cooling towers, river water pump house, and other miscellaneous structures. The plant under consideration was a PWR. A BWR would have probably included other structures.

In making its final calculation of the allowable annual expense for decommissioning the Commission started with the estimated cost of entombment ($23,600,000) and established an annual annuity ($74,000) such that at a 5.5% annual yield, tax free, over the remaining life of the plant (31 years) the total estimated cost would be accumulated.

Further, the annuity and its accumulated interest were to be placed in an escrow fund, with payments made monthly. The fund was to be invested in Commonwealth of Pennsylvania bonds, with the highest possible yields. Strict accounting would be maintained, so that if any change in methodology or cost estimates are made the account may be adjusted. Any over or under collections would be refunded to or collected from the ratepayers.

Using these provisions, not only should the interest on the escrowed fund be free of state and federal income taxes, but the annuity itself may be excluded from taxable income. IRS rules are but one of the many constraints upon the regulator.

The Commission has stayed with this policy since 1978, with the minor modification permitting investments in state and municiple bonds, thus raising the anticipated annual yield to 6.5%.

The Commission has considered other financing options including the posting of decommissioning bonds prior to the plant's operations and allowing the company to use the accumulated funds, among others. To date, however, the Commission has chosen the annuity method held in escrow as a proper balance between cost and the assurance that the dollars will be available. Two areas which we as a regulatory commission feel better information is needed are better estimates of the costs of decommissioning and better estimates of the expected lines of nuclear plants. Since we have yet to receive a case on decontamination expenses, no Commission policy has been developed.

TMI has underlined the possibility of premature decommissioning. Equity and cost could now take a backseat in the issue of decommissioning with assurability being paramount. The NRC in its workshop last week in South Carolina was to address this among other issues. Although the possibility of premature decommissioning may be an argument for up front financing, the decommissioning a dirty or damaged plant is likely to cost much more than normal decommissioning and may render the upfront funds inadequate. A possible answer may lie with insurance. I am anxious to see what the NRC workshop produces especially in their attempt to balance cost, assurability and equity. An avenue which we believe may have potential is the recommissioning of plants.

Decommissioning costs are only one of many costs which should be allocated over time so that rates can be set equitably and efficiently. Other inter-temporal allocations include depreciation, deferred income taxes, investment tax credits, construction work progress and others. In Pennsylvania we hope to examine all of these items which must be allocated over time under one cover in the near future.

Commission's Response to the Accident

On March 28 the TMI incident occurred rendering Unit 2 inoperable. The Commission ordered prompt reports by the technical staff.

On March 29 the technical staff informed the Commission that the situation was more serious than first reported and a full investigation should be instituted. Safety of public was the paramount concern and the Commission offered to act as support to the Pennsylvania Emergency Management Agency.

On March 30 the Commission task force was formed; meetings of the task force continued each day through Monday, April 9. The task force mandate was to determine the seriousness of the accident and the potential impact on ratepayers.

On April 4 the Commission instituted an investigation into financial consequences incurred by Metropolitan Edison Company, Pennsylvania Electric Company and their ratepayers as a result of TMI incident.

Interrogatories were issued by the Commission's Law Bureau requiring answers by Met-Ed and Penelec.

On April 19 the Commission instituted complaints against the rates of Metropolitan Edison Company and of Pennsylvania Electric Company.

Metropolitan Edison Company was ordered to file tariffs establishing temporary rates for a period of six months (those rates which were in effect prior to the most recent rate case), to approximate elimination of TMI-2 from base rates.

The Commission instituted investigations into the management practices of Metropolitan Edison Company and of Pennsylvania Electric Company during construction of TMI-2.

On April 24 a prehearing conference was held before the Commission sitting en banc in which GPU presented its position.

On April 25 the Pennsylvania Electric Company was ordered by the Commission to file tariffs establishing temporary rates for a period of six months, reducing Pennsylvania Electric Company's annual revenues by $25 million.

On April 27 a prehearing order was issued granting petitions to intervene and establishing hearing schedules.

On April 30 Chairman W. Wilson Goode presented testimony before the Subcommittee on Nuclear Regulation of the Federal Senate Committee on Environment and Public Works.

On May 2 the initial hearing was held in Harrisburg. A total of ten days were involved with the Commission sitting en banc.

On May 7 arrangements for purchased power between the GPU subsidiaries and the adjacent subsidiaries of the Allegheny Power System become effective. Terms were reviewed by the Commission staff.

On May 8 a non-evidentiary hearing was held in Harrisburg.

Also on that day, petitions were filed by Metropolitan Edison Company and Pennsylvania Electric Company requesting permission to change their net energy cost rates to begin the immediate recovery of the costs of purchased power.

On May 10 the Commission deferred decision on increasing net energy cost rates until the conclusion of proceedings.

On May 14 a hearing was held in Reading. Non-evidentiary hearings were also conducted.

On May 29 a hearing was held in Johnstown. Non-evidentiary hearings were also conducted.

On June 1 the formal hearings were concluded in Harrisburg.

On June 13 an oral argument was presented by GPU officials and the other parties before Commissioners in Harrisburg.

Finally just 10 weeks after the accident, June 15, the Commission issued its order.

This short period to a decision was made possible by the Commission sitting en banc and directly addressing the issues they needed to make a decision. This process eliminated most of the procedural friction which often occurs in an administrative agency.

The TMI Decision

After the hearings were completed, the Commission set about attempting to sort out the facts--to separate the emotions from the clear rational thought processes. We went in search of a clear rational regulatory response to a very real situation for which many felt there was no regulatory solution.

After an examination of all the facts the Commission reached its decision.

In its decision the Commission concluded in the final result

> "The ratepayers of Metropolitan Edison Company and Pennsylvania Electric Company should be no worse off - and no better off - because of the incident. The ratepayers should not pay for the costs of the incident; nor should they benefit from it. They should not pay the costs of a plant rendered useless through no fault of their own, nor should they receive needed electric power without payment."

To do this the Commission made two very critical findings: (1) we found that TMI Power Station No. 2 was not used and useful in the public service because of its prolonged shutdown; and all costs associated with that unit would be removed from based rates, as follows. This reduced the rates of Metropolitan Edison by $52.2 million and those Pennsylvania Electric by $26.6 million. (2) The Commission levelized the energy clauses of both companies for an eighteen month period starting in July 1979, under the assumption that TMI-1 would be back in service by January 1, 1980. Further, it was concluded that the question of whether TMI-1 should be considered used and useful and in the public service would be examined if it were not back in service by January 1, 1980. These findings were crucial to the Commission's conclusion that utilities could not expect to automatically pass costs along to the consuming public.

But, the Commission did not stop there, in addition, we required the companies to take steps to further reduce costs by:

(1) Requiring that Met-Ed and Penelec undertake an "Aggressive, imaginative program of encouraging conservation in order to reduce its costs of purchasing power." This included plans to urge all customers to conserve electricity along with the promotion of energy saving rate structures.

(2) We directed Met-Ed and Penelec to petition FERC and to negotiate with other members of the PJM Power Pool to eliminate split

savings during emergency conditions, suggesting that during such periods power should be sold at cost.

(3) We directed the companies to pursue purchased power arrangements which would reduce the cost of energy below normal interchange rates.

(4) The Commission would of its own accord petition the legislature to enact legislation removing sales and gross receipts taxes from increased revenues recognized during emergencies.

In addition, the Commission expressed the view that ratepayers should not pay for clean-up costs. It was our expressed view that such costs should appropriately be borne by the companies through their insurance companies. Though this expressed view is not legally binding, it was intended to send a very definite message to the company, that ratepayers should not pay for clean-up costs. Accordingly, the company should pursue all legal means to collect from the insurance company. Simply ratepayers should not be the insurer against nuclear accidents. It is estimated that Metropolitan Edison will incur non-insurable costs of $100,000,000 to restore TMI-2 to service and $5,500,000 to make the necessary modifications to TMI-1. The economics of nuclear generation, at least in Pennsylvania, will therefore be greatly affected by the TMI accident. In developing any cost comparisons, it is important to consider the statistical likelihood of possible future accidents of the TMI magnitude and their affect upon the economics of nuclear generation.

The Commission went on to express its concern about the management of the company. We expressed concern that the company had not aggressively and persuasively explored the pertinent issues in the case. We wanted to know whether or not our perception of management's actions during the entire proceeding was correct, and therefore instituted a study to determine whether management acted prudently, wisely and efficiently in managing its utility operations both during and after the crisis.

As we look beyond the accident itself, it must be pointed out, to date, that neither Metropolitan Edison nor Pennsylvania Electric Company has requested an increase in base rates as a result of the accident.

No rate of return witness in major electric cases has revised his request upward as a result of TMI. In short, other than the proceeding detailed above, to date no other proceeding before the Pennsylvania PUC has been affected by TMI.

REVIEW OF PLANT DECONTAMINATION METHODS

John F. Remark
Babcock & Wilcox Co.
P. O. Box 1260
Lynchburg, VA 24505

Alan D. Miller
Electric Power Research Institute
P. O. Box 10412
Palo Alto, CA 94303

In these days of continuous pressure to lower the maximum man-rem exposure limits, the increasing emphasis on plant availability and the increasing number of repairs, backfitting and inspections, decontamination techniques are becoming increasingly important in the operation of a commercial nuclear plant. Operating experience indicates that contaminated components may need to be decontaminated to help plant personnel comply with their ALARA program. Decontamination techniques are also needed in order to remove a component to a non-radiation area for inspection or repair, or for shipment off-site to a nonlicensed vendor for refurbishment, repair or disposal.

Few commercial nuclear utilities have reported their decontamination experience in the open literature; therefore, needed decontamination information must be generated at each nuclear site. Much time could be saved by plant personnel and costly errors avoided if up-to-date decontamination information were available. Recognizing this need, the Electric Power Research Institute contracted with the Babcock & Wilcox Company to conduct a survey of decontamination techniques employed by light water reactor operators. This survey was intended to identify the decontamination techniques and processes employed successfully and unsuccessfully at operating light water reactors. Personnel from all operating LWRs in the United States as well as personnel at selected plants in Canada and Western Europe were contacted by telephone or mail for information. Personnel from decontamination vendors, research laboratories, and nuclear steam supply system (NSSS) vendors were also contacted to determine decontamination methods presently being investigated for future plant application.

This paper discusses industrial decontamination experience subdivided into three areas:

(1) Miscellaneous tools and small equipment
(2) Removable components
(3) Isolable components

A section is also presented which provides guidance on the selection of a decontamination technique. The final report concerning this program (Contract No. TPS 78-816) will be available from the Electric Power Research Institue in early 1980.[1]

SELECTION OF A DECONTAMINATION TECHNIQUE

Before a decontamination technique is selected for use at a nuclear site, a review of the decontamination considerations should be performed. Among the considerations that should be included in the review are:

(1) Reason for the decontamination
(2) Location of activity in the corrosion film, type of activity, and type of surface to be decontaminated
(3) Decontamination equipment on site
(4) Decontamination experience on site
(5) Compatibility of waste generated with radwaste facility
(6) Material compatibility

The reason for the decontamination will have an important bearing on the decontamination technique chosen. If the component is going to be decontaminated to reduce the man-rem exposure during an inspection, a nondamaging decontamination technique would have to be employed. If man-rem exposure reduction for repair or inspection is the reason for decontaminating a component, then the decontamination technique employed must be a low man-rem exposure job. In other words, a decontamination technique must not use more man-rem than it saves.

If the component is decontaminated in order to be shipped to a nonlicensed vendor for repair or refurbishment, a different technique may be needed in order to fulfill the requirements of off-site unconditional release. When a component is being removed for disposal and damage to the component is not a concern, more harsh decontamination methods can be employed.

The site decontamination equipment and experience should be evaluated when determining the decontamination technique to employ. Techniques for which the site has equipment available and experience in using obviously should be given first consideration.

The type of surface, type of activity, and location of the

activity in the corrosion layer also need to be considered in the proper selection of a decontamination technique. If the activity is associated with the loosely adherent outer layer of the corrosion film, a simple wipe with a damp rag is expected to remove most of the activity. However, if the activity is associated with the tightly adherent layer of the corrosion film or associated with a rough surface finish, a more harsh and possibly more damaging technique would need to be employed.

In all cases the radwaste generated during the decontamination must be compatible with available radwaste disposal facilities. This disposal may limit the type of chemicals employed and the quantity of chemicals used during a decontamination.

The economic aspects of the decontamination also need to be evaluated when deciding on a particular decontamination technique. If the decontamination is on critical path, the downtime could cost $400,000 to $500,000 per day. Therefore, increased cost due to radiation exposure must be weighed against decreased exposure during repair or inspection. The cost of the exposure should also be evaluated against the contaminated component's replacement cost. At times, replacement of a contaminated component may be the most economical alternative.

When a chemical decontamination technique is employed on site, the compatibility of the chemical with the materials of construction must be acceptable. If the materials of construction exhibit unsatisfactory corrosion behavior, either different chemicals or a different decontamination technique should be chosen or a corrosion inhibitor added to the decontamination reagent.

DECONTAMINATION OF MISCELLANEOUS TOOLS AND EQUIPMENT

Survey participants identified six decontamination techniques for use on miscellaneous tools and equipment. The two techniques mentioned most often were hand scrubbing and ultrasonics.

Hand Scrubbing

Almost every plant surveyed stated that hand scrubbing has a role in their decontamination program. This may vary from a simple wipe with a dry or damp cloth to a scouring action with an abrasive pad with a chemical assist. The type of chemicals employed depends upon the amount of decontamination required and the type of corrosion film deposited on the surface. If the film is loosely held, simply wiping the surface with a damp cloth may remove the contamination. If the contamination is incorporated into a tightly-held corrosion film, various chemical cleaning agents must be employed assisted with abrasive scrubbing. Some of the cleaning agents named most often are Radiac Wash,[2] Formula 409,[3] Amway products, "soap", organic acids, or trisodium phosphate. The abrasive pads employed for the scrubbing were emory cloth or steel wool. If an oil layer was associated with the contamination, the surface was first wiped with cloths wetted by acetone or isopropyl alcohol. The organic solvents tend to remove the oil layer with any associated contamination, thus exposing the corrosion film for further decontamination.

A number of survey participants indicated that hand scrubbing can create an airborne contamination problem. Methods suggested for coping with the airborne contamination problem include wearing respirator protection or performing the cleaning under water. Water would also shield a portion of the radiation from the technician, further reducing the man-rem exposure.

Ultrasonics

Ultrasonics has also been identified by a large number of survey participants as a decontamination technique employed for miscellaneous equipment and tools. Unlike hand cleaning, the respondents reported mixed success with ultrasonics. The reason for the mixed success appears to lie in the mode of operation. Ultrasonic cleaning is dependent upon a number of parameters. The most critical are frequency, power intensity, pressure, solution viscosity and composition, temperature, and particulate concentration.[4] The decontamination can be degraded by improper control of the above parameters.

The plant personnel that have reported unsatisfactory decontamination factors have a number of operational techniques in common. These operators employ pure water as the ultrasonic solvent; however, some may add a cleaning reagent such as Radiac Wash.[2] These technicians operate the ultrasonic baths at room temperature. The geometric relationship of the surface to be decontaminated and the transducers, the source of the ultrasonic waves, is not chosen to optimize cleaning nor is the concentration of particulates monitored during decontamination.

Based on the survey results, the best decontamination factors have been achieved with ultrasonic baths by employing water as the solvent with an addition of 2 to 5% cleaning agent and/or a wetting agent. The temperature of the solution should be controlled at 150-170°F. This temperature range will optimize the cavitational intensity. The solvent should be recirculated through a cleanup filter system to reduce particulate concentration. Some site procedures require handwiping of all equipment to be cleaned by ultrasonics to reduce the concentration of particulates in the solvent. The cleanup filter should be shielded to provide radiation exposure control to the operator. Recirculation should be less than 2% of the tank volume per minute. Higher flow rates may introduce solution turbulence which can interfere with the ultrasonic waves and reduce the decontamination effectiveness. The arrangement of components in the ultrasonic bath as well as the number should be carefully monitored during each cleaning. Each surface to be cleaned should be in an area of maximum solution cavitation at some time during the decontamination. A large number of components may screen surfaces from maximum solution cavitation. The possibility of each surface being exposed to maximum cavitation can be enhanced if the components are rotated in the bath. An effective frequency and power intensity in which to begin ultrasonic cleaning is 20-25 kHz and 0.75 W/cm^2.[5] Reported decontamination factors employing this ultrasonic procedure range from 10 to 100.

Other Techniques

Other effective decontamination techniques for miscellaneous tools and equipment that were identified by survey participants include acid cleaning, electrocleaning, spray booths, and steam cleaning. However, very few plant personnel employ these techniques. A number of plant personnel are evaluating electrocleaning as a possible decontamination technique for miscellaneous equipment and tools. A number of vendors offer electrocleaning as a service and others plan to offer this service in the near future. If a plant has a substantial number of small components and tools to decontaminate, this method is attractive. However, because of the initial setup time if just a few tools or pieces of equipment are to be decontaminated, a different method should be evaluated. A number of participants expressed concern for damage to equipment with close tolerances. If properly applied, components with close tolerances can be decontaminated.[6] However, there is a risk of destroying the tolerance and rendering the component useless. This risk of possible damage should be discussed with the vendor performing the decontamination and then evaluated by plant personnel. For additional information regarding operational procedures for these techniques, Reference 1 should be consulted.

DECONTAMINATION OF REMOVABLE REACTOR COMPONENTS

Ten effective decontamination techniques have been identified by the survey participants for use on reactor components. The decontamination techniques identified were:

- Acid Bath
- Electrocleaning
- Grit Blasting
- Hand Scrubbing

- High Concentration Chemicals
- Hydrolasers
- Steam Cleaning
- Strippable Coatings
- Turbulators
- Ultrasonics

The techniques that are employed most often for removable reactor components are high concentration chemicals, hydrolasers, and ultrasonics. The decontamination technique employed depends not only upon the type of film on the surface to be decontaminated, but also where the contamination resides in the film as well as the parameters discussed in a previous section on selection of a decontamination technique. However, as a simple rule if the contamination is associated with loose-held, smearable film, a hydrolaser or ultrasonic unit is most often employed. If the contamination is associated with a tightly-held corrosion film, high concentration chemicals are most often employed.

Ultrasonics

Ultrasonics are limited in use since the largest size of the tanks commercially available is approximately 5' x 3' x 2½'. Larger units can be obtained by special order. Some site personnel have designed their own ultrasonic units. Oyster Creek and San Onofre operators have employed ultrasonics to decontaminate a control rod drive mechanism. Oyster Creek technicians employed no chemical assist while San Onofre technicians employed a citric acid solution. San Onofre personnel reported a decontamination factor of 30. No decontamination factor was reported by Oyster Creek personnel. Other components that have been successfully decontaminated employing ultrasonics are pump seals, small valves, pump casings, reactor coolant pump coolers, and seal injection filters. Decontamination factors ranged from 2 to 100. Low decontamination factors appear to have resulted from improper use of the ultrasonic technique as described in the previous section.

Hydrolasers

Hydrolasers are also employed at a large number of nuclear sites as a decontamination technique. Hydrolasers are commercially available that will produce a liquid stream of up to 10,000 psi. Most plant personnel recommend initially employing a hydrolaser at lower pressures, 500 psi, since the technique at the lower pressures in most cases performs just as well as at the higher pressures, 3000 to 5000 psi. If needed, higher pressures can then be attempted. A few plant personnel reported adding a chemical cleaning agent to the water to increase the decontamination factor. Other nuclear site personnel state the added cleaning agent makes the decontamination factor more consistent. Experience at one site indicated a decontamination factor of 2-50 if water alone was employed on a reactor component with corrosion film present, but if Radiac Wash was added to the water, the decontamination factor was consistently 40-50.

Some components that have been identified as being successfully decontaminated by a hydrolaser are reactor coolant pump impellers, small and large valves, cavity walls, spent fuel pool racks, reactor vessel walls and head, fuel handling equipment, feedwater spargers, floor drains, sumps, interior surfaces of pipes, and various storage tanks. The decontamination factors obtained ranged from 1 on the reactor vessel walls to several hundred on the cavity walls. Even though the reactor vessel walls had a decontamination factor of 1, the decontamination was still considered a success since the airborne contamination was reduced.

High Concentration Chemical Reagents

Approximately ten plants identified high concentration chemicals as an effective decontamination technique. Most plants' personnel used the two-step process employing alkaline permanganate as the pretreatment step followed by citrox (a mixture of oxalic and citric acids) as the cleaning or decontamination step. Both steps should be performed at 170-190°F in order to obtain successful results within a reasonable time frame. Chromium-51 appears to be the predominant radioisotope removed during the pretreatment step while the remaining radioisotopes are removed in the citrox step. Some boiling water reactor operators have employed a "brew" of EDTA plus oxalic acid. The mixture of this solution depends on the composition and age of the corrosion film. The composition ranges from 5-20% oxalic acid and 0.1% EDTA. High concentration chemicals should not be employed indiscriminately as they can have a severe impact on the materials of construction of both the component and the radwaste system.

Some of the reactor components that have been successfully decontaminated by high concentration chemicals are reactor coolant charging pump components, reactor coolant pump impellers, pump shafts, valves, and regenerative heat exchangers. The decontamination factors obtained ranged from 2 to 100. The average decontamination factor ranges from 6 to 25.

Electrocleaning

Electrocleaning is also employed as a decontamination technique for reactor components.[7] Some of the components that have been successfully decontaminated by this method include pump impellers, valves, and piping. When successful, decontamination factors have been reported to range from 100 to several thousand. A few plant personnel stated that electrocleaning techniques did not perform successfully if the surface was porous or grooved. Electrocleaning operators state that these type surfaces should be easily cleaned. Perhaps the electrocleaning operator did not orient the electrodes in the optimum position or perhaps it was some other operational difficulty; however, these surfaces were not successfully decontaminated at a number of plant sites. Some plant personnel report that components with close tolerances have been destroyed by electrocleaning techniques. At other plants the close tolerances have been protected by insulating tape or spray before electrocleaning. Other site personnel do not employ electrocleaning for components with close tolerance but rather use high concentration chemicals. However, a number of these sites do use electrocleaning for components without close tolerances.

Abrasive Blasting

Abrasive blast techniques have been attempted at four or five nuclear sites as a decontamination technique. No plant identified this as a successful technique. A variety of abrasive grits have been employed including boric acid crystals and aluminum oxide. During the decontamination attempt airborne contamination was a problem even though a vacuum technique was used which trapped greater than 95% of the grit. The technique did not produce a significant decontamination factor, i.e., greater than 1.5.

It is the authors' opinion that abrasive blast techniques should be evaluated carefully before being employed as a decontamination technique. If nothing else works, then abrasive blast probably will not work. The problems associated with these techniques as presently applied versus the decontamination factors obtained do not merit its use.

Other Identified Techniques

Other reactor component decontamination techniques which have been employed effectively at a few nuclear sites are mineral acid cleaning, steam cleaning, strippable coatings, and turbulators.

The techniques have not been employed extensively at any site. Decontamination factors reported from the use of these techniques range from 5 to greater than 100. These techniques have specific application or require specific equipment. More information can be obtained by consulting Reference 1.

DECONTAMINATION OF ISOLABLE COMPONENTS

Two decontamination techniques have been identified by survey participants for use on the reactor coolant system or other isolable components. These two techniques are high concentration chemicals and low concentration chemicals. These two techniques involve circulating chemicals through the reactor coolant system at elevated temperatures approaching the boiling point of the solution. A few participants mentioned redox cycling as a possible decontamination technique. Recent reports indicate that redox cycling does indeed increase the particulate concentration in the coolant.[8] The increase in particulate level is thought to originate from large releases of corrosion products from the fuel. Thus this decontamination technique could lead to decontamination factors of less than 1 or contamination of out-of-core components.

Low Concentration Chemical Reagents

The Douglas Point reactor in Canada was decontaminated with low concentration chemicals following full-scale tests at the NPG CANDU-PHW reactor and Gentilly-1 CANDU-BLW reactor.[9,10] Nutek L-106 supplied by the Nuclear Technology Corporation, Amston, Connecticut was selected for this decontamination because it prevented copper redeposition while producing adequate decontamination factors. The contamination was removed by cation exchange resin and, after the decontamination phase was complete, the reagent was removed by a mixed bed resin. Decontamination factors obtained furing the Douglas Point cleaning ranged from 4 to 10. One year of operation after the decontamination the radiation levels increased only slightly and remained substantially below the predecontamination levels. The only radioactive waste generated during this decontamination was ion exchange resin.

Other nuclear sites are planning for a low concentration chemical decontamination. Commonwealth Edison Company is planning to employ low concentration chemicals to decontaminate one or more of their reactors in the next few years and Vermont Yankee is planning to employ low concentration chemicals to decontaminate a regenerative heat exchanger in late September, 1979.

High Concentration Chemical Reagents

A number of plants have employed high concentration chemicals to decontaminate the reactor coolant system or steam generators. The decontamination of reactor coolant systems and steam generators was reviewed up to 1969 by Ayres.[11] A more recent literature review concerning decontamination reagents was recently published by EPRI.[12]

The Plutonium Recycle Test Reactor (PRTR) at Hanford was decontaminated twice during the 1960s. The first decontamination performed in 1962 required a number of chemical reagents to remove both the fuel debris and the activated corrosion products.[11] During the removal of the corrosion products oxalic acid was employed. It was determined that extended exposure of the reactor coolant system to oxalic acid caused a ferrous oxalate precipitate to collect on the piping. This precipitate led to low decontamination factors in a number of areas. To alleviate this precipitation problem, ammonium citrate plus EDTA were employed in place of oxalic acid. Employing this latter reagent, DFs of 1.8 to 56.5 were obtained.

The second PRTR decontamination took place in 1965.[13] The technique employed was a two-step process employing alkaline permanganate (3% $KMnO_4$) followed by a solution of ammonium citrate plus ammonium oxalate (3 to 5%). The decontamination factors obtained ranged from 12 to 350.[14]

The Dow Chemical Company has developed a high concentration chemical called NS-1 which has been employed to decontaminate the regenerative heat exchangers at Peachbottom 2 and 3.[15] The decontamination factor obtained was 2 to 10. The high level radioactive waste was solidified in 55-gallon barrels. Thirty-eight of these barrels were needed for Unit 2, and 34 barrels were required for Unit 3. The Dow solidification process was employed for this waste.

Other reactors or isolable components that have been decontaminated include N-Reactor steam generator at Hanford,[11] Shippingport,[16] and the steam generators at Sena.[17] All of these decontaminations employed at least a two-step process. The first step was the addition of alkaline permanganate; the following step involved the addition of sulfuric acid, diammonium citrate, or ammonium oxalate plus diammonium citrate. In all cases the solutions contained a suitable corrosion inhibitor. In all cases the decontamination was successful with most reported decontamination factors being greater than 10.

Other nuclear sites are planning for reactor cooling system decontamination. Dresden 1 appears to be the next site scheduled for reactor coolant system decontamination. This is currently scheduled for fall of 1979. Dresden 1 personnel will employ the

Dow NS-1 reagent for this decontamination. The NS-1 solution and rinse will be processed through a forced circulation evaporator and the bottoms will be solidified with the Dow system. Other papers in this symposium will present an update on this decontamination.

Research in Progress

Battelle-Northwest[18,19] and General Electric Company with the Commonwealth Research Corporation have programs to develop low concentration chemicals for use in both BWRs and PWRs while General Electric Company is developing chemicals for BWRs. Results and updates from both programs will be presented at this symposium.

Westinghouse is developing a slurry abrasive grit decontamination technique for use on reactor components including steam generator tube sheets. This technique is expected to be available commercially in 1980. These are the only programs that the authors have identified concerning new decontamination techniques being developed in the United States.

Conclusion

A number of decontamination techniques have been found to be effective for removing contamination accumulated during operation of a nuclear power reactor. Components as small as miscellaneous hand tools and equipment up to the size of a reactor coolant system have been decontaminated; however, the reactor coolant system of a commercial LWR has not yet been decontaminated.

The majority of the decontamination experience is on the smaller components. For these removable components a number of effective decontamination techniques are available. Generally acceptable decontamination techniques for larger components or the more intricate components are lacking. A number of decontamination vendors have proprietary solvents which may be effective on BWR or PWR films; however, few of these solvents have been demonstrated on the larger components. Material compatibility testing of these reagents has not been reported in the open literature.

Decontamination at most nuclear power plants appears to be the responsibility of the person who created the contamination problem or who is responsible for inspecting, repairing, or replacing a component. This responsibility appears divided between the health physics, chemistry and maintenance departments. The survey results indicate that decontamination techniques may be handled more efficiently if centralized in one department.

The survey results indicate that research in progress appears to be aimed at the needs of commercial nuclear utilities except in the area of corrosion or material compatibility studies. High concentration and low concentration chemical decontamination for BWRs and PWRs that have been tested with the major materials of construction for material compatibility would be helpful to the industry.

Acknowledgement

This work was funded by the Electric Power Research Institute Contract No. TPS 78-816.

References

1 John F. Remark, "Plant Decontamination Methods Review", EPRI No. NP-1168, to be published in 1980.

2 Trademark, Atomic Products Corporation, Center Mariches, NY.

3 Trademark, Clorox Company, Oakland, CA.

4 E. A. Neppuras, Ultrasonics, 10, 9 (1972).

5 T. F. D'Muhala, "Ultrasonic Cleaning in Decontamination of Nuclear Reactors and Equipment", J. A. Ayres (Ed). Ronald Press Company, NY, 1970.

6 H. W. Arrowsmith and R. P. Allen, "New Decontamination Technique for Exposure Reduction", presented at U.S. DOE Environmental Control Symposium, Washington, DC, November 28-30, 1978. Battelle Pacific Northwest Laboratory, PNL-SA-7279.

7 R. P. Allen, H. W. Arrowsmith, L. R. Charlot, J. L. Hooper, "Electropolishing as a Decontamination Process: Progress and Application", Battelle Pacific Northwest Laboratory, PNL-SA-6858, April 1978.

8 S. G. Sawochka, P. S. Wall, J. Leibovitz, W. L. Pearl, "Effects of Hydrogen Peroxide Additions on Shutdown Chemistry Transients at Pressurized Water Reactors", EPRI Report No. NP-692, April 1978.

9 P. J. Pettit, J. E. LeSurf, W. B. Stewart, R. J. Strickert, S. B. Vaughan, "Decontamination of the Douglas Point Reactor by the Can-Decon Process", Paper No. 39, presented at Corrosion/78, Houston TX, March 1978.

10 J. E. LeSurf, "Control of Radiation Exposures at CANDU Nuclear Power Stations", J. Br. Nucl. Energy Soc., 16(1), 53 (1977).

11 J. A. Ayres (Ed), "Decontamination of Reactors and Equipment", Ronald Press, NY, 1970.

12 G. R. Choppin, R. L. Dillon, B. Griggs, A. B. Johnson, Jr., J. F. Remark, A. E. Martell, "Literature Review of Dilute Chemical Decontamination Processes for Water-Cooled Nuclear Reactors", EPRI NP-1033, Research Project 828-1, March 1979.

13 J. A. Ayres, L. D. Perrigo, R. D. Weed, "Decontamination of a PWR", Nucleonics 25(4), 58 (1967).

14 R. D. Weed, "Decontamination of the Plutonium Recycle Test Reactor (PRTR) Primary System", USAEC Report BNWL-711, Battelle Northwest Laboratories, March 1968.

15 G. E. Casey, "Peachbottom 2 and 3 Regenerating Heat Exchangers", presented at Pennsylvania Electric Association Power Generation Committee Meeting, February 16, 1978.

16 C. S. Abrams, E. A. Salterelli, "Decontamination of the Shippingport Atomic Power Station", Bettis Atomic Power Laboratory, WAPD-299, January 1966.

17 H. M. Couez, L. F. Picone, "Decontamination of a PWR Primary System, Sena Plant", presented at American Power Conference, Chicago, IL April 20, 1971.

18 J. F. Remark, R. L. Dillon, B. Griggs, "Investigation of Alternate Methods of Chemical Decontamination - Semiannual Report, July 1, 1976 to December 31, 1976", Battelle-Northwest Laboratories, BN-SA-703, August 1977.

19 R. L. Dillon, B. Griggs, J. F. Remark, "Investigation of Alternate Methods of Chemical Decontamination - Second Progress Report, January 1, 1977 to December 31, 1977", Battelle-Northwest Laboratories, BN-SA-703-2, June 1978.

CHEMICAL DECONTAMINATION OF KWU REACTOR INSTALLATIONS

R. Riess
H. O. Bertholdt

Kraftwerk Union AG, Erlangen
Federal Republic of Germany

SUMMARY

KWU has developed decontamination methods for both PWRs and BWRs. These methods can be applied either for single components or complete primary heat transport systems (PHT). They were developed giving consideration to the different materials and operating conditions of the systems to be decontaminated. The available KWU methods are:

- Two-step process:
  This method is a modified APAC process which is chiefly applied for high-alloyed materials. It was used several times in the field.

- One-step process:
  This method is used for isolated systems or components of low-alloyed materials. An application is scheduled for the main steam line of a BWR.

- One-step process for total PHT systems:
  This procedure can be applied to both PWRs and BWRs. The laboratory development has been completed.

The above processes demonstrate that KWU can meet a wide range of decontamination needs.

INTRODUCTION

There is a close relation between the production of energy in nuclear power plants and that of radioactive substances. These radionuclides are transported, distributed and deposited in the complete system and result in the personnel being exposed to radiation. Until the present day a constant increase in the level of radiation has been observed in all nuclear power plants. Furthermore, consideration must be given to the fact that, as the number of nuclear power plants being placed in operation increases, the number of persons available for actual operation, repair work and special assignments becomes more and more restricted. In order to keep the radiation exposure of these persons as low as possible, especially during repair work and outage periods, it is necessary to limit the radiation dose rate within the plants. In order to achieve this, there are several possible solutions, such as:

- the selection of suitable materials for systems and components (e.g. low Co content)
- suitable grouping of the components
- measures taken with regard to process engineering (e.g. optimum clean-up rate)
- shielding (fixed or movable shielding)
- careful operations scheduling resulting in time-saving during repair work
- decontamination.

This report is concerned exclusively with the latter possibility, namely decontamination.

Either physical or chemical methods can be applied for the decontamination procedure. The extent to which physical methods, such as blasting treatment of surfaces (using glass beads, steam, water, B_2O_3, etc.) may be applied is restricted since a precondition for these is the accessibility of the surfaces to be treated. In addition, these processes tend to abrade not only oxide films but also base material, to roughen the surface and to effect the penetration of the radioactive oxide film into the base material. For this reason, this report is concerned exclusively with chemical decontamination. Experience gained with this process, which is available to Kraftwerk Union in this field, will be explained and described in more detail.

PRESENT LEVEL OF KNOWLEDGE

The occupational radiation exposure of the personnel of pressurized light water reactors built by KWU has decreased considerably in the newer plants in comparison to the older ones.[1] Table I shows the development. Measurements of the radiation exposure of the auxiliary personnel were made as well. Their dose proportion varies between 35% and 75% as related to the total dose.

A breakdown of the total dose into different work procedures is shown in Table II. As can be seen from this table, the radiation exposure during normal operation amounts to only 30%. This means that efforts to reduce the dose during power operation do not yield a marked reduction in the total dose, but if one succeeds in saving only 20% of the dose during the major inspection phase, this would equal a 50% reduction in the power operation dose. In any case the figures in Table II indicate that inspection and repair operations contribute the bulk of radiation exposure. The decontamination procedures which KWU has developed until now are therefore predominantly aimed at minimizing the radiation exposure to repair and maintenance personnel. The areas in which this kind of work is carried out can be described with the aid of a classification of the decontamination methods. This classification can be seen in Table III.

This report is concerned in particular with Field 3, but Field 2 is constantly gaining in importance. In Field 3 the main areas of development and application existed with respect to the available decontamination processes. Hard decontamination constitutes the complete removal of oxide films from the material surfaces, for which KWU developed a modified APAC process. As indicated in Section 3, it has been possible to apply this process several times in actual operation. In addition, KWU developed a soft decontamination process, the aim of which is a partial removal of the oxide films. The procedure has until now only been tested under laboratory conditions.

Besides hard decontamination and the one-step process which effects partial decontamination, KWU has developed processes for the decontamination of special systems in nuclear power plants. These procedures are aimed in particular at the decontamination of carbon steel components, with the main-steam piping system of BWR plants being a typical example.

The present state of knowledge at KWU resulted in the fact that the company now offers, apart from the previously described decontamination processes, the hardware for future decontamination operations for the area of auxiliary systems, not only for PWR but also for BWR plants. Commencing with the Grohnde and Philippsburg II plants, this hardware, a prerequisite for future decontamination operations, has been incorporated.

Table I: Occupational Radiation Exposure to the Personnel of Pressurized Light Water Reactors Built by KWU. Mean Values are Set in Paranthesis

	Annual Dose, man-rem		
	Power operation including short shut downs	Refuelling (revision and repairs)	Total dose
Demonstration power plant	140-500 (330)	360-850 (530)	600-1300 (860)
Power plants of the first commercial generation	90-150 (120)	200-450 (320)	300-550 (440)
Power plants of the second commercial generation	30-60 (40)	150-400 (280)	175-450 (320)
Aim for the standard power plant	≤50	100-150	150-200

Table II: Breakdown of the Annual Dose for KWU Pressurized Water Reactors (Operation of Commercial Power Reactors Over Many Years)

Power operation (incl. small shut downs)	
● Actual operation	50-90 man-rem
● Inservice inspections	10-20 man-rem
● Repairs	10-50 man-rem
	Average value: approx. 120 man-rem
Revision phase	
● Refuelling	10-20 man-rem
● General auxiliary work for refuelling (checks, cleaning, radiation protection etc.)	15 man-rem
● Inservice inspections	110-170 man-rem
● Repairs	30-80 man-rem
● Constructional works	20-50 man-rem
● General auxiliary work for inservice inspections and repairs (radiation protection, cleaning etc.)	40-60 man-rem
	Average value: approx. 300 man-rem

Table III: Chemical Decontamination of Reactor Systems
Fields of Decontamination

Field 1	Field 2	Field 3
Daily Routine	Temp. $<200\,°C$	Temp. $>200\,°C$
Floors Walls Clothing Personnel	Auxiliary Systems BWR/PWR Water/Steam Cycle BWR	Primary System BWR/PWR
1) Washing 2) Wiping 3) Flushing	1) Conventional Cleaning and Pickling Procedures	1) Hard Decontamination 2) Soft Decontamination

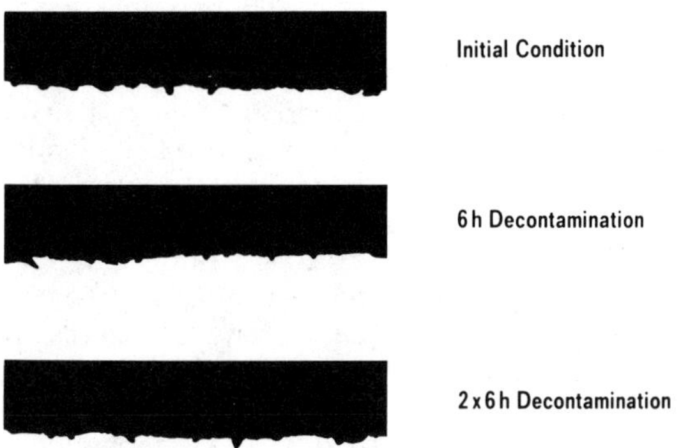

Figure 1: Decontamination of SS 304
The unstabilized and cold formed
Material was slightly pitted (10 years of operation)

It is evident from the description of the state of knowledge that KWU's strategy has until now been orientated towards the problem areas of plants and directed predominantly at the development and application of an efficient two-step process. However, it may be possible that in future the emphasis of decontamination process will be shifted to other areas. In this context the main objective is the application of the one-step process to the decontamination of complete loops, as well as the performance of decontamination in the so-called Field 2. This would mean that, above all, effective processes must be available in addition for inspection and maintenance work in the area of the auxiliary systems.

All developments in the field of decontamination obviously take the experiences described in the literature into consideration, special attention may be drawn to Schroeder, Ayres and the summary reports of the Ohio State University seminar on decontamination.[2,3,4]

KWU DECONTAMINATION PROCESS

DECONTAMINATION OF HIGH-ALLOYED MATERIALS (TWO-STEP PROCESS)

The KWU two-step process, which represents a modified version of the APAC process described in the literature, was given preference in the KWU development program, because the emphasis of the difficulties in the plants was on the decontamination of reactor coolant pumps (RCPs), pressurizer heater rods, secondary steam-to-steam heat exchangers, etc. In each case the KWU two-step process is based on the assumption that higher-grade materials are incorporated in the systems to be decontaminated. In KWU PWR plants the composition of the oxide films is determined by the materials used in the primary loop, namely Incoloy 800 or austenite. Thus it can be inferred that a high proportional content of nickel and chrome is present in the insulating layers. The proportion of nickel and chrome in the insulating layers of BWR plants is slightly smaller and their composition is determined essentially by the constituents of the austenites used. In the main, the elements iron, nickel and chromium are present in the following compound forms:

- as stoichiometrically combined spinel compounds, e.g. $NiFe_2O_4$
- as non-stoichiometrically combined spinel compounds, e.g. for cobalt, $CoNi_{1-x}Fe_2O_4$
- As an oxide, e.g. Fe_3O_4/NiO

At temperatures of less than 100°C the above metallic oxides are extremely insoluble, even in mineral acids and thus are not amenable to a conventional pickling or cleaning operation. The methods to be applied in such cases for the removal of metallic oxides are described in the literature.[2,3] On the basis of these findings, KWU developed a modified APAC process which provides adequately high decontamination factors and, in addition, avoids any specific attack on the base material.

It was possible to use the KWU decontamination process for the first time for the decontamination of the reactor coolant pumps (RCPs) of the Biblis A (KWB-A) and Biblis B (KWB-B) plants. Both plants were operated under the same operating conditions (1 - 2 ppm Li, 2 - 4 ppm H_2 and 1.220 ppm B) at an operating temperature of approx. 290°C in the area of the RCPs. Decontamination was necessitated to permit repair work and alterations to the pump inlet nozzles. The decontamination treatment was performed in immersion baths at temperatures of around 90°C. In the case of Biblis A the pump was dismantled prior to decontamination, whereas, in the case of Biblis B, the pump was immersed in the baths as a complete unit. Further decontamination work using the KWU two-step process was carried out in the Borssele and Biblis A plants for the replacement and disposal of defective pressurizer heater rod bundles. In both cases an external decontamination loop was set up and during decontamination the surfaces were given an additional brushing during the final water flushing. Further decontamination work on PWR plants were carried out on the RCPs of the Neckarwestheim, Biblis A and Biblis B plants. Experience was gained using the KWU decontamination processes on BWR plants for example in the Brunsbuttel plant. Here a decontamination of the internal recirculation pumps was necessary in order to permit the performance of maintenance and repair work. Similar to the decontamination work on the RCPs in Biblis A and B, the decontamination treatment was carried out in immersion baths. Furthermore, the decontamination of the primary channel heads of the secondary steam-to-steam heat exchangers, as well as of sections of the piping system was performed. These operations are described in the literature.[5]

In conjunction with the decontamination of sections of the primary system at the Gundremmingen plant, attention may be drawn to the fact that, from the point of view of material, this was the most difficult area in which decontamination had been performed. In Gundremmingen a non-stabilized austenite (SS 304) was used in the primary system. This cold formed material was during the course of its 10-year service life prior to decontamination slightly pitted. By means of a pre-approval test and procedure test coupons during manufacture, it could be demonstrated that an austenite pitted during operation suffers no further pitting during decontamination (Figure 1).

All the results achieved until now at KWU plants using the two-step decontamination process are summarized in Table IV. In

Table IVa: Applications of the KWU two-step-process

Unit	Operating Period	Component	Material	Decontamination Period	Dose Rate Initial m R/h	Dose Rate After Decontamination m R/h
KCB	28 Months	Heater Rod Bundle II	X 2 CrNiMo 1812	10,5 h	2000–3000	80–300
		Heater Rod Bundle III	X 2 CrNiMo 1812	20,5 h	2500–3000	45–300
		Heater Rod Bundle IV	X 2 CrNiMo 1812	10 h	3500–6000	50–200
KWB-A	20 Months	Heater Rod Bundle I	X 2 CrNiMo 1812	11,5 h	3000–5000	5–7
		Heater Rod Bundle III	X 2 CrNiMo 1812	6 h	500–2000	15–20
KRB	10 years	Connection Piece Loop I, Cold Side	SS 304	6 h	2000–4000	10–50
		Connection Piece Loop III, Cold Side	SS 304	6 h	2000–4000	10–20
		Connection Piece Loop III, Hot Side	SS 304	6 h	2000–4000	

Table IVb: Applications of the KWU Two-step-process

Unit	Operating Period	Component	Material	Decontamination Period	Dose Rate Initial m R/h	Dose Rate After Decontamination m R/h
KWB-A	12 Months	Impeller YD 10	G-X5 CrNi 13 4	11 h	7000	75
		Impeller YD 30	G-X5 CrNi 13 4	8,5 h	7000–10000	50–70
KWB-B	4 Months	Impeller YD 10	G-X5 CrNi 13 4	3 h	700	25
		Impeller YD 20	G-X5 CrNi 13 4	2 h	700	15–18
		Impeller YD 40	G-X5 CrNi 13 4	2 h	700	25
KKB	6 Months	Impeller P 2	X 6 CrNiMo 16 6	9 h	6000	50
		Impeller P 3	X 6 CrNiMo 16 6	10 h	1000	15
		Impeller P 5	X 6 CrNiMo 16 6	5 h	5000–15000	150–1500
		Impeller P 6	X 6 CrNiMo 16 6	3–5 h	20000–45000	200–3000

some cases very high decontamination factors may be determined on a purely arithmetic basis. On the basis of the results available, it would, however, appear more desirable no longer to talk of decontamination factors, but rather of achievable ultimate levels of activity. It is the aim of all practical applications of the KWU two-step process to achieve an ultimate level of activity of less than 50 mRem/h. If, in individual cases, this value is not achieved, it is merely a question of the duration of the treatment which, under certain circumstances, may be influenced by parameters peculiar to the particular plant.

Figure 2 shows a main coolant pump before and Figure 3 after a decontamination process.

DECONTAMINATION OF CARBON STEEL SYSTEMS

As already mentioned, the KWU two-step process described in the previous section can only be used for the decontamination of systems which have been manufactured in higher grade materials. In the case of decontamination of systems made of carbon steel, processes shall be applied which are used for conventional pickling. Each system to be decontaminated necessitates alterations to the process to be applied. In this section the chemical decontamination of the main steam piping system, as well as of the relief and auxiliary steam supply trains within the pressure suppression system of the Wurgassen nuclear power plant shall serve as a practical example. Decontamination was performed in June/July 1979. This was the first decontamination process in the Federal Republic of Germany, for the performance of which the licensing authority or an Authorized Inspector were called in. Prior to commencement of decontamination the Consultant (TUV) demanded that the following points should be explained in writing in a complete description of the procedure:

- Responsible persons (nomination of responsible representatives of both KWU and the Operator)
- Pre-approval, procedure qualification test
- Chemical Engineering
- Chemical treatment
- Supervision, checking, evaluation
- Accident considerations
- Flow chart

Figure 2: Decontamination of a Main Coolant Pump 1300 MW PWR
Pump before Decontamination Surface Dose Rate: ~ 30 R/h

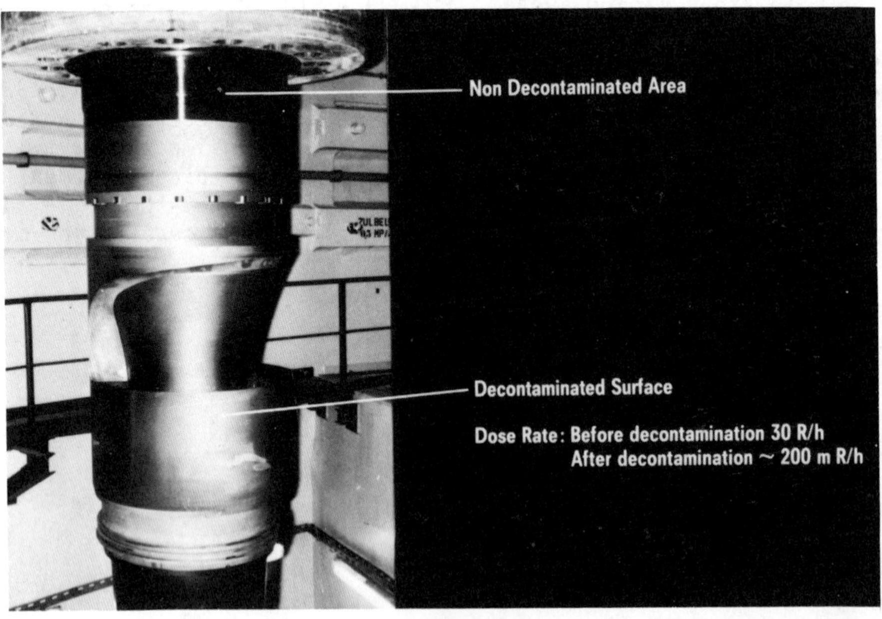

Figure 3: Decontamination of a Main Coolant Pump; 1300 MW_e BWR

- Assessment of radiation exposure

- Documentation.

The following explanations can be given to these points:

Responsible Persons

The representatives to be nominated by both the Operator and KWU were the main contacts of the Consultant both during the preparatory and implementation work for chemical decontamination.

Pre-Approval

Prior to the implementation of the chemical decontamination, the Consultant demanded a pre-approval of documents which had to include a description of material corrosion resistance, isolating valves, process technology, operations sequence, mandatory tests/ examinations, as well as of the surveillance of the decontamination procedure.

In order to evaluate the chemical decontamination process, a material corrosion resistance test was performed in the decontamination solution. During this test, selected material combinations were electrochemically monitored during treatment in the decontamination solution and examined metallographically after treatment.

Chemical Engineering

In a comprehensive documentation the operation sequences were described in detail in the form of flow charts and check lists. In addition, proof had to be established of corrosion resistance for all the organic materials used, e.g. flexible tubes. Proof was also required of the Manufacturer's and welders' qualifications, as well as of the performance of welding work.

Description of the Process

The chemical decontamination of the main steam lines was performed using an inhibited aqueous solution which was circulated through the system areas to be decontaminated with the aid of a temporarily constructed external decontamination loop. The areas of the Wurgassen plant to be decontaminated are shown in Figure 4. After the chemical decontamination treatment the system was post-treated using an alkaline oxidizing passivation solution. The delimitation of the system areas shown was accomplished by isolating equipment (see e.g. Figures 5 and 6). The drainage pipes leading out of the system were sealed for decontamination and were then flushed with demineralized water after decontamination. The spent decontamination solution was collected in the radwaste concentrate collection tank and then immobilized with Portland cement for disposal. The spent passivation and flushing solutions were deactivated in the evaporator feedwater tank.

Chemical Treatment

The chemical decontamination is divided up into the following operational steps:

- Trial operation using demineralized water
- Chemical decontamination treatment using an inhibited organic acid solution
- Flushing using demineralized water
- Chemical passivation treatment using an alkaline oxidizing solution
- Flushing using demineralized water
- Drying the system
- Lay up of the system.

Supervision, Check-up and Evaluation

The supervision, check-up and evaluation of the chemical decontamination was guaranteed by the following tests/examinations prior to, during and after performance:

- Procedure qualification test (material corrosion resistance test by means of simulation of the decontamination procedure)
- Analysis of stagnant pockets and leak test of the decontamination loop prior to decontamination
- Superveillance measures during the chemical decontamination
- Evaluation by means of the metallographic examination of procedure qualification surveillance coupons.

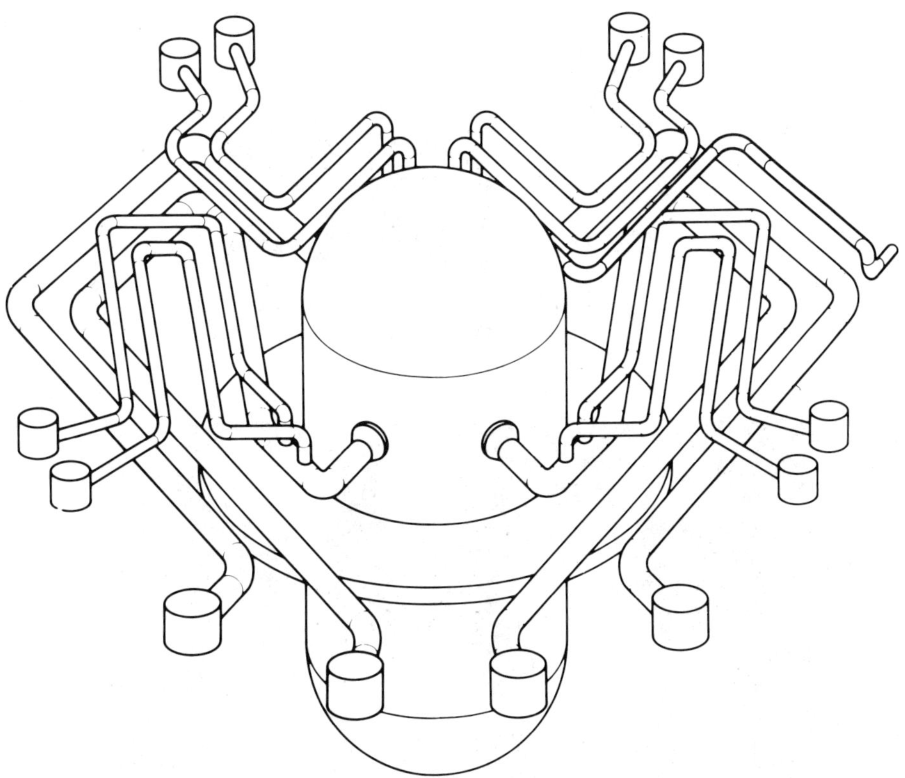

Figure 4: Decontaminated Areas in the Würgassen BWR

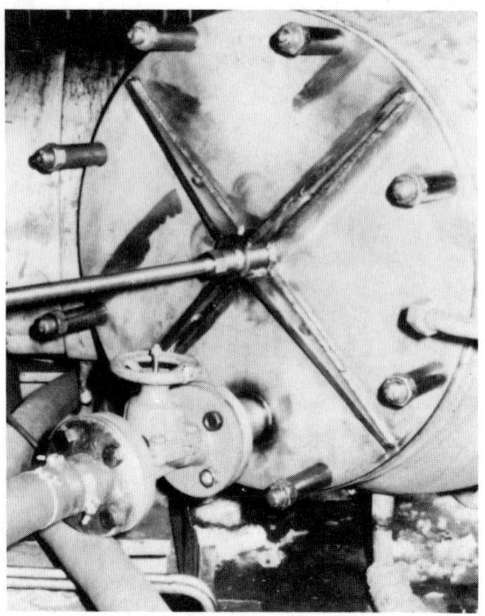

Figure 5: Decontamination of the Würgassen – BWR
Isolation Equipment of the Main Steam Line

Figure 6: Decontamination of the Würgassen – BWR
Isolation Equipment of the Main Steam Line

Accident Considerations

All conceivable accidents which could occur during decontamination were to be described in detail.

Flow Chart

All the details of the performance of decontamination in the field were to be shown in a flow chart. With the aid of check lists which had to be accepted by the Consultant, this operation sequence was monitored in detail.

Assessment of Radiation Exposure

The levels of radiation exposure of the decontamination personnel to be expected during the performance of decontamination were to be assessed in advance. The anticipated total radiation exposure to the decontamination personnel amounted to 10 man-rem. The number of staff employed on site was 12. The assessed total radiation exposure was not exceeded.

The results of the decontamination of the main steam lines in Wurgassen are shown in Figure 7. The curves shown in Figure 8 represent the four subsections of the main steam line system which were decontaminated. The diagrammatical representation indicates that the treatment was completed within 9 hours. The main criterion for the termination of the decontamination was the total activity in the decontamination solution. With initial levels of activity of 2000 - 3000 mRem/h, ultimate levels of activity were achieved of less than 200 mRem/h. This corresponds to decontamination factors of the order of 10. Decontamination factors of about 3 were obtained at areas with lower initial activities. These measured values were obtained by direct measurements within the decontaminated piping system.

ONE-STEP PROCESS

The development work on a one-step process for the decontamination of complete closed loops has been terminated. The aim of development was to make the application of this process possible in both PWR and BWR loops. The principle of this process is to add the decontamination chemicals to the reactor coolant at reduced temperatures (100 - 150°C). The coolant in the loops is recirculated in the reactor by coolant pumps (RCPs). The reactor coolant is simultaneously cleaned by the decontamination chemicals within the existing primary circuit pruification system. The ion exchangers of the primary circuit purification system are replaced

Table V: Chemical Decontamination of Primary Systems and their Components

Hard Decontamination	Soft Decontamination
Removal of Grown Oxide Layers Formed at Temperature >200 °C	
– High DF s	– Adequate DF s
– Total Removal of Oxide Layers	– Partial Removal of Oxide Layers
– Short Treatment Times	– Long Treatment Times
– Avoidance of Base Metal Attack	– Avoidance of Base Metal Attack
– Subsystem and Parts Decontamination	– Decontamination of Entire Systems
– High Waste Occurence	– Low Waste Occurence

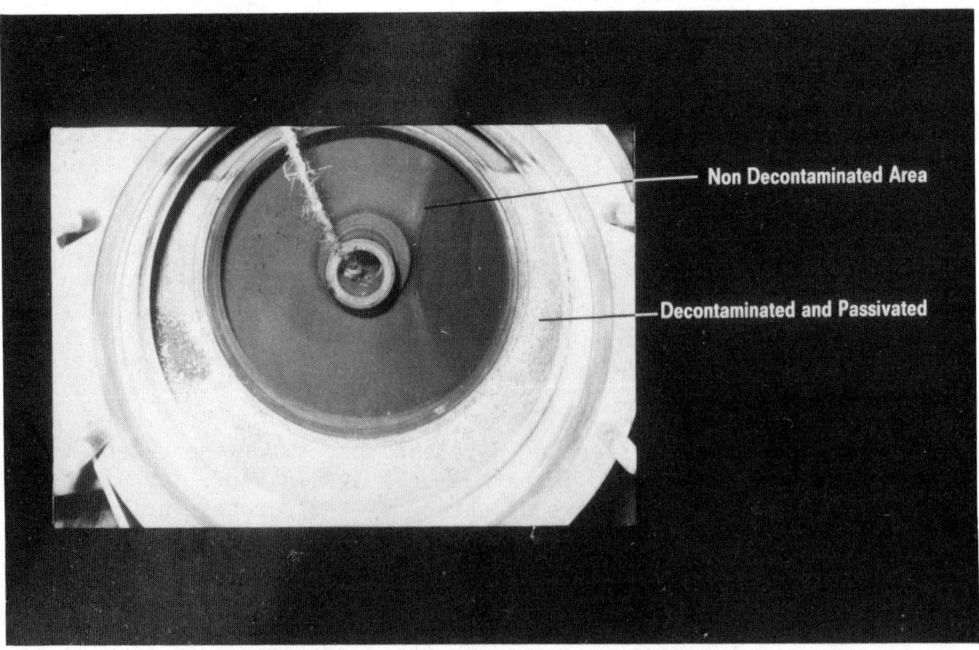

Figure 7: Decontamination of the Main Steam Line on the Würgassen – BWR

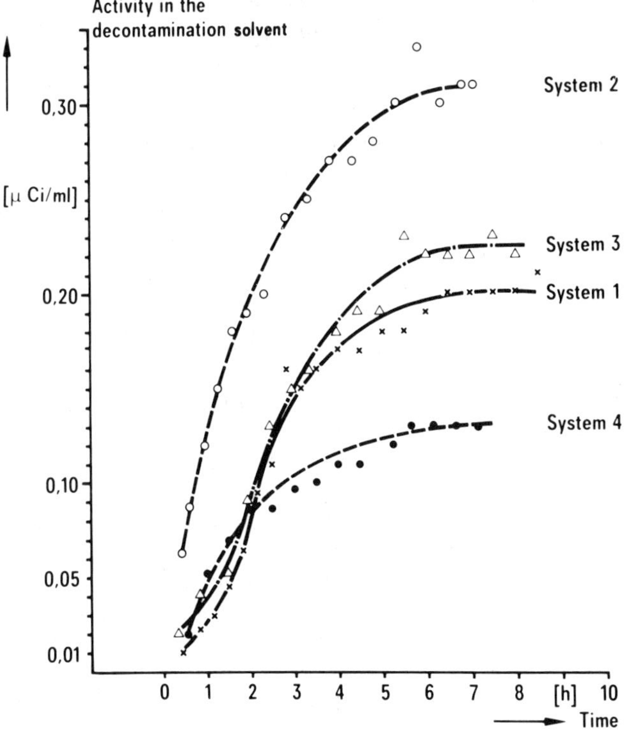

Figure 8: Decontamination of the Main Steam Line System of Würgassen BWR

during this decontamination process by special resins. The decontamination chemicals are regenerated after having passed through these ion resins, with the boron content of the reactor coolant remaining constant. The main principle of this process corresponds to the Can-Decon process. The advantages and disadvantages of the one-step procedure (soft decontamination) as opposed to those of the two-step procedure (hard decontamination) described in section 3 are listed in Table V. At present KWU is holding discussions with two operators on the possible practical application of this process to PWR plants.

PROSPECTS

The comments made in this report show that one process alone is not adequate to solve all the problems involved in decontamination, but rather be provided. The KWU strategy provides therefore that the following concepts be retained or developed further:

- Two-step process (modified APAC process for the decontamination of plant components or for the decontamination of systems of isolatable subsystems)

- One-step process for the decontamination of carbon steels, above all for practical application in BWR plants

- One-step process for the decontamination of complete reactor systems of PWR and BWR plant.

The processes previously developed and at present available complement each other. Furthermore they are to be further developed, including the associated process engineering.

Provisions are being made at present at KWU for the construction of a mobile automatic decontamination unit which is specially designed for the application of the two-step process.

An essential objective in the development for the future will be to design the auxiliary equipment in such a way that they are decontaminable. In new plants under construction, the necessary installations (piping of the connection nozzles, etc.) are all ready provided in the planning and actual construction stages.

REFERENCES

1. H. Untervossbeck and H. Weber, "BNES Meeting on Radiation Protection in Nuclear Power Plants and the Fuel Cycle," London 1978, page 167 - 175.
2. H. J. Schroeder, "VGB-Kraftwerkstechnik (Power Plant Technology), Vol. 56 (5), page 357, 1976. (VGB = Technical Association of Operators of Large Power Plants)
3. J. A. Ayres, "Decontamination of Nuclear Reactors and Equipment," The Ronald Press Company, New York, 1970
4. "Seminar on Decontamination of Nuclear Plants," 7-9 May, 1975, Columbus, Ohio, USA, Co-Sponsored by Electric Power Research Inst., American Society of Mechanical Engineers, The Ohio State University.
5. N. Eickelpasch and M. Lasch, "Atonwirtschaft," May, 1979 issue, page 247.

CANDIDATE REAGENTS FOR ACTIVITY REDUCTION IN BWR AND PWR PRIMARY SYSTEMS

A. B. Johnson, Jr., B. Griggs, and R. L. Dillon

Corrosion Research and Engineering Section
Battelle, Pacific Northwest Laboratory
Richland, WA 99352

R. A. Shaw
Nuclear Engineering & Operations Department
Electric Power Research Institute
Palo Alto, CA

INTRODUCTION

This paper is based on work sponsored by the Electric Power Research Institute (EPRI) to investigate alternatives to the traditional campaign-type decontamination process for reducing radioactive fields around nuclear reactor primary systems. The strategy is to identify reagents which effectively remove activity when injected into the primary coolant during a normal reactor outage. Removal of the suspended radioactive species by filtration and ion exchange minimizes the volume of radioactive waste which must be processed.

Procedures of this type have been applied successfully to Canadian reactors,[1] but have not been applied in U.S. reactors, which involve different materials and configurations.

The crux of the program has been the availability of radioactive sections from components filmed by exposure in reactor primary systems. We have investigated piping from three boiling water reactors (BWRs) and are developing comparisons of reagent effects on steam generator sections from seven pressurized water reactors (PWRs).

Prior papers[2,3] and reports[4-7] have summarized the course of the program.

REAGENT CRITERIA

Reagents for decontamination of nuclear reactor coolant systems must meet several stringent criteria:

- effective removal and suspension of radioactive species from coolant system surfaces
- minimal corrosion to reactor coolant system materials by the reagent and its decomposition products
- adequate thermal and radiolytic stability under decontamination conditions, but degradable to innocuous species at primary system conditions
- tractable to activity removal by ion exchange and filtration
- free from species which activate to offensive isotopes in a neutron flux

SELECTION OF REAGENT CANDIDATES

Selection of candidate reagents began with an assessment of prior reactor decontamination technology[8] and with a survey of properties of prospective reagents.[4] Prior decontamination reagents have comprised organic acids, chelating agents, and in a few cases, mineral acids. There is a strong motivation to work with classes of compounds already qualified for reactor service if possible. Therefore, the reagent studies began using organic acids and chelating agents, separately and in selected combinations. The first studies were conducted on preoxidized laboratory specimens in stirred beakers and in the Once-Through Test Facility (OTTF). When specimens from operating reactors became available, the studies moved into the radioactive phase.

EXPERIMENTAL APPROACH

The principal experimental tool in the program is the Once-Through Test Facility (OTTF) shown in Figure 1a. It is difficult to separate variables such as concentration, pH, etc., in a recirculating facility. The OTTF provides a method to investigate decontamination parameters on radioactive specimens. In the OTTF, reagent is heated, flows over the specimen, through a cool-down heat exchanger, through a filter and into a collection and sampling system. The OTTF operates at temperatures up to 180°C. Reagent flow rates are ∿120 ml/min; linear flow rates are ∿1 meter/min. The surfaces in the OTTF and RTF systems are inert (Zircaloy and Teflon) except for the radioactive test specimen. Figure 1b shows schematically the configuration used for BWR specimens,

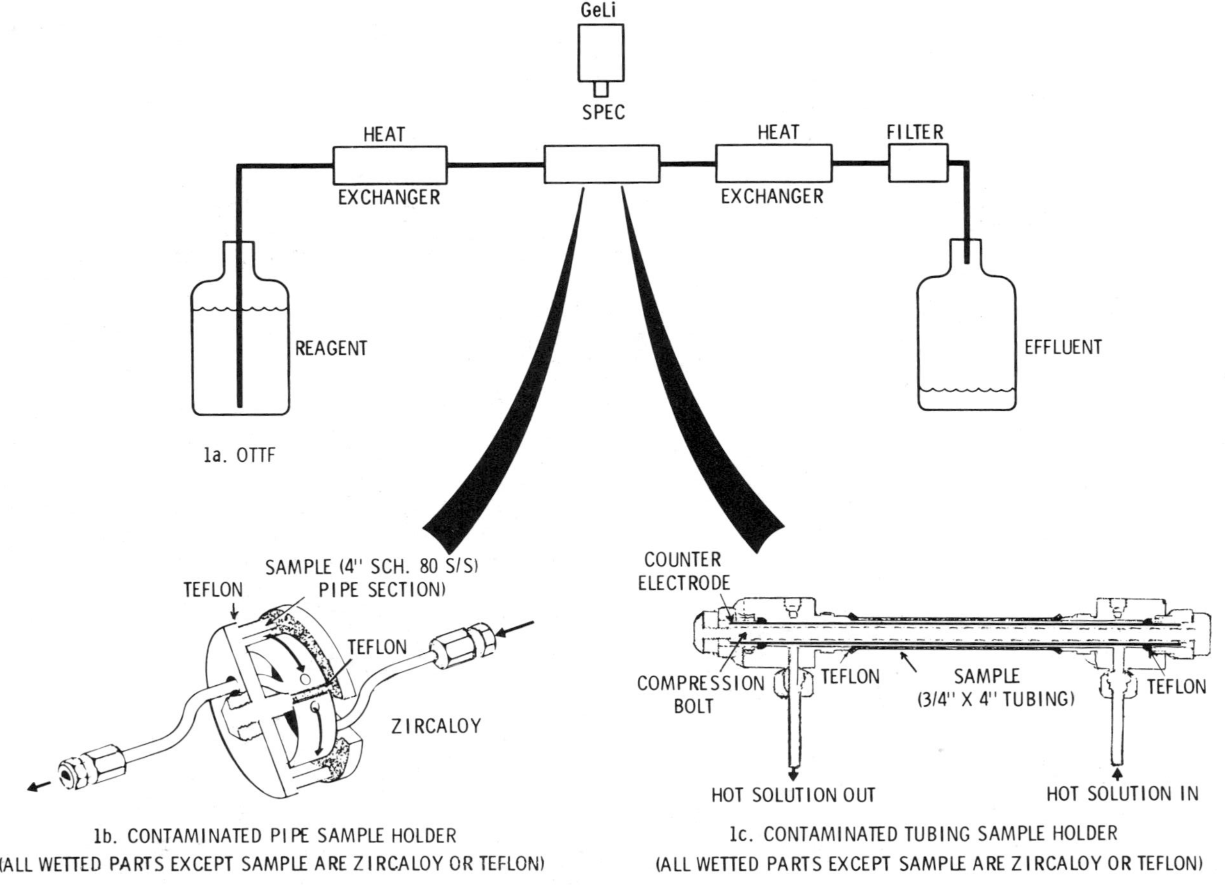

Figure 1. OTTF Configuration and Specimen Holders

involving a 2.5-cm wide section cut from 10-cm diameter 304 stainless steel recirculation bypass pipes. Figure 1c shows the PWR specimen configuration, involving a 10-cm length of Inconel 600 steam generator tubing, flared to seat in the specimen holder.

The course of a decontamination treatment is followed by:

(a) continuous radiation monitoring of the specimen using a Ge-Li detector
(b) periodic sampling of the OTTF effluent for both radioactive and nonradioactive (Fe, Cr, Ni) species
(c) inspection of selected specimens by scanning electron microscopy (SEM) to characterize effects of the decontamination on the crud/oxide layer and on the metal substrate.

A recent addition, the Recirculating Test Facility (RTF), permits testing under more prototypic recirculating conditions, also with radioactive specimens. Stainless steel recirculating autoclaves and stirred glass beakers have been used for tests with laboratory specimens.

REAGENT STUDIES - BWR SYSTEMS

Reagents tested on radioactive BWR specimens are summarized in Table I. The principal source of test specimens was a section of recirculation bypass piping from the Vermont Yankee reactor. Selected comparative runs were made on similar piping from the Millstone and Quad Cities reactors. Figure 2 compares the surface appearances of specimens from the three reactor pipes.

The compositions and morphologies for the three pipes differed. The Millstone deposits were duplex, with a high-iron (∿85%) outer layer and a lower-iron (∿60%) inner layer.[9] Copper was very low and zinc was 1 to 3%. Iron contents for the duplex Quad Cities deposit[10] was higher (90% and 68% respectively). Zinc and copper were not reported. The Vermont Yankee deposit consisted principally of faceted crystals and a compact layer, with some superficial particles and a shroud over the crystals. The outer layer was 85 to 93% Fe and the inner layer was 63 to 67% Fe. Copper was ∿1%; zinc was 9 to 19% in the compact layer. The principal radioactive isotope in the cruds from the three reactors ^{60}Co; ^{54}Mn was minor but detectable; ^{65}Zn also was detectable.

All three reactor pipes had depressions along grain boundaries, probably due to pickling during fabrication. The depressions were mild (up to ∿20 μm) on Millstone and Vermont Yankee pipe surfaces, but were much deeper (up to ∿250 μm) on the Quad Cities pipe surface.[10]

Table I. Reagents Tested on Radioactive 304 Stainless Steel BWR Pipe Specimens

Reagent	Range of Conditions: Concentration Molarity(a)	Temp °C	pH	Redox Conditions	Remarks
NTA	0.002	180	5.5	∿300 ppb O_2	Vermont Yankee
EDTA	0.00002 to 0.002	90 to 180	3.5,5.5, 7.5	∿300 ppb O_2	Vermont Yankee
EDTA	0.002	180	5.5	a) 43 ppm O_2 b) N_2H_4 (low O_2)	
EDTA Oxalic Acid Citric Acid	0.001 0.0005 0.0005	90	3.3 to 3.6	N_2H_4	Vermont Yankee (Beaker test only)
HEDTA Oxalic Acid Citric Acid	0.001 0.0005 0.0005	120 150	3.5	<50 ppb O_2 N_2H_4	Millstone, Quad Cities, Vermont Yankee
Oxalic Acid	0.002	120 to 150	3.5	<50 ppb O_2 N_2H_4 (MS/QC) NH_4OH (VY)	Millstone, Quad Cities, Vermont Yankee
HEDTA	0.002	120 to 150	5.5	<50 ppb O_2 NH_4OH	Vermont Yankee

(a) Conversions-molarity to ppm

NTA 0.002 M = 382 ppm
EDTA 0.002 M = 584 ppm
HEDTA 0.002 M = 556 ppm

Oxalic Acid 0.0005 M = 45 ppm
Citric Acid 0.0005 M = 105 ppm

Ms = Millstone-1
QC = Quad Cities-1
VY = Vermont Yankee

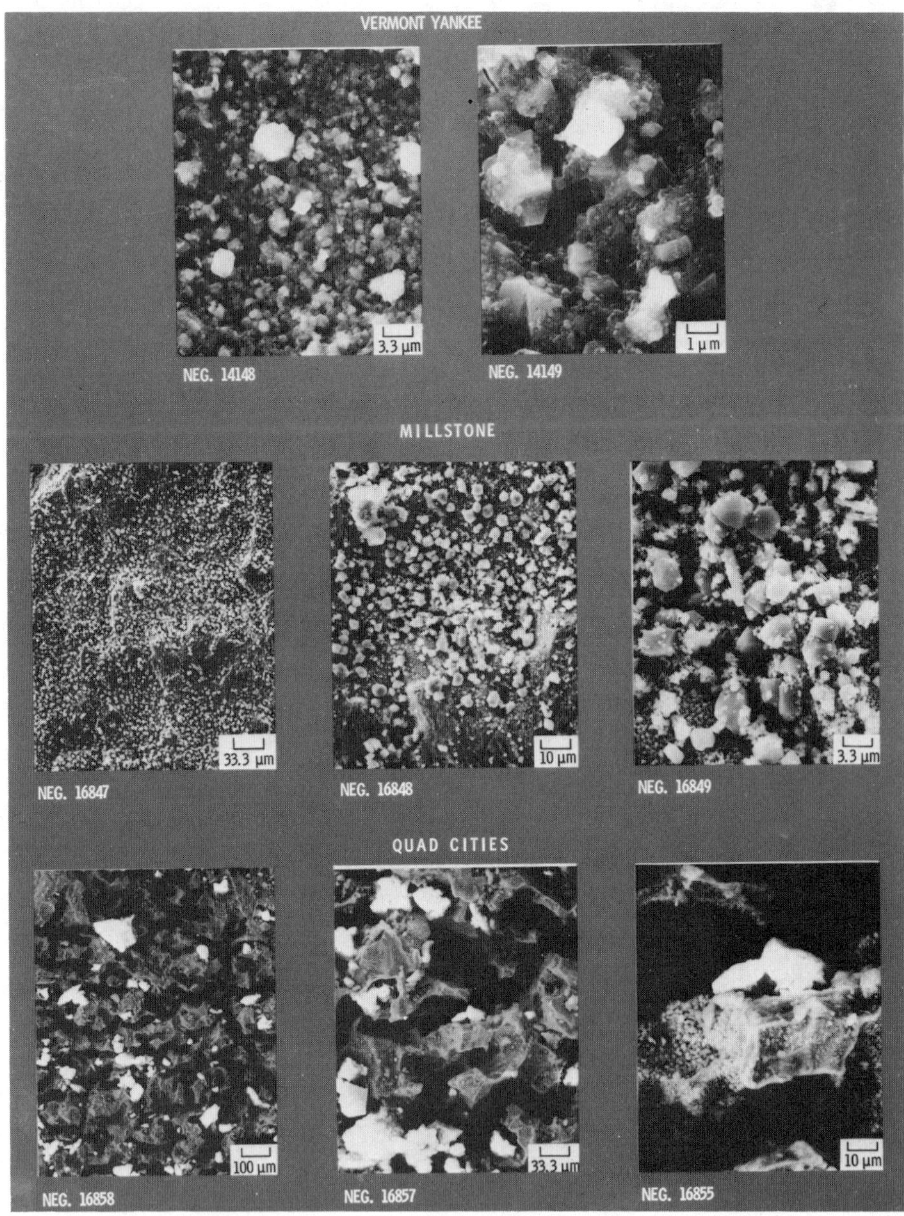

Figure 2. Surface Views (SEM) of Crud Layers on BWR Recirculation Bypass Pipe Specimens from 3 BWRs; no Reagent Treatment, Material is 304 Stainless Steel

Figure 3 compares ^{60}Co activity removal for specimens from the three reactors in a 3-component reagent. The Millstone and Vermont Yankee curves indicate that 94 and 88%, respectively of the ^{60}Co activity was removed after ∿300 min. By contrast, only ∿60% of the ^{60}Co was removed from the Quad Cities specimen. The lower activity removal from the Quad Cities specimen is attributed to more difficult transport of the radioactivity out of the deep depressions in the pipe surface.

Several important observations are derivable from Figure 3:

- Cobalt-60 removal by the 3-component reagent was similar to corresponding treatments by other reagents on Vermont Yankee specimens. Thus, several reagents are capable of effectively removing a large fraction of the activity from BWR pipe surfaces.
- Activity removal from Millstone and Vermont Yankee pipe surfaces was similar, despite differing crud morphologies and compositions, suggesting that a relatively large envelope of BWR cruds may respond similarly to a given reagent.
- The relatively poor activity removal from the Quad Cities surfaces is attributed to effects of the unusual surface morphology. This demonstrates the potential importance of characterizing surfaces to be decontaminated, where possible. The depth of surface depressions on the Quad Cities pipe appears to be unusual and probably represents a small fraction of the total surface area in the Quad Cities primary system.

Post-reagent inspections of Vermont Yankee surfaces by SEM indicated that the layer of faceted crystals remained. X-ray diffraction results suggest that the crystals are a form of nickel ferrite. Decontamination studies at General Electric Co. on pipe specimens from the Nine Mile Point reactor also identified a layer of undissolved crystals.[11] The crystals were easily detached in both sets of experiments. In both cases, 80 to 85% of the radioactivity had been removed by solubilizing the inner layer. Thus, the crystals contained a relatively small fraction of the total activity.

The detached crystal layer would almost certainly be removed under reactor flow conditions. Therefore, the decontamination process must be designed to deal with them, first by developing hydraulic conditions (not yet well-defined) which suspend the crystals; secondly, by providing effective filtration.

Results from this program and elsewhere[1,10,11] have demonstrated that several reagents can produce attractive decontamination factors on a variety of BWR pipe surfaces. Reducing conditions (e.g., using

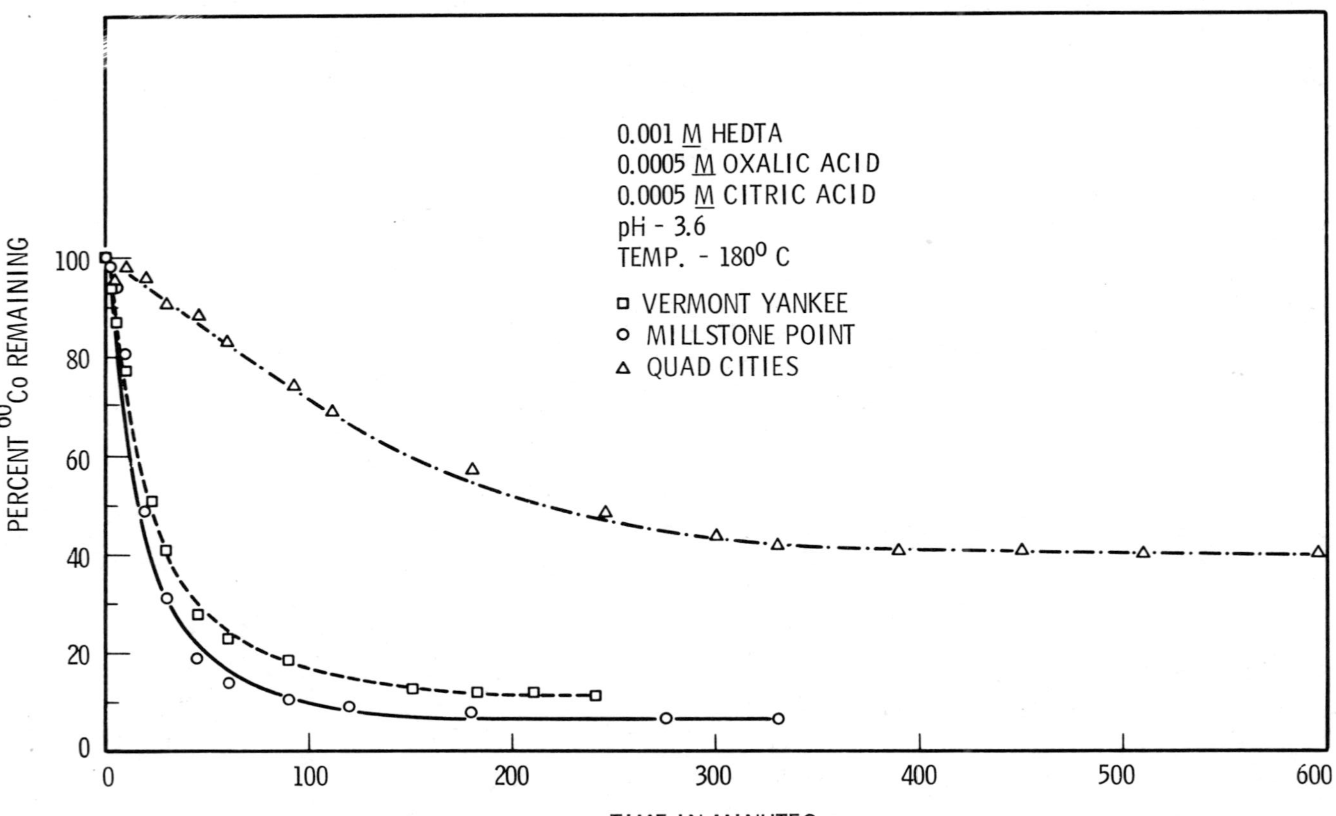

Figure 3. Decontamination of BWR Four Inch Pipe Specimens in OTTF; Material is 304 Stainless Steel

hydrazine additons) were most effective in activity removal from BWR surfaces.[7] Effectiveness of activity removal by ion exchange appears to favor HEDTA over EDTA because EDTA forms stronger complexes. Our studies on laboratory coupons suggest that NTA is more aggressive than EDTA toward stainless steel.[7] The 3-component (EDTA, NH_4 citrate, NH_4 oxalate) was more aggressive than NTA in stirred beakers at 90°C,[6] but was less aggressive than NTA in a flowing system (OTTF) at 180°C. HEDTA was less aggressive than NTA and EDTA on laboratory specimens in the OTTF.[6] On the grounds of system compatibility and tractability to ion exchange HEDTA appears to offer advantages over EDTA and NTA. Thermal and radiolytic decomposition of EDTA was indicated to be tolerable in a PWR below ∿140°C for an EDTA, oxalic acid, hydrazine reagent.[12] However, additional work appears necessary to better define radiolytic effects for candidate reagents under prototypic BWR decontamination conditions.

REAGENT STUDIES - PWR SYSTEMS

Early in the series of tests on radioactive PWR specimens it became evident that the reagents which readily removed activity from BWR pipe surfaces were less effective on PWR steam generator tube surfaces.

To investigate reagent interactions with PWR films, two types of runs were conducted in the OTTF:

- Screening Runs, involving sequential exposure of several reagents to a single PWR specimen; the contact time for each reagent was ∿2 hr. This approach developed information on effectiveness for several reagents without rapidly depleting the supply of radioactive specimens.
- Exclusive Runs, involving a single reagent for the full length of the OTTF run.

Table II summarizes the reagents tested in the PWR test series. to date it has involved Inconel 600 steam generator tube specimens from two PWRs.[(a)] To summarize results of the reagent studies on radioactive PWR steam generator tube specimens:

- For a common reagent (EDTA), activity removal rates were higher for BWR than for PWR specimens (Table III).

(a) Specimens designated I-4-30HL are from an 800 MWe Westinghouse PWR. Specimens designated IP2 are from the Indian Point 2 Reactor.

Table II. Reagents Tested on Radioactive Inconel-600 PWR Steam Generator Tube Specimens

Reagent	Range of Conditions: Concentration Molarity	Temp °C	pH	Redox Conditions	Remarks
EDTA	0.002	180 150 105	5.5,9.0	a) ∿0.3 ppm O_2 b) ∿40 ppm O_2 c) 0.004 M N_2H_4 d) 50 and 100 ppm H_2O_2 e) H_3BO_3/LiOH	Specimens for I-4-30HL and IP2(a) 1-4-30HL-2 IP2-3-2
NTA	0.002	180	5.5	∿0.3 ppm O_2	I-4-30HL-3
DTPA	0.002	180	5.5	∿0.3 ppm O_2	I-4-30HL-3
EDTA NH_4 Cit. NH_4 Oxal.	0.001 0.0005 0.0005	180 90	5.5,3.5	a) ∿0.3 ppm O_2 b) 100 ppm H_2O_2	I-4-30HL-4, IP2-3-3 I-4-30HL-4, IP2-3-3
EDTA/Lactic Acid	.001/.001	180	5.5	∿0.3 ppm O_2	IP2-3-4
DCTA/Citric Acid	.001/.001	180		∿0.3 ppm O_2	IP2-3-4A
Salacilic Acid	.002	180		∿0.3 ppm O_2	IP2-3-4B
Gluconic Acid	.002	180		∿0.3 ppm O_2	IP2-3-4C
HEDTA(b)	.002	180		∿0.3 ppm O_2	IP2-3-4D
Acetohydroxamic Acid	.002	180		∿0.3 ppm O_2	IP2-3-4G
Dihydroxybenzoic Acid	.002	130		∿0.3 ppm O_2	IP2-3-4H
Mendelic Acid	.002	180		∿0.3 ppm O_2	IP2-3-4I
EDTA	.002	180		O_3	
EDTA	.002	180	3.5, 5.5, 4.8	a) 50 ppm Fe^{++} b) 20 ppm Cr^{++}	IP2-3-4L Mod. Rate IP2-3-5 Rapid Rate
HEDTA(c)	.002	180	3.5	a) 0.8 to 1.2 ppm O_2 b) N_2H_4	IP2-3-7
H_3BO_3 LiOH(d)					

(a) 180°C, pH 3.5 and 5.5, ∿0.3 ppm O_2.
(b) OTTF run segments also attempted using HIMDA, TTHA, PDTA separately.
(c) Several combinations: HEDTA + O_2; HEDTA + H_3BO_3/LiOH; same with O_2; H_3BO_3/LiOH.
(d) Added to simulate PWR coolant: 2000 ppm boron as H_3BO_3, LiOH, 2 ppm.

TABLE III. Comparison of Activity Removal Rates for One BWR and Two PWR Specimens Filmed in Reactor Primary Systems. Reagent and Conditions: 0.002 M EDTA, pH 5.5 and 3.3, 180°C; Flow, 115 to 125 ml/min

Specimens	Activity Removal Rate % Per Min. pH 5.5	pH 3.5
BWR, Specimen Y-5 ^{60}Co	1.9[a]	
BWR, Specimen Y-13 ^{60}Co		2.4[a]
PWR, Specimen I-4-30HL-2 ^{60}Co	0.037	0.065
PWR, Specimen IP-3-2		
^{58}Co	0.020[b]	0.033[b]
^{60}Co	0.020	0.033

(a) Rates for the first ∿20 min; average rate over the first 150 min (to 85% activity Removal) is 0.57%/min for Y-5 and Y-13

(b) Corrected for Decay; initial separation of the ^{58}Co and ^{60}Co curves exceeds effect of decay.

- Activity removal rates differed by about a factor of two for steam generator specimens from two PWRs (Table III).
- Most of the reagents (Table II) produced low activity removal rates in the OTTF (Figure 4).
- The factors which accelerated activity removal from the PWR specimens were:
  - low pH (Table III and Figure 5)
  - H_2O_2 additions (temporary effect)
  - addition of Fe^{+2} and Cr^{+2} to a chelating reagent (EDTA); Figure 4 indicates a mild increase in ^{60}Co removal for Fe^{+2} addition and a dramatic increase for Cr^{+2} addition.

Inspection of PWR surfaces (from both reactors) prior to reagent treatments indicated grains in the oxide/crud layer (Figure 6). Post-reagent inspection indicated that the reagent attack occurred on all crud layer surfaces, but concentrated at the grain boundaries, which widened during the decontamination (Figure 6).

Suspension of activity from PWR steam generator tubing appears promising in low-pH chelating compounds and in reagents containing metal ion reducing agents such as Cr^{+2}. However, the development of a practical decontamination process is currently less well-defined for PWRs than for BWRs.

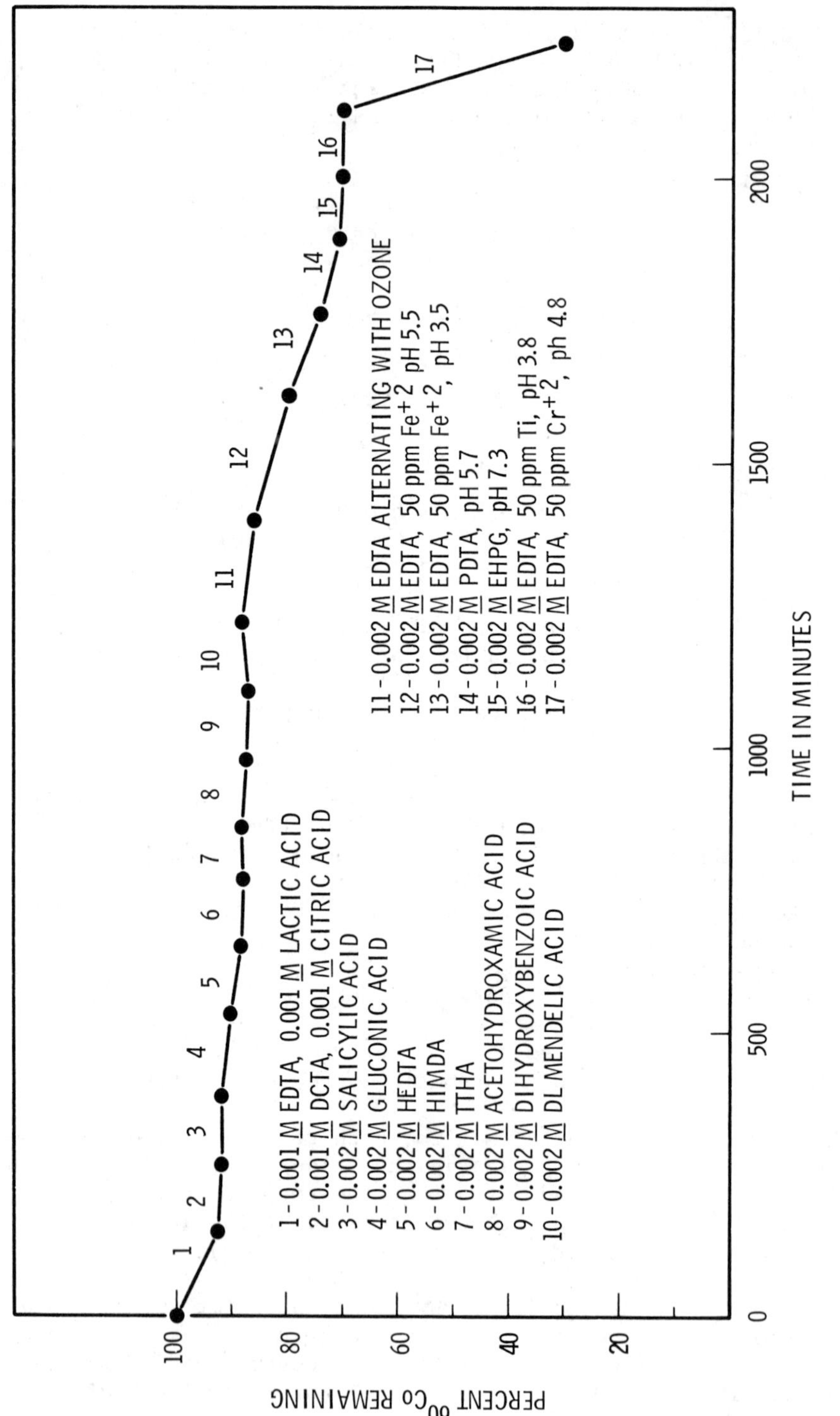

Figure 4. Decontamination of PWR Specimen IP2-3-4; Temperature 180°C; Material Inconel-600

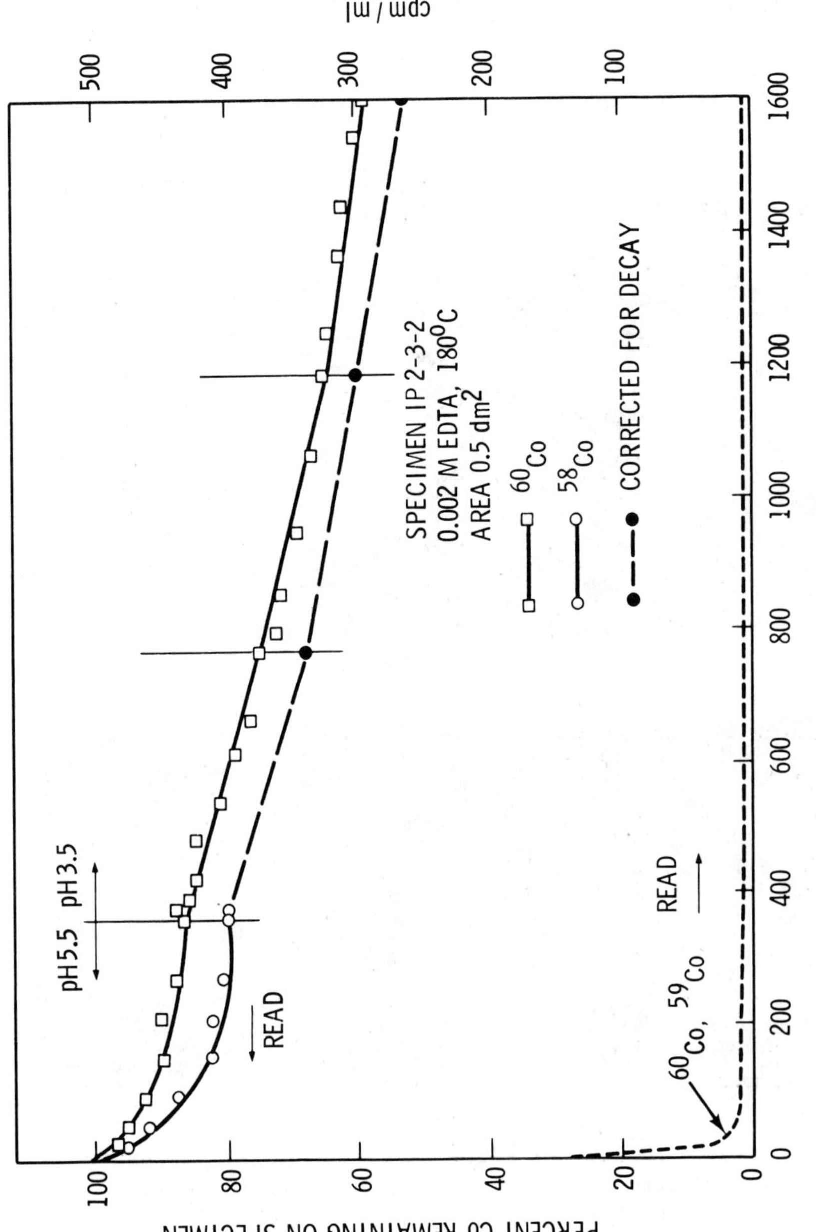

Figure 5. OTTF Decontamination of PWR Steam Generator Tube Specimen (Activity and Isotope Concentration vs. Time); Material - Inconel-600

NO REAGENT TREATMENT

IP2-3-5
REAGENT TREATMENT:
0.002 M EDTA WITH $Cr^{+}2$ AND $Cr^{+}3$
$180^{o}C$, pH ~5, ACTIVITY REMOVED ~65%

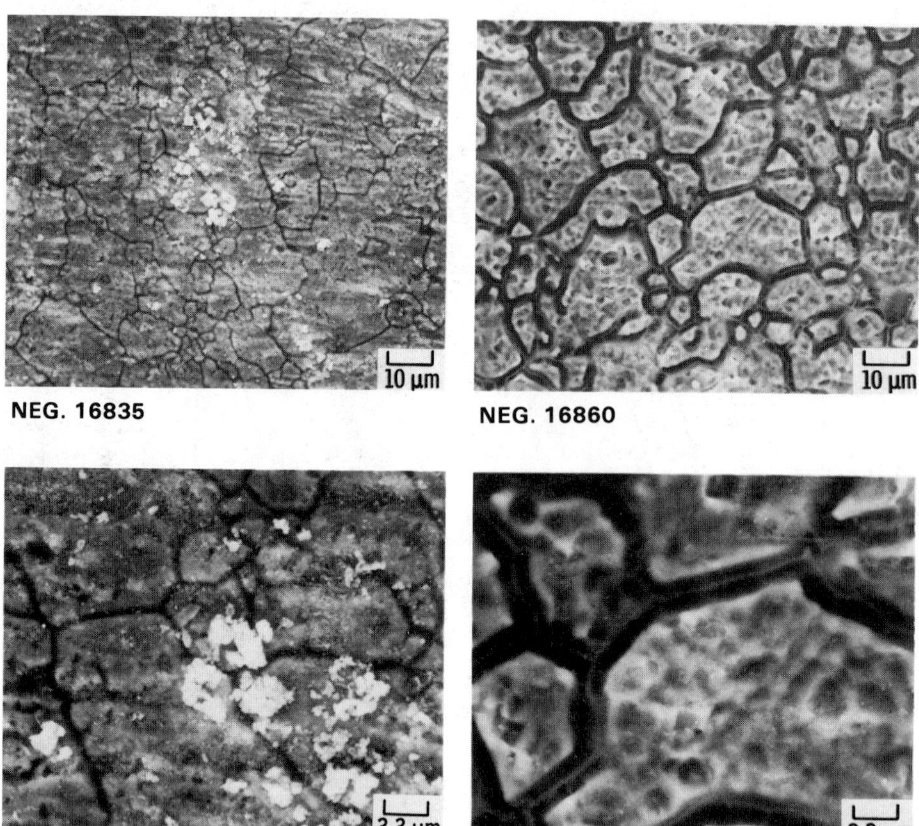

NEG. 16835

NEG. 16860

NEG. 16836

NEG. 16861

Figure 6. Surface Views (SEM) - Indian Point 2 Steam Generator Tube Specimens; In-Reactor Exposure ∿4.75 Years; Effects of Reagent Treatments; Material - Inconel-600

ACKNOWLEDGMENTS

The authors are grateful to the following for significant contributions to the work reported here: Consolidated Edison Co., General Electric Co., Westinghouse Electric Co. and Yankee Atomic Electric Co. for supplying radioactive reactor components; D. B. Mackey, R. L. McDowell, and R. D. Peters for assistance with the experimental work. The work was funded by the Electric Power Research Institute under RP 828-1.

REFERENCES

1. P. J. Pettit, J. E. LeSurf, W. B. Stewart and S. B. Vaughan, "Decontamination of the Douglas Point Reactor by the CAN-Decon Process," Paper No. 39 presented at Corrosion-78, Houston, TX, March 6-10, 1978, available from NACE, P.O. Box 986, Katy, TX 77450.

2. A. B. Johnson, Jr., B. Griggs and J. F. Remark, "Investigation of Chemicals and Methods of Dilute Reagent Decontamination for Potential Application in Light Water Reactors," Paper No. 32 presented at the Corrosion-78, Houston, TX, March 6-10, 1978, available from NACE, P.O. Box 986, Katy, TX 77450.

3. A. B. Johnson, Jr. and B. Griggs, "Comparison of Decontamination Characteristics of Films from BWR and PWR Primary Systems," Paper No. 161, presented at Corrosion-79, Atlanta, GA, March 12-16, 1979, available from NACE, P.O. Box 986, Katy, TX 77450.

4. Literature Review of Dilute Chemical Decontamination Processes for Water-Cooled Nuclear Reactors, EPRI NP-1033, Project 828-1, March 1979.

5. J. F. Remark, R. L. Dillon, and B. Griggs, Investigation of Alternate Methods of Chemical Decontamination - Semi-Annual Report, July 1, 1976 to December 31, 1976, BNSA-703, Battelle, Pacific Northwest Laboratories, Richland, WA 99352, August 1977.

6. A. B. Johnson, Jr., R. L. Dillon, B. Griggs and J. F. Remark, Investigation of Alternate Methods of Chemical Decontamination - Second Progress Report, January 1, 1977 to December 31, 1977, BN-SA-703-2, June 1978.

7. A. B. Johnson, Jr. R. L. Dillon, and B. Griggs, Investigation of Alternate Methods of Chemical Decontamination - Third Progress Report, January 1, 1978 to December 31, 1978, BN-SA-703-3 in press.

8. J. A. Ayres (ed.), Decontamination of Reactors and Equipment, TID-25450, Ronald Press, New York, 1970, p. 840.

9. Characterization of Corrosion Products on Recirculation and Bypass Lines at Millstone-1, EPRI NP-949, Project 819-1, December 1978.

10. L. D. Anstine, BWR Decontamination and Corrosion Product Characterization, NEDE-12665, March 1977.

11. J. Blok, Summary Report-Water Chemistry Program Extension, NEDC-21550, June 1977.

12. K. Oertel, et al., Decontamination of Primary Loop Equipment with Ethylenediaminetetraacetic Acid, BNWL-TR-290, January 25, 1978.

DESIGN CONSIDERATIONS FOR PHWR DECONTAMINATION

W.B. Stewart and T.S. Drolet

Ontario Hydro

Toronto, Canada

INTRODUCTION

Operation of early CANDU-PHW* reactors resulted in radiation fields around out-of-core heat transport system components and equipment which were higher than originally anticipated[1]. As a result of the accompanying increases in radiation exposures of maintainers and operators an intensive collaborative research and development program between scientists and engineers from Atomic Energy of Canada Limited (AECL) and Ontario Hydro (OH) was undertaken to identify the mechanisms of activity production, transportation and deposition and to recommend methods to reduce ambient radiation fields around the complete nuclear steam supply system. The objective being to reduce the radiation fields to levels that permit station staff to perform all operating and maintenance functions, without the necessity of additional staff. In addition all radiation exposures should be kept as low as reasonable achievable (ALARA).

As a result of these programs a number of options to control or reduce the growth of radiation fields were identified. The options include[2]:

* CANDU-PHW - Canadian Deuterium Uranium - Pressurized Heavy Water

(1) Tighter specifications for the materials of construction. This has reduced objectionable impurity levels in the materials. Therefore the source of potential radionuclides has been reduced.

(2) Increased coolant purification rate. This has been used to remove both active and inactive corrosion products[3].

(3) Improved coolant chemistry. This has minimized corrosion product production, deposition on the fuel, subsequent dissolution and out of core deposition.

For stations where radiation fields have risen to unacceptably high levels, only decontamination has the potential to quickly reduce these radiation fields. Designers therefore should consider that a decontamination may be required during the life of a station and necessary provisions should be incorporated in the design.

The prime objective of a nuclear steam supply system decontamination is to reduce the ambient radiation fields around the system. This will result in a reduction in the total radiation exposure associated with work performed on or in the vicinity of the system. It will also have an impact on the radiation dose, associated with the system for several years following the decontamination.

PROCESS DEVELOPMENT

Most conventional decontamination processes[4] have been developed primarily for stainless steel systems and has involved the addition of relatively strong (6-10 wt. %) concentrations of organic acids. For a heavy water system such processes would result in downgrading of the heavy water by hydrogen in the chemicals and would require either extensive upgrading or replacement of the heavy water coolant. This concentrated acidic process would result in the generation of several reactor volumes of acidic, radioactive liquid decontamination wastes with the associated processing and storage requirements. An extensive station shutdown would be required to prepare the reactor, perform the decontamination and return the reactor to operating condition.

The cooperative program between AECL and OH[1] resulted in the development of the dilute chemical decontamination technique known as CAN-DECON. Compared to conventional decontamination processes CAN-DECON has the following advantages for a CANDU-PHW reactor.

- Effectively reduces radiation fields.
- Can be performed by station staff.
- May be performed with the heavy water coolant left in the system.
- Requires relatively short outage times.
- Little corrosion occurs and hence there are no deleterious after effects.
- Produces conveniently managed radioactive wastes.

The CAN-DECON process involves the addition of chemical reagents to the heat transport system of a shutdown reactor, to give a maximum chemical concentration of 1 g/kg D_2O (0.1 w/o conc.). The solution of acidic reagents attacks and complexes a portion of the corrosion product oxide layer and releases both particulates and dissolved material to the coolant. The complexed corrosion products, some of which are activated, are removed in a side stream high flow purification circuit.

Filtration, in the form of submicron cartridge filters, is used to remove the particulate material. Cation resin removes the corrosion products from the complex and regenerates the reagent. The process is continued until the resin is spent or until the allotted time has expired. The reagents and remaining dissolved corrosion products are then removed by mixed bed resins in the final clean-up phase.

The process is particularly suited to heavy water reactors since the heat transport system is neither drained nor flushed. Compared to conventional decontamination processes, using strong reagents, this process requires little additional equipment.

The regeneration of the reagent minimizes the quantity of reagents while at the same time the radioactive wastes are collected on the resins and filters. This greatly simplifies subsequent disposal of the waste since there are essentially no liquid wastes to deal with.

The feasibility of the CAN-DECON process has been demonstrated by extensive laboratory and loop testing, test applications at the NPD generating station, a full scale heat transport system application at the Gentilly-1 Generating Station, and a full system decontamination at the Douglas Point Generating Station.

Details of the Douglas Point decontamination have been previously reported[5]. The highlights of the Douglas Point decontamination are as follows:

- significant reduction in radiation fields on carbon steel components was obtained (Table 1) and the rate of growth of fields, following the decontamination, has been slow
- approximately 230-250 Ci of Co-60 was removed from the heat transport system
- no failures or malfunctions of equipment have occurred which could be attributed to the decontamination
- the average leak rate from the heat transport system did not increase
- less than 10 rem were expended in performing the decontamination
- approximately 150-180 man rem were saved during a scheduled maintenance outage shortly after the decontamination
- the time from reactor shutdown to start the decontamination until the reactor was ready for start-up was 72 hours.

A CAN-DECON purification system is presently being designed and procured for the Pickering Generating Station. The Pickering Generating Station is a four unit CANDU-PHWR, each unit is rated at 540 MW(e). Figure 1 is a schematic diagram of the heat transport system of the reactor. Although not required at the present time, it is expected that decontamination of the heat transport system will be required during the life of the station. This may occur on short notice (a few months) or as a scheduled event.

DESIGN REQUIREMENTS

The Pickering CAN-DECON purification system has been designed to meet the following functional requirements:

(a) Permanent connections for a side stream purification system shall be provided on the heat transport system of each unit. The permanent piping is located within the reactor building. ASME Code Section III, Class 1 and Class II values are provided for isolation.

(b) Removal of the dissolved corrosion products released during the decontamination and simultaneous regeneration of the decontamination reagent.

(c) Removal of the decontamination chemicals and residual dissolved corrosion products at the completion of the decontamination.

(d) Removal of the particulate material released during the decontamination.

TABLE 1

SUMMARY OF RADIATION FIELD MEASUREMENTS

Location	Before CAN-DECON mR/h	Immediately After CAN-DECON mR/h	1 Year After CAN-DECON mR/h
Walkway at Feeder Cabinets	2000 - 3000	300 - 500	285 - 700
East Fuelling Machine Vault*	1900 - 7900	650 - 850	685 - 1470
West Fuelling Machine Vault*	1600 - 5800	420 - 970	515 - 1045
Lower Main Circulating Pump Area (Affected by Fields from Monel Boilers)	800 - 4000	500 - 1400	500 - 1750
Contact with Boiler Cabinets (Average)	520 - 1300 (810)	320 - 875 (540)	300 - 850 (530)

* Measurements taken vertically up the center of the reactor face at 1.2 metres from the end-fittings.
Fields increase with height as the feeder cabinet region is approached.

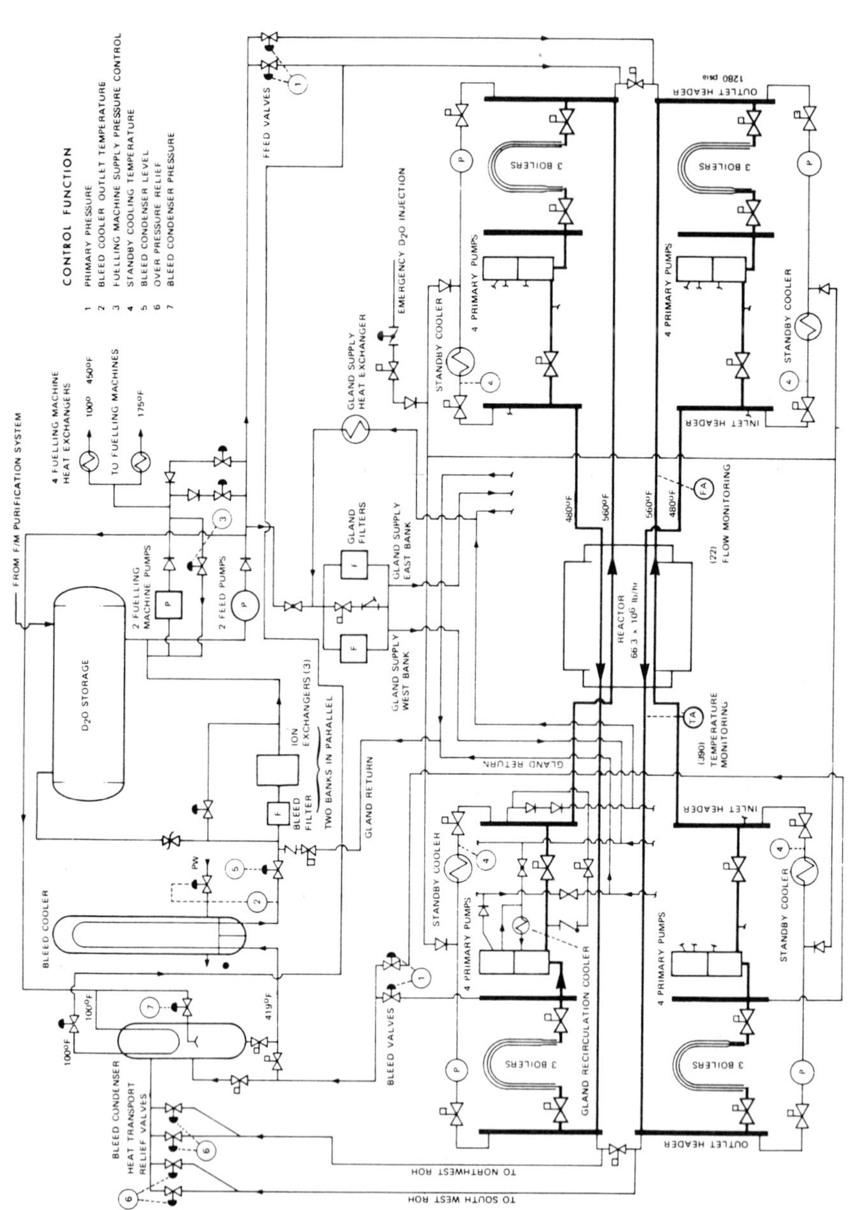

Figure 1. Pickering Heat Transport Systems

(e) The purification system shall include cation bed and mixed bed ion exchange columns and disposable filters. The equipment shall be sized to permit decontamination of a whole reactor. Internal surfaces of the ion exchange columns shall have a protective coating. The coating shall minimize corrosion and retention of spent resin.

(f) Each ion exchange column and the filter must be capable of handling full purification design flow. At Pickering the design flow is limited to the maximum available from the heat transport bleed and feed system and at the same time provide a reasonable head for the CAN-DECON system.

(g) The design of the ion exchange columns shall ensure that in situ deuteration and dedeuteration of the resin can be performed. The ion exchange resin must be deuterated prior to use to prevent downgrading the heavy water. After the decontamination dedeuteration is performed to recover the heavy water held up in the ion exchange columns.

(h) Loading of the ion exchange resin shall be done manually. Complete removal of the spent resin shall be by hydraulic means. The spent resin can be slurried to either the station spent resin storage tanks or a bulk resin transportation flask for transportation to Ontario Hydro's waste storage site. Ion exchange resin is not solidified.

(i) The system components, including the piping, shall be grouped, skid mounted and self supporting. Minimum inter-connecting joints, to facilitate disassembly and reassembly of the system and transportation between units or to storage, shall be employed. Since the system is transportable and permanent connections are provided for each unit only one purification system is necessary for the station.

(j) The design of the system and in particular its piping shall enable complete draining of the fluid, to minimize the potential downgrading of the heavy water by light water after hydraulic testing and to enable efficient removal of heavy water from the system after the decontamination prior to equipment disassembly.

(k) Maximum leak tightness of the system, in order to avoid heavy water losses and the release of tritium and other radioactive materials shall be ensured by design of components and selection of materials. Except for flanged interconnecting piping between the skids, filter and heat transport system, all piping throughout the system is welded.

(l) For economy, the principal material of construction shall be carbon steel. Sampling lines shall be stainless steel.

(m) Corrosion shall be measured during the decontamination. Corrosion coupons shall be used to evaluate system corrosion.

(n) A dedeuteration clean-up filter shall be included. It is provided for removal of any radioactive particulate from the dedeuteration effluent water. The heavy water is transferred to the collection or downgraded heavy water systems.

(o) Control of the CAN-DECON purification system valves shall be manual.

(p) The CAN-DECON purification system need not be seismically qualified. Should an incident occur the level of radioactive material released from it would be low. The purification system can be located outside containment since it operates at low temperature and pressure.

(q) Shielding shall be provided to protect personnel from radiation fields emanating from the system. The design radiation level, in accessible areas, during and after the decontamination, shall be 0.6 mR/h.

SYSTEM DESCRIPTION

The Pickering CAN-DECON purification system, shown in Figure 2, consists essentially of a cartridge type disposable filter connected in series with a set of four ion exchange columns connected in parallel. The CAN-DECON system is connected to the heat transport system via permanent connections penetrating the containment wall of each unit. To lower the temperature of the heat transport system coolant to that appropriate for ion exchange resins, the coolant is passed through a bleed condenser and then a bleed cooler. The purification flow is taken downstream of the bleed cooler and returned to the HT system feed pump suction or heavy water storage tank.

The filter is provided with isolating and bypass valves. Each ion exchange column is also provided with isolation and there is a bypass valve between the ion exchange column inlet and outlet headers. This arrangement enables operation of the CAN-DECON system in a filtration mode, ion exchange mode or filtration/ion exchange mode. Although all ion exchange columns

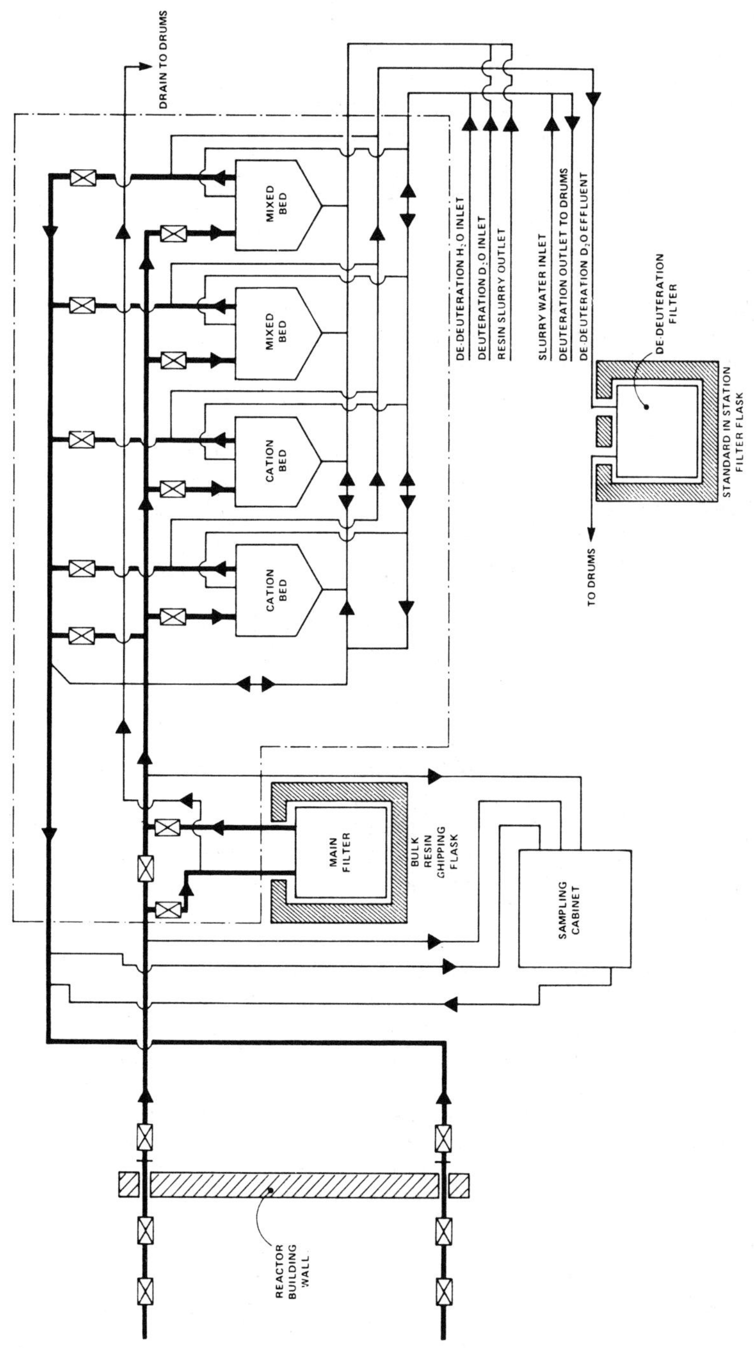

Figure 2. Candecon Purification System Flow Diagram

are identical in design, two of the columns contain cation resin, for use during the initial phases of the decontamination for regeneration of the CAN-DECON chemicals and removal of the dissolved corrosion products. The other two columns contain mixed bed resin which are used for removal of the reagent and for system clean-up. The filter and each ion exchange column are designed to handle full system flow (400 Igpm). The CAN-DECON decontamination reagent is added to the heat transport system via slurry techniques.

Auxiliary facilities and connections to plant services (eg. instrument air, heavy water collection, downgraded heavy water transfer, demineralized water, etc.) are provided for resin deuteration/dedeuteration, slurrying of the resin and displacement of the heavy water from the spent filter.

Sampling lines from strategic points in the purification system to a sampling cabinet are provided to enable monitoring of system performance and control of chemistry during the decontamination. Sample connections are also provided for on-line pH, conductivity and gross instrumentation.

The CAN-DECON system, heavy water handling areas and sampling cabinet are equipped with exhaust hoods connected to the active ventilation system, to minimize the tritium hazard, should spills or leaks occur.

The system is designed for manual control and has simple instrumentation consisting of temperature measurement, located upstream of the filter inlet and pressure measurement, for monitoring pressure losses across the filter, ion exchange columns and strainer. Flow meters are provided for resin deuteration/dedeuteration and slurrying.

The entire system is shielded to reduce the radiation fields, in accessible areas, during and after the decontamination, to less than 0.6 mR/h. Where necessary valves are provided with extension handles into accessible areas.

The purification system located outside containment will operate at low pressure and temperature and is designed to meet ASME Code Section III, Class 3 requirements.

CONCLUSION

Because a CAN-DECON decontamination is performed with the fuel in place in the core both radioactive and non-radioactive corrosion products deposited on the fuel are removed before they have been dissolved from the fuel and redistributed on out of core surfaces. Since the portion of activity deposited on the fuel can be significant this reduces the rate of growth of fields after the decontamination.

Decontamination factors achieved by the CAN-DECON process are less than those obtainable with more agressive decontamination methods but it is relatively quick, inexpensive and can be performed by station staff. These features suggest that it could be used more frequently before radiation fields become so high that agressive decontamination is required.

Although the CAN-DECON process was developed for CANDU-PHW reactors which have carbon steel heat transport circuits, the principle of using dilute regenerable decontaminants is applicable to light water reactors with stainless steel circuits.

REFERENCES

1. J.E. LeSurf, "Control of Radiation Exposures at CANDU Nuclear Power Stations", J. Br. Nucl. Engineering Society, 16, 53 (1977).

2. W.B. Stewart and T.S. Drolet, "Design Considerations for Radiation Field Control in CANDU-PHW Reactors", Paper Presented at Corrosion/78, Houston, Texas (1978).

3. B. Montford, "Decontamination by Cycling Techniques at the Douglas Point NGS", Proc. Am. Power Conf., 35, 902 (1973).

4. Ayres, J.A. (Ed.), "Decontamination of Nuclear Reactors and Equipment", The Ronald Press Co., New York, 1970.

5. P.J. Pettit, J.E. LeSurf, W.B. Stewart, R.J. Strickert and S.B. Vaughan, "Decontamination of the Douglas Point Reactor by the CAN-DECON Process", Paper Presented at Corrosion/78, Houston, Texas (1978).

DECONTAMINATION EXPERIENCE IN ONTARIO HYDRO

C.S. Lacy and B. Montford

Ontario Hydro
Toronto, Ontario
Canada

ABSTRACT

This paper describes the decontamination of a Bruce Nuclear Generating Station heat exchanger. The heat exchanger, tubed with Incoloy-800 was contaminated with both neutron activated corrosion products and fission products. Sequential decontaminations using the CANDECON process, followed by alkaline permanganate (AP) and Decon-Turco 4521A, were used. Decontamination factors of between 8 to 12 and 10 to 50 were achieved on the tube bundle and tubesheet, respectively. The cost effectiveness of these and their effects on system materials will be discussed.

INTRODUCTION

The Bruce Nuclear Generating Station[1] is a typical CANDU reactor operated by Ontario Hydro. Each of the four units produces 740 MWe net. As in all CANDU reactors, the heat transport system is purified in a bypass circuit called the bleed system.

To lower the temperatures of the heat transport system to that appropriate for ion exchange resins, it is first passed through the bleed condenser and then the bleed cooler. The temperature of the bleed flow is reduced from 250°C to 204°C in the bleed condenser. It is then reduced to 38°C to 66°C in the bleed cooler.

The bleed cooler is a conventional tube-in-shell type heat exchanger with two tube side passes for the bleed flow and two shell side passes for lake water. The heat transfer area consists of 900 U-tubes made of Incoloy-800.

During service, a heat exchanger from one nuclear electric generating unit failed and was removed from the system for inspection and repair. When this work started, it was found that both loose contamination in, and radiation fields around, the heat exchanger was high. The replacement for this exchanger was one removed from the fourth unit (at the time in its early commissioning phase). It was quickly apparent that decontamination and repair should be done as expediently as possible, so as not to delay Unit 4 startup.

It is the objective of this paper to describe:

1. Solvent selection and qualification for contaminated Incoloy-800.
2. The decontamination factors achieved on this heat exchanger.
3. The cost and benefits accrued.
4. The problems encountered with the chemical decontamination and their potential solutions.

REAGENT AND PROCEDURES

In the Canadian Decontamination Program, we have considerable experience with conventional decontamination reagents for the removal of fission products from experimental loops.[2] However, this was our first attempt to decontaminate a power reactor component contaminated with fission products. Since the heat exchanger material, Incoloy-800, is the same as steam generators in future reactors, there was additonal impetus to gain as much information as practical.

Problem

The problem was to decontaminate the heat exchanger to as low a level as practicable to permit its reinstallation in the radiologically unzoned construction area of the fourth unit.

After the failed bleed cooler had been removed from containment, it was placed inside a plastic tent to avoid the spread of contamination. A radiation survey of the heat exchanger was carried out. The results are shown in Figure 1. Contact fields varied from 100 mR/h on the tubesheet to 15 mR/h around the tube bundle. The radionuclides contributing to the gamma fields were determined using a portable gamma spectrometer. These are given in Table 1.

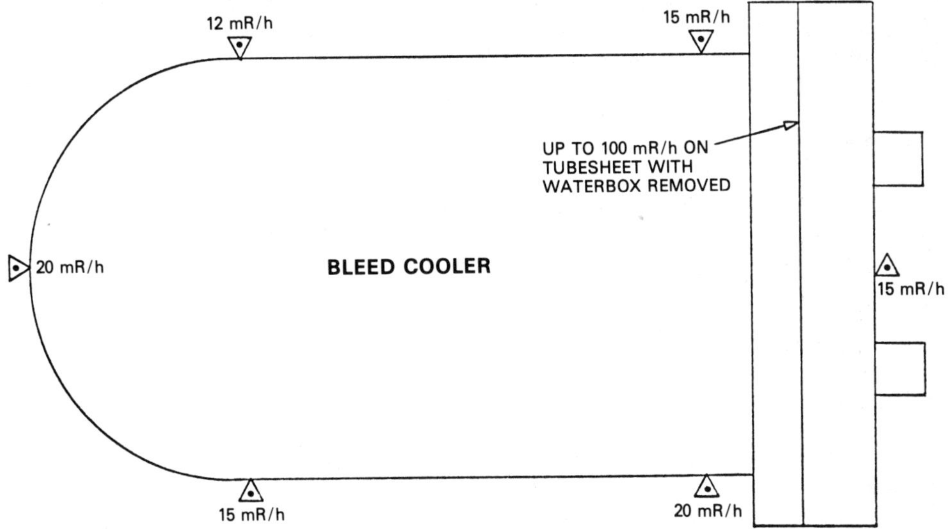

Figure 1: Radiation Fields

Table 1. Radionuclide Contribution to Gamma Radiation Field

Radionuclide	Percent Contribution
^{124}Sb	6
^{95}Zr	14
^{95}Nb	29
^{65}Zn	51

As can be seen, the radiation field emanates from fission products and the neutron activated product ^{65}Zn in equal proportions. Contribution from ^{60}Co was insignificant. The reasons for this lower ^{60}Co contribution than observed in older reactors are:

1. The close attention paid to alloy specifications, such that only alloys with very low concentrations of ^{59}Co were used.

2. Activation of that ^{59}Co present was still small as the reactor had only operated for about one year.

Swipes, taken from inside the tubes, indicated that gross loose contamination was present.

The heat exchanger was to be: eddy current tested, the failed tubes removed for metallographic examination, repaired and used in the fourth unit - a radiologically unzoned construction area.

The problem was therefore to decontaminate the heat exchanger with the following objectives:

1. Remove all loose contamination.

2. Reduce radiation fields to as low as reasonably achievable.

3. Evaluate the CANDECON process[3] for the removal of fission products from Incoloy.

Reagent Selection

Because of the advantages and simplicity of the CANDECON process, we decided that this process should be compared to the conventional processes. However, before proceeding with the bleed cooler decontamination, beaker tests were performed on three reagents and their effectiveness evaluated by Ontario Hydro Research Division. The reagents examined were:

1. For CANDECON, LND 101*, an acidic reagent, containing organic complexing agents.

2. Alkaline permanganate, AP.

3. Decon-Turco 4521A, a corrosion inhibited mixture of ammonium oxalate and citrate.

These were applied as solutions (0.1% LND 101, and 6% 4521A) to contaminated corrosion coupons. These coupons of carbon steel and Inconel-600 had been removed from the generating unit for these tests. We did not have contaminated Incoloy coupons. However, it was assumed that the type of oxides and mode of contamination would be similar for Inconel and Incoloy.

The test solutions were applied in two different sequences and their effects examined. The sequences were:

Step	Sequence 1	Sequence 2
1	LND 101	4521A
2	AP	AP
3	4521A	4521A

* Proprietary reagent of London Nuclear Decontamination Limited

Radionuclides on the coupons were determined before and after exposure to each reagent and the decontamination factor calculated. The decontamination factors were:

Sequence 1

Radionuclide	LND 101 Carbon Steel	LND 101 Inconel 600	LND 101 - AP-4521A Carbon Steel	LND 101 - AP-4521A Inconel 600
^{95}Zr	3-5	2-4	8-15	10-20
^{65}Zn	14-25	4-11	50-80	35-100

Sequence 2

Radionuclide	4521A Carbon Steel	4521A Inconel 600	4521A - AP-4521A Carbon Steel	4521A - AP-4521A Inconel 600
^{95}Zr	2-3	2-5	4-8	20
^{65}Zn	3-4	4-9	7-15	40-70

As well as radiochemical analysis, the Inconel-600 coupons and uncontaminated Incoloy-800 (also exposed to the reagent) were examined metallographically[4] for intra and intergranular cracking. None was evident.

Sequence 1, LND 101 followed by AP-4521A, was preferred to Sequence 2 for two reasons:

1. The bulk of the contamination on the heat exchanger could be transferred to ion exchange resin in the CANDECON step, thus avoiding the handling of highly active waste solutions.

2. Because in CANDECON the solution is constantly regenerated, it never becomes highly active. Therefore, any leaks to the shell side would not cause subsequent contamination problems.

We therefore proceeded with Sequence 1 and determined the effects of each step on the radiation fields.

Equipment

A schematic of the equipment used for this decontamination is given in Figure 2. A decontamination skid was rented from the Chalk River Nuclear Laboratories of Atomic Energy of Canada Limited. The skid consisted of a stainless steel mixing tank and pump. The remainder of the system comprized: two 110 litre ion exchange columns and a 0.2 μm filter assembly and was formed into a closed loop by hose and quick disconnect fittings. The temperature of the decontamination solutions was raised to 85°C by injecting steam into the water-filled shell side of the heat exchanger.

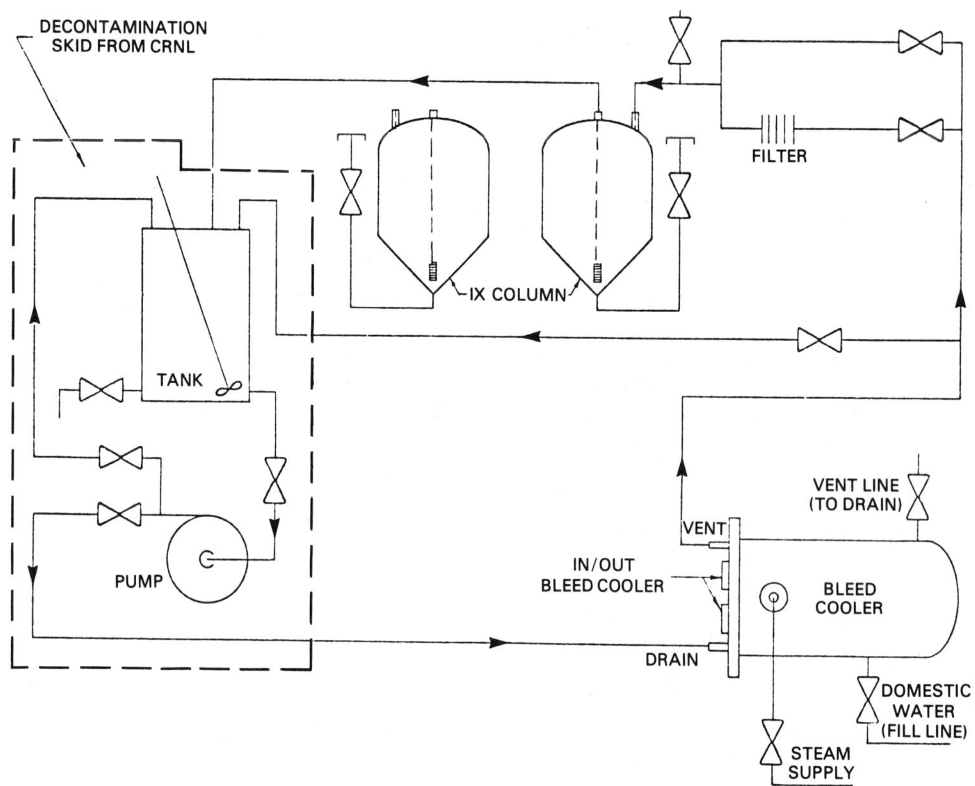

Figure 2: Equipment Flowsheet

Decontamination Procedure

The decontamination procedure is summarized in Table 2. Initially, the loop was filled with demineralized water, heated to 85°C and maintained there by manual control of the steam supply. To start the first stage of the decontamination, the reagent LND 101 was added to the hot circulating demineralized water. The reagent was constantly regenerated by passing the solution through cation ion exchange resins. When radionuclides ceased to be removed by this solution, the reagents and trace radionuclides were removed by mixed bed ion exchange resin. This stage took four hours and the radiation fields on the cation exchanger reached 4 R/h.

The second stage of the decontamination was started by adding liquid 50% NaOH and $KMnO_4$ powder to the hot demineralized water. This solution was circulated for three hours, but not ion-exchanged. At the end of this time it was drained and the loop rinsed. Remaining chemicals were removed on mixed bed ion exchange resins.

The third stage was started by reheating the circulating water to 85°C. Decon Turco 4521A powder was added to the water and recirculated for three hours. Again the solution was drained together with one rinse.

At a later stage the AP and the 4521A solutions together with their rinses were solidified in 200 L drums.

Table 2. Decontamination Procedure

Time	Temperature	Chemicals	Comments
CANDECON:			
4 hours	85°C	0.1% LND 101	Chemical full flow regeneration with cation resin followed by chemical removal with mixed bed resin.
AP:			
3 hours	85°C	10% NaOH 3% $KMnO_4$	All liquid and one rinse drummed for solidification. Second rinse purified with mixed bed resin.
4521A:			
3 hours	85°C	5% 4521A	As for AP.

RESULTS OF DECONTAMINATION

During each step of the decontamination, the concentrations of: crud, dissolved iron, dissolved nickel, and dissolved chromium were determined every 30 minutes. The results are shown in Figure 3.

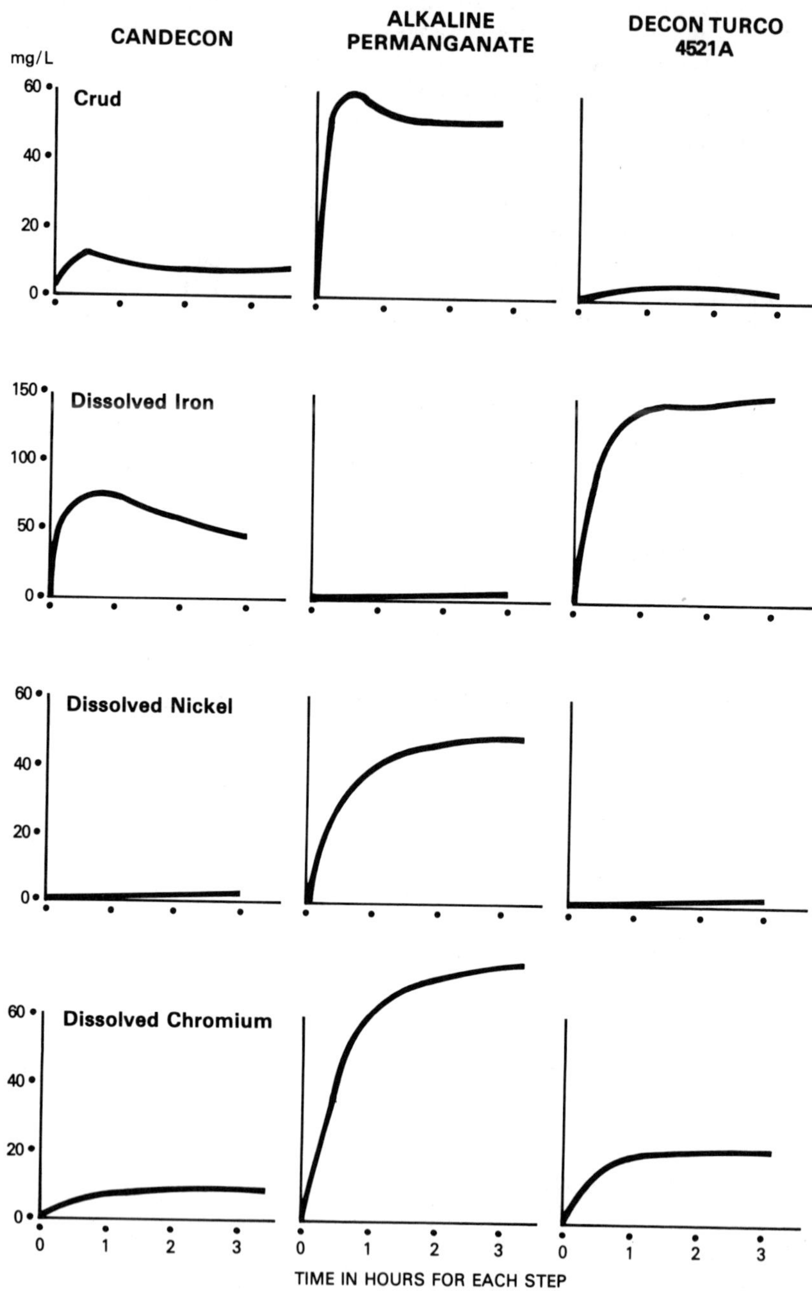

Figure 3: Crud, Iron, Nickel and Chromium Concentrations

The significant effects observed were:

1. Iron concentrations peaked at 70 mg/kg during the CANDECON step, dropped to essentially zero during the AP step and reached a maximum of 150 mg/kg during the 4521A step.

2. Nickel and chromium concentrations were low (< 1 mg Ni/kg and < 5 mg Cr/kg, Ni and Cr respectively) during the CANDECON step but both increased to 50 mg/kg at the start of the 4521A step.

To determine the efficiency of each step, radiation field measurements were continually taken. These are shown in Figure 4. The average decontamination factors are summarized in Table 3:

Table 3. Decontamination Factors

	Tube Bundle	Tubesheet
CANDECON	4-6	5-20
AP-4521A	2	2-2.5
Total	8-12	10-50

Swipes taken from inside the heat exchanger tubes indicated that essentially all loose contamination (> 98%) was removed by the CANDECON step. Even though at this stage the heat exchanger was considered clean to be handled in the unzoned area, we decided to continue with steps 2 and 3 to compare the effectiveness of the two decontamination procedures.

Carbon steel and Incoloy-800 corrosion coupons were installed in the recirculation tank. These together with a heat exchanger tube removed subsequent to the decontamination showed:

1. Insignificant general corrosion.

2. No intra or intergranular cracking.

DISCUSSION

Costs

The costs of the decontamination are given in Tables 4 and 5. They are subdivided into:

1. Those costs associated with equipment - 17.5 k$. These would be the same for CANDECON or the AP-4521A decontamination.

2. Those costs associated with the individual decontamination techniques - 6.5 k$ for CANDECON and 46 k$ for AP-4521A.

Table 4. Cost Associated With Equipment

Planning and Engineering	10 k$
Operation and Maintenance	2
Cleanup	2
Decontamination of Equipment	2
Rental of Tank and Pump	1.5
	17.5

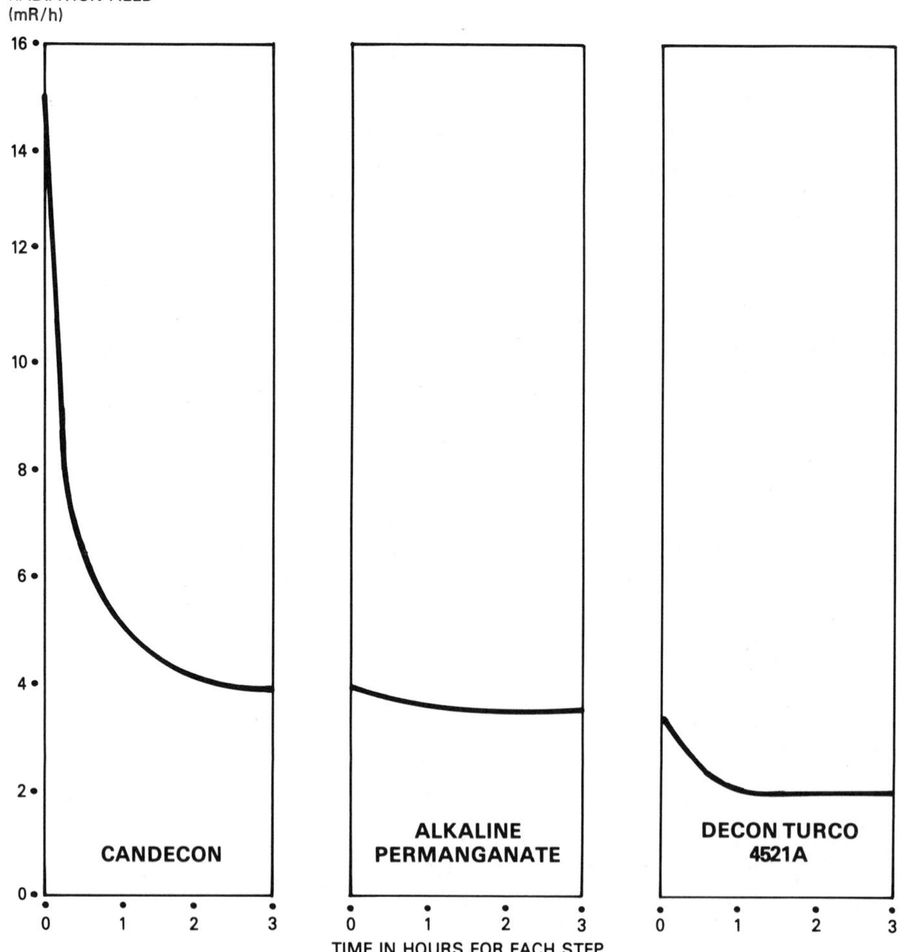

Figure 4: Radiation Field Reduction

Table 5. Costs Associated With Decontaminations

	CANDECON	AP-4521A
Chemicals		
- and Ion Exchange Resin	.8 k$	Insignificant
Equipment		
- Ion Exchange Columns	3	
- Solidification Equipment		10 k$
Solidification of Waste	NA	6
Waste Storage	2.5	30
	6.5 k$	46 k$

Therefore, to complete the CANDECON alone would cost 24 k$. To complete the AP-4521 type decontamination alone would cost 63.5 k$. However, it should be explained:

1. We were able to rent the tank and pump. To have built this would have cost 20 k$ making costs for CANDECON 42.5 k$ and AP-4521A 82 k$.

2. The ion exchange columns used during the CANDECON, while purchased for this job, were not disposed of and are still available for reuse.

Decontamination Factors

The decontamination factors achieved by CANDECON were 4-6 on the tube bundles, and 5-20 on the tubesheet. AP-4521A improved these by a factor of 2 to give overall decontamination factors of 8-12 on the tube bundle, and 10-50 on the tubesheet.

Radiation Dose

Radiation dose expended to do the job was 4 rem of which 3.5 rem were involved with the solidification process. Estimated dose saved to do the subsequent cleaning, inspection and maintenance was 10 rem. This 10 rem does not include any estimate for extremity dose the inspection crew would have received. Without the removal of the gross loose contamination, eddy current testing would have been a very difficult task. It would also be difficult to contain the loose activity.

Although the dose saving was only 6 rem, as mentioned earlier, the prime reason for the decontamination was to place the bleed cooler in the radiologically unzoned construction area, for use in Unit 4. This objective was achieved.

Operating Problems

During the decontamination, several operating problems were encountered worth detailing. The filling and draining of the horizontal heat exchanger presented unforeseen problems. Filling was accomplished using the 3 l/s circulating pump, while draining was by compressed air. Although the heat exchanger was laid on a flat surface, it was not perfectly level. Because of this, when the heat exchanger was filled, it is now believed that air pockets remained in some tubes. Therefore, it is unlikely we fully filled or drained the heat exchanger. In future, careful attention would be required to purging the heat exchanger free of air. In addition, during draining, the use of compressed air was unsuitable as contaminated spray was generated which could not be fully contained by the tank. For any future jobs, we plan to use an eductor to assist filling and draining the heat exchanger.

CONCLUSIONS AND RECOMMENDATIONS

Neither decontamination caused any corrosion damage to the carbon steel or Incoloy-800 surfaces. The CANDECON decontamination steps of the bleed cooler achieved a DF of 4 to 6 on the tube bundle and effectively removed all the loose contamination. The cost was 24 k$. We knew CANDECON worked on carbon steel surfaces. This work has shown that the same technique is equally applicable to Incoloy-800 surfaces contaminated with fission products and is cost effective. The additional AP-4521A cost 46 k$ to improve the DF by a factor of 2. AP-4521A was not cost effective.

Portable skids, containing all the equipment for CANDECON, should be procured to be available for future decontaminations, and must overcome all mechanical problems associated with this decontamination.

REFERENCES

1. L.J. Ingolfsrud, "The Bruce Generating Station", AECL Report 3370 (1969).

2. G.M. Allison, S.P. Gibson, J.A. Atherley and D. McLaughlin, "Decontamination of the X-3 Loop Pressure Tube Using Alkaline Permanganate and Ammonium Citrate Solutions", AECL Internal Report (1960).

3. P.J. Pettit, J.E. LeSurf, W.B. Stewart, R.J. Strickert and S.B. Vaughan, "Decontamination of the Douglas Point Reactor by the CANDECON Process", Paper Presented at Corrosion 78, Houston, Texas.

4. E. Ho, Private Communication, Ontario Hydro Research Division, 1978.

DECONTAMINATION OF THE HANFORD N-REACTOR IN SUPPORT OF CONTINUED OPERATION

William K. Kratzer

Chemistry and Waste Treatment Technology
UNC Nuclear Industries
Richland, WA 99352

INTRODUCTION

N-Reactor is a graphite moderated, light water cooled, horizontal pressure tube pressurized water reactor operated by UNC Nuclear Industries for the United States Department of Energy (DOE). The reactor has two functions, producing plutonium for DOE and byproduct steam used for generating 860 MWE by the Washington Public Power Supply System Hanford Generating Project.

Since the start of operation in 1964, radiation dose rates of piping and components in contact with reactor coolant have gradually increased, requiring a continuing program to mitigate the effects of the increasing activity. The program to minimize personnel exposure involves increasing the efficiency of radiation zone operations, shifting to remote operation where possible, shielding, and decontamination.

The purpose of this paper is to describe some of the decontamination activities at N-Reactor applied in support of continued reactor operation.

BASIS FOR DECONTAMINATIONS

Decontamination decisions at N-Reactor are based on cost-benefit analyses and on maintaining exposure as low as practicable. Analyses include impact of the decontamination on program objectives, personnel exposure costs and savings associated with the decontamination, effect of the decontamination on continued operation, waste disposal costs and problems, and risks entailed in the decontamination.

TYPICAL N-REACTOR DECONTAMINATION ACTIVITIES

Reactor Coolant System Decontamination

Since 1967, a portion of the reactor coolant system including the piping to and from the reactor and the in-core fuel charge has been decontaminated ten times. The purpose of the decontaminations was to reduce dose rates for the refueling operation, maintenance work on reactor cooling system valves, instrumentation, and other components, and necessary inspections and tests. Specifically excluded from the decontamination were the reactor coolant pumps and the steam generators.

The decontaminations were performed by placing the reactor in the once-through backup cooling mode using demineralized water and injecting inhibited concentrated phosphoric acid into the coolant stream. The dilute spent acid was routed from the reactor to a waste storage tank from which it was subsequently shipped for evaporation and storage.

In contrast to commercial United States PWRs, N-Reactor has carbon steel reactor coolant piping and components out of the reactor core. Pressure tubes and fuel surfaces inside the core are Zr-2. Reactor coolant system parameters are shown in Table 1.

Table 1. Reactor Coolant System Parameters.

Volume: 586 m^3

Surface Area:

Carbon Steel	8 920 m^2
Zr-2	8 450 m^2
Inconel 600	14 860 m^2
304 SS	2 970 m^2

Coolant: Demineralized water with NH_4OH added to give a pH of 10.0 - 10.2.

Circulating flow rate: 12.6 m^3/s

Temperature: 205 C inlet, 288 C outlet.

Pressure: 11 MPa

The eight weight percent phosphoric acid applied to the carbon steel piping at 85C for about 20 minutes is effective in removing activated corrosion products from the carbon steel and in removing nonadherent deposits from the Zr-2 fuel surfaces. During the 1979 decontamination, 62 m^3 of concentrated phosphoric acid applied to the reactor coolant system removed about 476 curies of activated corrosion products, Table 2, generating 1700 m^3 of waste solution. The effect of the decontamination on dose rates on the front and rear work platforms are shown in Table 3. Corrosion of the carbon and low alloy steels in contact with the solution ranged from 0.8 to 1.6 μm, Table 4. Total annual carbon steel piping wall thickness reduction from decontamination, film reformation, and normal high temperature operation is generally less than 25 μm.

Table 2. Radionuclides Removed During the 1979 N-Reactor Decontamination.

Isotope	Ci Removed
^{51}Cr	50.6
^{54}Mn	32.4
^{58}Co	21.5
^{59}Fe	69.2
^{60}Co	271.9
^{65}Zn	10.2
^{95}Nb	4.7
^{103}Ru	7.5
^{124}Sb	4.1
^{141}Ce	4.1

The net saving achieved by the 1979 reactor coolant system decontamination was estimated to be 182 Rem personnel exposure savings and $590,000.

Removal of the radioactive source from the reactor core also benefits surfaces that are not routinely decontaminated. For example, the steam generator tube sheet dose rates are about ten times lower than on the other PWRs at comparable operating times.

Steam Generator Decontaminations

N-Reactor has horizontal U tubed steam generators with a central cylindrical tube sheet. The design and operating parameters for the generators are shown in Table 5. The generators are located in pairs in isolation cells, with one reactor coolant pump in each cell. The plant was built with ten steam generators, five cells; two additional steam generators were added later.

The original ten steam generators procured for N-Reactor were tubed with sensitized 304 stainless steel. Stress cracking of the

Table 3. Effect of Decontamination on Work Platform and Piping Radiation Levels.

Measurement Date	EFPD Since Previous Decon	Front Platform, mR/hr				Rear Platform, mR/hour			
		Center	Apron	Nozzle	Connector	Center	Apron	Nozzle	Connector
7-19-77 before decon	367	173	235	292	418	176	308	633	683
7-21-79 after decon	0	52	140	107	82	73	238	252	153
5-20-78 before decon	180	100	105	170	203	189	183	583	550
5-25-78 after decon	0	45	71	80	78	60	94	98	106
5-8-79 before decon	188	88	152	175	95	213	407	575	667
5-11-79 after decon	0	59	98	99	61	74	220	205	87

tubes led to retubing one of the generators with Inconel 600 before it was placed in service. The remaining steam generators were scheduled for tube replacement on a continuing basis, retubing two generators at a time while the others were operating. This approach permitted operation while the retubing was taking place, but required working on steam generators that had been contaminated, with dose rates on the secondary side up to 4 R/hr.

Table 4. Corrosion of Low Alloy Steels During the 1979 Reactor Decontamination.

Material	Corrosion, μm
A-216	1.6
A-105	0.9
SAE 4140	0.9
A-515	0.9
A-516	0.8
A-234	1.4
A-106	1.0

To reduce dose rates in the steam generators for retubing, the reactor coolant side of the generators was decontaminated by an alkaline permanganate, citrox procedure which was extremely effective, reducing dose rates to as low as two to twenty-five mR/hr. Results of the decontaminations are shown in Table 6.

The decontamination procedure was refined as additional generators were decontaminated and the final approach involved the application of a six weight percent solution of citrox at 80 C to remove nonadherent film, a ten weight percent sodium hydroxide, four weight percent potassium permanganate, solution at 100 C for oxide conditioning, followed by a six weight percent solution of citrox at 80 C for adherent film removal. Total waste generated was about 800 m^3. Corrosion of stainless steel was about 0.5 μm.

External Decontamination of Carbon Steel Piping

In 1972 a campaign was started to remove 1004 valves from the reactor outlet piping. The valves were a major source of leakage and their function was no longer required for reactor operation. Reactor coolant from the leaking valves had badly contaminated the surrounding piping, providing an airborne contamination problem for the valve removal work. To reduce the surface contamination, a procedure was devised in which a six weight percent solution of citrox at 80 C was sprayed over the valves and surrounding piping for about 30 minutes, followed by a demineralized water spray for rinsing.

Table 5. N-Reactor Steam Generator Parameters.

Tubing

Material: Inconel 600 and 304 stainless steel.

Size: 15.9 mm OD, 1.24 mm wall.

Number of tubes: 1916 x 12 units.

Surface area: 1486 m^2 x 12 units.

Reactor Coolant Side Conditions

Design pressure: 12.58 MPa.

Design temperature: 316 C.

Operating temperature:

Inlet 271 C
Outlet 205 C

Secondary Side Conditions

Design pressure: 5.52 MPa.

Design temperature: 316 C.

Operating pressure: 900 kPa.

Operating temperature: 180 C.

Normal steam generation, each unit: 150 kg/s.

Shell

Material: A 212-B carbon steel.

Length: 17.4 m

Diameter: 3.35 m

The decontamination reduced smearable contamination on the piping surfaces by about a factor of ten. Dose rates in the area were also reduced by 100 to 300 mR/hr.

A similar spray application of citrox was tried in 1974 to reduce external contamination on the reactor front face nozzle caps. Dose rates were reduced about 35 percent.

Reactor Test Loop Electropolish Decontamination

As part of the long term corrosion surveillance program at N-Reactor, coupons of the materials in the reactor coolant system are exposed to reactor coolant at normal operating conditions in a carbon steel bypass loop located in one of the steam generator cells. The loop test section is a 67 mm ID, 5.8 m long heavy walled A 106 Gr B carbon steel pipe.

As a demonstration of electropolishing applied to an in-place reactor component, the loop test section was decontaminated in 1978 by electropolishing. This was a demonstration of the technique, and not required for continued operation.

Electropolish decontamination is a method of removing films and radioactivity from metallic surfaces by electrolytic dissolution at high current densities, 1000 to 2000 A/m^2. The advantages of electropolish decontamination over normal acid or chelant dissolution decontamination are increased effectiveness, greater speed, and reduced waste volumes. Disadvantages include initial equipment requirements and the need for more precise control. The technique had been tried on small components removed from the reactor; it had not previously been applied to a large component on the reactor.

The decontamination was accomplished by passing 1130 A/m^2 through the phosphoric acid electrolyte between the inner test section pipe wall as the anode and a centered stainless steel tube in the pipe as the cathode. The current was timed to remove 50 μm from the wall, or 1.356×10^6 C/m^2.

The electropolish decontamination reduced the average radiation readings in the loop from 6.7 R/hr to 2.2 R/hr. Much of the final radiation field came from surrounding hot components that could not be shielded out. Test section wall thickness measurements confirmed the minimal wall reduction, and a final pressure test at 18.3 MPa demonstrated the suitability of the loop for continued operation as a part of the reactor coolant system pressure boundary.

Graphite Cooling System Heat Exchanger Decontamination

N-Reactor has a supplementary moderator cooling system in which demineralized water is pumped through cooling tubes in the graphite moderator and then through four heat exchangers. These heat exchangers are horizontal shell and tube exchangers in a four pass configuration. Each exchanger has 622 15.9 mm OD U tubes with 302 m^2 heat transfer surface. When numerous leaks began to occur in the 304 stainless steel tubing, the heat exchangers were removed from service one at a time for retubing with Inconel 600. To reduce dose rates and to control contamination spread, the heat exchangers were decontaminated by the same procedure as used for the steam generators. Dose rates were reduced from 30 to 50 mR/hr before decontamination to 300 to 500 counts per minute after decontamination.

Table 6. Steam Generator Decontamination Results.

Cell No.	Steam Generators Decontaminated	Year Work Done	Radiation Readings (mR/hr) Before Decontamination	After Decontamination
1	1A	1969	4000	2 to 7
	1B	1969	4000	5 to 18
2	2A	1965	3500	3 to 8
	2B	1965	3500	5 to 10
3	3A	1967	3000	7 to 10
	3B	1967	3500	22 to 25
4	4B	1968	1100	3 to 5

Spent Fuel Storage Basin Decontamination

When N-Reactor fuel is discharged, it falls into the spent fuel storage basin. The first, isolatable, section of the basin has underwater equipment for sorting and packaging the fuel elements for storage or shipment. A breakdown of the underwater equipment requires draining that section of the basin for access to the equipment for repair. After draining, the 6 m high painted concrete walls are a source of airborne contamination as they dry, as well as a source of radiation exposure to repairmen in the basin.

Cleaning the walls with a stream of hot 5 weight percent sulfamic acid from a fire hose removes the loose activity and reduces the dose rate for basin work. The sulfamic acid is readily neutralized and disposed of with the contaminated basin water.

Pressure Tube Nozzle Cap Decontamination

Fuel in each of the 1003 pressure tubes in N-Reactor is discharged and replaced about once a year. Pressure caps at each end of the 68.6 mm tubes are removed and refurbished each time the tube is refueled. Caps and cap components have radiation dose rates of several hundred mR/hr, requiring decontamination to minimize personnel exposure during cap reclamation. Caps have been routinely decontaminated with phosphoric acid, reducing the dose rate by a factor of 3 to 5. Development work with a vibratory finisher, a mechanically agitated bin of ceramic or steel shapes in which the cap is immersed, has shown dose rate reductions of a factor of 100, with greatly reduced waste volumes and little personnel exposure to operate the equipment. A vibratory finisher is being procured for cap cleaning, fuel spacer decontamination, and handtool decontamination, and is just now being placed in service for routine operation.

Small Component Decontamination

A variety of techniques have been used to decontaminate small components removed from the reactor for maintenance. Such components can be valve parts, instrument parts, sample line coolers, fittings, pipe sections, or tools. The method chosen depends on the degree of decontamination and the dimensional control required. Techniques that have been used include water rinsing, detergent application, scrubbing, water and sand blasting, acid cleaning, both inorganic and organic, chelant cleaning, ultrasonic cleaning, electropolishing, and vibratory finishing.

CONCLUSIONS

In a reactor plant as large and complex as N-Reactor, decontamination is mandatory both to reduce personnel exposure to a low as reasonably practicable and to minimize costs. We are continually searching for improved decontamination procedures with improved effectiveness, reduced corrosion, and reduced waste.

EXPERIENCE IN DECONTAMINATION AND OPERATION OF RADWASTE EVAPORATORS AT COMMONWEALTH EDISON'S ZION GENERATING STATION

T. P. Hillmer and W. Breen

Commonwealth Edison Company

INTRODUCTION

Zion Nuclear Generating Station is a 2 Unit, 1100 Megawatt Westinghouse fourloop pressurized water reactor. Unit I went critical June 19, 1973 and Unit II on December 24, 1973.

Zion has had a past history of relatively low activity in its primary systems and minimal primary coolant leakage. The liquid waste disposal system has had to process an average of 16 GPM of low activity in leakage throughout its operating history.

In 1978 Zion Station created an engineering group to handle radwaste problems. One of the major goals of the group was to improve the efficiency of Zion's two radwaste evaporators. Plans for modifications and repairs to the Westinghouse and Aqua Chem Radwaste evaporators were developed which we believed would improve their operating performance. The chemical cleaning of the evaporators was decided upon to improve the heat transfer capability of the Westinghouse Radwaste evaporator and to reduce the estimated 200 man-rem exposure involved in the repairs to the Aqua Chem Radwaste evaporator.

The Aqua Chem 12 GPM evaporator was started up in late 1972 and operated satisfactorily with only minor problems through 1974. These minor problems centered around the continual clogging of the concentrates discharge line, the erratic operation of the boiling point instrument (OPIC-102) which automatically discharges the concentrates at a predetermined concentration, and the continual replacement of the concentrate pumps.

In March of 1975 a forced outage of the evaporator occurred. This outage centered around problems with the vapor condenser and plugging of the concentrates discharge line. A cleaning of the vapor condenser and the addition of a flush connection on the concentrates line placed the evaporator back in service.

During this outage a service representative from Aqua Chem was at Zion Station to make recommendations that would improve the operation of the evaporator and to check on the three recurring problems. He recommended the change-over from "Chem Pumps" to "Gould" pumps. He commented on the Station's plan to increase the concentrates discharge line from a 1/2" to a 1" line. He stated that the boiling point instrument was designed to work only on a uniform feed solution and that our mixed feed with its varying concentrations of constituents would not allow it to function properly.

The change-over of the pumps was scheduled to take place as soon as new pumps could be obtained. The boiling point elevation instrumentation would be abandoned and a sampling program developed to determine the concentration of the evaporation.

In November 1976 another forced outage occurred. The major problem at this time was the control of the steam to the evaporator. Repairs were made to the pressure regulating valve and two tubes in the steam tube bundle were plugged.

In November 1977 a forced outage occurred once again. This was due to extensive failure of the tube bundle. Of the 135 tubes, 29 were involved in the failure. A new tube bundle that had been purchased with the original evaporator package was installed. This tube bundle was constructed of 304L stainless steel.

Discussions with Aqua Chem and the station's engineering department over the tube bundle failure resulted in the recommendation that the next tube bundle ordered be made of Inconel 825. It was further recommended that a chemical cleaning and a series of repairs and modifications should take place 1) as soon as the waste inflow rate was low enough so that the Westinghouse evaporator could handle the input or 2) as soon as the use of portable demineralizers on Zion's waste streams was proven.*

In November 1978 a chemical cleaning was scheduled for both radwaste evaporators. After the use of portable demineralizers to process radwaste had been proven, we then had the means to shut down the evaporators with enough time to perform the needed maintenance.

* See Appendix I, Outage Plans.

Both evaporators were flushed with demineralized water on numerous occasions with little success in dose rate reduction. A solvent was apparently needed. Scale samples were obtained from rashing rings in the gas stripping column and the steam tube bundle of the Westinghouse evaporator. Chemical analyses were performed by Dow Nuclear Services. Dow recommended a dilute solution of strong acid with an inhibitor to remove the scale. Dow had tried unsuccessfully to clean the specimens with EDTA, NSL, organic and several inorganic acids.

The recommended solution, along with several others, were tested at Zion to verify decontamination. Commonwealth Edison's Production Systems Analysis Department and Operational Analysis Department were contacted to help develop the procedures to be used and to determine the effects of the cleaning solution on the 304L stainless steel evaporators. Upon recommendations from these departments the proposed procedure for the Dow solution was modified because it was felt that the deleterious effects of stress, of sensitization of the material, or the presence of crevices, were not originally considered.

The chemical cleaning was scheduled to last four days on each evaporator and the cleanings were to be separated by one week. The Westinghouse was flushed and all minor leaks repaired prior to chemical addition. The Aqua Chem was flushed, filled with demineralized water and placed in recycle. It was to be used as a backup to the portable demineralizers only if needed. During the cleaning chemistry samples were taken at 30-minute intervals for the first four hours of solvent contact, then taken every two hours for the remainder of the chemical cleaning operation. The samples indicated the amount of radioactivity being removed and the amount of corrosion taking place.*

Both evaporators and all associated components were filled with a 10% inhibited sulfuric acid solution supplied by the Dow Chemical Company. The solution was recirculated for 15 hours and the temperature was allowed to rise as high as 120°F.

At the completion of the acid phase a 1% solution of NaOH was added to the evaporator to neutralize any remaining acid. The evaporator was filled and the caustic was circulated by the evaporator recirculation pump and the distallate discharge pump for approximately 3 hours. When all the caustic was removed, the evaporator was filled with demineralized water to the normal

* See Graph 1, Activity Removal Rates.

TABLE I

AQUA CHEM RADWASTE EVAPORATOR

I. Modifications or repairs that must be completed during a prolonged shut down of the evaporator.

1. Increase discharge line size between concentrates cooler and concentrates tank. (Increase size to 1-1/2-inch or as large as possible.)

2. Install feed inlet flow totalizer.

3. Unplug or reroute plugged drain lines.

4. Reroute sample sink lines and increase sample line size.

5. Pull feed nozzle header, clean header and replace plugged spray nozzles.

6. Install new demister pads in the absorption tower.

7. Inspect, replace, or clean vapor separator lines and pads.

II. Repairs that can be made during short shut down periods.

8. Finish installation of Gould pumps that are replacing Chem pumps.

9. Separate common demister low flow alarm into 3 individual alarms.

10. Replace all diaphragms on diaphragm valves that have a history of failure.

TABLE II

WESTINGHOUSE RADWASTE EVAPORATOR

I. Modifications or repairs that must be completed during a prolonged shut down of the evaporator.

1. Install Vent Condenser sight glass with a drain valve.
2. Install check valve on line between vent condenser and eductor.
3. Change WEV-34 from a manual valve to an AOV that works in conjunction with the concentrates discharge valve.
4. Install second Gould concentrate pump.
5. Relocate sample sink lines and sample sink.
6. Inspect tube bundle for potential problems.
7. Chemically clean evaporator.
8. Replace demister pads.
9. Check vent condenser, feed preheater, and evaporator condenser for tube leaks.
10. Inspect vent condenser trap.
11. Replace gaskets on numerous leaking flanges.

II. Repairs that can be made during short shut down periods.

12. Install new pressure gauges or repair old ones on concentrates and distillate pumps.
13. Change all diaphragms on valves which have a failure history.

operating level. The evaporator was then placed in operation on recycle. The water was recycled for 6 hours then drained out. This demin flush was repeated until no change in the pH occurred during recirculation.

The Westinghouse 15 GPM radwaste evaporator was installed in 1973 to handle liquid waste because the actual inflow to the radwaste system had exceeded the capacity of the Aqua Chem evaporator. However, it was never able to operate at greater than 3 to 7 GPM and produce acceptable quality water of 10 micromhos or less. Due to its inability to consistently produce acceptable quality water the Westinghouse was used only to process excess waste when all available tankage was full. Numerous tests were conducted to determine the cause of the problem. From these tests several modifications and the need for cleaning to improve heat transfer were recommended.

The cleaning of the Westinghouse evaporator was very successful and proceeded without major problems. The dose reduction on the Westinghouse evaporator showed a decontamination factor of 6.25:1 on contact reading on the evaporator shell and a 38.9% reduction in general area room dose rates.* The evaporator capacity increased to the design flow rate after cleaning, but it is unable to produce acceptable quality water unless it is operated at 10 GPM or less.

The Aqua Chem evaporator was cleaned the week following the Westinghouse cleaning. The procedure was similar to that used on the Westinghouse evaporator. The cleaning resulted in a decontamination factor of 1.7:1 on contact readings. This was far less than hoped for and the estimated presonnel exposure for the maintenance work was still unacceptable.

The steam tube bundle had a minor leak prior to the chemical cleaning. The bundle was eddy current inspected and pressure checked after the chemical cleaning and 56 of the 135 tubes were found to leak. The location of the cracked tubes suggested that the cracking was caused by stress-corrosion from chlorides. One side of the tube bundle showed greater cracking than the other. This suggested that the cracking may have been accelerated by uneven spraying of the tubes caused by pluggage of one or more spray nozzles. To further arrive at the cause of the scale build-up, scale samples from the Aqua Chem tube bundle are now being analyzed by Westinghouse Analytical Laboratories. Tube samples were taken and are being analyzed for the exact cause of the tube failures by Battelle Columbus Laboratory.

* See Table III, Westinghouse Evaporator Dose Rates.

TABLE III

WESTINGHOUSE EVAPORATOR DOSE RATES

Location or Equipment Component	Before Cleaning	After Cleaning	%Reduction In Exposure	Decon Factor
Evaporator Side, Location #1 NE Corner	120 MR/HR	50 MR/HR	58.3%	2.4:1
Evaporator Side, Location #2 Middle North	100 MR/HR	40 MR/HR	60.0%	2.5:1
Condenser Side, Location #4 West Side	45 MR/HR	30 MR/HR	33.3%	1.5:1
Condenser Side, Location #5 SW Corner	100 MR/HR	60 MR/HR	40.0%	1.67:1
Condenser Side, Location #6 East End	120 MR/HR	80 MR/HR	33.3%	1.5:1
Evaporator Shell Man-way (contact reading)	250 MR/HR	40 MR/HR	84.0%	6.25:1

The tube bundle was plugged so further decontamination of the evaporator could be attempted. A new tube bundle, previously ordered, of Inconel 825 would not be available for 6 months so a new 316L stainless steel tube bundle was ordered and would be available within a few weeks.

A new plan was developed whereby flushes were made with heated demineralized water with a controlled pH. Initially the flush water had the pH adjusted to a value of 3. The evaporator was then heated to normal operating temperature and left in the recirculation mode until the pH exceeded 7. The flush solution was then drained and the process repeated. By this procedure the decontamination factor was increased to 7.8:1 on contact readings.*

The dose rates still pose a problem, therefore, the current plans are to make repairs in segments so that the man-rem will be spread over several quarters of exposure reporting.

In conclusion, it was apparent that the interdepartmental cooperation played a major role in the development of the decontamination procedures for radwaste equipment and that other utilities could use a coordinated approach to similar problems in decontamination.

The chemical cleaning of the Westinghouse was successful and a sufficient amount of scale was removed to improve its heat transfer and reduce the radiation levels. The cleaning of the Aqua Chem was not as successful. The suspect cause of the poor decontamination factors are attributed to the possibility that the scale build-up on the Westinghouse had a different composition than the scale on the Aqua Chem.

Chemical cleaning resulted in a recommendation that a controlled pH flush-type cleaning of the evaporators take place once a year. This type of cleaning, if done on a regular basis, will improve operation of the evaporators and will reduce the radiation levels during maintenance.

Numerous problems still face Zion Station in the radwaste area. However, the experience gained and the procedures developed during chemical cleaning will play an important role in the future successful operation of Zion's radwaste evaporators.

* See Table IV, Aqua Chem Evaporator Dose Rates.

TABLE IV

AQUA CHEM EVAPORATOR DOSE RATES

Equipment	Before Cleaning	After Cleaning	After Flush	%Reduction In Exposure	Decon Factor After Flush
Distillate Cooler	600 MR/HR	400 MR/HR	220 MR/HR	63.3%	2.7:1
Concentrates Cooler	450 MR/HR	390 MR/HR	400 MR/HR	11.1%	1.2:1
Distillate Condenser	150 MR/HR	120 MR/HR	50 MR/HR	66.7%	3.0:1
Evaporator Shell	2500 MR/HR	1450 MR/HR	320 MR/HR	87.2%	7.8:1
Vapor Separator	610 MR/HR	300 MR/HR	120 MR/HR	80.3%	5.1:1

CHART I

AQUA CHEM RADWASTE OUTAGE SCHEDULE

Week 1	Week 2	Week 3	Week 4	Week 5	Week 6	Week 7	Week 8
Reroute drain lines			Install sample taps & reroute sample line			Replace diaphram valves	Start-up & testing
Pull Spray Header and inspect spray nozzles and replace as needed				Install spool piece to bypass concentrates cooler			
Inspect vapor separator, replace demister pads and unplug demister spray lines			Install new demister pads on absorption tower				

CHART II

WESTINGHOUSE RADWASTE EVAPORATOR OUTAGE SCHEDULE

Week 1	Week 2	Week 3	Week 4	Week 5	Week 6 (Float week)	Week 7	Week 8
Chemically clean evaporator							
Decon evaporator	Pull Stripping column rings & clean	Pull tube bundle & inspect bundle and evaporator internals for corrosion (Weeks 3–5)			Install diaphram valves	Start-up & testing with Chem/Rad systems (Weeks 7–8)	
	Replace demister pads (Weeks 2–3)		Check vent condenser, feed preheater and evaporator condenser for leaks (Weeks 4–5)				
	Install vent condenser site glass, drain valve, inspect trap and install check valve (Weeks 2–3)		Relocate sample sink (Weeks 4–5)		Install conductivity cell permanently		

DRUMMING STATION UPDATE

The best time would be during the Westinghouse Radwaste Evaporator outage due to work crews passing in and out of the shipping area.

APPENDIX I

OUTAGE PLANS

The planned outage on the waste evaporators and the drumming station could be handled in either of the following ways:

Plan I

Shut down the Westinghouse Radwaste Evaporator for approximately six weeks. During this time all inflow to radwaste could be handled by the Aqua Chem Evaporator since its 12 GPM output nearly matches the normal inflow rate to radwaste. If any excess inflow during the outage was encountered, it would be stored in one of our overflow waste storage tanks. If an unexpected increase in inflow occurred and the available tankage was full, portable demineralizers could be used to clean the waste water. Available overflow waste storage tankage would be about 200,000 gallons or twelve days of full operating capacity of the Aqua Chem Evaporator.

After the Westinghouse Evaporator was back in service we could clean up any stored water that accumulated during its outage. Chem/Rad Systems feels that the Westinghouse Evaporator will be able to run at closer to design capacity after the recommended changes have been made. The outage to the Aqua Chem Evaporator then would be handled in the same manner as the Westinghouse Evaporator outage.

The drumming station could be worked on during either or both evaporator outages or after outage completion. During an evaporator outage the concentrates could be solidified in liners. The addition of one valve and a T-connection is all that would need to be added to our present system. This outage should coincide with the date of the installation of the new Stock Equipment Company cement handling equipment. In April of 1978 the silo section of the Stock Equipment is to arrive on site. Should Unit I have to be shut down, the silo can be errected during this outage. The construction plans require an outage on Unit I so that crane can be used near the power lines. The remainder of the equipment should be ready for construction in July of 1978.

Plan II

This plan calls for the total shut down of the Radwaste System for a period of ten to twelve weeks. All inflow would be handled by portable demineralizers. If sample data that indicated a tank of waste water could not be handled or would drastically shorten the life of the demineralizer, it would be solidified in a shipping liner by a contractor and transported to burial. Depleted resins would be solidified in their liners and shipped off site.

In the past, prior to AVT steam generator blowdown, the blowdown demins were successfully used to periodically clean radwaste water. The life of the beds was limited but clean effluent was produced. The life expectancy of portable demin beds would be in the area of one week. Thus, eight to ten portable demin beds would be used. A T-connection and isolation valve would be needed on the common feed header to the waste evaporators. This line would be used as the inlet line to the portable demins.

The outlet of the portable demins would be connected to an existing T-connection in the Aux. Building on the 592' elevation. This total shut down of Radwaste would allow all equipment and panels to be worked on and would free the stationman assigned to Radwaste to start a clean-up and decon program in Radwaste.

As with the first plan, a definite time limit would be placed on the total outage. A priority list of tasks to be performed would be used. A minimal amount of work that would have to be completed before the outage ended would have high priority. Other work that could be finished or completed without shut down of more than a day would have a low priority, and if this work was not finished, it would not delay the return of the system. A list of needed modifications and work for each major area is listed in Tables 1 and 2.

DECONTAMINATION FOR CONTINUED OPERATION:

AN INDUSTRY-WIDE APPROACH

J.E. LeSurf and H.E. Tilbe

London Nuclear Decontamination Limited, Niagara Falls, Ontario
London Nuclear Services Inc.
Niagara Falls, New York

ABSTRACT

Several reasons are discussed why decontamination is desirable now, and will become increasingly necessary in future years. It is recommended that an industry-wide strategy be developed to cope with radiation problems in general and decontamination requirements in particular.

Considerations are discussed which will be important in evolving such a strategy for both the near term and the long term.

Three different approaches are presented for estimating the cost of a rem, and thereby appraising the cost/benefit of any radiation reduction procedure.

It is recommended that standards for decontamination be prepared and adopted as part of the industry-wide strategy.

The evolution, adoption, and practice of a radiation reduction strategy will help to improve the public image of the nuclear industry. We should be seen to be taking a morally commendable and professionally competent approach to a problem which is inherent to nuclear power production.

INTRODUCTION

The 1973 OPEC oil crisis and subsequent events have emphasized the need for better use of the world's resources. These events have resulted in increasing attention being devoted to improving the efficiency and availability of power stations. Generating stations operate at an overall efficiency of not much more than 30 percent. The availability of many power stations is not much more than 70 percent. Utilities, equipment suppliers, consulting engineers, and architect-engineers have all increased their attention to providing better services and improved maintenance of power station equipment.

Because of such attention to the provision of improved services, the need to decontaminate radioactively contaminated equipment has become increasingly apparent during the past three years.

This need will increase as nuclear power stations age, because of both the increased requirement for servicing and the growth of radiation fields with time. A well defined strategy is required to guide the nuclear industry on when and how to decontaminate, so as to minimize radiation exposures of workers.

In this paper, we attempt to formulate such a strategy, considering the near term, the long term, and a cost/benefit appraisal of radiation prevention measures. We hope that publication of this paper will stimulate others to improve on our approach, so that an industrial strategy will be formulated <u>and adopted</u> in the near future, with consequent reductions in radiation exposures.

WHY DECONTAMINATE?

Man-Rem and ALARA

The fundamental reason for decontaminating a piece of equipment or a reactor system is to reduce the exposure of people to radiation.

Most radiation problems can be categorized as either chronic (i.e., lasting a long time) or acute (i.e., intense, but of limited duration). An example of a chronic condition was the rapidly rising radiation fields around the Douglas Point Nuclear Generating Station in the period 1969 to 1971. All work on the reactor was hampered by radiation. How this problem was successfully overcome has been reported elsewhere.[1,2]

An example of an acute condition is the repair carried out on a steam generator at Indian Point-1 Nuclear Generating Station in 1970.[3] Seven hundred skilled workers were required to make the repairs because of the high radiation fields present. No decontamination was performed in this case.

The ALARA principle (As Low As Reasonably Achievable)[4] is now widely accepted by the nuclear community throughout the world, requiring the industry to avoid radiation exposures of people as much as possible. The application of this principle will frequently lead to decontamination as a means to achieve low radiation exposures.

There are other pressing reasons why a utility should decontaminate:

Power Generation

Reducing the radiation fields around a reactor lowers the overall man-rem burden of both scheduled and unscheduled maintenance. This in turn enables more preventative maintenance to be carried out within a given man-rem budget, thereby improving the reliability and availability of the reactor to produce power. Increased availability of nuclear power reactors reduces the use of fossil-fueled power stations, reducing power production costs for the utility and conserving scarce fossil fuel.

Ethical and Social Considerations

Controversy still rages on the health consequences of low levels of radiation exposures.[5,6] However, whether or not the effects are linear with dose, the need to minimize exposures to station workers is clearly established as an ethical and social consideration:

- ethical, because there is a moral responsibility on all industries to provide a safe working environment for their workers;

- social, because industrial harmony between workers and management is enhanced when workers are aware that efforts are being made for their protection.*

These moral considerations are equally as important as the pragmatic ones discussed earlier. Indeed, an apparent lack of concern of an industry for ethical and social considerations can quickly lead to a public outcry and worker discontent, which can cause very real, practical consequences.

*"Social cost" has been defined in terms of the burden on society of providing health services to counteract illness induced by radioactivity. This second definition is discussed later in this paper.

The case for decontamination, then, is clearly established for several reasons. What is needed is a strategy for the industry so that decontamination can be applied easily, with a frequency and ease that will minimize disruption of power production and maximize station operation.

NEAR-TERM CONSIDERATIONS

In the near term, a strategy has to be devised which can be applied to stations which are now operating or are in an advanced stage of construction. It is assumed that no structural changes can be made which will have a significant effect on radiation field growth or ease of decontamination.

Choices have to be made between (i) decontamination of a component (such as a pump, heat exchanger, or steam generator) or a reactor subsystem (such as the reactor water cleanup system or the ringheader of a BWR), and (ii) decontamination of the complete reactor circuit. In this paper, we refer to the component or subsystem choice as a "partial decontamination", and the complete reactor circuit as a "whole system decontamination".

Another consideration is whether to apply a dilute chemical decontamination (DCD, or "soft decontamination"), or a concentrated chemical decontamination (CCD, or "hard decontamination").

Some considerations in making these choices will now be discussed.

Partial versus Whole System Decontamination

There are many considerations which have a bearing on the decision to restrict decontamination to an isolated piece of equipment or subsystem, or whether to decontaminate the whole reactor; for example,

- the ease of application,
- total cost,
- cost/benefit appraisal,
- requirements for subsequent decontaminations,
- the amount of work to be done,
- the need for repassivation,

and so on. All these factors have to be taken into account. In practice, the decision will often be governed by whether the problem is chronic or acute, and also upon the nature and extent of maintenance work to be done.

Table 3-1. Comparative Advantages and Disadvantages of Partial and Whole System Decontaminations

Advantages	Disadvantages
Partial Decontamination	
Less expensive Quicker Less waste produced	Radiation fields may be present from surrounding equipment which is not decontaminated Radiation dose used in isolating the equipment Recontamination may be rapid Fuel is not decontaminated
Whole System Decontamination	
Greater immediate savings in dose Benefit lasts longer	More expensive May take longer Produces more waste

A partial decontamination was applied to the regenerative heat exchangers at Peach Bottom BWR.[7] The decontamination enabled work to be done to modify the seal between the head and the body of the heat exchanger; however, when the unit was reconnected to the rest of the reactor circuit, which had not been decontaminated, the cleaned surfaces rapidly became recontaminated. In this case the decontamination achieved what was required, but no long-lasting field reduction was obtained.

A whole system decontamination was applied at Douglas Point NGS in 1975.[8] Benefits were achieved by reduced radiation exposure in the shutdown which followed the decontamination. In addition, the fields have remained low during continued operation for the four years since the decontamination, giving additional savings in man-rem at each shutdown.

Some advantages and disadvantages of partial and whole system decontaminations are listed in Table 3-1. The considerations listed in Table 3-1 are not of equal importance in different situations. A whole system decontamination may give the greatest saving in radiation exposure, but other considerations may make a partial decontamination the preferable choice in a specific case.

Table 3-2. Comparative Advantages and Disadvantages of Hard and Soft Decontaminations

Advantages	Disadvantages
Hard Decontamination	
Higher DF (10-100)	Large volume of liquid waste created
Has been used since the 1950s	Corrosion damage to sensitive surfaces (e.g., seals, bearings)
Soft Decontamination	
Little corrosion	Low DF (2-10)
Smaller amount of waste (mostly in solid form)	Large IX columns and filters required
Can be used on fuel bundles	Highly active resins and filter cartridges requiring disposal
No need to drain	

Hard versus Soft Decontamination

A hard decontamination may be defined as one which uses concentrated chemicals, generally in the range from 5 to 25 weight percent. The major advantage of a hard decontamination is the large reduction in radiation fields which can be achieved, with decontamination factors* up to 100 being possible. A major disadvantage is the large volume of waste produced. Draining and flushing steps are required between and after addition of the chemicals, producing several reactor volumes of waste to be disposed of. In addition, residual metallic oxides have been found to promote the failure of Zircaloy-clad fuel in a reactor loop which had been previously decontaminated by a hard method.[9]

Soft decontaminations may be defined as using relatively dilute concentrations of chemicals, say, less than 1 weight percent, with corresponding reduction in corrosion. The reagents contain no metallic anions, such as the chromates or permanganates which are

*Decontamination factor, DF = $\frac{\text{radiation field before decontamination}}{\text{radiation field after decontamination}}$

frequently used in hard decontaminations. The reagents used in soft decontaminations are completely destroyed by high-temperature water, reactor radiation, or both, so that any residual reagent will be converted to harmless products on reactor start-up. Reagent removal is still performed at the end of the decontamination procedure, but the necessity to avoid any residual reagent is less stringent than with a hard decontamination process. A disadvantage of soft decontamination procedures has been the smaller reduction in radiation fields which was achievable (generally, decontamination factors less than 10, and usually in the range 2 to 6). Recent laboratory developments have resulted in improved reagent compositions and application techniques, so that decontamination factors greater than 10 are likely to be achievable on austenitic stainless steel surfaces.

Soft decontamination techniques are particularly suitable for application to the whole reactor system, including the fuel bundles. They are equally attractive for subsystems which will be reconnected to the reactor circuit. The recent developments resulting in improved decontamination factors improve the attractiveness of these processes for isolated components as well.

Some advantages and disadvantages of soft and hard decontaminations are listed in Table 3-2.

LONG-TERM CONSIDERATIONS

A long-term industrial strategy to minimize radiation exposures must consider design changes which will reduce the rate of growth of radiation fields and other changes that will facilitate decontamination.

Specifying that alloys be used which have a low level of cobalt impurity is an important step in reducing the rate of growth of fields,[10] thereby delaying the need for decontamination.

Other important design improvements concern the use of components which have longer periods between maintenance and are more easily maintained than existing components.[11]

The layout of stations is important, to ensure that components which need routine, regular maintenance (such as valves and pumps) are not exposed to the fields from components where activity will collect (such as heat exchangers, ion exchange resins, or filters, all of which tend to collect radioactivity). Experience in station operation and feedback to the designers of later stations is important here. Equally important is a rigorous assessment of the design for radiation expenditures (man-rem audit[12]), which should be reviewed by managers of other disciplines, to ensure that an optimum station layout is obtained.

The judicious use of shielding can be valuable in reducing radiation exposure. However, its indiscriminate use can result in more radiation exposure rather than less, because the reduced accessibility of some components may require longer time in the field to carry out the maintenance.

When all of these measures have been taken to reduce the rate of growth of radiation fields and to reduce radiation exposures during station operation, decontamination will still be a requirement at some time in the station's operating life.

There is a need for design modification of reactors to facilitate decontamination. Some of the features required were incorporated in the design of prototype or early commercial reactors. As power reactors became bigger and the design evolved, less attention appears to have been paid to ease of decontamination. Some items to be considered in the design are:

- ways to isolate major components or subsystems so that they may be decontaminated easily without treating the whole reactor circuit;

- provision of access points around the system to inject and remove reagent;

- adequate space to install temporary decontamination equipment and its shielding;

- ways of obtaining access to the station active waste disposal system.

There is a need for standards by which to judge proposed methods of decontamination before committing any particular job. An industry-wide approach to improvements in station design to cover all aspects of man-rem control is necessary. Such modifications are likely to be cost-beneficial when considering the economic value of the man-rem saved over a station's operating life. Some ways of assessing the value of a man-rem are presented in the next section.

ECONOMIC VALUE OF REDUCED RADIATION EXPOSURE

There have been many attempts to evaluate a man-rem, using different ground rules and reaching different conclusions.[13,14,15,16] Some values which have been cited are quite arbitrary, with little attempt at an economic justification. In 1975, the U.S. Nuclear Regulatory Commission recommended that a figure of $1,000 be used as the budgetary value of 1 man-rem.[17] Also in 1975, a group of authors from Ontario Hydro and Atomic Energy of Canada Limited cited[18] a range of from $650 to $18,600/man-rem/a (in 1985 dollars) for design purposes. The range covered different classes of employee, from normal station complement to additional permanent staff employed solely because of radiation problems.

In this paper, three different approaches to estimating the value of a man-rem will be described briefly. They have all three been printed before, but do not seem to have received much recognition. We summarize them here in the hope of stimulating further efforts to estimate the cost of radiation exposure.

The Operator's Viewpoint

The first method comes from an unclassified Ontario Hydro report.[19] Costs are evaluated as they apply at an operating nuclear power station. The costs are considered under four categories:

- social costs,
- exposure costs,
- temporary labour costs,
- permanent replacement labour costs.

Social Costs. In this study, social cost was defined as the increased burden on society, such as ill health or death, due to radiation exposure. Using 1971 health and census data for the Province of Ontario, a value of $200/man-rem was derived in 1971 dollars. (Note: In reference 18, a social value of $650/man-rem/a in 1985 dollars was assigned.)

Exposure Costs. In this category are included the costs to the station of providing such radiation services as:

- dose accounting,
- training/awareness of radiation safety,
- dosimetry.

In the original report these costs are evaluated in great detail, considering items such as the cost of buying and processing film for dosimeters, and the cost of analyzing bioassay samples.

Temporary Labour Costs. This is the cost of bringing in additional temporary staff to perform special radiation duties or supplement permanent staff because the fields are too high to complete the work with the normal station complement.

Permanent Replacement Labour Costs. This cost is based on the need to increase the permanent station complement because of radiation.

In the original report the methodology was applied to an actual case, viz, the replacement of pressure tubes at Pickering Unit 3 in 1974. The author computed a value of \$620/man-rem (1974). Because some data were not well established and assumptions had to be made, the author recommended that a value of \$1,000/man-rem be used for planning purposes.

The Designer's Viewpoint

The second method under review considers the station lifetime benefit to be derived from changes introduced at the design stage.[20] The authors recommend that a basic man-rem target be assigned when starting to design a station, and that no financial cost for man-rem reduction be levied against the original design efforts to achieve that target.

Cost assessments for man-rem reduction should be introduced under defined guidelines, such as:

- when the dose target of the station (or of a subsystem) is difficult to achieve, particularly in the later stages of design when changes are undesirable;

- when a cost/benefit decision of dose reduction has to be made independently of other factors;

- when the station design is a repeat of earlier designs, and any changes have to be justified on a cost/benefit basis;

- when it can be shown that radiation exposures are expected to limit station operation, unless the improvements are introduced.

The authors consider the value of a rem based on the salary costs of hiring temporary or additional permanent staff. They compute, for a station coming into service in 1981,

- cost of using temporary staff = \$ 4,400/rem (1981 dollars);
- cost of additional permanent staff = \$22,800/rem (1981 dollars).

They then apply a capitalization factor over a station life of 30 years, assuming historical values for inflation and interest rates. No credit is allowed for the first five years of operation, during which period fields are relatively low. From this methodology, the authors compute:

- justifiable capital expenditure to save 1 rem/a (temporary staff) = \$ 34,000;

- Justifiable capital expenditure to save 1 rem/a (permanent staff) = $175,000.

The absolute values may differ for each utility or reactor vendor doing the assessment, depending on the values assumed for average staff salaries, inflation, and interest charges. However, using this method of assessment, a range of values may be computed within a spectrum of assumptions, giving a basis on which to make judgement design decisions.

The Institution's Viewpoint

The final example of a way to compute the value of radiation exposure comes from studies done for the International Atomic Energy Agency,[21] which were based upon recommendations of the International Commission on Radiological Protection.

The evaluation in this case starts from the biological consequences of radiation dose. It distinguishes between effects on the individual receiving the dose (somatic effects) and those on his descendents (hereditary effects), and also between those consequences which occur with a random probability (stochastic occurrences) and those for which the occurrence is well established (nonstochastic occurrences).

"Risk" is defined as the probability that an individual will incur a specific deleterious effect as a result of a given radiation exposure.

Ways are given of assessing severity factors for different consequences of radiation exposure ("effects") in terms of loss of working time or of life expectancy. Fatalities are regarded as equivalent to a loss of 6,000 working days.

Account is taken of the amount of adsorbed radiation and the quality factor Q, to distinguish between the damage caused by different types of radiation (X-rays, gamma rays, neutrons, alpha particles, etc.). Also considered is the tolerance of different body organs to radiation damage, leading to the concepts of Effective Dose Equivalent H_E, and Collective Effective Dose Equivalent S_E. Further discussion considers rate of exposure, and the summation of effects of successive exposures.

By proceeding mathematically in this way, a cost/benefit appraisal of an operation which incurs radiation exposure may be obtained from:

$B = V - (P + X + Y)$, where:

B = the net benefit;
V = the gross benefit;
P = the basic cost of the operation, excluding protection;
X = the cost of achieving the selected level of protection;
Y = the cost assigned to the detriment involved in performing the operation.

The exposure in doing the job cannot be justified if B is negative, and becomes increasingly justifiable at increasing positive values of B.

The author points out the difficulty and uncertainty of making absolute assessment of the values needed in this computation, but emphasizes the value of the method in comparative assessments of different courses of action, where many of the assumptions are the same for each analysis and disappear on subtraction.

The paper goes on to discuss (and compute) the optimization of radiation protection, and the "cost of a statistical life". From such considerations, the value of a man-rem is a few hundred dollars (the actual value depending on the assumptions used), and may be equated with the "social cost" assumed in the earlier papers discussed. (Note that this assessment makes no allowance for increased labour costs.)

APPLICATION OF MAN-REM EVALUATIONS

In the previous section, three different approaches to estimating the value of radiation exposure were illustrated. These, and other papers cited, suggest that the value of a rem differs with the viewpoint taken; thus:

- the cost to the station operator is generally in the range \$1,000 to \$1,000/rem;

- the cost to the reactor designer is generally in the range \$30,000 to \$200,000/rem/a saved;

- the cost to the population as a whole (social cost) is generally in the range \$100 to \$700/rem.

London Nuclear recommends that an industry-wide study be started into the methods to be used to evaluate the necessity and economic desirability of radiation reduction practices, such as decontamination. We strongly urge all sectors in the nuclear industry to take part in this exercise so that a rational situation may be reached in the shortest practicable time.

The nuclear industry now has enough know-how and data to prepare standards for decontamination. These should be prepared and adopted as a further guide to station staff when deciding whether or not to decontaminate and how it should be done.

The nuclear industry is adopting a moral, professional, and practical approach to radiation protection. This work should be publicized as another step towards improving the image that the public has of the nuclear industry.

SUMMARY

In this paper, we have attempted to specify reasons why decontamination should be carried out. We have suggested strategies that the nuclear industry might follow in the near term and the long term, to reduce man-rem exposures. Finally, we have presented three ways of assessing man-rem, applicable to different segments of the nuclear industry.

We strongly recommend the development of methods to quantitatively evaluate radiation protection measures. A concentrated effort by the industry will result in a rational approach to radiation protection. Industrial strategies for man-rem control for the near term and the long term will follow naturally from such a study.

References

1. J.E. LeSurf, Control of Radiation Exposures at CANDU Nuclear Power Stations, J. Brit. Nuclr. Energy Soc., 16, No.1, pp. 53-61 (Jan. 1977)
2. B. Montford, Decontamination by Cycling Techniques at the Douglas Point Nuclear Generating Station, Proc. Am. Pwr. Conf., 35, pp. 902-914 (Mar. 1973)
3. A. Flynn, W. Nelson, W. Warner, D. McCormick, Thermal Sleeve Failure and Repairs - Indian Point #1 Nuclear Unit (285 MW), Nuclear Technology, 25, pp. 13-31 (Jan. 1975)
4. Recommendations of the International Commission on Radiological Protection, ICRP Publication 26, Pergamon Press, 1, No.3 (1977)
5. Biological Effects of Ionising Radiation, III Report, prepared by the BEIR Committee for the National Academy of Sciences (May 1979)
6. Risks Associated with Nuclear Power: A Critical Review of the Literature - Summary and Synthesis Chapter, prepared by the Committee on Literature Survey of Risks Associated with Nuclear Power for the NAS Committee on Science and Public Policy (May 1979)

7. M. Rohner, G.E. Casey, Decontaminating Heat Exchange Equipment Enhances Personnel Safety for Maintenance, Power, pp. 82-85 (Dec. 1978)
8. P.J. Pettit, J.E. LeSurf, W.B. Stewart, R.J. Strickert, S.B. Vaughan, Decontamination of the Douglas Point Reactor by the CAN-DECON Process, CORROSION/78 Conference, Paper No.39 (Mar. 1978)
9. A. Van der Linde, A.C. Letsch, E.M. Hornsveld, Some Observations on Pitting Corrosion in the Zircaloy Cladding of Fuel Pins Irradiated in a PWR Loop, Netherlands Energy Research Foundation Report No. ECN-51 (Nov. 1978)
10. G.F. Taylor, Heat Exchanger Tubing Materials for CANDU Nuclear Generating Stations, Proc. Nuclr. Congress of Rome, pp. 251-266 (Mar. 1978)
11. P.A. Ross-Ross, E.J. Adams, D.F. Dixon, R. Metcalfe, Performance and Reliability of Primary Circuit Components in CANDU Reactors, Proc. 1st European Nuclr. Conf. (Apr. 1975), Progress in Nuclear Energy series, Pergamon Press, 6 (1976)
12. G.G. Legg, Reducing Radiation Exposure in CANDU Power Plants, Technical Paper No. 5/22 at NUCLEX 75 (Oct. 1975)
13. A. Hedgran, B. Lindell, PQR - A Special Way of Thinking?, Health Physics, 19, p. 121 (Jul. 1970)
14. H.J. Dunster, A.S. McLean, The Use of Risk Estimates in Setting and Using Basic Radiation Protection Standards, Health Physics, 19, p. 121 (Jul. 1970)
15. J.J. Cohen, Plowshare: New Challenge for Health Physicist, Health Physics, 19, pp. 633-639 (Nov. 1970)
16. R. Wilson, Man-rem, Economics and Risk in the Nuclear Power Industry, Nuclear News, pp. 28-30 (Feb. 1972)
17. New NRC Ruling Provides Appendix 1 Compliance, Nuclear News, pp. 38-39 (Oct. 1975)
18. R. Wilson, G.A. Vivian, C. Bieber, D.A. Watson, G.G. Legg, Man-rem Expenditure and Management in Ontario Hydro Nuclear Power Stations, presented at the 20th Anniversary Meeting of the Health Physics Society (Jul. 1975)
19. C. Bieber, A Methodology for Costing Man-rem, Ontario Hydro Report (Unclassified), HPD-76-4 (Mar. 1976)
20. R. Collins, R.A. James, Design Worth of Radiation Exposure Reduction, Proc. of the 5th Intnl Conf. on Reactor Shielding, pp. 309-316 (Apr. 1977)
21. D. Beninson, Modern Trends in Radiation Protection, presented at the Topical Review Meeting on Water Chemistry and Materials in Nuclear Reactors, organized by the Argentine Atomic Energy Commission (Buenos Aires) (Apr. 1979)

DECOMMISSIONING OF COMMERCIAL

SHALLOW-LAND BURIAL SITES

E. S. Murphy
G. M. Holter
Pacific Northwest Laboratory*
Richland, Washington 99352

INTRODUCTION

This paper describes the results of a study to conceptually decommission commercial low-level waste (LLW) burial grounds.[1] The study was sponsored by the U.S. Nuclear Regulatory Commission to provide information on the available technology, the safety considerations, and the probable costs of decommissioning LLW burial grounds after waste emplacement operations are terminated. This information is intended for use as background data in the development of regulations pertaining to decommissioning activities. It is also intended for use by regulatory agencies and site operators in developing improved waste burial and site maintenance procedures at operating burial grounds.

CHARACTERISTICS OF THE REFERENCE BURIAL GROUNDS

Two generic burial grounds, one located on an arid western site and the other located on a humid eastern site, are used as reference facilities for the study. The characteristics of these postulated facilities are based on real characteristics of the six commercial burial grounds that have operated in the United States. The reference burial grounds are assumed to have the same site capacity for waste, the same radioactive waste inventory, and

*Operated by Battelle Memorial Institute.

similar trench characteristics and operating procedures. The climate, geology, and hydrology of the two sites are chosen to be typical of real western and eastern sites.

Each site has an area of 70 hectares (7×10^5 m^2), of which about 50 hectares contain burial trenches. The total site capacity for waste is about 1.5×10^6 m^3 contained in 180 burial trenches. The trenches are 150 m long, 15 m wide at the top sloping to 10 m wide at the bottom, and 7.5 m deep. A space of at least 3 m separates adjacent trenches at the ground surface. Each trench is filled with waste to within 1 m of the ground surface; the top 1 m of trench is reserved for fill soil.

The reference waste inventory in the burial trenches is comprised of 40% (by volume) non-fuel-cycle waste and 60% reactor fuel-cycle waste. An average byproduct specific activity of 9.0 Ci/m^3 is assumed for the waste at the time of burial.

At some commercial sites, high activity beta-gamma waste is buried separately from other waste. To evaluate cost and safety requirements for exhumation of this waste, the reference burial ground is assumed to include 10 slit trenches. Slit trench dimensions are 150 m long by 1.2 m wide by 6 m deep; each trench can accommodate ninety 0.76-m-diameter by 3.6-m-long waste canisters. Typical activities at the time of waste burial range from 1,000 to 5,000 Ci/m^3.

The western site is semi-arid. Summers are marked by very low precipitation and high temperatures, resulting in soil moisture deficiencies. Occasional periods of high winds are accompanied by blowing sand. Additional characteristics include 1) low annual precipitation, with evaporation greatly exceeding precipitation, 2) great depth to ground water, 3) soil with moderate permeability, 4) relatively great distance to the point of groundwater discharge into surface streams, and 5) no farming in the immediate vicinity of the site.

The eastern site has a continental climate with a wide range of temperature throughout the year. Summers are characterized by intense heat and high humidity, and winters by extreme cold with occasional heavy snowfall and moderate-to-high winds. Additional characteristics include 1) high annual precipitation, 2) shallow depth to ground water, 3) soil with low permeability, 4) relatively short distance to the point of ground water discharge into surface streams, and 5) farming and recreation in the vicinity of the site.

DECOMMISSIONING ALTERNATIVES

Decommissioning may be defined as the measures taken at the end of a facility's operating life to ensure that future risk to public safety is within acceptable bounds. Decommissioning an LLW burial ground normally involves engineered procedures (site/waste stabilization procedures) to ensure the adequate confinement of radioactive wastes left buried at the time of site closure. Burial ground stabilization is followed by a period of long-term care designed to maintain and verify the confinement capability of the site. Long-term care continues until the wastes no longer pose a significant radiological hazard, or until additional actions are taken to reduce the potential consequences of unrestricted site usage.

Relocation of part or all of the waste from the burial trenches may be necessary in some situations. This decommissioning mode is therefore also considered in the study.

Site/Waste Stabilization

Potential site/waste stabilization activities include:

- engineered routing/flow control of ground and surface water
- modification of trench caps to minimize water infiltration into trenches
- stabilization of the land surface and erosion control
- grouting or injection of chemicals into the waste matrix to reduce the mobility of the waste
- control of plants or animals that might disrupt surface stabilization or transport radioactivity from the trenches
- erection of physical barriers to control human activities at the site.

To select an appropriate stabilization plan, radionuclide transport mechanisms capable of initiating a release of radioactivity (i.e., "release agents") are identified, and suitable stabilization techniques for dealing with these release agents are catalogued and evaluated. Plans are then formulated based on the techniques selected. Release agents of importance for a particular site are identified by critical pathway analysis techniques. For the reference western site of this study, the release agents of concern are human activities (excavation and agriculture) and wind erosion. For the reference eastern site, the release agents of concern are human activities (excavation and agriculture), hydrological releases (percolation and overflow), and water erosion.

In the Reference 1 study, three stabilization plans are described and evaluated for each reference site. The stabilization plans include a minimal plan, a relatively modest one, and a more complex one; they correspond to varying levels of effort that may be required to properly stabilize a site. Brief descriptions of each plan are given in Table 1.

At each reference site, the minimal plan assumes that necessary site stabilization activities have been performed as part of burial ground operating procedures. Therefore, extensive decommissioning measures are not required when burial operations cease. The minimal plan includes site inspections and the performance of any repairs needed to ready a site for long-term care.

The main feature of the modest plan at each site is an increase in the soil thickness over the buried waste (increased capping thickness). In addition, at the eastern site, a system

Table 1. Descriptions of Stabilization Plans for the Reference Sites

Plan	Western Site	Eastern Site
Minimal	Site inspection, Preparation for long-term care.	Site inspection, Preparation for long-term care.
Modest	Increased capping thickness, Revegetation, Vegetation management.	Increased capping thickness, Modification of capping soil, Improved capping drainage, Revegetation, Vegetation management.
Complex	Subsurface rock layer with hard top, Increased capping thickness, Revegetation, Vegetation management.	Peripheral drainage and diversion, Sump pumping with treatment, Subsurface hard layer, Increased capping thickness, Revegetation, Vegetation management.

of drainage ditches is installed to allow precipitation to be directed away from the trenches. Both sites are revegetated with shallow-rooted plants to control vegetation and reduce surface erosion.

The complex plan for each site includes installation of a subsurface layer in addition to an increase in capping thickness. The subsurface layer reduces moisture percolation and plant-root infiltration into the wastes, acts as a deterrent to animal or human penetration, and provides a secondary control against erosion. At the eastern site, a series of drainage/diversion ditches is constructed around the site perimeter to intercept runoff that might encroach on the site. The sites are then revegetated.

Long-Term Care

After site decommissioning is completed and the operating license is terminated, access to a site may need to be controlled for a time until the radioactivity in the buried waste has decayed to levels that permit unrestricted site release. Long-term care procedures are employed during this time to maintain and verify the continued capability of the site to adequately contain the buried radioactivity.

Long-term care includes administrative control, environmental surveillance, and site maintenance. Administrative control includes control of public access/activities at the site, coordination of site surveillance and maintenance programs, and record keeping. Environmental surveillance includes the collection and analysis of environmental samples and the analysis and preservation of environmental data. Site maintenance includes inspection and repair of fences and equipment, erosion control, trench cap repair, water infiltration control, and vegetation management.

Waste Relocation

Waste relocation involves the exhumation of buried waste, repackaging of the waste if necessary, and reburial of the waste at another disposal site or in another trench on the same site. Exhumation of waste originally buried without any intent of later retrieval is an expensive and time-consuming operation, with a potential for significant radiation exposure to decommissioning workers. Therefore, waste relocation would likely be considered only if other decommissioning procedures are inadequate to ensure that future risk from the buried waste is within acceptable bounds.

Relocation of the waste from part or all of a particular trench or trenches may be more likely than waste exhumation from an entire burial ground. Partial waste relocation may be necessary if individual trenches are known to contain very high levels of radioactivity, high concentrations of transuranic or other long-lived wastes, or waste in a form that makes it particularly susceptible to migration (i.e., waste mixed with organic complexing agents). Partial waste relocation cases evaluated in the study include 1) relocation of high beta-gamma activity waste from a slit trench, 2) relocation of a transuranic (TRU) waste package buried in a regular burial trench, and 3) relocation of all the waste from one burial trench. In addition, an estimate is made of the manpower, time, and cost of relocating all the waste from the reference burial ground.

Slit trench excavation and waste retrieval require personnel protection from high radiation dose rates. Most operations are performed remotely. Several equipment options are available for remote excavation of the trench. Canister retrieval is accomplished with a boom crane. Sheet piling is used along the sides of the trench as a safety measure and to limit the width of the trench.

To reduce the possibility of airborne release of TRU contamination, all excavation and waste retrieval operations related to TRU waste exhumation take place inside a metal enclosure equipped for control and filtration of the air leaving the building. If a second level of protection from airborne releases is required, the metal enclosure can be located inside a plastic air-support weather shield.[2] All personnel operating within the confines of the metal enclosure wear plastic bubble suits for protection from airborne contamination.

After selective exhumation of the high beta-gamma and TRU wastes, removal of the waste remaining in a regular trench can be accomplished by relatively simple earthmoving techniques. Wastes are exhumed by bulk excavation of the trench, using conventional equipment. Because of the difficulty and added cost of sorting soil, it is assumed that all of the soil in the bottom 6.5 m of a trench is exhumed and repackaged with the waste. Two exhumation alternatives are evaluated. One utilizes a backhoe operating from above the trench, permitting most of the operating crew to be relatively remote from the exposed waste. The second involves use of a front-end loader operating from the floor of the trench, with laborers assisting in grappling and excavation of large containers and loose waste. The second alternative requires less time but results in a higher worker exposure to radiation than the first alternative.

DECOMMISSIONING COSTS

Costs for the decommissioning alternatives described previously are calculated assuming efficient performance of the work. Direct costs are estimated, including labor, material, equipment, surveillance, and, where applicable, packaging, transportation, and disposal of wastes. A 25% contingency is added to account for work delays and equipment cost escalations. All costs are in 1978 dollars.

Costs of Site/Waste Stabilization

Estimated costs of site/waste stabilization are summarized in Table 2. Total site stabilization costs for the western site are \$0.4 million for the minimal plan, \$2.5 million for the modest plan, and \$7.7 million for the complex plan, while total costs for the eastern site are \$0.5 million, \$3.8 million, and \$5.3 million for the minimal, modest, and complex plans, respectively. For both the modest and complex plans, site stabilization is performed by a contractor hired by the site operator. Therefore, a contractor's fee is included in the total cost.

Manpower costs include both support staff and decommissioning worker costs. The support staff includes those persons involved in planning and supervising decommissioning operations, in preparing documentation, and in carrying out the accounting, quality control, health physics, and site security functions related to decommissioning.

Table 2. Estimated Costs of Site Stabilization

	Cost (\$ millions)[a,b]					
	Arid Western Site			Humid Eastern Site		
Cost Category	Minimal Plan	Modest Plan	Complex Plan	Minimal Plan	Modest Plan	Complex Plan
Manpower	0.395	1.033	1.586	0.424	1.391	1.467
Equipment and Material	0.034	1.262	5.473	0.035	2.107	3.430
Contractor's Fee[c]	--	0.184	0.565	--	0.280	0.391
Miscellaneous Owner Expense[d]	0.002	0.009	0.010	0.002	0.010	0.010
Environmental Monitoring	0.011	0.027	0.031	0.011	0.034	0.034
Records Maintenance	0.001	0.001	0.001	0.001	0.001	0.001
Total (rounded)	0.4	2.5	7.7	0.5	3.8	5.3

[a]Number of figures shown is for computational accuracy only and does not imply precision to the nearest thousand dollars.
[b]Costs include 25% contingency.
[c]Based on 8% of the sum of manpower, equipment, and material costs.
[d]Includes utilities, insurance, and taxes.

The complex plan for the western site is calculated to be more costly than that for the eastern site. Both plans include an increase in capping thickness over the trenches. For the western site, the additional soil must be brought from offsite by truck. For the eastern site, a large portion of the required backfill is provided by digging the peripheral drainage/diversion ditches. This also results in somewhat reduced labor and equipment costs for the eastern site.

Costs of Long-Term Care

Long-term care costs for the reference sites are summarized in Table 3. To provide a basis for calculations, long-term care is assumed to continue for 200 years following stabilization of a site.

Annual costs are assumed to be highest during the first few years because of greater site maintenance and environmental surveillance requirements. After an initial site "maturation" period of about 5 years, annual maintenance costs are assumed to decrease significantly. Environmental surveillance is maintained at a significant level for 25 years to evaluate the effectiveness of site stabilization procedures. Environmental surveillance is then reduced to about one-fourth of the original level by reducing the number of sample locations and/or the sampling frequencies. Long-term care requirements and costs following stabilization of

Table 3. Summary of Estimated Long-Term Care Costs

Stabilization Plan That Precedes Long-Term Care	Annual Costs (in millions of 1978 dollars) for Time Period:[a,b] 0-5 Years After Stabilization	6-25 Years After Stabilization	26-200 Years After Stabilization	Total Cost for 200 Years (in millions of 1978 dollars)[a]
Minimal and Modest Plans for the Arid Western Site	0.181	0.125	0.081	17.6
Complex Plan for the Arid Western Site	0.250	0.120	0.075	16.8
Minimal and Modest Plans for the Humid Eastern Site	0.254	0.196	0.134	28.6
Complex Plan for the Humid Eastern Site	0.382	0.201	0.139	30.2

[a]These costs include contingency costs of 25%.
[b]Number of figures shown is for computational accuracy only.

a site by the minimal or modest plans are postulated to be identical. Maintenance requirements and costs following stabilization by the complex plans are somewhat greater because of the need to maintain and repair subsurface layers and diversion and drainage systems.

Total long-term care costs at the eastern site are almost twice as high as they are at the western site, partly because of higher environmental surveillance costs at the eastern site. However, the cost differential is mainly due to the additional costs of maintenance of stabilization features needed to reduce water infiltration into the trenches at the eastern site.

Costs of Waste Relocation

Costs of partial waste relocation at the western site are summarized in Table 4. Costs are shown separately for exhumation (including core drilling and sampling, sheet piling installation and removal, trench excavation, waste exhumation, and site restoration) and for waste management (including packaging, shipping and disposal). Offsite waste repositories are assumed to be located 2,400 km from the LLW burial ground, with all waste shipments made by truck.

Table 4. Estimated Costs of Partial Waste Relocation at the Western Site

	Cost ($ millions)[a]				
	Slit	TRU Waste Package		Entire Burial Trench	
	Trench Exhumation	Single Enclosure	Double Enclosure	Excavate from Above	Excavate from Within
Deep Geologic Disposal					
Exhumation	0.411	0.361	0.801	0.590	0.471
Waste Management	2.336	0.024	0.024	43.240	43.240
Total (rounded)	2.8	0.38	0.82	43.8	43.7
Shallow-Land Burial					
Exhumation	0.411			0.590	0.471
Waste Management	1.121			7.190	7.190
Total (rounded)	1.5			7.8	7.7

[a]Number of figures shown is for computational accuracy only.

Some of waste exhumation activities are postulated to require a 20% longer time for completion at the eastern site than at the western site because of the greater potential for adverse weather at the eastern site. Therefore, the costs of waste exhumation are higher at the eastern site than at the western site. However, total waste relocation costs are not significantly changed for the eastern site because waste management costs are the same at both sites.

Waste management controls the cost of relocation of slit trench waste and the waste from an entire burial trench. Based on deep geologic disposal costs in Reference 3, unit costs of \$2,100/m^3 for burial of low-level waste (waste that does not require shielding for transport or handling), and \$7,100/m^3 for intermediate-level waste (waste that arrives in a shielded container) are assumed for the study. A basic cost of \$168/m^3 (\$4.75/ft^3) is assumed for disposal at a shallow-land burial ground. Enclosure costs control the cost of TRU waste removal.

Costs of total waste relocation from the reference burial grounds are summarized in Table 5. It is assumed that 10 slit trenches and 10 TRU waste packages must be exhumed in addition to the complete exhumation of the 180 burial trenches. Waste relocation at the reference sites is estimated to require about 20 years (plus 1.5 years for planning and preparation) and to cost about \$1.4 billion. About 94% of this cost is related to waste management (packaging, shipment, and disposal of exhumed waste).

PUBLIC AND OCCUPATIONAL SAFETY

Radiological and nonradiological impacts from normal decommissioning operations and potential accidents are identified and evaluated for the site stabilization and waste relocation options described previously. The safety evaluation includes estimates of radiation dose to the public from normal decommissioning operations and postulated accidents, radiation dose to workers from normal decommissioning operations, and deaths and injuries to decommissioning workers from industrial-type accidents.

The radiation dose to the public is evaluated by determining the 50-year committed dose equivalent to the populace within 80 km of the site from airborne releases. The same population distribution is assumed for both reference sites. This allows a direct comparison of safety effects that are related to the physical characteristics of the two sites. For decommissioning workers,

Table 5. Summary of Costs for Total Relocation of the Waste from the Reference Burial Grounds

Cost Category	Cost ($ millions)[a] Western Site	Eastern Site
Manpower	46.0	54.2
Equipment and Material	30.5	36.0
Contractor's Fee[b]	6.1	7.2
Work Enclosures	0.5	0.5
Environmental Surveillance	1.1	1.4
Records Maintenance	0.1	0.1
Miscellaneous Owner Expenses	0.4	0.4
Subtotals	84.7	99.8
Waste Management	1 318	1 318
Totals	1 403	1 418

[a]Number of figures is for computational accuracy only.
[b]Based on 8% of the sum of labor, material, and equipment costs.

the dose from direct exposure is determined. Workers are presumed to wear respiratory protection devices as needed to provide protection from inhalation of airborne radionuclides.

Safety Evaluation of Site/Waste Stabilization

Table 6 shows radiation dose to decommissioning workers from normal site stabilization activities, based on an assumed external exposure rate of 1 mrem/day. Ingestion and inhalation are not considered important contributors to occupational dose for site stabilization.

Accident data from a U.S. Atomic Energy Commission study[4] provides a basis for estimating lost-time injuries and fatalities from industrial-type accidents to decommissioning workers. It is estimated that only about one lost-time injury might be expected to occur during site stabilization. The number of fatalities from industrial-type accidents during site stabilization is estimated to be about 1×10^{-3}.

Table 6. Radiation Dose to Decommissioning Workers from Normal Site Stabilization Operations

Stabilization Plan	Radiation Dose (man-rem)	
	Western Site	Eastern Site
Minimal	0.2	0.2
Modest	1.2	2.0
Complex	2.5	2.2

Stabilization of a burial ground involves modification of and addition to surface soils, but no intentional uncovering or exhumation of buried wastes. There is no transportation of radioactive waste. Therefore, routine site/waste stabilization operations are not expected to contribute any significant radiation exposure to the general population.

Most accidents during stabilization are not expected to contribute a significant radiation dose to any member of the public. Trench void-space collapse resulting from the passage of heavy equipment over a trench is calculated to result in a first-year dose of 0.61 mrem and a 50-year committed dose equivalent of 5.2 mrem to the bone of the maximum-exposed individual.

Safety Evaluation of Waste Relocation

Because partial waste relocation is much more probable than relocation of the waste from an entire burial ground, and because complete waste relocation is estimated to take more than 20 years, it is unlikely that one person would be subjected to the radiological effects of an entire burial ground exhumation. Therefore, public and occupational radiological safety effects are described on the basis of partial waste relocation activities. The results are shown in Table 7. The table shows the 50-year committed dose equivalent to the populace within 80 km of the site from airborne releases and the total dose to decommissioning workers from direct exposure. Occupational doses for exhumation of a complete trench are calculated on the basis of excavation from within the trench to maximize the dose numbers. The dose numbers shown in Table 7 indicate that waste exhumation can be very costly in terms of worker exposure.

Table 7. Radiological Safety Evaluation of Normal Waste Relocation Operations

Source of Safety Concern	Radiation Dose (man-rem)[a]		
	Slit Trench Exhumation	TRU Waste Package Exhumation	Exhumation of Entire Burial Trench
Public Safety[b]			
Decommissioning Operations	6×10^{-5}	1×10^{-1}	1×10^{2}
Transportation	2×10^{-1}	2×10^{-3}	3×10^{0}
Occupational Safety			
Decommissioning Operations	5.2×10^{1}	1.1×10^{2}	2.4×10^{2}
Transportation	1.8×10^{1}	2.0×10^{-1}	9.9×10^{1}

[a]Radiation doses from postulated accidents are not included.
[b]The 50-year committed dose equivalent to the population residing within 80 km of the site is reported for routine operations. The organ of reference is the bone. Transportation doses are from external exposure to the population along the transport route.

Table 8 summarizes the results of an analysis of postulated decommissioning accidents during waste relocation operations. A wide spectrum of accidents is considered, with appropriate assumptions leading to calculated airborne releases of radioactivity and resulting radiation doses to the maximum-exposed individual. Results are shown for the seven accidents that result in the highest dose to an organ of the maximum-exposed individual. Table 8 shows that for each of the decommissioning accidents studied, the 50-year committed dose equivalent to the maximum-exposed individual is very small compared to the 50-year dose from natural background radiation.

Table 9 shows estimated occupational lost-time injuries and fatalities from industrial-type accidents and from decommissioning transportation accidents for waste relocation. Except for transportation accidents associated with waste relocation from an entire trench, the number of nonradiological injuries and deaths is estimated to be low.

Table 8. Summary of Radiation Doses to the Maximum-Exposed Individual from Decommissioning Accidents[a]

Accident	Airborne Release (μCi)	Estimated Frequency of Occurrence[b]	50-Year Committed Dose Equivalent (mrem) Total Body	Bone
Severe Transportation Accident (TRU)	3.0×10^{3}	Low	4.2×10^{0}	9.6×10^{1}
Exhumation of Undetected TRU Waste	1.6×10^{3}	High	2.2×10^{0}	5.1×10^{1}
Waste Package Handling (TRU)	5.3×10^{2}	Low	7.5×10^{-1}	1.7×10^{1}
Undetected TRU Core Drilling	1.6×10^{2}	High	2.3×10^{-1}	5.2×10^{0}
Failure of HEPA Filters (TRU)	6.6×10^{1}	Low	9.4×10^{-2}	2.1×10^{0}
Spontaneous Combustion of Wastes	1.7×10^{1}	Medium	1.2×10^{-2}	1.9×10^{-1}
Minor Transportation Accident (TRU)	3.0×10^{0}	Medium	4.3×10^{-3}	9.6×10^{-2}

[a]Inhalation doses only.
[b]Frequency of occurrence: High $>1.0 \times 10^{-2}$; Medium 1.0×10^{-2} to 1.0×10^{-5}; Low $<1.0 \times 10^{-5}$ events per year.

Table 9. Estimated Occupational Injuries and Fatalities from Nonradiological Sources for Partial Waste Relocation

Waste Relocation Operation	Waste Exhumation: Number of Injuries	Waste Exhumation: Number of Fatalities	Transportation: Number of Injuries	Transportation: Number of Fatalities
Slit Trench Exhumation	5.2×10^{-2}	1.0×10^{-3}	2.4×10^{-1}	1.4×10^{-2}
TRU Package Exhumation	1.3×10^{-2}	6.7×10^{-5}	2.7×10^{-3}	1.6×10^{-4}
Exhumation of Entire Trench	1.6×10^{-1}	7.3×10^{-4}	2.6×10^{0}	1.5×10^{-1}

SUMMARY AND CONCLUSIONS

Estimated costs and safety considerations for decommissioning LLW burial grounds have been evaluated. Calculations are based on a generic burial ground assumed to be located at a western and an eastern site. Decommissioning modes include 1) site stabilization followed by long-term care of the site, and 2) waste relocation.

Site stabilization is estimated to cost from $0.4 million to $7.5 million, depending on the site and the stabilization option chosen. Long-term care is estimated to cost about $100,000 annually, with somewhat higher costs during early years because of increased site maintenance and environmental monitoring requirements. Long-term care is required until the site is released for unrestricted public use. Occupational and public safety impacts of site stabilization and long-term care are estimated to be small.

Relocation of all the waste from a reference burial ground is estimated to cost more than $1.4 billion and to require more than 20 years for completion. Over 90% of the cost is associated with packaging, transportation, and offsite disposal of the exhumed waste. Waste relocation results in significant radiation exposure to decommissioning workers.

REFERENCES

1. E. S. Murphy and G. M. Holter, Technology, Safety and Costs of Decommissioning a Reference Low-Level Waste Burial Ground, NUREG/CR-0570, U.S. Nuclear Regulatory Commission Report by Pacific Northwest Laboratory, 1979.

2. D. H. Card, Early Waste Retrieval Interim Report, TREE-1047, EG and G Idaho, Inc., Idaho Falls, ID, February 1977.

3. K. J. Schneider and C. E. Jenkins, Technology, Safety and Costs of Decommissioning a Reference Nuclear Fuel Reprocessing Plant, NUREG-0278, U.S. Nuclear Regulatory Commission Report by Pacific Northwest Laboratory, Vol. 1, p. 7-129, October 1977.

4. Operational Accidents and Radiation Exposures Experienced Within the USAEC 1943-1970, WASH-1192, U.S. Atomic Energy Commission, Washington, DC, 1972.

SOIL DECONTAMINATION AT ROCKY FLATS

R. L. Olsen, J. A. Hayden, C. E. Alford,
R. L. Kochen, J. R. Stevens

Rockwell International
Energy Systems Group
Rocky Flats Plant
Golden, Colorado 80401
Under Contract DE-AC04-76DP03533
for United States Department of Energy

INTRODUCTION

Soil Contamination

During the last few years, many articles have appeared in newspapers and journals concerning radioactive contamination around Rocky Flats.[1,2] The amount of plutonium in the soil has been of particular interest. The Draft Environmental Impact Statement[3] on Rocky Flats lists the yearly alpha releases from the Plant since it was opened in 1952. The largest release was from "contaminated oil leakage." In 1958, a drum storage area was established on virgin ground just inside the present-day east gate to Rocky Flats. The drums contained cutting oil and carbon tetrachloride contaminated with plutonium and uranium cuttings from the machining of nuclear weapon components. Deterioration of and leakage from the drums was first observed in 1964. This leakage resulted in the soil being contaminated. In January of 1967, the last drums were added to the area. Shortly after this time, drum removal commenced and continued until all drums were removed by January 1968. The level of contamination in the soil was measured at this time. The results showed levels from 2,000 to 300,000 dpm/100 cm^2 and penetration depths of 3 to 20 cm. In April of 1969, a gravel cover of approximately 15 cm was applied and by July the area had been covered with a 7.5 cm asphalt pad.

Form of Contamination

The plutonium at this site was originally in the form of plutonium metal from the machining process. In the environment, the plutonium metal oxidized to PuO_2. Actual PuO_2 particles have been identified in the cutting oil used for machining. The average diameter of these particles is 0.2 microns. Besides the particulate form of the plutonium in the soil, there exists also a dispersed form. The dispersed form of the plutonium will pass through a 0.01 micron pore filter. Up to 50% of the total contamination may have been in this form. As the barrels sat outside, hydrochloric acid was probably generated. This acid reacted with the plutonium giving a dissolved plutonium species. The exact nature of the dispersed form is not known. However, the plutonium is probably absorbed onto the clay or organic material in the soil or precipitated as Fe_2O_3-PuO_2 coatings on the mineral surfaces. When evaluating potential decontamination processes, both the particulate and dispersed forms of the plutonium must be considered.

Guidelines

The EPA has issued a guideline for maximum levels of radioactivity in soil. The level of 200 mCi/km^2 for plutonium in soil was calculated from a maximum permissible air concentration of respirable size dust assuming reasonable values for the soil density and mass loading. This is approximately 10 to 15 pCi/g or 20 to 30 dpm/g. Because of instrumental limitation during rapid survey of soil, DOE has indicated that soil containing less than 1,000 dpm/g does not have to be excavated, but can be left in place. However, if the soil is removed, it cannot be returned to the ground unless it is less than 30 dpm/g. Therefore, any process employed to decontaminate soil must reduce the level of contamination to below 30 dpm/g.

Alternative Solutions

The pad area is 113 m wide and 120 m long. Approximately 80 to 90 grams of plutonium are dispersed in 2×10^7 kg of soil beneath the pad. Water transport has not been observed under the pad since monitoring at the four corners began in 1969. Therefore, the contaminated soil is effectively isolated from the environment. In spite of this apparent stability, there was concern about long-term diffusion of plutonium from the site. Cost estimates to remove and ship the contaminated material were made and alternative solutions investigated. The present-day alternatives are as follows:

1) Stabilize in place.
2) Remove and transport all of the soil to a permanent

disposal site.

3) Decontaminate.

The first alternative has been eliminated as a viable long-term solution because of the concern about diffusion. The second alternative has been used at Rocky Flats extensively during the last few years on small areas of contaminated soil. The decontamination alternative offers improved socio-political impact and will reduce any potential long-term effect. That is, the soil is actually being "cleaned up," not just moved to another site. Moreover, the decontamination option is also more economical. The actual cost to dig, package, and ship soil to a permanent disposal site in Nevada is $255 per 1,000 kg. Projected costs to decontaminate 90% of the soil and ship 10% is $123 per 1,000 kg. This second cost estimate includes manpower and chemicals, but excludes the initial capital cost of the processing facility. This cost was excluded because the process facility will be used only two years at Rocky Flats and then moved to other sites. The cost would increase proportionally if the decontamination was less than 90% effective.

Because of the projected savings of the decontamination alternative, a soils decontamination project was initiated. The objective of the project is to develop and demonstrate decontamination processes to concentrate or remove actinides associated with contaminated soils at Rocky Flats. The bulk of the soil would be returned to its natural environment, while the remaining small fraction would be packaged for shipment. Approximately nine man-years of effort were expended in FY-1979. The ultimate goal of the project is the construction of a $5-7 million mobile process facility.

EXPERIMENTAL AND RESULTS

Laboratory Studies

Several soil conditions exist at Rocky Flats that are advantageous to decontamination processes:

1) The soil is very rocky, 2) the contamination exists on the surface of the minerals, and 3) the surface-contaminated soil contains only approximately 20% clay and organic matter. A typical soil profile in the Rocky Flats area contains three distinct horizons. The dark top soil layer is usually 15 to 25 cm thick. This horizon is usually followed by a rocky zone rich in limonite and hematite-coated minerals. This zone runs from 25 to 40 cm in depth, but is missing under the pad. Finally, a layer of tan bentonite clay is

observed. This layer varies in thickness and is located at a depth of 40 to 75 cm below the surface. The plutonium contamination does not penetrate the clay layer. The total cover over the Rocky Flats area consists of this rocky alluvial material.

Because of this rocky material, a physical scrubbing process is very effective. Four such processes have been investigated:

1) Wet screening at high pH.
2) Attrition scrubbing with Calgon® at elevated pH.
3) Attrition scrubbing at low pH.
4) Cationic flotation of clays.

Wet Screening. The first process is a simple wet screening process with the pH adjusted to 11 with NaOH. Both the Na^+ and the OH^- ions help disperse the clay particles. In fact, a colloidal suspension is formed. This suspension is due to the hydroxide ions attaching to the surface of the clay, creating negatively charged particles which repel each other. Details of one run are shown in Figure 1. The +35 mesh (greater than 420 micron) fraction of soil is decontaminated to less than 30 dpm/g. Typically, this material represents 60 to 70 wt % of the soil and can be returned to the environment. The remaining 30 to 40 wt % is radioactive and would be packaged and shipped for permanent disposal.

Attrition Scrubbing at High pH. Calgon solutions at high pH also effectively decontaminate soil. The soil is scrubbed in a rotary-type attrition scrubber (jar mill) four times. The fine material is decanted each time. In a typical experiment, 1 kg of soil is scrubbed with 1,000 ml of solution for 5 to 7 minutes. The fines are decanted removing most of the contamination. The scrub is repeated three more times with 250 ml of solution each time. A total of 99.9% of the activity is removed with the fine fraction. This fraction represents about 20 wt % of the total soil. The remaining 80 wt % is below the EPA guideline and can be returned to the environment.

Two processes are taking place that decontaminate the soil. One process is the physical grinding action due to the large rocks present in the soil. The tumbling action in the jar mill actually grinds away the outer surface of the particles. The second process is the dispersion of the clay and the fines being generated by the high pH Calgon solution. The two processes work together to decontaminate the soil.

If an additional blender-type scrub is incorporated in the process, the amount of material decontaminated can be increased. The material less than 4 mm is scrubbed at 800 rpm in a special

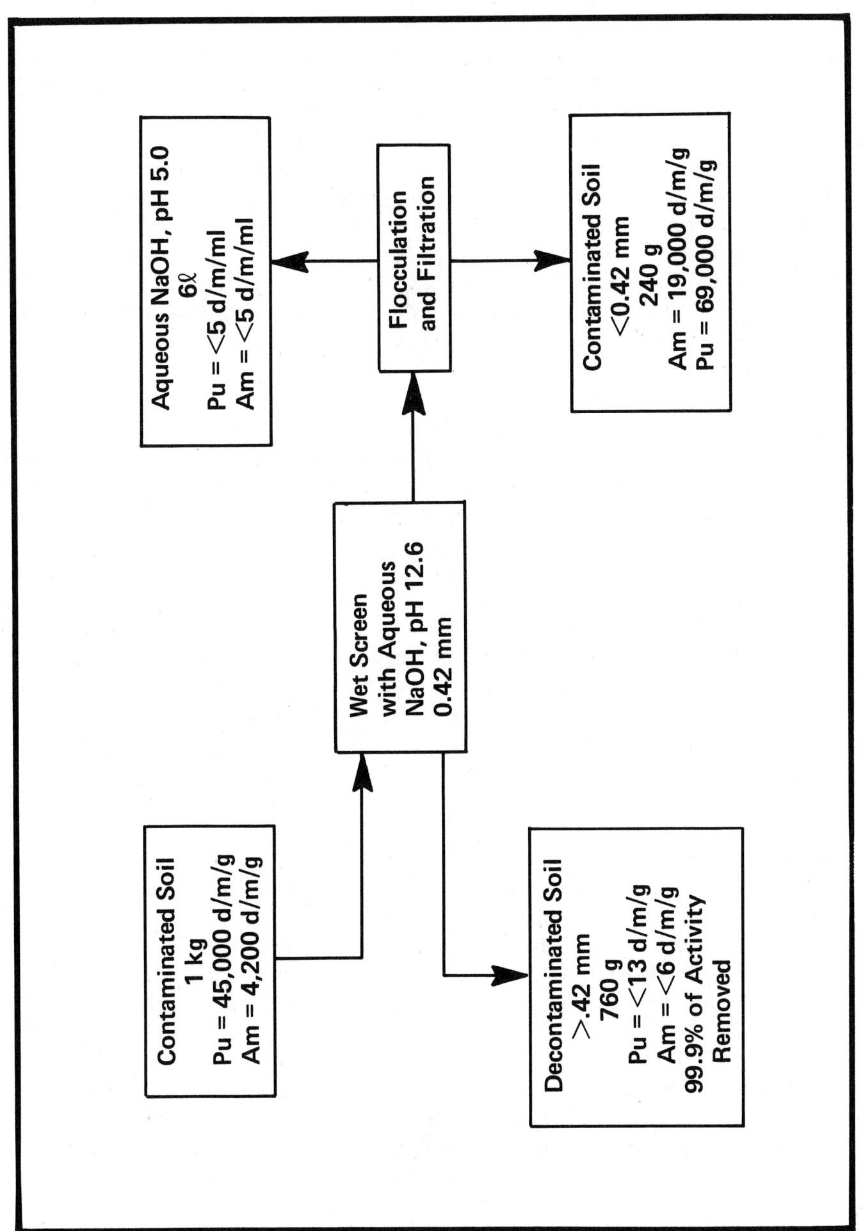

Figure 1. Wet Screening at High pH.

container. This process imparts a high shear to the particles thus liberating more of the contamination into the fine fraction: The outer surfaces of the particles are also effectively removed.

Attrition Scrubbing at Low pH. Attrition scrubbing at low pH employs the same process as attrition scrubbing at high pH except that the scrubbing solution has been changed. An aqueous solution of 2% HNO_3, 0.2% HF, 2% pine oil, and 5% Calgon is most effective. The soil is scrubbed five times in the rotary-type scrubber: three times for seven minutes each and two times for one minute. After each wash, the fines are removed. The first wash removes 88.1% of the contamination, while the second, third, and fourth washes remove 7.1, 3.5, and 1.3% of the contamination respectively. A typical experiment decontaminated 84% of the soil to less than 5 dpm/g. Originally, the soil contained 45,000 dpm/g.

The acid attacks the surface of minerals and aids in the grinding process. Smaller sized particles are scrubbed effectively, thereby increasing the total amount of soil that is decontaminated. Using the acid scrub solution, approximately 2% of the decontaminated material is less than 200 mesh (74 microns) in size.

When an acid solution is used in scrubbing, no colloidal suspension is observed. However, some of the plutonium does dissolve. This plutonium must be removed from the water so it can be recycled. The removal can be accomplished by: 1) Co-precipitation of the plutonium with $BaSO_4$ or $Fe(OH)_3$, or 2) adsorption by the hydroxide form of an anion exchange column. The ion exchange process is actually a precipitation of $Pu(OH)_4$ on the resin in the column.

In some experiments, the larger material (-5 +35 mesh) did not decontaminate to an acceptable level. However, by removing the magnetic fraction, the level was reduced to less than 30 dpm/g. Close examination of the magnetic fraction revealed that all the contamination was contained in several small particles. These particles could have been pieces of the rusted drums.

Cationic Flotation. The fourth decontamination process takes advantage of the anionic surface of the clay particles. A cationic flotation agent such as an amine can be used to float the clay material in a conventional flotation process. By adding a quartz suppressor, the soil can be scrubbed at a high speed (greater than 1,000 rpm). Usually at such high speeds, the larger rocks are abraded extensively, causing an increase in the weight of the fine fraction. However, with the addition of the suppressor, these abraded particles will not float with the clay fraction. Further research would be needed on this method.

Methods to extract the plutonium from the fine clay fraction are also being considered. These processes include leaching with ceric solutions in HNO_3 and contact leaching with HF, HNO_3, and Na_2CO_3 solutions.

Pilot Plant Studies

The attrition scrubbing process at high pH is the most feasible process to scale up. The process flow diagram shown in Figure 2 was generated based on the high pH scrub. A feed rate of 10 tons per hour (9,000 kg/hr) was selected for the full scale facility. The soil would pass through a 4-inch (10 cm) grizzly to eliminate the large rocks. The material would then enter a rotary Trommel scrubber. A screen attached to the end of the scrubber would separate out the material greater than 1/4 inch (6 mm). This fraction would be sent to landfill because it would contain less than 30 dpm/g. The fines material would then be washed and screened at 35 mesh (0.420 mm). The material greater than 0.420 mm would be sent to landfill. The fines fraction would be further processed using three stages of 1-inch (2.5 cm) liquid cyclones. The cyclones would separate the soils fractions at 10 µm. The smaller fraction would be packaged for shipment while the larger fraction would be decontaminated by further treatment. This process would provide a weight reduction of 88%. As the final plant must be mobile and self-contained, a water recycle system is shown on the diagram.

Results. Pilot Plant testing was performed on "cold" material fed at a rate of approximately 275 kg/hr. Testing on "hot" material was accomplished at a rate of 70 kg/hr. The mass balance for the cold test circuit is shown in Table I. Calculations showed that 4-inch (10 cm) cyclones were better suited for the desired flow rates. Three stages of cyclones were employed to separate the soil at 10 µm. However, as indicated in Table I, the underflow (the desired decontaminated product) still contained 17% of the minus 10 micron material.

If the cut is made at 400 mesh (37 micron) after two stages of cyclones, the underflow contains only 1.6% and 0.4% of the -38 micron and -10 micron material respectively. These numbers represent 0.36% and 0.086% respectively of the total mass. This cut is made very efficiently. Therefore, a separation could be made at 400 mesh (37 microns) instead of 10 microns. The weight reduction is lowered to 83.8%.

FUTURE WORK

Modifications

If the circuit were modified to cut at 100 mesh (149 microns),

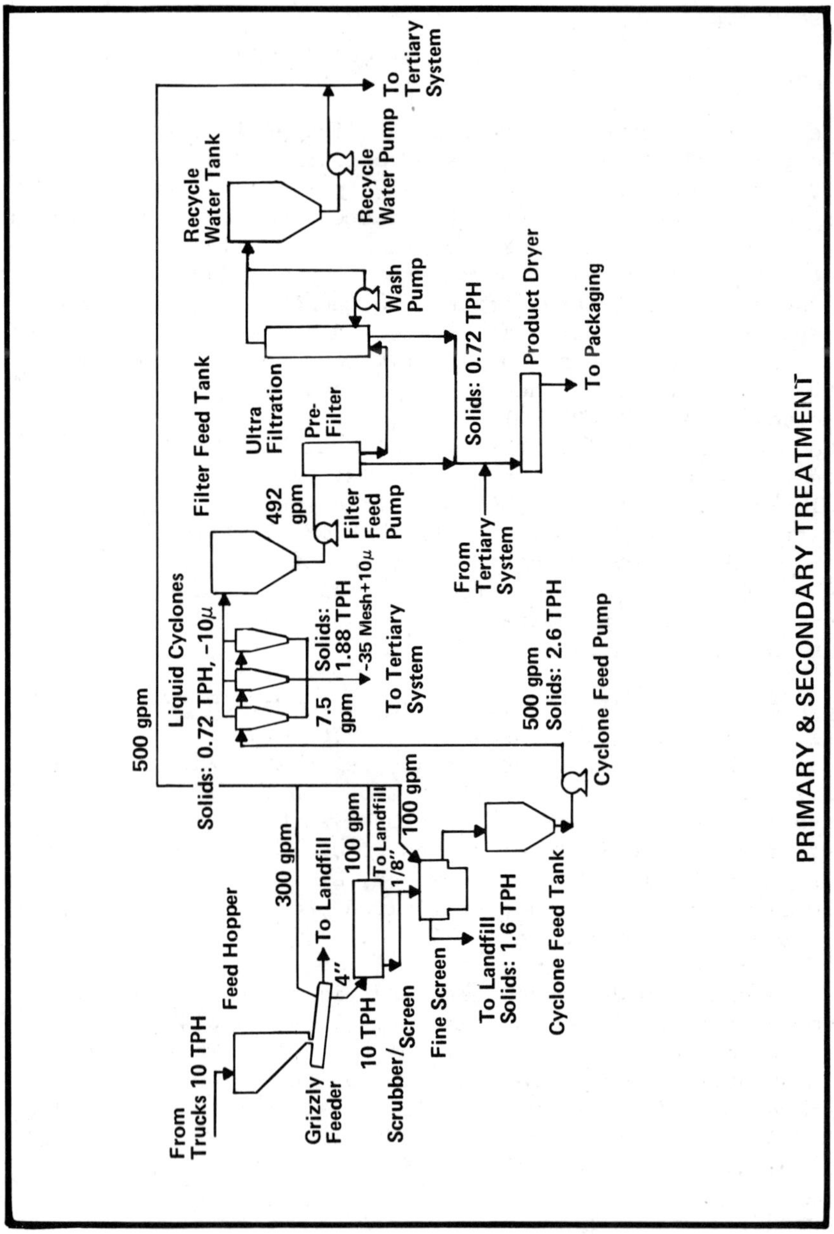

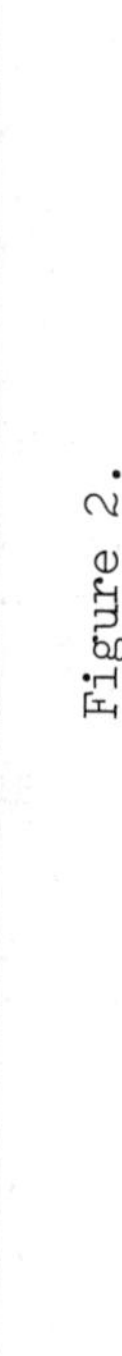

PRIMARY & SECONDARY TREATMENT

Figure 2.

TABLE I

Product	Pulp Density Weight (%)	Pulp Density Solids (%)	Particle Size Distribution -37μm (%)	Particle Size Distribution -10μm (%)
Grizzly				
Feed	100.0	100.0	100.0	100.0
+4 in. (10 cm) rock	15.0	100.0	0.0	0.0
-4 + 1-1/2 in. rock (100-38 mm)	25.0	100.0	0.0	0.0
	40.0	Removed by Grizzly, 60% Sent to Scrubber		
Scrubber				
Feed (-1-1/2 in.)	60.0	100.0	100.0	100.0
Discharge	60.0	63.0	100.0	100.0
-1-1/2 in. + 1/4 in.	26.5	70.0	.6	.4
	26.5	Removed by Screen on End of Scrubber, 33.5% Sent to Vibrating Screens		
Sweco Vibrating Screen				
Oversize (-1/4 in. +35M) (6-.42 mm)	10.9	77.3	.06	.04
Undersize (-35M) (-.42 mm)	22.6	8.5*	99.3	99.6
	10.9%	Removed, 22.6% Sent to Cyclones		
1st Stage Cyclone				
Overflow	15.1	18.0	91.1	92.6
Underflow	7.5	68.0	8.2	7.0
	15.1%	Sent to 3rd Stage, 7.5% Sent to 2nd Stage		
2nd Stage Cyclone				
Overflow	1.1	5.0	6.6	6.6
Underflow	6.4	71.0	1.6	0.4
	6.4%	Removed, 1.1% would be Treated Further		
3rd Stage Cyclone				
Overflow	10.0	9.5	0.0	75.6
Underflow	5.1	33.1	0.0	17.0

*This product was thickened to 25% solids prior to 1st stage cycloning.

the cyclones and screens could be replaced with spiral classifiers. This modification would result in a weight reduction of only 80% compared to 84% by the previous circuits. However, the advantages are numerous. Spiral (or screw) classifiers are almost maintenance free while cyclones are not. The spiral classifiers are also excellent dewatering devices leaving a product with a pulp density of 80% solids. Furthermore, the screen is eliminated making the circuit simpler. Spiral classifiers for a 10-ton per hour plant would have a spiral diameter of 16 inches (400 cm) with a total machine length of 12 ft (3.6 m). Four of these units operating in a counter-current configuration would make a very precise cut producing a clean product with a low moisture content.

Pilot Plant runs based on the circuit in Figure 2 using hot soil revealed that sometimes the -5 +35 mesh (4 to 0.42 mm) fraction contained over 100 dpm/g. This level is not acceptable. The probable cause of this situation is that the clay is lubricating the rocks in the Trommel scrubber and efficient grinding action is not being achieved. The solution is to simply deslime (remove the clay) the material prior to scrubbing. One method of achieving this removal is to have two Trommel scrubbers. The clay would be removed in the first scrubber, therefore allowing effective scrubbing action in the second scrubber.

Further Research

Several areas need more research before final design criteria can be issued. These areas include water recycle and clay dewatering. Dissolved solids in the water must be controlled carefully if the water is to be continually recycled. The product that is to be packaged and shipped is the clay fraction. This clay is present in the water as a colloid; therefore, it must be flocculated. This flocculated clay then must be dewatered so that the volume to be shipped is minimized. Probably the best method to accomplish dewatering is with an automatic filter press. Such presses are relatively new, but have been shown to eliminate 20% more water than previous presses. The possibility of using a vacuum brick extruder is also being considered.

Research is also planned using the modified circuit with an acid solution. The acid solution was not selected originally because of its corrosive nature. However, tests indicate that the abrasion of the equipment by the rocks and not the corrosion by the solutions is the main factor in equipment wear.

The Total Concept

As previously mentioned, the total process must be mobile so that it can be moved from site to site. Conceptual designs have been generated for mounting the process in semi-trailers. Three

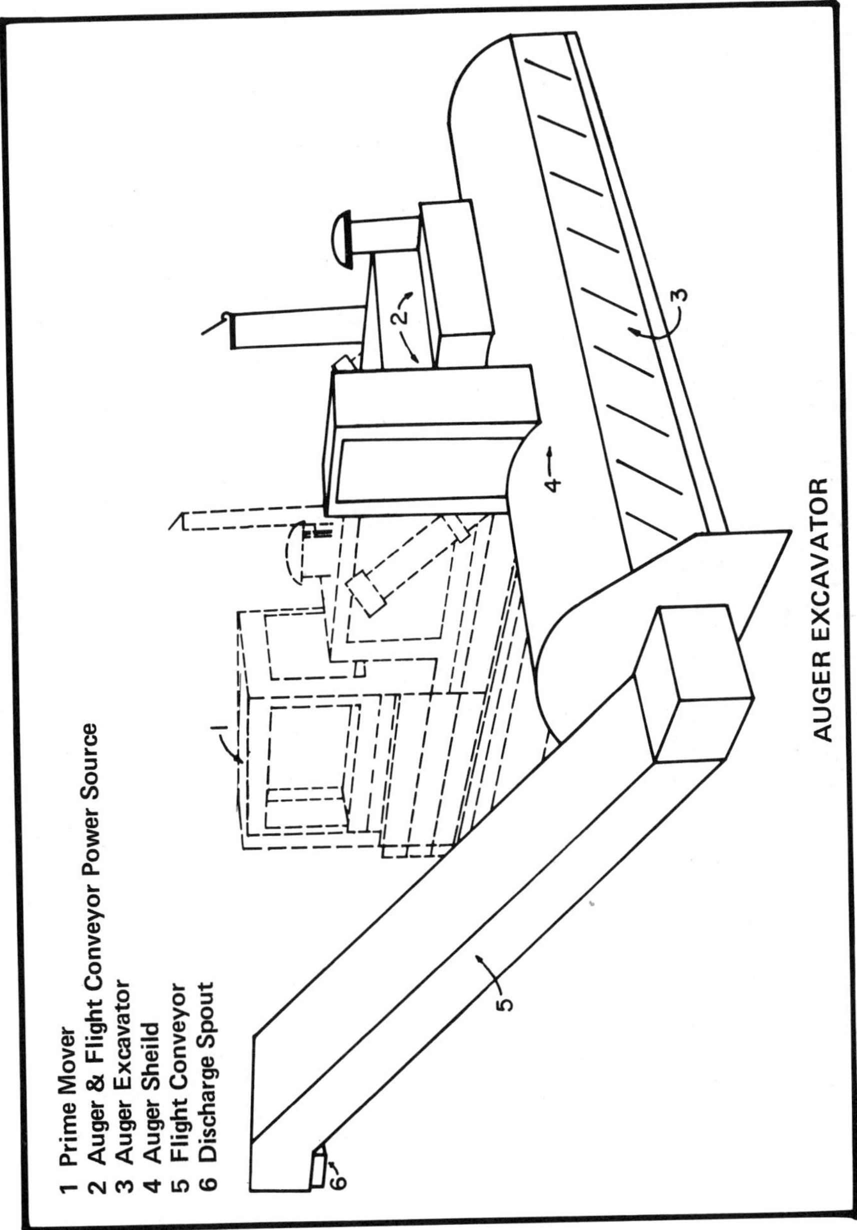
1 Prime Mover
2 Auger & Flight Conveyor Power Source
3 Auger Excavator
4 Auger Sheild
5 Flight Conveyor
6 Discharge Spout
AUGER EXCAVATOR
1
2
3
4
5
6

Figure 3.

trailers will probably be required: One containing the process equipment, one with two stages of HEPA air filteration, and one with water recycle and power generation equipment.

The total concept also involves excavating the contaminated soil. To accomplish this excavation, a dustless mining machine has been conceptually designed (Figure 3). This machine has many advantages over the typical front-end loader method generally used. One of the main advantages is that the rotary blade is continually against the face of the excavated bank; therefore, the contaminated soil is not exposed to the open. The machine is also very maneuverable, capable of being adjusted precisely (within ± 2 inches). As the contaminated material is excavated, it is moved up the enclosed conveyor into a 20-ton dumpster. The dumpster is then moved to the process facility.

Other DOE Sites

Soils from four other Department of Energy sites are currently being evaluated: Hanford, INEL, LASL, and Mound. Preliminary decontamination tests are encouraging. Results will be available later this year.

References

1. C. J. Johnson, R. R. Tidball, R. C. Severson, Plutonium Hazard in Respirable Dust on the Surface of Soil, Science, 193:488 (1976).
2. J. A. Hayden, Size Fractionation Methods: Measuring Plutonium in Respirable Dust, Science, 202:754 (1978).
3. Draft Environmental Impact Statement, Rocky Flats Plant Site, Golden, Colorado, United States Energy Research and Development Administration, ERDA-1545-D, 1977.

ANALYSIS OF SOIL SAMPLES FROM OMRE DECOMMISSIONING PROJECT

O. D. Simpson, J. A. Chapin, R. E. Hine,
J. W. Mandler, M. P. Orme and G. A. Soli

EG&G Idaho, Inc.
Idaho National Engineering Laboratory
Idaho Falls, Idaho 83401

INTRODUCTION

The decontamination and decommissioning (D&D) of retired nuclear reactor plants and the restoration of the landsite to its original state is of primary importance to the nation. As of December 1977, there have been fifty-two reactors constructed at the Idaho National Engineering Laboratory (INEL). Seventeen of these are currently operable; the others are in various stages of retirement ranging from defueled and placed under protective confinement to completely decontaminated and open to the public. Several reactors at the INEL have undergone D&D in the past, but these projects were done individually without an integrated plan. An integrated, site-wide, long-range D&D plan[1] has been developed at the INEL, and the Organic Moderated Reactor Experiment (OMRE) facility is the first reactor to be addressed by this plan.

Criteria are being developed at the INEL to be used as guidelines for D&D of land, structures, and equipment at the INEL in order that such land, structures, and equipment may safely be returned to unrestricted use whenever or wherever practicable. In order to accomplish this for the land, the radioactive contamination in the soil must be reduced below a maximum permissible concentration. Criteria suggested in the INEL D&D long-range plan for maximum permissible concentration are 100 pCi/g of soil for transuranic alpha-emitting radionuclides and 1 nCi/g of soil for beta-gamma-emitting radionuclides. The above values assume natural backgrounds have been subtracted. These criteria are still under review, and as data are developed, they may be revised.

The D&D long-range plan also requires a cost/benefit analysis to be performed on each project to evaluate the feasibility of decommissioning to the unrestricted levels listed above. The planned use of the INEL, in the near term, may preclude the expenditure of large sums of money and manpower for D&D when the site will be reused for nuclear applications. A different set of criteria would be considered if the site is to be released for activities such as farming or housing. Each end use may require a separate, specific set of criteria.

Until a critical pathway analysis can be performed for all possible alternatives, specific criteria cannot be established for each facility. Current criteria will be influenced most heavily by safe access considerations, near-term site utilization, radioactive storage space, and cost/benefits.

In order to establish that the present OMRE site does not exceed the above criteria for radioactive contamination, samples obtained from the remainder of the facility that was not removed such as soil, concrete pads, various structural materials, and the leach pond area were analyzed to determine their radioactive content. This paper presents the results of the analyses performed on soil samples.

REACTOR HISTORY AND DESCRIPTION OF REACTOR AREA

The general purpose of the OMRE facility was to study feasibility, economics, and behavior of organic moderator and coolant under power conditions. The nominal reactor power was from 2 to 12 MW thermal. The reactor was designed by Atomics International, constructed by Wadsworth and Arrington and Fluor Maintenance Co., and owned by the United States Atomic Energy Commission. Construction was started in June 1956 and completed in May 1957. The reactor went critical in September 1957, attained full power in February 1958, and was operated until 1963. The facility remained in a deactivated condition from 1963 until October 1977, at which time D&D of the facility was begun. The reactor vessel, buildings, and other structures were removed from the site during 1978.

The reactor core was of rectangular shape having dimensions of 57 cm x 69 cm x 91 cm high. It had a critical mass of 16.4 kg ^{235}U. Fuel elements were constructed from alloyed plates containing 25 wt.% fully enriched UO_2 and 75 wt.% stainless steel. Cladding made of 0.013-cm-thick 304 stainless steel was metallurgically bonded to the fuel plate. The moderator consisted of 380 liters of Santowax OM (diphenyl 16.0%, ortho-terphenyl 46.1%, meta-terphenyl 31.8%, para-terphenyl 6.1%). The mixture is partially solid at 21°C. Melting begins at approximately 38°C and is completely liquified slightly below 93°C, depending on irradiation history. Total volume of moderator and coolant was 19,000 liters.

A picture of the OMRE facility before D&D was started is shown in Figure I. The reactor was located in the corrugated silo shown on the left side of the picture. Figure II shows how the reactor pit area appears today. All structural materials have been removed including the silo building. The bottom of the pit is where the base of the reactor vessel rested.

SOIL SAMPLING TECHNIQUES

Four sampling techniques (vertical augering, vertical punching, vertical coring, and trenching with horizontal coring) were tested in order to determine a method which will allow radionuclide

Figure I. The OMRE facility before decontamination and decommissioning was started. The reactor is housed in the corrugated silo shown at the left. The reactor core is ~5.5 m below ground level.

Figure II. The OMRE reactor pit area as it appeared in August 1979. Since this time an additional 1 m of basalt and soil has been removed from the bottom of the pit in order to meet soil contamination criteria at INEL. The basalt at the bottom of the pit is where the concrete pad rested that supported the base of the reactor vessel. The Ge(Li) detector used for an in situ measurement is shown on the right side of the reactor pit near the center of the picture. The motor home that houses the electronic equipment can be seen in the background.

characterization of the soil with a minimum amount of cross contamination from level to level yet minimize cost. The amount of time required for sampling and analysis, personnel exposure, and the complexity of equipment necessary to perform the sampling process were considered.

Vertical augering was done using a 15.2-cm-diameter, two-man, motor-driven post-hole digger. Samples were collected in 15-cm-thick increments as they were expelled by the digger. The auger was brushed clean between samplings to minimize cross contamination. These samples were sent to the laboratory for radionuclide analysis.

Vertical punching was accomplished by using an 8-cm-diameter by 91-cm-long steel punch driven by a sledge hammer. A beta-gamma radiation detector was lowered into the hole and total radiation levels recorded as a function of depth.

Vertical cores were taken by driving a core-sampling probe into the soil. The probe was retracted through the center of the coring tool, the coring tool then driven to the depth of the sample required. Samples were removed and sent to the laboratory for radionuclide analysis.

Trenching and horizontal coring were done as follows. A trench was dug down to the basalt level. The wall face was gently scraped to remove surface contamination that may have been deposited during the digging process. Horizontal coring was then accomplished by driving a 5-cm-diameter, 30-cm-long hollow pipe into the wall. Core samplings were normally spaced 15 cm apart starting at the base of the trench.

Core sampling of the basalt regions was done as follows. After cleaning of the surface area of the basalt, an air hammer was used to first spall the top 2-3 cm of basalt and then to drill a hole in 15-cm increments to a depth of up to 76 cm. Samples were taken from each 15-cm-thick segment of crusted rock and sent to the laboratory for radionuclide analysis.

In all of the above sampling procedures great care was used to minimize cross contamination. Most soil samples ranged from 50-500 g in size.

ANALYSIS METHODS

Laboratory and *in situ* analyses were performed to establish the concentrations of the alpha-, beta-, and gamma-emitting isotopes in the soil samples.

Alpha Analysis

Three different techniques were used: (1) A general survey of the area was done using a hand-held proportional counter. (2) Samples were prepared for gross counting by sieving and leaching 10 g of soil. The sample was then gross counted using a gas-flow proportional counter. (3) Ten grams of soil were fused so that plutonium and americium isotopes could be extracted from the sample. Isotopic identification was performed using electrodeposition techniques and surface barrier spectrometry.

Beta Analysis

General mapping of the OMRE area was accomplished with hand-held end-window G.M. detectors. Laboratory analysis of the samples was carried out as follows: (1) The samples that were prepared for gross alpha analysis were also counted for beta activity by changing the operating voltage of the gas-flow proportional counter. (2) The ^{90}Sr activity was determined by fusing a 10-g soil sample, separating the strontium with wet chemistry, letting the sample decay for seven days, separating the ^{90}Y that grew into the sample from the decay of ^{90}Sr, and then counting the ^{90}Y in the gas-flow proportional counter. Yields were established by using a ^{85}Sr gamma-emitting tracer and yttrium as a carrier.

Gamma-ray Analysis

A shielded beta-gamma survey G.M. instrument was used to measure the gross gamma activity at the site area. A mobile Ge(Li) spectrometer was used to collect gamma-ray spectra at the site area. This spectrometer was calibrated to give reliable relative answers only. Quantitative measurements with this spectrometer were not attempted. Three types of laboratory analyses were performed on the samples obtained from the OMRE site: (1) Gross count using methane flow chambers, (2) Na(I) spectrometry, and (3) Ge(Li) spectrometry. Ge(Li) spectrometry was used to establish the concentration of the various gamma-ray emitting isotopes. The spectrometers were calibrated using National Bureau of Standards samples which were distributed uniformly in various soil samples. From these calibrations absolute numbers could be obtained. The samples were counted as received with no sieving or drying procedures being used. Uncertainties in sampling procedures, uniformity, soil dryness, etc., can cause activity uncertainties to be as large as 200-300%. Variations of this amount are considered to be acceptable.

Natural Backgrounds

The thorium (^{232}Th) and uranium (^{238}U) atoms' natural radioactive decay set off a series of nuclear transformations which are called the Th-U daughter series. Emission of an alpha particle from ^{238}U transforms the atom into ^{234}Th which is the beginning of a long chain of naturally occurring radioactivities. Likewise, the natural emission of an alpha particle from ^{232}Th produces ^{228}Ra which also starts the production of a long chain of other radioactive isotopes. Potassium (^{40}K) is also a naturally occurring radioactive isotope. The ^{40}K produces about 85% and the Th-U daughters 15% of the natural gamma-ray background in the vicinity of the OMRE site. There are also naturally occurring beta emitters such as tritium (^{3}H), ^{14}C, and ^{87}Rb that add to the natural background. At the INEL, the natural background radioactivity in the soil is approximately 63 pCi/g (24 pCi/g due to alpha-emitting and 16 pCi/g due to beta-gamma-emitting isotopes in the Th-U chain, 19 pCi/g due to ^{40}K, and about 4 pCi/g due to pure beta-emitting isotopes). There is also a man-made background, mostly ^{137}Cs and ^{90}Sr, that is produced from weapons tests which amounts to about 2-3% of the naturally occurring background at the INEL.

These naturally occurring backgrounds sometimes make measurement for the man-made isotopes difficult to perform.

RESULTS

Reactor Area

Since the OMRE reactor was located below ground with little shielding between the core and the surrounding soil, this soil became radioactive by neutron-gamma reactions. The majority of this soil was removed to the waste storage area. Approximately 150 samples taken from the remaining soil and from the basalt base were analyzed for radionuclide concentrations. Table I shows a summary of the concentrations observed. The primary radionuclides observed in the soil samples taken from the reactor area were ^{60}Co, ^{152}Eu and ^{154}Eu. Fission product concentrations were established to be < 3 pCi/g of soil around the reactor pit perimeter and in the basalt that supported the base of the reactor. Small quantities of ^{137}Cs (< 3 pCi/g) and ^{90}Sr (< 7 pCi/g) were observed in some of the soil that was collected above the basalt base. Off-site backgrounds for these two isotopes are approximately 1.0 and approximately 0.5 pCi/g, respectively.

Additional soil and basalt is being removed from the reactor pit area so that soil activities will meet the criterion of < 1 nCi/g. When the < 1 nCi/g criterion has been met, the pit will be backfilled with clean soil. The area will then be restored to its original state by grading and seeding with native grass.

Table I. A Summary of the Radionuclide Concentrations Observed in Soil Sources Taken at the OMRE Reactor Area Site

Location	pCi/g					Beta-Gamma Natural Background
	^{60}Co	^{90}Sr	^{137}Cs	^{152}Eu	^{154}Eu	
Perimeter	0.6-2	0.4	0.4-3	0-5	<0.5	~39
Soil Above Basalt (0-61 cm)	20-370	6.4	0-3	9-960	6-70	~39
Basalt (0-61 cm)	10-210	---	0-0.2	26-650	2-50	~39
Off Site	<0.1	~0.5	~1	<0.3	<0.03	~39

An _in situ_ survey of the reactor pit area and the fill dirt was carried out to verify that the gamma-ray-emitting radionuclides measured in the soil samples were the only ones present in the area (excluding natural radioactivity). This was done utilizing a motorized laboratory containing a high-resolution Ge(Li) gamma-ray detector. Figure II shows the mobile laboratory with the Ge(Li) detector positioned on the edge of the pit and viewing the general pit area.

Results of the _in situ_ measurements (see Table II) indicate that 99% of the gamma-ray activity in the fill dirt is due to natural backgrounds of ^{40}K and Th-U daughters and 1% is due to ^{137}Cs. Approximately 75% of the gamma-ray activity in the reactor pit was found to be from ^{152}Eu and ^{154}Eu, 18% from ^{60}Co, 1% from ^{134}Cs (the ^{134}Cs activity came from the neutron-gamma reaction on ^{133}Cs), and 6% from ^{40}K. The Ge(Li) mobile spectrometer was also used to measure the activities in other areas: 3 m back from the edge of the reactor pit, the cement dirt pile, and an area approximately 2 km from the OMRE site (at the OMRE-PBF road junction). These results are also shown in Table II.

Leach Pond Area

The OMRE leach pond is a trench situated on natural basalt bedrock. The pond size at the berm (brim of the trench) is approximately 8 m wide by 22 m long. The sides slope to a base approximately 5 m wide by 15 m long. The depth of the soil above

Table II. Results of the *in situ* Ge(Li) Gamma-ray Spectral Measurements Taken at the OMRE Area

	Percentage of Activity					
	Full View Reactor Pit Base	3 m Back Reactor Pit Edge	Cement-Dirt Pile	Edge Leach Pond	Fill Dirt Pile	OMRE-PBF Road Junction
^{40}K	6	11	71	24	83	86
Th-U Dau.	--	--	27	3	16	14
^{60}Co	18	20	--	21	--	--
^{134}Cs	1	--	--	--	--	--
^{137}Cs	--	--	1	52	1	--
^{152}Eu	69	63	1	--	--	--
^{154}Eu	6	6	--	--	--	--

the basalt base varies from 30 to 46 cm. Figure III shows the leach pond with the mobile Ge(Li) detector being set into position to measure the gamma-ray field.

Approximately 200 samples were obtained from the leach pond area and analyzed for their radionuclide content. Only four radionuclides (^{60}Co, ^{90}Sr, ^{137}Cs, and traces of ^{152}Eu) were detected in samples from the pond. The small traces of ^{152}Eu were observed only in surface samples. A compilation of the radionuclide concentrations measured in the soil samples is given in Table III. The ^{60}Co concentrations ranged from approximately 1 pCi/g at the pond perimeter to 1500 pCi/g at the basalt surface. The ^{137}Cs concentrations ranged from approximately 1 pCi/g at the perimeter to 6700 pCi/g at the basalt surface. The average concentrations of the four radionuclides and an estimate of their total activities in the leaching pond are given in Table IV.

Selected samples were analyzed for alpha-emitting radionuclides. The results indicated that the gross alpha-emitting radionuclide concentration (including the natural Th-U chain) is

Figure III. The OMRE leach pond. The pond is approximately 8 m wide and 22 m long. The Ge(Li) gamma detector is shown in the background at the doorway of the motor home where the electronic equipment is housed.

Table III. A Summary of the Radionuclide Concentrations in Soil Samples Taken from the OMRE Leach Pond

	pCi/g				Beta-Gamma
Location	^{60}Co	^{90}Sr	^{137}Cs	^{152}Eu	Natural Background
Perimeter	0.3-3.0	--	0.2-4.0	0-0.5	~39
Berm	2-220	2-43	1-425	0-3	~39
Base Soil	3-1500	6-650	3-6700	0-19	~39
Basalt (0-15 cm)	0.4-9.0	--	0.3-26.0	<0.3	~39
Off-Site	<0.1	~0.5	~1.0	<0.3	~39

Table IV. The Average Concentration of the Four Radionuclides and an Estimate of their Total Activity for the OMRE Leach Pond

		Activity			
Isotope	Half Life	Total (mCi)	Berm Area (mCi)	Pond Base (mCi)	Average (pCi/g)
^{60}Co	5.27 yrs	22	9	13	151
^{90}Sr	29.1 yrs	~18	~7	~11	~123
^{137}Cs	30.0 yrs	36	14	22	246
^{152}Eu	13.33 yrs	0.3	0.2	0.1	1.4

Note: 220,000 kg soil in berm
62,000 kg soil in base

< 20 pCi/g. Concentrations for specific radionuclides are < 0.04 pCi/g for ^{238}Pu, < 0.04 pCi/g for ^{239,240}Pu, and < 0.01 pCi/g for ^{241}Am.

Results of the _in situ_ measurements obtained using the mobile Ge(Li) spectrometer (see Table II) indicate that about 50% of the gamma-ray activity in the leach pond is due to ^{137}Cs, 20% is due to ^{60}Co, and almost 30% is due to Th-U daughters.

Figure IV shows how the average concentrations of the detected radionuclides will drop off as a function of time. These concentrations are compared to the INEL D&D criterion (< 1 nCi/g), INEL natural background (mainly ^{40}K and daughters of Th-U), ^{137}Cs and ^{90}Sr background (from weapons tests), and ^{60}Co and ^{152}Eu INEL backgrounds.

CONCLUSIONS

Results of this study indicate that the activity at the OMRE decommissioned area is confined to localized areas (i.e., the leach pond area and reactor area). Comparisons of radionuclide concentrations measured in soil taken from the lip of the leach pond with concentrations in soil obtained outside the INEL site boundaries indicate that the concentration in the soil at the edge of the leach pond is at background levels.

The vertical augering technique was determined to be the best approach for obtaining shallow soil samples at the INEL. Selection of this technique was based on ease of operation and analytical results. Less area is disturbed per sample than with the horizontal trenching and coring techniques.

The radionuclide analysis of the samples presented in this report shows the existence of a few regions in the reactor and leach pond areas that were still above INEL release criteria. These regions have been or are being further decontaminated.

The following specific conclusions are worthy of note.

Reactor Area

1. Major radionuclides came from neutron activation of the soil (^{60}Co, ^{152}Eu and ^{154}Eu). Highest concentrations were in the soil samples taken above the basalt base.

2. Concentrations of the above isotopes are less than INEL D&D criterion (< 1 nCi/g).

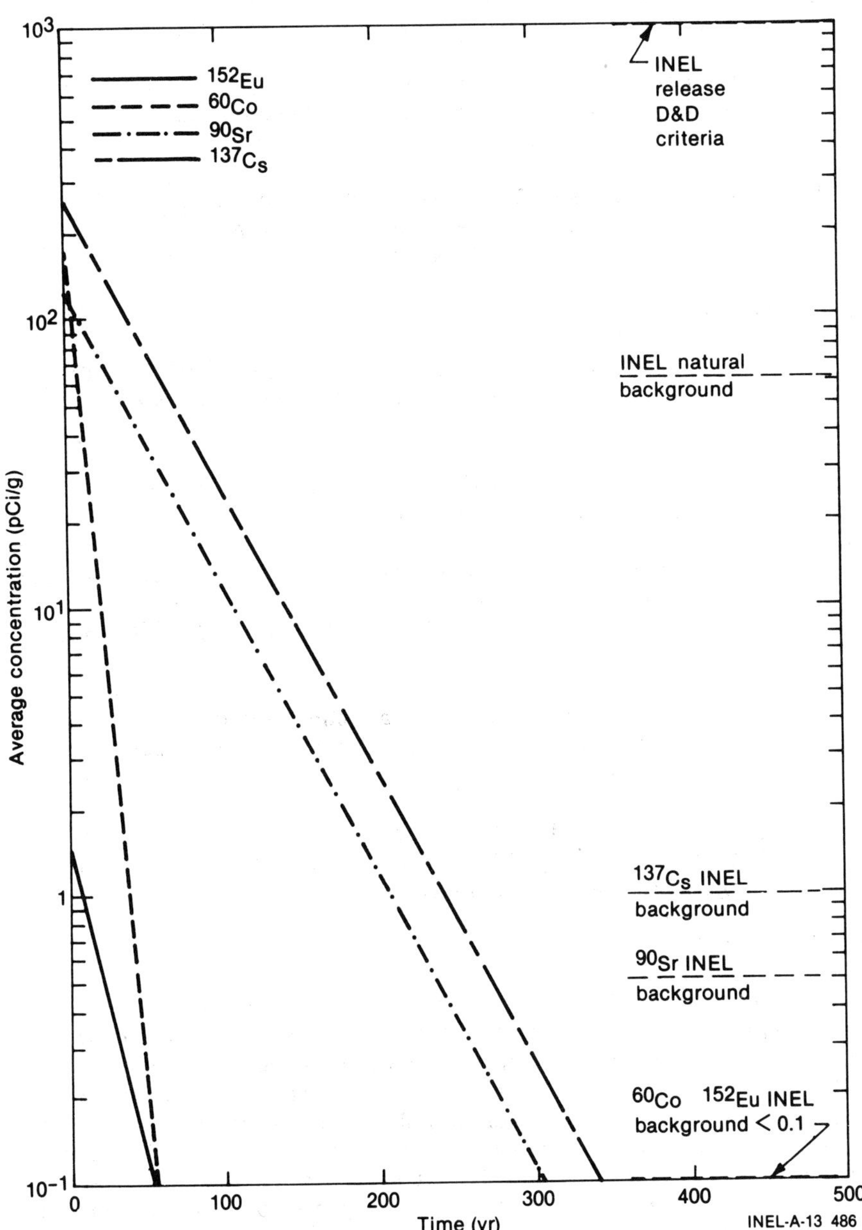

Figure IV. The OMRE leach pond average radionuclide concentrations versus decay time. These values are compared to the INEL D&D criteria.

3. No significant amounts of fission products were observed in the reactor area (< 3 pCi/g).

4. The reactor perimeter area including fill dirt had radionuclide concentrations at essentially background levels.

5. Basalt core samples taken from a depth of 30-60 cm showed ^{60}Co concentrations < 10 pCi/g, ^{137}Cs < 0.2 pCi/g, ^{152}Eu < 28 pCi/g, and ^{154}Eu < 3 pCi/g.

Leach Pond Area

1. The major radionuclides observed were ^{137}Cs, ^{60}Co, and ^{90}Sr. Only trace quantities of ^{152}Eu were measured.

2. The highest contamination was located in the soil next to the basalt surface.

3. Radiation readings taken on a few basalt core samples indicate extremely low level contamination in the upper 15 cm of the basalt (< 26 pCi/g of ^{137}Cs and < 9 pCi/g of ^{60}Co).

4. Analysis of samples taken at the perimeter of the pond show activities at essentially background levels.

5. The concentration of transuranic radionuclides was measured to be < 0.1 pCi/g. This is three orders of magnitude below the INEL D&D criterion.

ACKNOWLEDGEMENTS

The authors would like to express their appreciation to E. D. Cadwell and C. L. Rowsell for their assistance and direction of the Radiation Measurements Laboratory personnel who did the Ge(Li) gamma-ray spectral analysis of the many soil samples. Thanks are also given to R. A. Stevenson, J. L. Alvarez and J. F. Sommers for their support and recommendations. Appreciation is also expressed to C. W. Filby who directed the beta and alpha analyses that were performed by the Radioanalysis Group of Exxon Nuclear Idaho Corp.

REFERENCES

1. J. A. Chapin and R. E. Hine, Decontamination and Decommissioning Long Range Plan, Idaho National Engineering Laboratory, EG&G Idaho, Inc., TREE-1250 Vol. 1, June 1978.

LOW-LEVEL RADIOACTIVE WASTE FROM RARE METALS PROCESSING FACILITIES

Jeanette Eng, New Jersey Department of Environmental Protection
Donald W. Hendricks, ORP-Las Vegas Facility, U.S. EPA
Joyce Feldman, Radiation Branch, U.S. EPA Region II
Paul A. Giardina, Radiation Branch, U.S. EPA Region II

INTRODUCTION

The federal government has recognized that companies which process thorium and uranium ores require regulatory controls in order to protect man and the environment from unnecessary radiation. The recent passage in November, 1978 of the Uranium Mill Tailings Bill (H.R. 13650) demonstrates the government's recognition that the front-end of the uranium fuel cycle, i.e., mining and milling of uranium, had been neglected. The bill defines procedures for a remedial action program at inactive mill sites and regulations for active mill sites.

Companies which provide titanium, phosphorus, rare earths, and rare metals for industrial and chemical use are not normally regarded as possessors of large quantities of radioactive materials. In fact there appears to be a historical laxity in documenting the processing and waste disposal activities of these industries. A recent EPA publication reviews the available literature on technologically enhanced natural radiation due to mineral extraction industries (Bliss, 1978). It is only recently that phosphate industrial wastes have been listed as hazardous radioactive wastes in the U.S. Environmental Protection Agency's proposed Hazardous Waste Guidelines and Regulations (Costle, 1978). This paper will review the situations at the existing Teledyne Wah Chang Co., Inc. located at Albany, Oregon, and the former Carborundum Corp./Amax Specialty Metals, Inc., facilities located at Parkersburg, West Virginia, and Akron, New York, in order to show the extent of the radioactivity problem at rare metals processing facilities and the need to identify for radiological review other rare metal and rare earth processing sites.

As shown in Figure 1, the unusual grouping of rare earth and rare metal processing industries stems from their common ore origin. The ores used in rare earth and metals processing are byproducts of mining for titanium ores, since the ores for the specific processing are seldom found in economically mineable rock. The principal domestic areas for raw materials are Florida and Georgia, although mining has occurred in western and other southeastern states. Outside of the U.S., major deposits are located in Australia, Canada, Brazil, South Africa, Sri Lanka, India, and Mexico. Often ores with higher specific mineral content were imported, such as Nigerian zircon sand for hafnium processing and Australian zircon sand for zirconium processing.

The beach and fluvatile sand deposits from these areas are rich in marketable, ilmenite, rutile, monazite, and zircon. Monazite commonly incorporates thorium and uranium as well as rare earths due to similarity in geochemistry and electronic structure. Ilmenite and rutile are ore materials for titanium processing, monazite is the principal ore for rare earth processing, and zircon the principal ore for zirconium and hafnium processing. Generally, these beach or placer sands are treated to produce heavy mineral concentrations containing the zircon, rutile, ilmenite, monazite, and other marketable minerals. The concentrates may then be treated by various combinations of gravity, electrostatic or electromagnetic methods to separate the individual minerals. Monazite being slightly magnetic can be separated from zircon by electromagnets. The purity of the zircon product (or conversely the degree of monazite contamination of the product) is obviously a function of the degree of separation effort. Initial treatment usually is provided at or near the mine site. As a rule of thumb, the sand deposits are usually but not always processed primarily for the titanium content in the form of rutile and ilmenite. The zircon and monazite fractions are then byproducts which are treated separately to extract zirconium and rare earths, respectively. Thorium is then a further byproduct of the rare earth processing of the monazite portion. This has been the major source of thorium up to the present.

For zirconium metal production, zircon sand is usually processed to minimize the monazite content since the phosphate content of the monazite has a deleterious effect on the metallurgical process. This in turn should mean a lower thorium and uranium content in the metallic zirconium wastes than in foundry wastes where the monazite content of the zircon sands should be of less importance to the process. However, Wagstaff has reported levels of radium-226 from the uranium decay chain to be about 100 pCi/g in incoming zircon sands at both foundries and metallic zirconium production facilities (Wagstaff, 1978). As Table 1 shows, the uranium and thorium content of monazite concentrates varies depending on where the ore is mined. The amount of monazite in the

TABLE 1: THORIUM AND URANIUM COMPOSITION IN MONAZITE CONCENTRATES (WEIGHT PERCENT)

	ThO_3	U_3O_8
Australia	7-8	1
Brazil	6-7	0.2
India	9-10	0.3
Madagascar	9	0.4
South Africa	6	0.1
United States	4-5	0.4

TABLE 2: RADIOISOTOPIC CONCENTRATES IN SOILS FROM THE AKRON, NEW YORK, RARE METALS PROCESSING FACILITY

Radioisotope	Range of Concentrations (pCi/g Dry)
Ra-226	2.1 - 35
Pb-214	0.66 - 7.0
Bi-214	0.47 - 2.7
Ac-228	1.3 - 140
Pb-212	1.1 - 150
Tl-208	1.1 - 120
K-40	8.8 - 120

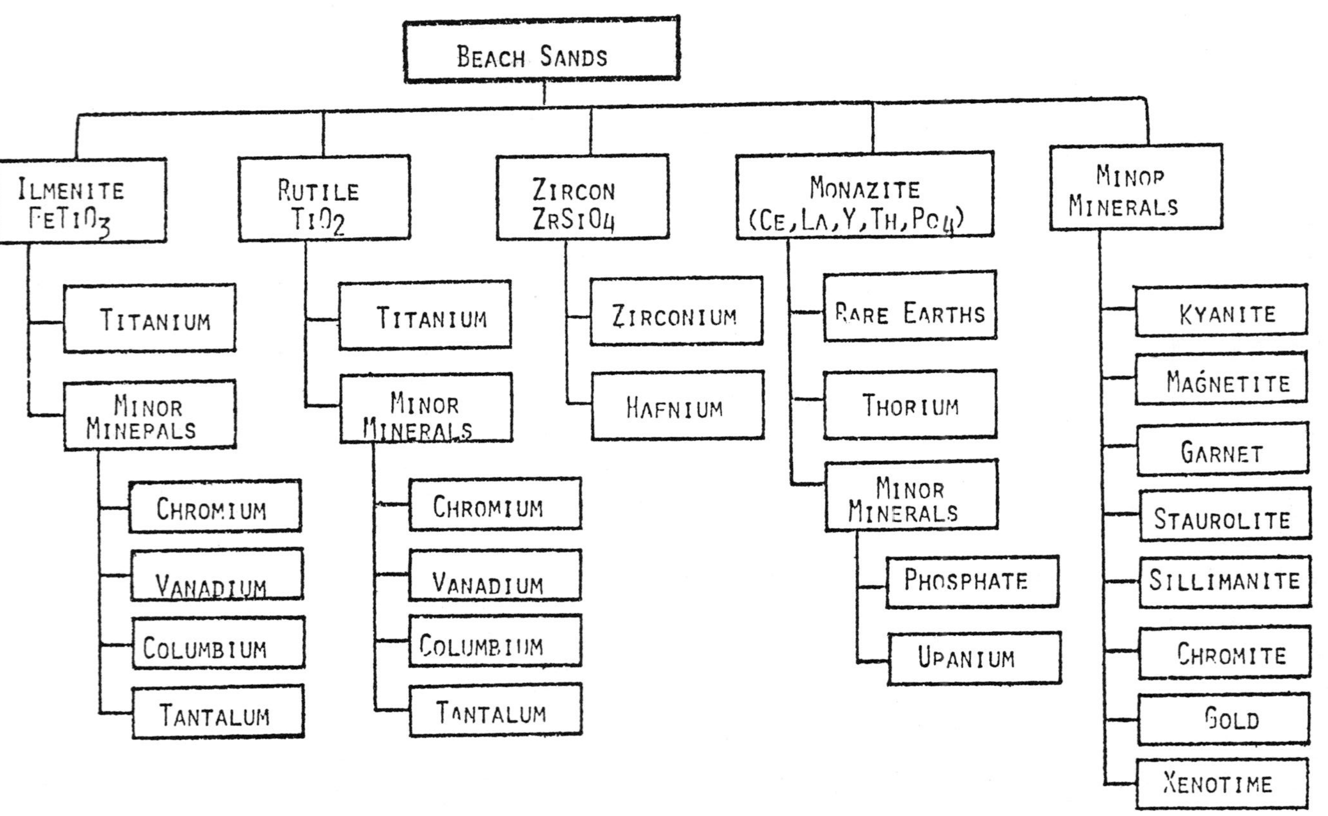

Figure 1: Potential Beach Sand Composition

zircon sand also depends on how well the separation facility removed the monazite before shipping to the use point. At the use point (such as a zirconium metal manufacturing plant), the manufacturer may find it necessary to further separate monazite from the sand. Low-level radioactive wastes may be generated at each separation point. The natural concentration of uranium and thorium decay series products in the sands are low but the industrial processing to obtain the specific minerals concentrates radioactivity in the waste residues. The disposition of these waste residues is the subject of this paper.

CASE STUDIES OF THREE FACILITIES

The zirconium and hafnium processing method was developed by W. J. Kroll for the U.S. Bureau of Mines. The bureau established a pilot plant in 1947 at Albany, Oregon, to extract zirconium and hafnium using the Kroll process and a purification plant in 1951 at Oak Ridge, Tennessee, to produce high purity, low hafnium, reactor grade zirconium. The zircon sand is mixed with graphite or coke and is fused in an electric furnace to produce a mixture of carbonitrides of zirconium and hafnium. The carbonitrides are chlorinated in a vertical shaft furnace and the gaseous chlorides of zirconium and hafnium are collected in a nickel condenser. The zirconium and hafnium chlorides are reduced in the Kroll process to the metals by reaction under an inert atmosphere with magnesium. The end product, commercial grade zirconium sponge, will contain about 2% hafnium suitable for non-nuclear use. Current industrial practice uses zirconium tetrachloride produced by chlorinating zircon directly instead of the carbonitride (Minerals Yearbook, 1975). In order to produce reactor grade zirconium, i.e., that containing about 0.3% hafnium, the commercial grade zirconium sponge is dissolved and the hafnium is solvent extracted to hafnium thiocyanate using methyl isobutyl ketone. The hafnium is precipitated as an hydroxide, calcined to about 99% hafnium oxide. The resulting zirconium sponge is crushed, compacted into consumable electrodes, and vacuum melted in an inert atmosphere to ingot. Further product purity is achieved by applying the deBoer-vanArke refining process. A similar procedure is applied to the hafnium solvent extraction in order to obtain high purity hafnium metal. The residues generated by the extraction processes contain graphite, coke, unreacted silicates, and non-volatile silicates.

Wah Chang Corp. began operating the Bureau of Mines' Albany, Oregon, facility in 1955. Today it is one of the major producers of reactor grade zirconium and hafnium metals. Concern over the environmental and health safeguards at the facility grew when explosions were encountered during digging operations near the facility's industrial waste piles. Apparently the explosions were caused by rapid combustion of the zirconium in the waste

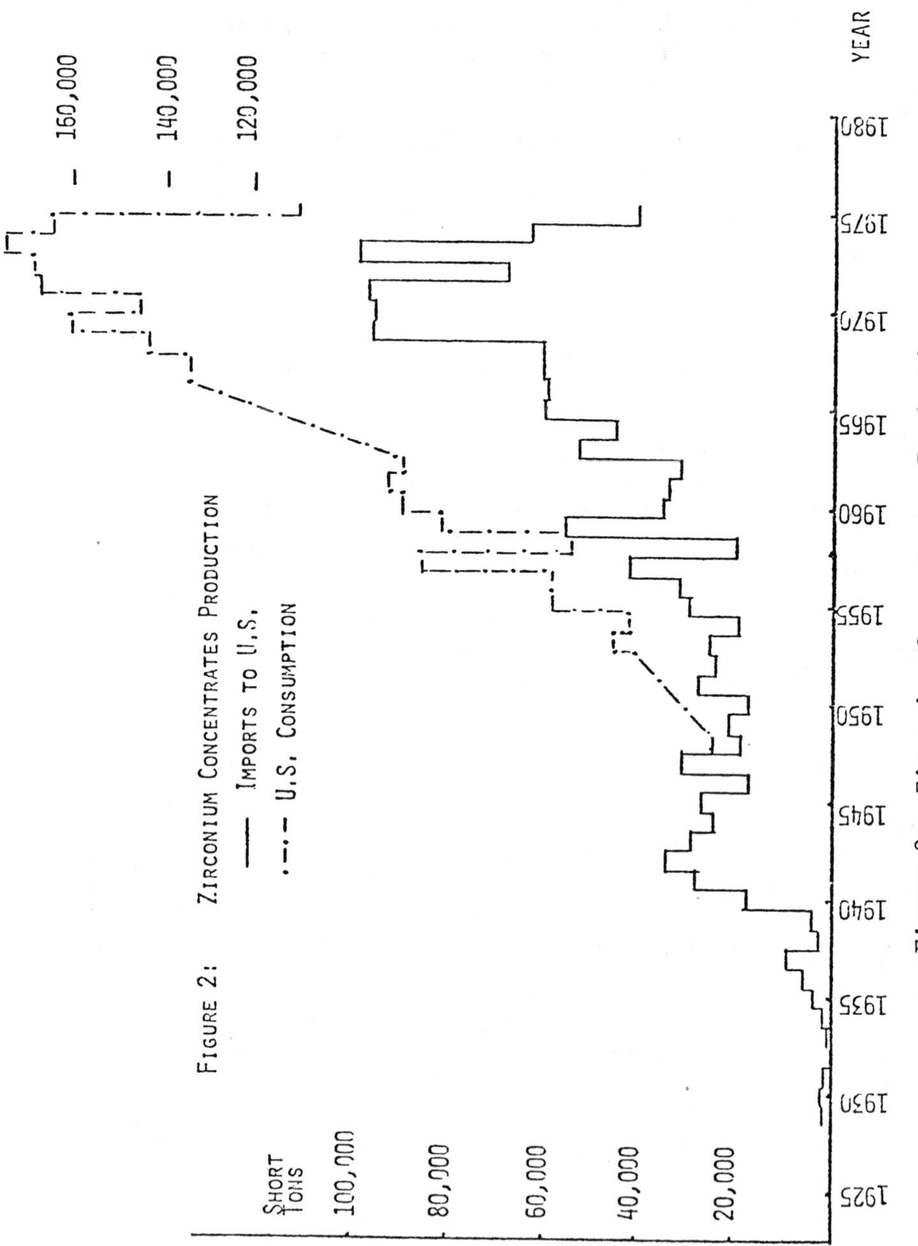

Figure 2: Zirconium Concentrates Production

piles. At the same time the Radiation Control Section of the Oregon Department of Environmental Quality (DEQ) became concerned that the large chlorinated residue piles may be a radiological problem. A gamma radiation survey showed maximum reading of 1200 uR/h. When the Oregon DEQ checked the radium concentration of the piles, it found that the Ra-226 ranged from the original zircon sand concentration of about 60 pCi/g to over 1300 pCi/g. One water sample taken within the residue pile showed Ra-226 concentration of 45,000 pCi/L, hence the concern of a potential ground water contamination. These radiological parameters for the rare metals chlorinated residues can be compared with those for uranium mill tailings.

Most uranium mills in the U.S. typically processed an average uranium ore grade of about 0.15-0.35% uranium which would give expected radium-226 concentrations in the mill tailings ranging from 420-980 pCi/g. Individual tailings samples at a given mill may have concentrations that are more or less than these values by as much as a factor of five or so.

Due to Oregon DEQ's work at the Wah Chang facility, Oregon limited the volume and radium content of chlorinated residue which the facility may accumulate onsite before disposal in an out-of-state facility is required and mandated all users of zircon sand to file an application for a radioactive materials license. The criteria for release to an unrestricted area are 57 uR/h, 30 pCi/L of Ra-226 in effluent, and 0.03 WL of radon* (Wagstaff, 1978). Of the twenty-six potential users of zircon sand, the state estimates only four will require specific licensing. Similarly, Utah has restricted onsite accumulations to no more than 100 tons of chlorinated residue and no more than 3 curies of Ra-226.

Efforts to determine the extent of possible radiological problems are more difficult for sites which have ceased rare metals processing activities for several years either due to changes in site ownership or unfavorable economic climate. Locating residue piles and sludge ponds, estimating amount and origin of ores processed, and determining processing and waste disposal activities must rely on historical records which are vague or nonexistent.

In the mid-1950's the responsibility for zirconium production was shifted to private enterprise when the civilian nuclear power program was established. In order to meet the increased

---

* One Working Level (1 WL) is a unit describing any concentration of short-lived decay products of radon-222 in one liter of air which results in the release of 1.3 x 10^5 MeV of potential alpha energy.

demands for reactor grade zirconium, Carborundum Corp. which operated a facility in Akron, New York, expanded its production capacity by building a facility in Parkersburg, West Virginia. The plant's designed capacity was 600 tons annually; it began operations in 1957. In the mid-1960's, Amax Specialty Metals Co., Inc., became a partner and in 1967 obtained full ownership of the company. The Parkersburg site was sold in 1977 to L. B. Foster Company, a steel pipe fabrication plant. As a result of Foster's plan to expand its buildings, pyrophoric waste materials were encountered during backhoe operations. In investigating the causes of the explosion, it was discovered that zirconium and thorium may have been buried onsite, and that Amax Specialty Metals had not adequately terminated its license with the U.S. Nuclear Regulatory Commission (NRC) for possession of radioactive materials. The NRC estimates that two million pounds of zircon ore, mainly from Nigeria, were processed at the Parkersburg plant since 1957. A radiological survey of the site shows gamma radiation to range from a background level of 10 uR/h to 150 uR/h. Soil samples show concentrations of thorium-232 and its decay products to range from background level of 1 pCi/g to 10,000 pCi/g. The thorium contaminated area is limited to a few acres of the 100 acre site. The NRC's tentative clean-up goal of 5 pCi/g above background of thorium-232 with a three to four foot overburden and deed restrictions on excavation was developed based on an assessment of the long-term hazard due to thoron (radium-220). A radiologal survey of Parkersburg by a contractor to Amax Specialty Metals estimates 50,000 cubic yards of soil may need to be removed. Some disposal alternatives being considered are burial at a disposal facility, burial onsite in a clay lined cavity with land use restrictions provided for the burial area, and ocean disposal. Whether there are other locations onsite where zirconium and/or thorium are buried may never be known since records on waste disposal and processing activities are incomplete.

The Akron, New York, zirconium and hafnium processing facility was the pilot plant for the Parkersburg, West Virginia, facility and presently is owned by Amax Specialty Metals Co., Inc. Processing activities by Carborundum Metals Co., Inc., began in 1953 at the Akron site to produce hafnium free zirconium under an Atomic Energy Commission (AEC) contract.

The plant's designed processing capacity was 162 tons of zirconium annually. Although the plant had a contract with the AEC to produce zirconium metal, there does not appear to be any AEC, NRC, or NYS (an agreement state) license for byproduct material other than for research and development purposes. An industrial license with the NYS Department of Labor (DOL) for x-ray and gamma sources was in effect from 1960 through 1978.

During the summer of 1978, the EPA Region II radiation office queried the NYS DOL and Department of Environmental Conservation about the status of the Akron, New York, plant. The EPA was concerned that a situation similar to the Parkersburg, West Virginia, plant may exist at the New York plant due to their operating history. EPA was informed that Amax Specialty Metals Co., Inc., had contracted Atcor, Inc., to perform a radiological survey of the Akron site as the initial step in terminating its industrial license with the NYS DOL. The June, 1978, survey showed radiation levels ranging from a background level of 7 uR/h to 1500 uR/h outside, with some building measurements up to 40 uR/h (Levesque, June, 1978). The extent of processing activities at the site is not well known due to incomplete records. There were areas where magnesium and zirconium residues were found but no pyrophoric incidents occurred. A single soil sample from an area with an external radiation reading of 1500 uR/h was analyzed and showed the soluble portion contained the radioactive material, but no further radiochemical analyses were performed. The elemental composition of the sample indicates it may be Nigerian ore, the principal ore processed at the Parkersburg plant. Surface soil samples were taken at locations with above background gamma radiation and were spectroscopically analyzed. The range of concentrations of Ra-226, Pb-214, Bi-214, Ac-212, Tl-208, and K-40 are shown in Table 2 (Levesque, November, 1978). For these isotopes, the background concentrations are less than 1 pCi/g except for K-40 with concentration of 12 pCi/g. The limited results in Table 2 indicate levels of thorium and uranium chain nuclides elevated above expected background levels.

The monazite fraction of zircon ores typically runs about 3-10% thorium dioxide (ThO_2) content with a tri-uranium octoxide (U_3O_8) content up to 0.41%. Zircon sands of 96.7% pure zircon are currently imported and quoted on a minimum basis of 65% zirconium oxide (ZiO_2). Hence the maximum monazite content of incoming zircon sands should be less than 3.3%. Assuming a 10% ThO_2 content in the 3.3% monazite fraction of the zircon sand, the overall ThO_2 content of the zircon sand should be less than 0.33% or less than 300 pCi/g. Nigerian sands are reported to range from 0.4 to 7% ThO_2. Based on this and on the assumption that the monazite is some lesser fraction than 3.3% of the non-zircon portion of the sands, then the values of 120-150 pCi/g for the thorium chain nuclides do not seem unreasonable. Similarly, using a 0.41% U_3O_8 content in the monazite fraction and a maximum 3.3% monazite content, one would estimate a maximum uranium or radium-226 concentration, assuming equilibrium, of about 38 pCi/g, which seems to be in probably fortuitous agreement with the maximum measured values of 35 pCi/g for Ra-226. The higher K-40 values are certainly higher than the expected background values of about 12 pCi/g. However, one stage of the hafnium purifying process uses a potassium chloride molten mixture. If this plant used this process and if some of the molten mixture were spilled, it could account for the higher K-40 values since the potassium chloride probably contains about 400 pCi/g.

Between June and September, 1978, Amax Specialty Metals removed soil from areas with high gamma radition levels. About 25 cubic yards of soil were shipped to the commercial low level burial site in Barnwell, South Carolina. Two of the three lagoons or infiltration ponds were excavated to a three feet depth and the material was disposed in a nearby hazardous chemical landfill. The radioactivity of the excavated material is not known since no analyses were performed prior to disposal.

In September, 1978, Atcor resurveyed the site after clean-up efforts and made recommendations for additional decontamination to reduce levels to "as low as reasonably achievable." However, it is not known to what extent these recommendations were pursued by the site owner. It is not known whether any attempts were made to identify the source of the slightly elevated levels of radioactivity in the buildings.

In December, 1978, the NRC performed a survey of the Akron site (Stohr, 1978). The survey identified one area near a ridge with gamma radiation of twenty times background. Areas near the lagoons and the tube mill building had levels ranging from background to ten times. Analyses of the soil samples from these three locations indicate thorium concentrations in the range 6.0-19 pCi/g with background concentration of 1.2 pCi/g. The one air sample showed no Rn-220 daughters above background levels. Gamma radiation levels in the buildings were within twice background, indicating little contamination after removal of the gamma gauges.

The site could have been released for industrial use with little clean-up necessary in order to meet the NYS DOL Industrial Code Rule 38 that no gamma radiation levels exceed 250 uR/h at the surface and that source material in soil be less than 0.05% by weight, which is 5,000 pCi/g of Th-232. Due to experience at rare metal processing facilities in Albany, Oregon, and Parkersburg, West Virginia, Amax considered a more stringent clean-up program to meet the goal of "as low as reasonably achievable." In general, the clean-up program has been successful, although EPA Region II would have liked to have seen the levels reduced to twice background and to 5 pCi/g for Ra-226 and Th-232. A record of processing and disposal activities at the site would have greatly assisted in answering questions concerning the possibility of any buried radioactive materials.

TABLE 3: STATES WITH RARE METAL AND RARE EARTH PROCESSING ACTIVITIES

Activity	States
Producers of zirconium oxide, zirconium and hafnium sponge metal, ingot, and alloy	NJ, MA, AL, MI, OR, WV, NY, OH, PA, CA, NH
Refractory firms using zircon in products	KY, NY, PA, MO, OH, WV, MI
Producers of zirconium compounds and chemicals	NJ, NY, MA
Producers of zirconium oxide for other than metal production	NY, AL, OH, SC, NJ, WV
Milled and sold ground zircon	NJ, NY, OH, SC, DE, CA, PA
Producers of rare earth compounds and chemicals	NY, PA, CA, CO
Producers of high purity rare earth metals	NJ, PA
Processed rare earth concentrates	IL, TN, NJ
Processed Canadian uranium mill solutions for rare earths	MI

DISCUSSION

It appears from Figure 2 that the amount of zircon ore imported into the United States has been increasing steadily since 1930. The imports account for approximately 50% of the annual U.S. consumption of zircon concentrates; the remainder is attributed to domestic production and stock piles. Australia supplied about 60% of the imports before 1950, and over 95% after 1950. Brazil contributed about 20% during the period 1930 to 1950. In total, the U.S. consumed about 300,000 short tons of zircon before 1950 and 1.3 million short tons thereafter.

The potential radiological problem can be likewise divided into two periods. Prior to 1950, most of the Australian zircon was imported as a black sand mixture containing zircon (40-75%), ilmenite (14-43%), rutile (7-18%), and monazite (2-8%)ores (Minerals Yearbook, 1936). No attempts were made to separate the ores until the sand mixture reached the processing facility. In 1948, the Commonwealth Government declared its intent to purchase and stockpile monazite ore. As a result, future shipments of sand had most of the monazite ore removed in Australia before export.

In order to provide a conservative estimate of the radiological content in the sands imported before 1950, the sand mixture is assumed to be composed of 8% monazite ore. Assuming the ThO_2 content in the monazite fraction to be as high as 10%, then the ThO_2 content in the beach sands could reach 0.8% or about 800 pCi/g. Similarly, assuming the U_3O_8 content in the monazite fraction to be as high as 1%, then the U_3O_8 content in the beach sand could reach 0.08% or about 200 pCi/g. After 1950, the Australian ore had the monazite fraction separated to some extent, hence the sands contain a minimum of 96.7% zircon. Assuming the ThO_2 content in the monazite fraction remains 10%, then the ThO_2 content in the beneficiated sands could be as high as 0.33% or about 300 pCi/g, the U_3O_8 content could reach 0.03% or about 100 pCi/g. Subsequently, the rare metals processing facilities which operated prior to 1950 may have greater radiological problems with their chlorinated residues than those which use ore obtained after 1950.

We would expect that any user of zircon sands would receive some monazite in the zircon sands, since the separation of ilmenite, rutile, zircon, and monazite, as indicated in Figure 1, is often incomplete. Hence some small fraction of monazite, containing thorium and uranium will be present in any industry which extracts titanium, chromium, tantalum, etc., from beach sands. The monazite and hence the amount of thorium and uranium decay chain products will vary with the sand origin and the degree of ore beneficiation. In fact, facilities which need only the zircon, ilmenite, or rutile fraction of the beach sands and insist on high purity ores may not have as great a radiological problem as facilities which need only the monazite fraction or use sands with little ore separation.

Producers of reactor grade zirconium have rigid specifications for thorium and uranium content and generally require high purity zircon which implies low radioactivity content. To achieve this, the zircon is either purchased as high purity material or further processed at the rare metal producing plant to achieve purity. For metal production, then, the radioactivity remains in the wastes while very little goes with the metal product. On the other hand, the purity of zircon sands consumed at foundries is not critical, hence these sands may have the highest radioactivity content. Manufacturers of refractory materials, producers of milled or ground zircon, and ceramics manufacturers will most likely have some portion of the radioactive content incorporated into the products due to the manufacturing process.

In reviewing information from the annual Minerals Yearbook for 1929 through 1975, over twenty states were identified to have some facility which processed beach sands in zirconium, hafnium, and rare earths production or used beach sands in foundry processes. Table 3 provides a breakdown of the states according to the type of processing or use activity. Facilities presently operating in these areas of activity can be fairly easily identified and evaluated to determine where these facilities dispose their chlorinated wastes and whether the sands or concentrates used by the facilities have any appreciable monazite fraction. For facilities which have ceased operating or changed their owernship or their products, such an evaluation is more difficult.

Since only limited data and in some cases no data are available for radioactivity and exposure levels associated with industries such as discussed above, it seems apparent that considerable work needs to be done to assess the environmental and health impact of such industries.

REFERENCES

1. Bliss, J. D., 1978, "Radioactivity in Selected Mineral Extraction Industries - A Literature Review," U.S. Environmental Protection Agency, Office of Radiation Programs - Las Vegas Facility, November 1978, Technical Note ORP/LVF-79-1.

2. Costel, D., 1978, "Hazardous Waste Proposed Guidelines and Regulations, and Proposal on Identification and Listing," Federal Register, Vol. 43, No. 243, December 18, 1978, pp. 58946-59028.

3. Levesque, R. G., 1978, "Results of ATCOR's Gamma Scan Survey of June 9, 1978," letter to H. Kall (AMAX Specialty Metals Corporation), June 20, 1978.

4. Levesque, R. G., 1978, "Results of Soil Samples - Analysis by Teledyne Isotopes," letter to H. Kall (AMAX Specialty Metals Corporation), November 16, 1978.

5. Minerals Yearbook, Annual Publication 1929 through 1975, U.S. Department of Interior, U.S. Bureau of Mines.

6. Stohr, J. P., 1978, "Results of NRC Radiation Survey at AMAX in Akron, New York," letter to F. Bradley (NYS Department of Labor), December 26, 1978.

7. Wagstaff, D. G., 1978, "NORM - Problems in Oregon," paper presented at the Region X Radiation Control Meeting, September 26, 1978.

DECOMMISSIONING A TRITIUM GLOVE-BOX FACILITY*

C. L. Folkers, S. G. Homann**, A. S. Nicolosi,
S. L. Hanel, and W. C. King

Lawrence Livermore Laboratory
University of California
Livermore, California 94550

ABSTRACT

A large glove-box facility for handling reactive metal tritides was decommissioned. Major sections of the glove box were decontaminated and disassembled for reuse at another tritium facility. To achieve the desired results, decontamination required repeated washing, first with organic liquids, then with water and detergents. Worker protection was provided by simple ventilation, which was combined with careful monitoring of the work areas and employees. Several innovative techniques are described.

INTRODUCTION

A major tritium research facility at Lawrence Livermore Laboratory (LLL) was obsolete for current programs and was decommissioned. This facility included:

- A nine-section glove box with a volume of about 8 m^3.
- An enclosed 500-ton hydraulic press.

---

*This work was performed under the auspices of the U.S. Department of Energy by Lawrence Livermore Laboratory under contract No. W-7405-Eng-48.

**Present address: Measurex Corporation, 1 Results Way, Cupertino, California 95014.

- A gas purification system with vacuum pumps and monitoring equipment.

- Three flush hoods with access to the glove-box air locks (pass-throughs) and gas purification system, the hoods having once-through air flow.

- Numerous specialized tools and equipment.

Examples of the glove box and its contents are shown in Figs. 1 and 2.

This glove-box facility had been used for the synthesis and processing of metal tritides (and deuterides and hydrides), including compounds of Li, U, Ti, Zr, and Ce. Gram quantities of tritium were processed frequently. Processing operations included crushing, grinding, sieving, and weighing. Experimental compacts were formed in a hydraulic press, and some samples were machined for special tests. Other operations included microscopy, thermal treatments, welding and brazing of containers.

These operations created tritiated dust that spread throughout the glove box. Some of this dust reacted with traces of water vapor and oxygen in the glove box to form tritium gas (HT) or tritiated water (HTO). These were scavenged by a gas-purification system and stored on a large molecular-sieve bed as HTO. The gas-purification

Fig. 1. Glove box for handling reactive metal tritides in an inert atmosphere.

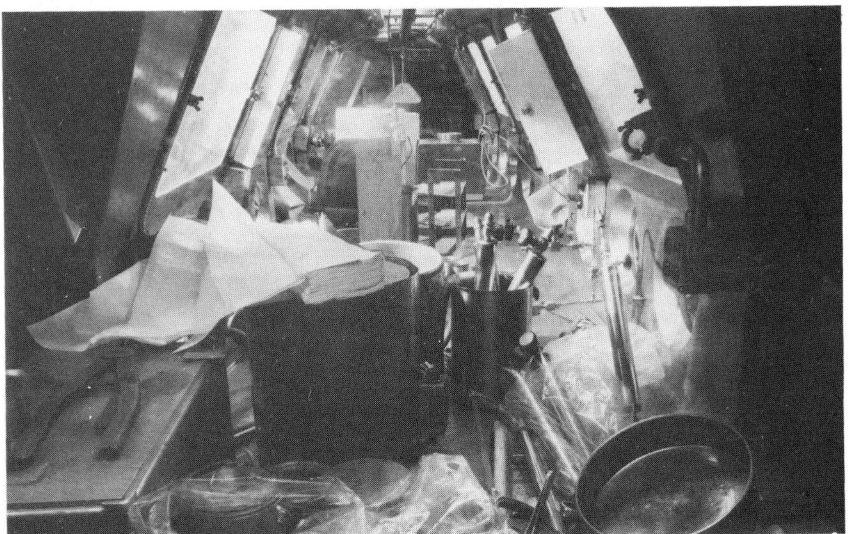

Fig. 2. Interior view of glove box, showing typical equipment to be disposed of as waste.

system included a Pd-catalyst bed and a 140-kg bed of 5A molecular sieve, with a blower, vacuum pumps, piping, and monitoring equipment. We estimated the molecular-sieve bed held about 9 g of tritium as HTO.

The molecular-sieve bed had been last regenerated nearly 12 years previously. The original equipment for regeneration was inadequate, and all temporary equipment was removed at that time. Even now the purification system maintained the glove-box atmosphere at about 1 to 2 ppm water (containing all hydrogen isotopes) and about 0.2 Ci/m^3 (combined HT and HTO). However, without the ability to regenerate the molecular-sieve bed, any large additions of water during decontamination would partially displace the existing HTO, thereby increasing tritium in the glove-box atmosphere and thus increasing worker exposures as well as effluent releases to the atmosphere. This greatly influenced the decontamination processes we used.

We knew that tritium-contaminated facilities at several laboratories, including ours, had been decommissioned, but we were unaware of any reports on the procedures used or the amount of effort needed for specified goals. After making repairs on a similar glove box, Folkers and Johnson[1] reported decontamination of nearby areas with methyl alcohol and trichloroethylene and an effective use of local ventilation for worker protection. Harris, Kokenge and Marsh[2] and Gilbert, Wright and Madding[3] reported that procedures for decontaminating facilities for ^{238}Pu and ^{210}Po, respectively, required repeated washing with detergents to get the desired results.

Because of this paucity of information, we had to develop innovative techniques for tritium decontamination while minimizing worker exposures and atmospheric releases. This report describes the methods used and our results.

PLANNING

The two key issues identified[4] were:

- To adopt standards for residual tritium contamination that are acceptable for subsequent use of recovered equipment.
- To minimize the radiation dose to workers and the public during the operation.

Normal procedures for operation of our tritium facility provided the basis for most of our planning.[5]

A safety assessment[6] provided guidance for the entire operation, especially for unusual or high risk operations. Key recommendations included:

- Use of protective clothing while handling contaminated equipment.
- Provision for adequate ventilation (0.75 m/s) while working on open contaminated equipment.
- Monitoring of workers by taking daily bioassay samples during high hazard operations and at least weekly samples at other times.

The principal tritium-exposure pathways were associated with the initial breaking of tritium lines, decontamination efforts, and the handling of tritium-contaminated vacuum-pump oil.

Primary protection for the workers consisted of maintaining a positive flow of air away from personnel and the wearing of protective clothing, that is, 5-mil polyvinyl chloride gloves and cotton lab coats. While working in the glove box, 1.75-mil shoulder-length polyethylene gloves were worn to prevent skin contact with the rubber gloves on the box; these polyethylene gloves were replaced four times per hour. The bioassay program recorded actual exposures.

The room containing the glove box was continuously monitored for airborne tritium using two, Overhoff Beta-Tech*, 2-litre,

*Reference to a company or product name does not imply approval or recommendation of the product by the University of California or the U. S. Department of Energy to the exclusion of others that may be suitable.

gamma-compensated, ion chambers. The average tritium concentration in the room was $<10\ \mu Ci/m^3$ throughout the decommissioning project. On two separate occasions, the tritium room concentration approached $350\ \mu Ci/m^3$. Normal room ventilation reduced the level in about 10 min.

Maximum allowable surface contamination levels are shown in Table 1. As we expected to ship the glove-box sections to Sandia Laboratory Livermore (SLL) for use in their tritium facility, special contamination levels were adopted for this purpose. Surface-contamination levels were assessed, using standard swiping techniques and a liquid scintillation unit located in the same building.

The maximum off-site exposure from a postulated accidental release of 1 g of tritium as HTO was estimated[6] to be less than 30 mrem. This was unlikely to occur since most of the tritium here was bound on the molecular-sieve bed.

Table 1. Maximum Levels of Tritium Surface Contamination Allowed for Various Uses.

Item	Maximum Allowable Level ($dis/s \cdot cm^2$)
Equipment for unrestricted use, including general public[a]	1
Equipment or furniture, use limited to offices within our tritium facility[a]	2
Equipment or furniture, use limited to laboratories within our tritium facility[a]	20
External surfaces of equipment to be sent to SLL for use in their tritium facility	2
Internal surfaces of equipment to be sent to SLL for use in their tritium facility	2000[b,c]

[a]Taken from Ref. 5.

[b]Each glove-box section also limited to total (absorbed plus swipable) tritium of 1×10^{12} dis/s ($\sim$25 Ci).

[c]Enclosed during shipment; required continuous ventilation except during shipment; required protective clothing during work.

Both HT and HTO outgas continuously from contaminated surfaces, even after decontamination. All steps were planned to retain the gas purification system or to flow air continuously, so that unexpected concentrations of tritium would not build up in enclosed spaces or piping.

A preliminary survey was made of the glove box, its contents, and related equipment. We anticipated the hazard for various operations from the data in Table 2. For example, we know from experience that surface contamination levels of <20 dis/s·cm^2 can be handled in the open by workers appropriately monitored, while levels up to about 8×10^5 dis/s·cm^2 can be safely handled with plastic or rubber gloves inside a flush hood. Vacuum-pump oil with at least 30 mCi/ℓ of tritium contamination and no water content can be handled with plastic gloves and modest ventilation. The hazard is severe when the oil also contains water.

Table 2. Tritium Contamination Levels in a Research Glove-Box Facility After Exposure of Glove Box to Solid Tritium Compounds for Over 11 Years.

Item	Measured Contamination Level
Molecular-sieve bed, 140 kg type 5A, tritium stored as HTO	$\sim 9 \times 10^4$ Ci (estimated)
Equipment and surfaces inside glove box	0.3 to 30×10^7 dis/s·cm^2 (1 to 100 Ci/m^2)
Equipment and surfaces inside flush hoods	0.7 to 800×10^3 dis/s·cm^2 (0.2 to 200 mCi/m^2)
External surfaces of glove box, purification system, nearby walls, floor, and furniture	0.4 to 8 dis/s·cm^2 (0.1 to 2 μCi/m^2)
Vacuum pump oil, pump used to recover hydrogen isotopes	$\sim$150 mCi/ℓ
Vacuum pump oil, pump used for glove-box atmosphere	1 to 30 mCi/ℓ
Argon atmosphere inside glove box	$\sim$0.2 Ci/m^3

DECONTAMINATION AND DISASSEMBLY PROCEDURES

Some metal tritides react spontaneously with H_2O or O_2 to liberate tritium. For example:

$$LiT\ (s) + H_2O\ (\ell \text{ or } g) \rightarrow LiOH\ (s) + HT\ (g) \quad .$$

Other tritides such as TiT_2 are relatively stable and as contaminants must be physically removed. Water-based detergents used to remove stable particulates would react with less stable materials, giving excessive HT in the glove-box atmosphere and thus increase worker exposures. Use of water was also limited by the capacity of our molecular-sieve bed, as described earlier. Therefore, our procedures used mineral oil and kerosene to remove "loose" contamination before using any water.

Glove Box Contents

All quantities of metal tritides were removed, including dust and debris vacuumed from the glove box. Equipment inside the glove box was packaged as contaminated waste. The following procedures were generally followed:

- Tritium contamination was estimated from past experience.
- Equipment was placed in "lard" cans (∿45-ℓ volume). Pieces too large to fit were disassembled, broken, or crushed.
- Outsides of the lard cans were wiped off with mineral oil, and closed cans were moved from the glove box to the flush hood.
- Lids were sealed with silastic sealant and taped.
- Cans were sealed in plastic bags and placed in 55-gal barrels. These also had to be promptly sealed or some contamination from the lard cans would permeate the plastic bags.

Glove Box Decontamination

Decontamination of the inside surfaces of the glove box was as follows:

- Washed with mineral oil.
- Washed with kerosene, followed by kerosene with detergent.

- Water spray to decompose reactive tritides.
- Washed twice with water and detergents.
- Converted the glove box from argon atmosphere to once-through air flush, followed by two water rinses.
- Removed windows and gloves, washed adjacent areas with water-based detergent.
- Washed entire glove box with water-based detergent.
- Separated glove-box sections and washed flanges with water-based detergent.

The air locks were cleaned by a modified procedure, which eliminated the use of kerosene with detergents.

Results of the decontamination procedures for the glove box are shown in Fig. 3. Repeated scrubbings were clearly necessary to achieve the desired decontamination, even though the final contamination levels were well below the target level of 2000 dis/s$\cdot$cm^2.

When the glove-box interior was washed or rinsed, this included spraying hard-to-reach corners with a pressurized nozzle and an external source of fluid. Dishwashing swabs with extended handles provided scrubbing action. Fluids and dirt were wiped up with paper products and sealed in paint cans.

The first use of water as a "spray" was by a hand-held spray bottle similar to those used for household detergents. Reaction with tritide salts caused an effervescence, increasing the tritium in the glove-box atmosphere. When the level exceeded 1 to 2 Ci/m^3, we stopped spraying until the level came back down so as to minimize worker exposures. Joints were saturated repeatedly to decompose everything possible before converting to once-through air flow. This decomposition was much slower than expected--this was true even on flat surfaces after repeated applications of water, as evidenced by increasing tritium in the glove-box atmosphere. We later estimated that decomposition of salts liberated about 2 to 12 Ci/m^2 of surface as HT, although these surfaces were swiping only a few percent of this just before the water spray (see Fig. 3).

In order to use water, we valved out our large molecular-sieve bed and installed an _in-situ_ dryer in the glove box. The regular purification-system blower and catalyst were maintained so that HT would be converted to HTO; while the glove-box atmosphere was still basically argon, it contained several hundred ppm O_2 for this purpose. This effectively gave us an unlimited capacity for water. The _in-situ_ dryer is described elsewhere.

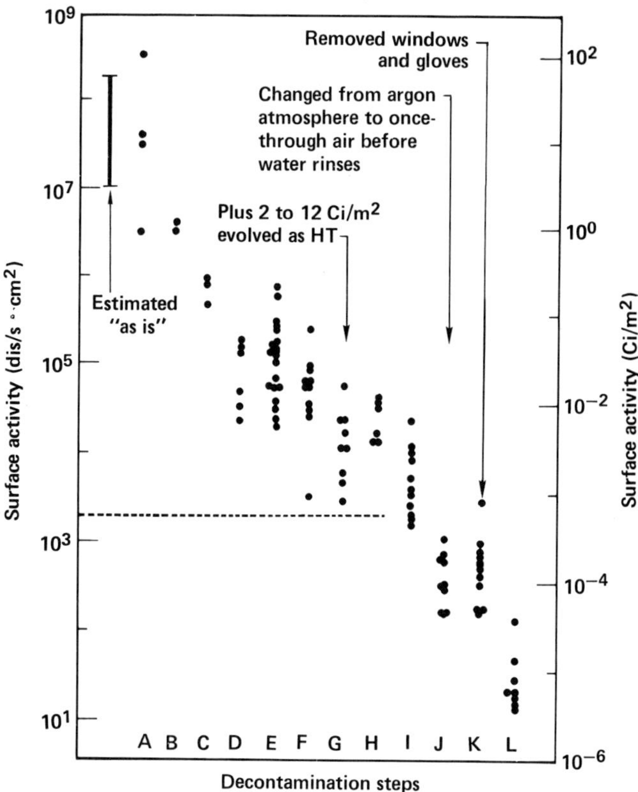

Fig. 3. Glove-box internal-surface contamination as reduced by decontamination procedures. Area below dashed line shows target level of activity for shipment off site. The decontamination steps are:

A measured "as is"
B mineral oil
C kerosene
D kerosene
E various detergents with kerosene
F 2% Triton CF-10 with kerosene
G water spray
H, I 2% X-100 with water
J water rinses (2)
K air flush 30 days
L swish; water and methanol rinse

We tried using several detergents in kerosene to lift off visible grime. These were anhydrous nonionic surface active agents, one example being a 2-vol% solution of Triton CF-10 (trademark of Rohm and Haas Co.), a benzyl ether of octyl phenol. Detergents in kerosene did not appear to be effective beyond the ability of kerosene to wipe away loose particulates, as shown in Fig. 3.

With water, we used a general purpose, nonionic surface active agent as a 2-vol% solution of Triton X-100 (trademark of Rohm and

Haas Co.), a type of alkylaryl polyether alcohol (see Fig. 3). Another detergent we used effectively was Swish (trademark of Haviland Products Co.), a proprietary water-base amphoteric surfactant containing alkaline silicates, which is packaged in a spray can. This product foams upon application, and after standing it was rinsed off with water or methyl alcohol (see Fig. 3). At this point, we conclude that any good detergent in water would be effective.

After most of the glove box decontamination was done, the box was purged with room air for about 30 days. The windows and gloves were removed during this time. We then swiped the inside surfaces again to see if there was any tritium accumulation from outgassing. Results are noted in Fig. 3. There was no increase of tritium activity inside the glove box and only a modest increase on the air-lock surfaces.

Windows, Gloves, and Fittings

After contamination levels dropped below 2000 dis/s·cm^2, we removed windows and gloves. We used a simple, portable, acrylic plastic hood with an exhaust blower over the work area and with the glove box under negative pressure so that air would flow into it. The hood is illustrated in Fig. 4.

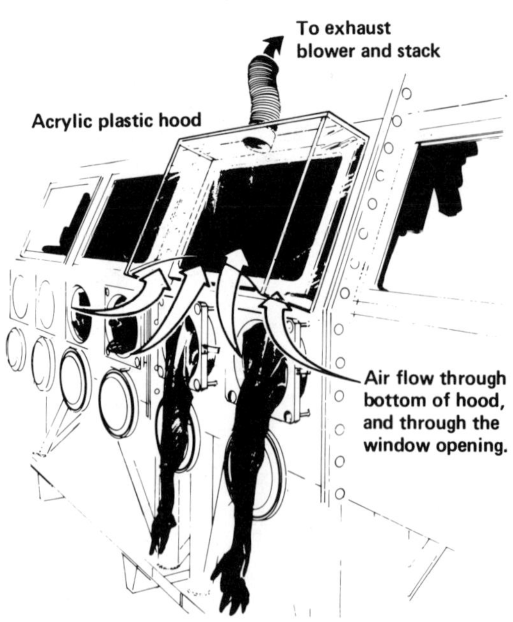

Fig. 4. Portable hood to provide ventilation while removing windows and cleaning adjacent areas.

Purification System

The purification system was removed in a manner analogous to that used for the windows. Piping was cut into sections and capped for disposal.

Separation of Glove Box Sections

We separated the flanges on the glove boxes by 10 to 15 cm and kept the air flowing into the glove box as before. A plastic sheet was taped beneath the flanges. Rubber gaskets were removed for disposal, and the flange areas were cleaned. Air flow could be increased in a given work area by taping plastic across other areas.

SPECIAL PROBLEMS

Alternate Swipe Techniques

Regular smear tabs would have been useless for swipes inside the glove box, because the gloves would contaminate the tabs. As an alternate technique we used cotton swabs.

The stem of a cotton swab was prescored close to the cotton; the swab was placed in a glass bottle, which in turn was placed inside a plastic bag. After a swipe was made inside the glove box, the now contaminated swab handle was broken off and discarded. The sample was sealed in the plastic bag, taken outside the glove box, and counted in the same way as a regular smear tab. A background sample for each occasion was obtained by waving a swab in the glove-box atmosphere for a corresponding time period. We usually moistened the cotton with glycerol before use.

In-Situ Packed-Bed Dryer

We used an inexpensive, temporary desiccant bed that could be operated inside the glove box, eliminating regular use of our purification-system molecular-sieve bed. This bed was regenerated by simply pouring out the expended desiccant and replacing it with fresh desiccant.

The *in-situ* dryer was made from a lard can, a small blower, and a wire screen, as shown in Fig. 5. Joints were sealed with silastic adhesive. The blower provided about 20 ℓ/s (42 cfm) gas flow through a bed of 5A molecular sieve about 5 cm deep.

A molecular-sieve charge of 2.3 kg gave a maximum capacity of about 460 g of water. A fresh charge of molecular sieve in the

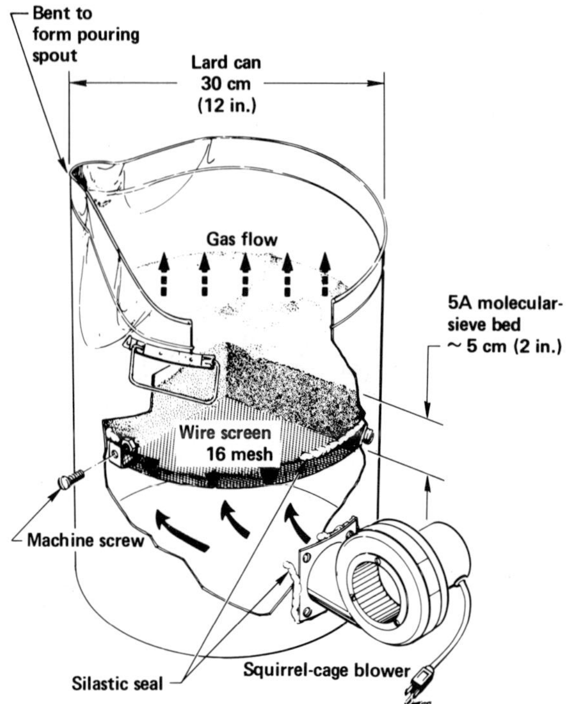

Fig. 5. Construction features of in-situ pack-bed dryer for scavenging water containing some HTO.

dryer could maintain the glove-box atmosphere at 250 ppm H_2O, with HTO levels typically 0.2 to 0.3 Ci/m^3. It could also reduce the glove-box atmosphere from about 30,000 to 2000 ppm H_2O in less than 2 h. This performance was adequate for our purposes.

Bioassay Program

Workers submitted urine samples daily during heavy work loads and at least weekly at other times. The typical range of tritium concentration was 1 to 5 μCi/ℓ, and the highest value was 8 μCi/ℓ. The maximum worker exposure was 110 mrem, and the total integrated dose for all workers was 420 mrem for the 6-month period to date. These values are within our guidelines.

Rubber Gaskets

We recovered several gaskets from air-lock doors and other locations to verify their condition after more than 11-years service exposed to tritium. Most of these were made of Buna N with known hardness.

Analyses of tritium concentrations in the gasket materials ranged up to 2 mCi/g. There was some indication that Buna N has hardened, but this could be because of other factors as well as tritium exposure. We judged the gaskets to be generally suitable for further service.

WASTE DISPOSAL

Requirements of our waste disposal program were largely limited to packaging waste in containers that conformed to current regulations[7]. Ultimate disposal was handled by the LLL Waste Disposal Group, and commercial burial sites were used.

Bagged or canned waste was loaded into containers, and the tritium content was estimated; when appropriate, the item was weighed. A log of these data was kept. From this point on, the Waste Disposal Group processed the containers.

CONCLUSIONS

A major tritium research facility for handling reactive metal tritides was decommissioned. A large glove box was decontaminated and disassembled for reuse at another tritium facility. Tritium contamination inside the glove box was reduced by six orders of magnitude to a level of about 10 to 100 $dis/s \cdot cm^2$. Decontamination required repeated washing, first with organic liquids and then with water and detergents, to achieve the desired results. Special swipe techniques were used to monitor progress inside the glove box. Careful planning before and during our operations played a major role in our success. Adequate worker protection was provided by the use of simple protective clothing in conjunction with continuous monitoring of the room atmosphere and regular bioassays of the workers. Adequate local ventilation gave protection to workers while opening contaminated equipment or breaking contaminated plumbing lines. Radiation exposures were well within our guidelines; the highest individual exposure being 110 mrem for a 6-month period.

An inexpensive *in-situ* packed-bed dryer was developed for temporary service during decontamination operations. This effectively gave us unlimited capacity to adsorb water containing some HTO.

Rubber gaskets exposed to tritium service for over 11 years were soft and pliable and were generally suitable for further service.

ACKNOWLEDGMENTS

Any project of this magnitude involves many talents beyond those of the authors. We gratefully thank all who have generously helped us, especially these individuals: F. L. Vanderhoofven for his intimate understanding of the whims of this system; R. L. Stark for his many novel suggestions; H. H. Miller for the tritium assays of rubber gaskets; E. E. Volk for assistance with waste disposal; and R. M. Alire for counsel and encouragement and for reviewing this manuscript.

REFERENCES

1. C. L. Folkers and K. A. Johnson, "Repairing a Large Tritium-Contaminated System," UCRL-50850, Lawrence Livermore Laboratory, Livermore, Calif. (1970).
2. W. R. Harris, R. R. Kokenge, and G. C. Marsh, "Termination of the Special Metallurgical (SM) Building at Mound Laboratory," MLM-2381, Mound Facility, Miamisburg, Ohio (1976).
3. K. V. Gilbert, E. M. Wright, and R. M. Madding, "A Report on the Decontamination and Decommissioning of the Technical (T) Building at Mound Laboratory," MLM-2239, Mound Facility, Miamisburg, Ohio (1976).
4. "Environmental Development Plan (EDP), Decontamination and Decommissioning," DOE/EDP-0028, U. S. Department of Energy, Washington, D. C. (1978).
5. "Operational Safety Procedures, Tritium Facility, Procedure 331," Revised March 1, 1978, Lawrence Livermore Laboratory, Livermore, Calif. (1978).
6. M. A. Matthews, D. H. Denham, and S. G. Homann, "Safety Assessment Document for Decommissioning of Building 331 Hydride Operations," Internal Report, Lawrence Livermore Laboratory, Livermore, Calif. (1979).
7. Code of Federal Regulations, Title 49, Subtitle B, Chapter 1, Parts 170-189, Department of Transportation, Washington, D. C. (1978).

DECONTAMINATION AND REFURBISHMENT OF THE HOT FUEL EXAMINATION FACILITY SOUTH (HFEF/S) ARGON CELL

J. P. Bacca, R. L. Brookshier, J. C. Courtney*, K. R. Ferguson, M. F. Huebner, and J. P. Madison

Argonne National Laboratory
P. O. Box 2528
Idaho Falls, Idaho 83401

INTRODUCTION

The HFEF/S argon cell[1] is a large hot cell at the Argonne National Laboratory-West (ANL-West) Site of the Idaho National Engineering Laboratory (INEL). From the start of operations in 1964, until the fall of 1977, this hot cell was operated without personnel entry in support of United States breeder reactor programs. During the first four years it was used for remote pyrometallurgical reprocessing and refabrication of about 35 000 uranium-fissium† metallic driver-fuel elements for EBR-II. Thereafter, the cell was used for nondestructive and destructive examinations of fast-reactor fuel and material experiments irradiated in EBR-II or in the Transient Reactor Test Facility (TREAT).

In late 1977 the cell was shut down to prepare it for refurbishment of important in-cell systems. To date, remote removal of in-cell equipment, and remote decontamination of the cell interior have been completed, and contact decontamination is about 25% complete.

Objective

The objective of the decontamination phase of the refurbishment project is to reduce the radioactive contamination so personnel can

*Louisiana State University, Baton Rouge, Louisiana.
†Fissium (Fs) is a mixture of fission-product alloying elements, principally molybdenum and ruthenium.

work in-cell for substantial periods while remaining well within DOE and ANL radiation-exposure guidelines. Cell entries are necessary to facilitate refurbishment of the in-cell systems including: 1) modification, repair, and maintenance of two bridge cranes and six bridge-mounted electromechanical manipulators; 2) modification of the present mercury-vapor lighting system to increase the light-level; 3) modification of 18 lead-glass viewing windows to improve light transmittance; and 4) maintenance and repair of three equipment-transfer locks, 10 master-slave manipulator wall-feedthrus, and 65 utilities-service cell-penetrations.

After refurbishment, the cell will again provide a large inert-gas-atmosphere hot cell for DOE breeder reactor programs.

Summary and Status

Cleanup and decontamination of the HFEF/S argon cell was started in the Fall of 1977 with the removal of process and examination equipment and irradiated fuel using remote handling methods. The cell floor and other accessible surfaces were dry-vacuumed remotely, then a foaming agent was applied to the floor and picked up by wet-vacuuming. For the first time since 1964, a personnel entry into the cell was made on June 23, 1978, to determine in-cell radiological conditions. Hot-spots were identified and smears were taken to determine the nature of the contamination.

Contact decontamination, involving personnel entries, was started in September 1978. Initially, conventional contact-decontamination methods were used. The interior was dry-vacuumed to pick up loose debris in areas difficult to reach using remote handling equipment. Wet swabs and mops, then manual and motor-powered brushes were used.

None of the above decontamination methods proved adequately effective; nevertheless, the waste generated indicated that significant quantities of contaminants had been removed. Next, a strippable latex coating was tested. It was moderately effective, but it proved excessively difficult to apply, then peel the coating from interior projections.

A high-pressure water-spray is now being used. The liquid waste is vacuumed up, collected in a drum, and pumped to a receiving tank. Over 250 entries into the argon cell have been made. Usually, two persons at a time enter to perform specific assigned tasks. To date, this high-pressure water-spray decontamination technique has been used over about 40% of the cell interior surface, and the work is continuing.

Facility Description

The HFEF/S facility, illustrated in Fig. 1, comprises two large, heavily shielded hot cells; the air-atmosphere "air cell", and the inert-gas atmosphere "argon cell".

In plan view, the argon cell (illustrated in Fig. 2) is a 16-sided polygon with a control room at the center. The argon cell contains 16 trapezoidal bays, each 4.9 m (16 ft) deep, and 3.7 m (12 ft) across the widest end. The cell is 6.7 m (22 ft) high inside. On radial lines between bays there are 13 sets of utilities floor-feedthrus. The internal volume is approximately 1700 m^3 (60 000 ft^3), and the internal surface area is approximately 1140 m^2 (12 300 ft^2). The cell has 14 lead-glass viewing windows through the outer walls, and four through the walls of the central control room. The in-cell overhead handling system is comprised of two 4.5-Mg (5-ton)-capacity bridge cranes and six 340-kg (750-lb)-capacity bridge-mounted electromechanical manipulators. There are two small equipment-transfer locks between the argon cell and the adjacent air cell, and one large transfer lock through the floor into a subcell that communicates with the air cell. All internal surfaces of the cell are lined with carbon steel, the surface of which is zinc-metallized and shot-peened. During normal operations,

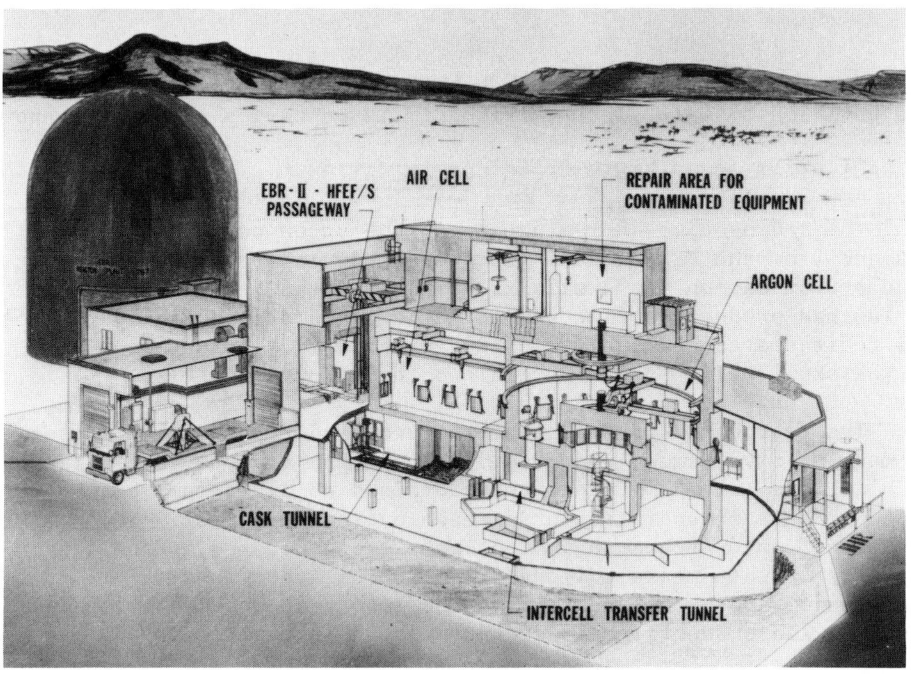

Fig. 1. HFEF/South Facility

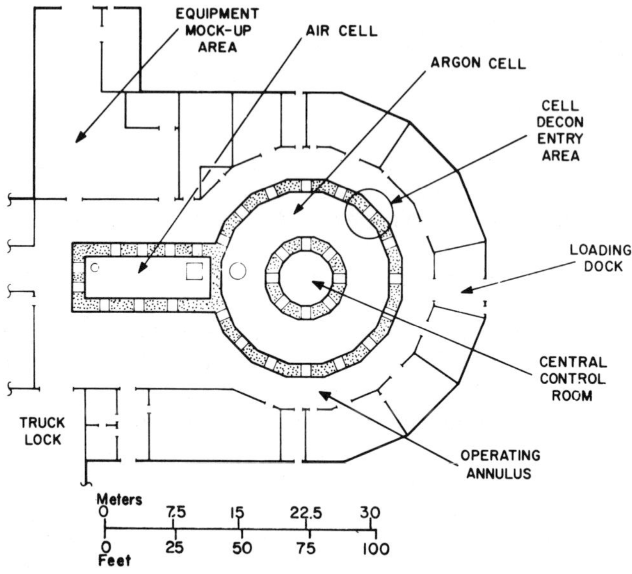

Fig. 2. Plan View of HFEF/South Hot Cells

argon gas is recirculated through the cell and continuously cooled and purified to maintain specified limits on oxygen and water vapor.

CHARACTERIZATION OF RADIOACTIVE CONTAMINATION

The radioactive contamination in the argon cell was principally residue from the EBR-II fuel reprocessed during the 1960s. In particular, oxidation of "skulls" remaining in the melt-refining crucibles had produced a fine powder. The argon atmosphere in the cell was recirculated at high velocities without filtration, so spills of the skull-oxide powder were blown throughout the cell.

The reprocessed EBR-II fuel was recycled through the reactor as many as five times, and received about one atom-percent burnup each cycle. Initially, the fuel contained no plutonium, but about 0.06% ^{239}Pu was generated during each cycle. Only negligible amounts of other transuranic isotopes were generated. From the analysis of the smears collected, plutonium averages 0.2% of the uranium and, consequently, was an important consideration in planning and conducting all decontamination operations.

The significant long-lived fission products in the samples are as expected after more than 10 years decay; namely, ^{90}Sr-^{90}Y and ^{137}Cs-^{137m}Ba. Trace quantities of ^{125}Sb, ^{134}Cs, ^{144}Ce, and ^{155}Eu have also been detected. The calculated ^{90}Sr concentration is 0.02 Ci/g uranium; the concentration measured in a number of samples averaged 0.06 Ci/g. During the melt refining process, the cesium was volatilized from the fuel and collected on a fume trap. Nevertheless, significant amounts of cesium are present. The ^{137}Cs concentration is more variable than the ^{90}Sr, measuring from 0.005 to 0.05 Ci/g uranium.

Beta radiations from ^{90}Sr, ^{90}Y, and ^{137}Cs have the following maximum energies (MeV) and probabilities of emission: 0.55 (100%), 2.27 (100%), 1.18 (7%), and 0.51 (93%). The average range of the ^{90}Y betas is 3 m in air; the maximum range is 9 m. All other betas of significance have a maximum range less than 1.4 m. Gamma radiation from the ^{137m}Ba is 0.66 MeV with an 89% probability of emission.

REMOTE DECONTAMINATION

Methods developed in 1976 (when the HFEF/S air cell was decontaminated and refurbished[2]) were used to prepare in-cell equipment for out-of-cell maintenance, storage, or disposal. Compressed air was blown over the contaminated equipment to remove loose particulates within the cell and minimize contamination dragout. Equipment was then transferred through the equipment lock and adjacent air cell to an enclosed spray chamber. The equipment was further decontaminated as necessary for its intended disposition using high-pressure water and manual wiping. Afterwards, the equipment was packaged for on-site storage or maintenance, or for disposal at the INEL Radioactive Waste Management Facility.

Dry Decontamination

The cell floor was brushed remotely using the electromechanical manipulators. The floor and all other accessible horizontal surfaces were then vacuumed using special tools and fixtures.

Criticality and Accountability Considerations. Specific procedures were developed to insure nuclear criticality safety and fissile-materials accountability during the dry decontamination operations. The quantity of material collected in a batch, either by remote brushing into a pan or by vacuuming, was limited to a safe weight of 2.5 kg. Collected materials were passed through coarse and fine screens, one batch at a time; materials retained on either screen were classified by viewing through a periscope. Materials positively identified as either fissile or nonfissile were separated and packaged. All other materials collected on the screens were packaged and identified as suspect fuel. The suspect materials are temporarily stored, and will be assayed for fissile content before their disposition is determined. The criticality weight assigned to the fines material that passed through both screens was based on analysis of representative samples vacuumed from the cell floor. The highest concentration of ^{235}U found in these samples, 5.2%, was used to assign criticality weights; the average concentration, 3.4%, was used to assign accountability weights. The fines material was disposed of in HFEF high-level-waste cans along with other nonfissile waste. About 200 g fuel, 3000 g unidentified materials, and 7700 g fines material (assigned 400 g fissile weight) were collected during the remote dry decontamination operations.

Representative samples of material remaining on the cell floor after the dry decontamination operations were collected and analyzed. Based on these samples, the quantity of ^{235}U remaining on the entire floor was determined to be less than 260 g--an amount safe under any conditions.

Wet Decontamination

After dry decontamination, the cell floor was wetted with Turco* #5865, a foam-type decontamination agent. The liquid waste was collected using remote wet vacuum cleaners, transferred to drums, and solidified by mixing with Safe-T-Set†, all within the cell. The waste-filled drums were subsequently transferred to the INEL Radioactive Waste Management Facility. The entire floor area was decontaminated twice using this wet procedure.

CONTAMINATION CONTROL AND PERSONNEL SAFETY

All refurbishment project operations are controlled by written, approved procedures to assure that all contamination and personnel safety hazards are considered, and that adequate precautions are specified. To meet project objectives, personnel must enter the cell despite the relatively high in-cell radiation levels. Therefore, each step is carefully preplanned to develop techniques that minimize personnel exposures. In-cell operations during the previous week are reviewed to assess their effectiveness, to identify problems, and to assign responsibility for developing solutions. When practical, procedures are mocked-up and tested out-of-cell before they are performed in-cell. New techniques are also subjected to limited in-cell trials before committing to full-cell operations. In-cell radiation is surveyed as necessary to control personnel exposures and to evaluate the effectiveness of decontamination operations.

---

*Turco Products Division, Purex Corp., Carson City, California.
†Oil Center Research, Inc., Lafayette, Louisiana.

Administrative Controls and Procedures

The safety of in-cell workers is optimized by mandatory administrative controls and procedures that detail specific measures to be taken prior to and during each cell entry. Responsibilities for normal and emergency actions are clearly delineated. For each cell entry one person is designated as the person-in-charge (PIC). All support personnel work under his direction. The PIC directs the in-cell workers by radio. In an emergency, he is in charge until relieved by the HFEF Emergency Director. Technicians are stationed at the cell windows to maintain constant visual surveillance of each in-cell worker. Closed-circuit television is used to maintain the surveillance when the work is out of the observer's line of vision.

If a cell worker becomes ill or is injured, other in-cell personnel will assure that the worker's breathing-air supply is functioning and, if he can walk, will assist the worker to the exit. A rescuer, suited and equipped with full-face mask, stands by at all times ready to enter the cell in case a worker were to need immediate assistance. An unconscious worker would be rescued by firemen, on site and equipped for emergency rescues.

The administrative procedures include checklists to assure the readiness of all required support and safety-related systems, and that hazardous equipment has been tagged out-of-service prior to each entry. Supervisors of all organizations that provide support services in normal or emergency circumstances are notified prior to the start of an entry.

Special radiation safety requirements are documented and approved for each entry. An approved procedure describes each specific task to be accomplished, emphasizing safety precautions.

In case of an out-of-cell emergency, such as a fire, in-cell workers would be summoned to the exit by blasts from an air horn. Emergency procedures for unsuiting workers have been prescribed, and a special vehicle is provided in case a site evacuation should be necessary.

Each worker's eligibility to enter the cell is verified in advance. The worker's medical history and work restrictions are reviewed by the ANL-West staff nurse. Persons with open wounds are ineligible. The worker's radiation-exposure history is reviewed by Radiation Safety. During an entry, the PIC periodically questions the workers concerning their need for rest. An in-cell bench, enclosed for shielding from nonpenetrating radiation, is provided for breaks.

Personnel Training

HFEF and Radiation Safety personnel involved in the refurbishment project are qualified in basic procedures, with emphasis on prescribed safety precautions and emergency actions. Outside-contractor personnel, and ANL-West personnel not previously trained, are given radiation-worker instruction that includes radiation-exposure control, contamination control with emphasis on plutonium safety[3], and site emergency responses. Refurbishment Group personnel train the workers to use specialized equipment and instruct them in the specific tasks to be performed.

Personnel Protective Clothing

For cell entries, personnel wear the following anticontamination clothing (listed in order from the body outward): 1) shorts, T-shirt, socks, and safety shoes provided by the Laboratory; 2) a pair of sack-type cotton coveralls; 3) two pairs of low-quarter polyethylene shoe covers; 4) one pair of high-top polyethylene shoe covers; 5) two pairs of low-quarter shoe covers; 6) one pair of cotton glove liners; 7) one pair of rubber gloves; 8) one TYVEK* surgeon's cap; 9) one pair of safety glasses; 10) one pair of TYVEK coveralls; 11) a polyethylene supplied-air breathing hood; 12) a second pair of rubber gloves; 13) a two-piece plastic wet-suit; 14) one pair of rubber boots; 15) a lead-loaded apron (0.5 mm Pb); and 16) one pair of lead-loaded gloves (0.35 mm Pb).

Before use, the breathing-air hoods are modified to include a short section of 25 mm (1 in.)-dia polyvinyl chloride (PVC) tubing clamped to a Scott HEPA-filter approved for use in radioactive mists and fumes. The filter is covered with duct tape folded over to provide a pull-tab. If the air-supply to the worker's hood were to be interrupted, the worker would bite the stub of the PVC tubing inside his hood, pull the tape from the HEPA filter, and breathe through the filter while making an emergency exit from the cell. Integrity of the HEPA filter and its installation in the breathing-air hood is pretested with stannic-chloride fumes. Each in-cell worker carries a pair of heavy-duty shears with which to sever his air-supply hose if it becomes entangled.

Breathing-air Supply

Certified-safe breathing air is supplied to in-cell workers from redundant sources. In order of priority and backup the sources are: 1) a large-capacity, two-compressor plant-air system in the EBR-II facility; 2) a standby bank of breathing-air cylinders; and

*TYVEK is a fabric by DuPont Corp., Wilmington, Delaware.

3) a low-capacity breathing-air system in the HFEF Complex. At 30-day intervals, certification of each breathing-air system is reviewed. The operational readiness of each system is confirmed prior to every cell entry.

The breathing-air supply is connected out-of-cell to a NIOSH-approved* breathing-air manifold† that can provide 2.8 L/s (6 cfm) air to each of four workers. From the manifold, breathing air is supplied through a continuous length (about 30 m) of 200 mm-OD, heavy-wall hose which passes through a cell-wall penetration without intermediate connections. A quick-disconnect fitting, protected from contamination by a plastic sleeve, connects the air hose to the worker's hood.

Personnel Dosimetry

Each in-cell worker wears three pairs of thermoluminescent dosimeter (TLD 700) chips--one chip in the pair is unshielded, and the other is shielded with 2 mm aluminum. One TLD pair is included in a standard INEL badge worn, along with a self-reading pocket dosimeter, in the breast pocket of the cotton coveralls. The second TLD pair is taped to the worker's forehead, and the third pair is taped near the back of the worker's thigh. TLD finger rings are worn on the middle finger of both hands. All TLD chips and rings are processed, and the dosimetry data are reported by the DOE Radiological Environmental Services Laboratory at the INEL.

The worker's whole-body exposure to penetrating radiation is based on the highest reading of the three shielded TLDs. Skin exposure is based on the sum of the whole-body penetrating-radiation exposure and the highest nonpenetrating-radiation reading of the three TLD pairs. Nonpenetrating radiation for a given TLD-pair is determined by subtracting the shielded TLD reading from the unshielded TLD reading.

The exposure to extremities is monitored by the TLD finger rings. During the initial cell entries, TLD pairs were also attached to the worker's forearm, ankle, and foot. However, data from these extra TLDs showed that exposures to the extremities were not significantly greater than the whole-body exposure. Hence, the hands were considered the most exposed extremities, and the most conservative ones to monitor for extremity exposures.

During the entry, a digital dosimeter, taped to the front of the worker's lead-loaded apron, is used to monitor his exposure to

*NIOSH is the National Institute of Occupational Safety and Health.
†Mine Safety Appliances Co., Evans City, Pennsylvania.

penetrating-radiation. By radio, the worker is periodically requested to read his digital-dosimeter by a Radiation Safety technician who monitors and logs the accumulated exposure. When a worker's exposure approaches a preestablished control value, the person-in-charge of the entry is advised that the worker should start to exit the cell. In-cell working times are now limited by the workers' whole-body skin exposures. The present 100 mR control value has limited exposures to the skin and extremities safely below 2000 mRem, the ANL-West administrative limit for any four-week period.

Cell Entry Rooms

To control contamination and air flow, temporary rooms were built inside and outside the argon cell, at the window port used for worker's ingress and egress. Periodic cleanup, particularly in the room closest to the cell, has successfully controlled migration of contamination. A plan view of the rooms is shown in Fig. 3.

The large room outside the cell is 9.1 m (30 ft) long by 2.4 m (8 ft) wide. The dry-wall interior surfaces are attached to metal

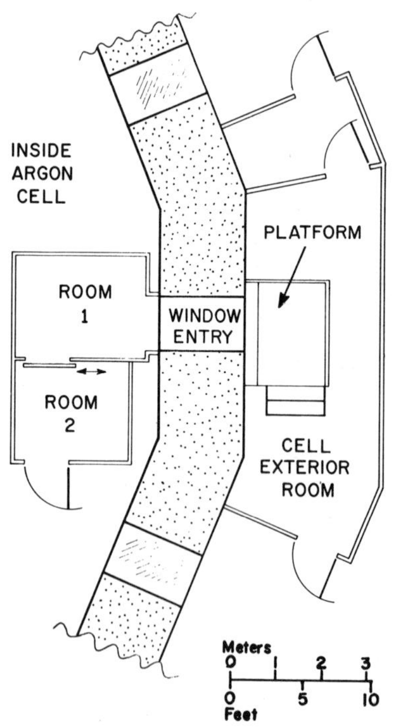

Fig. 3. Argon Cell Entry Rooms

studs. The interior walls, floor, and ceiling are lined with reinforced plastic sheeting for ease of decontamination. A segment of the room is partitioned to provide an isolated area in which to monitor personnel exiting the cell. The remainder of the room is used for final checkout of protective clothing and radio communications.

Inside the cell, at the window aperture, is a 1.8 m (6 ft) by 2.4 m (8 ft) room (Room No. 1) assembled from prefabricated plywood modules. The 20-mm (3/4 in.)-thick plywood provides shielding against nonpenetrating radiation. Plastic booting connects Room No. 1, through the window opening, to the room outside the cell. The second in-cell room, Room No. 2, is adjacent to Room No. 1 and is 2.8 m (6 ft) by 2.1 m (7 ft). Rooms 1 and 2 also are lined with reinforced plastic sheeting.

The argon-cell exhaust system maintains a flow of 280 L/s (600 cfm) through the window opening. A beta-gamma constant-air monitor samples the air just above the opening between Rooms 1 and 2.

To enter the cell, workers crawl through the window opening into Room No. 1, then proceed into Room No. 2. Their breating-air hoods are connected to the in-cell hoses by an assistant standing in Room No. 1. After entering the cell, the workers put on their boots and their lead-loaded aprons and gloves, tape on their digital dosimeters, and begin assigned tasks.

Before exiting the cell, the workers damp-wipe their lead-loaded aprons and gloves and leave them in containers. They also damp-wipe 3 m (10 ft)-lengths of their air hoses, and the exterior of each other's air hoods. The rubber boots and the (outer) wet-suit are left in the cell, and the outermost pair of shoe covers is removed as Room No. 2 is reentered. In Room No. 2, the next layer of anticontamination clothing is removed. As the workers progress from the "dirtier" to cleaner areas, their anticontamination clothing, including the forced-air supply hood, is systematically removed, and radiation surveys are made by Radiation Safety personnel.

CONTACT DECONTAMINATION

Following completion of remote decontamination and radiation surveys in the argon cell, contact decontamination began in September 1978. Because of the size and complexity fo the cell, and the number of in-cell systems that could be damaged by liquids, dry methods, or methods using very limited quantities of liquid, were preferred for the in-cell contact decontamination. Furthermore, ANL-West does not have facilities for processing large volumes of radioactively contaminated liquids.

Vacuuming, and Wet Methods

Initial contact decontamination consisted of sweeping and vacuuming the cell floor and areas inaccessible during the remote decontamination operations (the inner area of the central control room roof, bridge-rail offsets high on the walls, and rails and bridges of the cranes and electromechanical manipulators). To facilitate decontamination of the rails and bridges, a mobile platform was installed on the bridge rails. This platform enabled workers to move easily from one end of the bridges to the other using a hand-powered drive wheel. The crane and manipulator bridges were wiped with water-soaked rags.

The floor of the cell was scrubbed with a water-Radiacwash* mix using hand-operated brushes. Abrasive household cleansers were sometimes applied. Powered floor scrubbers were also used with a Turco foaming agent (#5865). Liquids from these operations were picked up by vacuuming, accumulated in 0.24 m^3 (55 gal.) drums, solidified using Safe-T-Set, and disposed of as solid contaminated waste. Hot-spots on the floor were repeatedly treated, using grease-solvents when appropriate. Such conventional decontamination methods were only moderately effective.

Latex Strippable Coating

To develop a more effective method, applicable to the cell wall and floor areas, a strippable coating was tested. Using an airless spray gun, test areas of the cell floor and walls were sprayed with a Turco water-based latex, strippable coating (#5931). After it cured, the coating was stripped away to remove contamination. Although the test results were encouraging, operational problems discouraged the use of strippable coatings for general decontamination of the cell. It proved very difficult to control over-spray or the thickness of coating applied (important for ease of removal). Stripping the coating from projections on the cell wall and floor proved exceedingly difficult and inefficient.

High-pressure Water Spraying

Because of the problems mentioned above, tests of high-pressure water-spray decontamination methods were begun despite the waste-disposal problem inherent in such operations. These tests showed the high-pressure-spray method to be both effective and efficient; accordingly, it was selected as the method to be used for general in-cell decontamination.

---

*Atomic Products Corp., Center Moriches, New York

To minimize the volume of contaminated liquid generated, tests were made to determine the maximum nozzle-to-wall distance at which the spray would be effective. Based on these tests, a high-pressure water-pump and spray-nozzle system capable of delivering 0.25 L/s (4 gpm) of water at 14 MPa (2000 psi) was selected. To control and confine the water spray, a shroud was installed around the nozzle and has proven very effective. The shroud, illustrated in Fig. 4, also positions the nozzle to produce a spray pattern about 150 mm wide. The spray is moved horizontally with the patterns overlapping about 50%. The rate of coverage is controlled so about 19 L (5 gal.) water is applied per m^2.

The cell windows, master-slave manipulator penetrations, and mineral-insulated electrical cable ends are covered to protect them from water damage. The metal-framed plastic window covers are temporarily sealed to the cell wall. The space between the window and the window cover is slightly pressurized with air to prevent in-leakage of water. Duct tape, plastic bags, and rubber gloves are used to cover the ends of electrical cables.

In regular decontamination operations now in progress, one worker operates the shrouded high-pressure water spray while a second worker vacuums and operates a sump pump to collect

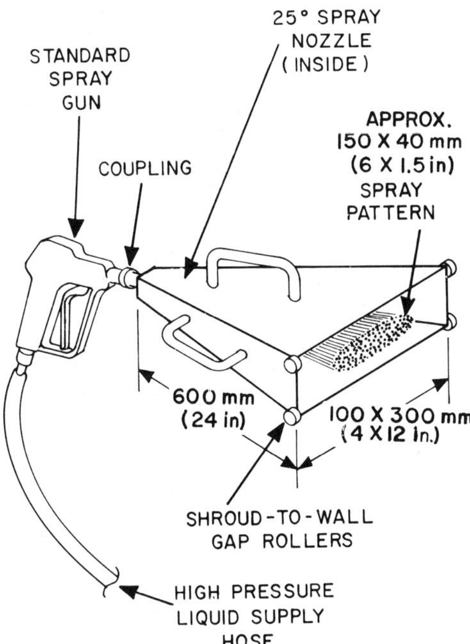

Fig. 4. High-pressure Spray Gun and Shroud

and transfer waste water from the floor to an in-cell drum. Temporary dams on the floor contain, and facilitate collection of the water. Accumulated water is pumped from the drum through a cell-wall penetration, through all-welded piping, to a 3.8 m^3 (1000 gal.)-capacity, trailer-mounted tank parked outside the facility. The tank-truck hauls the contaminated water to the INEL Idaho Chemical Processing Plant for disposal.

ASSESSMENT OF PROGRESS

Thermoluminescent dosimeters (TLDs), unshielded and shielded, together with ionization-type radiation meters, have been used to measure decontamination progress. Standard surface-smear techniques also have been used to determine the level and nature of loose surface activity.

The TLDs have been particularly useful because the results correlate directly with the personnel dosimetry. The shielded TLD (2 mm Al) responds to none of the ^{137}Cs beta radiation, to only 10% of the beta radiation from ^{90}Sr-^{90}Y, and to only the energetic bremsstrahlung from both. Nonpenetrating beta-, X-, and gamma-radiation is measured by the difference between responses of the unshielded and shielded TLDs. The ionization meters also were shielded in all directions with 2 mm aluminum to measure penetrating radiation, and in all directions except one to measure both nonpenetrating and penetrating radiation.

TLD dosimeters were inserted through floor penetrations to monitor progress during the remote decontamination. Before dry vacuuming, the average radiation at four locations about 0.5 m above the floor was 6 R/hr penetrating and 30 R/hr nonpenetrating. Dry vacuuming reduced the penetrating radiation by a factor of seven, but the nonpenetrating radiation was reduced by a factor of only two. Remote wet decontamination of the cell floor removed a substantial quantity of activity, but penetrating radiation near the floor was reduced by a factor of only two, and the nonpenetrating was reduced even less. The penetrating-radiation intensity was estimated to be under 500 mR/hr, and the nonpenetrating radiation was estimated to be under 10 R/hr, making possible a personnel entry to survey the radiation fields.

The survey located numerous hot spots in which penetrating plus nonpenetrating radiation ranged from 1 to 4 R/hr at one foot. The average personnel exposure rate during this first entry (30 min) was 270 mRem/hr penetrating and 2.6 Rem/hr nonpenetrating, giving a skin dose-rate of 2.9 Rem/hr. The floor was remotely washed a second time before further entries. Based on the average exposure received during the following four entries, the penetrating radiation was unchanged, but the nonpenetrating radiation was reduced by a factor of almost two. From these data, it was concluded that significant radiation was coming from areas impractical to decontaminate further by remote means, so contact decontamination was initiated.

The first quadrant of the cell to be washed with the high-pressure spray was monitored with Juno* ionization-type meters (both shielded and unshielded) held in contact with the wall at selected locations. Clean plastic was placed on the wall to prevent contamination of the meters during the contact readings. The ratio of the nonpenetrating radiation readings before and after washing was used to evaluate the washing technique. The nonpenetrating-radiation decontamination factor, averaged over 28 wall areas, was nearly six. The resulting average nonpenetrating-radiation intensity in contact with the walls was 40 mR/hr; the penetrating radiation in the center of the work zone was about 30 mR/hr. Hot spots of nonpenetrating radiation up to 700 mR/hr were found on the walls. The lower wall, up to about one meter from the floor, gave detectably higher-than-average nonpenetrating-radiation readings. Smears from the walls after washing showed a very low level of loose contamination.

Because all activity washed from the walls had to be picked up from the floors, the average nonpenetrating-radiation intensity from the floor was essentially unchanged after the first wash (500 to 700 mR/hr, with hot spots of 1 to 2 R/hr penetrating and nonpenetrating). Also, smears from the floor indicated considerable residual contamination. During the survey made after the first cell quadrant was washed, personnel exposures averaged 320 mRem/hr nonpenetrating and 35 mRem/hr penetrating radiation.

Analysis of the decontamination wash water affords an indirect measurement of process effectiveness. The first 3.8 m^3 (1000 gal.) contained 6 Ci beta-emitting radioactivity, of which 3.4 Ci was ^{137}Cs with associated gamma radiation. The liquid samples contain about 0.01% suspended solids which emit most of the alpha activity (about 640 μCi ^{239}Pu in the first 1000 gal.).

Diminishing in-cell radiation intensity, as indicated by the dosimeters, confirm the progress of the decontamination process. The general trend downward is shown in Fig. 5. Although a given one-month average may include work in cell areas differing markedly in radiation levels, individual exposure rates usually are within a factor of two of the average. The ratio of nonpenetrating-to-penetrating radiation exposure has ranged from four to seven with

*Juno meters were manufactured by Technical Associates, Canoga Park, California.

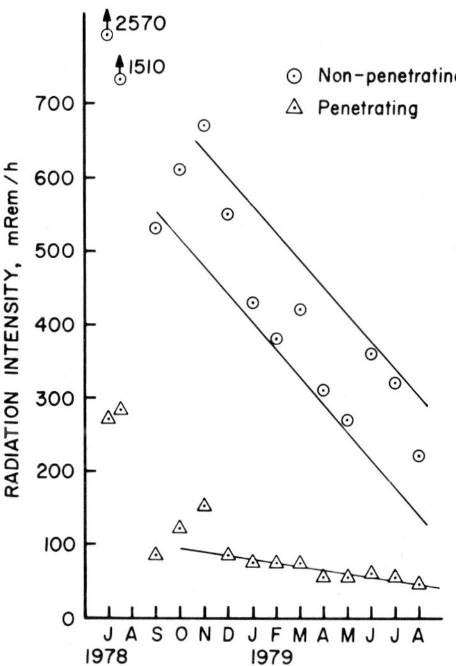

Fig. 5. Radiation Intensities During Contact Decontamination

an average of 5.5. At any ratio larger than three, the nonpenetrating exposure controls the allowable in-cell working time for multiple entries.

After completion of the high-pressure water-spray decontamination now in process, in-cell radiation will be surveyed to determine 1) the need for water spray-down of the cell roof (not yet sprayed because data regarding its contribution to overall radiation levels is inconclusive); and 2) the need for a second (and, possibly, additional) complete spray-down of the cell interior. In-cell radiation levels must be further reduced by a factor of 10 or more to meet ANL-West exposure-control guides (1000 mRem penetrating radiation, and 4000 mRem penetrating plus nonpenetrating, per quarter-year). Hopefully, in-cell radiation intensity can be reduced to 6 mR/hr or less penetrating, and 20 mR/hr or less nonpenetrating plus penetrating.

ACKNOWLEDGMENTS

The authors acknowledge the important contributions to the project of the HFEF Refurbishment Group, especially P. Wayne and R. A. Kifer; the Radiation Safety personnel, including F. R. Hunt and R. S. Peterson; and R. Villarreal of the ANL-West Division.

The authors further acknowledge contributions by M. F. Adam and B. W. Haff of the HFEF organization, and by D. K. Jackson of the ANL-West Division, during preparation of this paper. Our special thanks go to Carol Anderson who typed the manuscript.

REFERENCES

1. J. C. Hesson, M. J. Feldman, and L. Burris, "Description and Proposed Operation of the Fuel Cycle Facility for the Second Experimental Breeder Reactor (EBR-II)", ANL-6605, April 1963.
2. L. Larsen and R. L. Brookshier, "Decontamination and Refurbishment of the Hot Fuel Examination Facility/South Air Cell", Proc. 25th Conf. on Remote Syst. Technol., 1977.
3. J. C. Courtney and V. N. Thelen, "A Plutonium Safety Training Program", Health Physics, Vol. 35, December 1978.

A DECOMMISSIONING PLAN FOR PARTICLE ACCELERATORS

J. H. Opelka, M. J. Kikta, G. J. Marmer,
R. L. Mundis, J. M. Peterson, B. Siskind

Argonne National Laboratory
9700 South Cass Avenue
Argonne, Illinois 60439

Nuclear facility decommissioning and waste disposal has been addressed by the Comptroller General in a June 2, 1977, report to the Congress entitled "Cleaning-Up the Remains of Nuclear Facilities--a Multi-billion Dollar Problem".[1] The primary thrust of the report was towards the nuclear power industry; however, other aspects of nuclear work, including isotope usage and particle accelerator facilities, were recognized as sources of potential problems. In order to assess the magnitude of the problem presented by particle accelerator decommissioning, several tasks have been accomplished.

Task 1. To compile a census of the United States particle accelerators that have a likelihood of involving radioactive material at the time of decommissioning. There are perhaps as many as 1,200 accelerators in the United States, ranging in size from the very small Cockroft-Walton and electron linear accelerators to the multi-GeV research synchrotrons. At least fifty accelerators produce significant induced activation, and several hundred more are capable of producing fluxes of neutrons which could result in non-negligible activation of various components of the accelerator. The census is not presented in this paper because of space limitation.

Task 2. To review the history of past particle accelerator decommissionings with regard to technological, environmental, health and economic aspects. The decommissionings of four AEC-funded accelerators were examined: the synchrocyclotrons operated by the University of Rochester[2] and Carnegie-Mellon University,[3] the Cambridge Electron Accelerator,[4] and the Yale Heavy-Ion Linear Accelerator.[5] Some information was obtained via personal communication about the decommissioning of the Brookhaven Cosmotron.[6] Components ranging from electronics to shielding to magnet frames and coils, even those exhibiting induced radioactivity, were generally used again at another experimental laboratory rather than disposed of, because of the expense of obtaining new steel and copper. Other radioactive components were shipped to commercial burial grounds for disposal. Details of these five past decommissionings are not reported in this paper due to space limitation.

As an alternative to dismantlement, some accelerators, especially smaller machines such as particle injectors, have been transferred in a relatively intact form to other accerator sites. For example, the 50-MeV proton linac injector for the Alternate Gradient Synchrotron was moved from Brookhaven National Laboratory to Lawrence Berkeley Laboratory for use as an improved injector to the Bevatron; the 2.2-GeV Cornell Electron Synchrotron was sent to Argonne National Laboratory for use as a proton booster; the 450-MeV University of Chicago synchrocyclotron was shipped to Fermi National Accelerator Laboratory, where it has served as a particle spectrometer in the muon laboratory; and the 3-GeV Princeton-Pennsylvania Accelerator was shipped to Argonne National Laboratory.

The shutdown of an accelerator, especially if not anticipated, often results in a termination of technical and professional staff positions. For example, in the case of the Cambridge Electron Accelerator, 83 staff positions were terminated.

In general, the regulation of accelerator decommissionings is a function of state, not federal, governments. An important exception to this is for those accelerators located on federal lands such as national laboratories. At past decommissionings, the involvement of state and federal regulatory agencies was financial rather than regulatory in scope. A review of state policies with regard to accelerator decommissioning revealed that the states have not carefully considered this specific area.

Task 3. To develop a generalized plan for decommissioning a particle accelerator. As part of this generalized plan, the cost for dismantling, the radiation dose to dismantling workers and the radioactive waste generated can be estimated. The alternative methods of decommissioning, as they relate to an accelerator, are:

(a) Dismantle and remove the entire accelerator and its building, followed by waste disposal, temporary storage, or immediate recycle of its parts.

(b) Remove the accelerator only to a waste burial site, temporary storage or for immediate recycle, leaving the building for research or office space.

(c) Mothball or entomb the accelerator in place with performance of (a) or (b) above, possibly at a future time.

Since accelerators, especially the largest ones, are unique in design and application, it is not possible to develop a precise cost prediction, a nuclear waste management plan, and a radiation dose assessment associated with each one. A generalized planning guide was developed and applied specifically to four prototypic accelerators located at Argonne National Laboratory,[7-9] with suggestions for extension of the prototypic results to a general case. The four prototypic accelerators are the Zero Gradient Synchrotron (ZGS), the 60" fixed frequency cyclotron (built by Cyclotron Corporation), the 22-MeV electron linac (built by ARCO), and a 9-MV Tandem Van de Graaff (pressure tank by High Voltage Engineering, accelerating system by National Electrostatics Corporation). The ZGS was chosen as the prototype for large proton synchrotrons, since it will be the next major accelerator to be decommissioned (October, 1979). The 60" cyclotron is an average sized cyclotron, although it is perhaps small for planning a synchrocyclotron dismantling. The electron linac was chosen as typical of the wave guide components of a linac and also to emphasize the different radiation environment at an electron or ion accelerator. The Van de Graaff was chosen because electrostatic accelerators form the largest fraction of operating research accelerators, and also because a Van de Graaff has a massive pressure tank despite generally low beam energy (10's of MeV). The choice of the Van de Graaff also serves to emphasize the difference in radiation environment at high and low energy ion accelerators.

The cost analysis presented in this paper is limited to the dismantling and radioactive waste disposal of the entire accelerator. All costs are based upon an engineering evaluation of the dismantling procedures. The level of health physics effort during dismantling assumes radiation levels that would exist immediately after final shutdown. The cost of dismantling allows for conveyance of all components out the facility door and onto railroad cars or trucks or into a temporary storage building. Disposal activities are the packaging, transporting, and burial of the radioactive wastes generated. Costs associated with the disposal of nonradioactive materials are not considered. The nonradioactive wastes could be disposed of as ordinary trash in a local landfill. The options of entombment, mothballing, and interim storage were examined but are not discussed in this paper. All costs presented are in $1978.

The estimated dismantling costs have been divided into activity- and period-dependent costs. Activity costs are those costs incurred in performing specific dismantling and removal activities. Examples of activities are unbolting magnet segments and scarfing concrete. Period costs are those costs which occur throughout a portion of the dismantling period. Examples of period costs are health physics and administrative management.

All shipments of radioactive wastes to a burial site are assumed to be by truck from Chicago, Illinois to Richland, Washington, a one way distance of approximately 2000 miles. For this study, it is assumed that all shipments will be of legal weight except for those involving very heavy objects such as sections of magnets. The contaminated material from the accelerators that requires packaging would be packaged in 1.2-m x 1.2-m x 2.13-m fiberglassed wooden boxes and the tools, protective clothing, and rags that have become contaminated during the decommissioning of the accelerator would be packaged in 55-gallon drums.

In some instances, the dismantling of the experimental area could be a more complex task than the accelerator itself. However, the dominant task will generally be the dismantlement of the accelerator proper. In addition to overall size and radiological concerns, two factors principally determine the magnitude, complexity and cost of the dismantling. First, the method of treatment of the radioactive portions of the permanent shielding concrete will greatly influence the cost estimate. The decision among the alternatives for the shielding depends upon the potential reuse of the building. Second, the cost of dismantling is generally sensitive to the method of removal of the magnetic or electric components proper. The cutting of the magnets or other custom-made components limits their reuse potential and should be avoided unless disposal (or bulk reuse as shielding or counterweight) is the only option available. If cutting is considered, the optimal cut sizes are such that the resulting sections are the largest manageable under the decommissioning plan.

The estimated period- and activity-dependent costs for the ZGS dismantling are shown in detail in Tables 1 and 2. Only the large accelerator complexes such as ZGS will require extensive planning and close-out periods. The entire process for the ZGS will cover a period of approximately 30 months. From an examination of the engineering evaluation of the dismantling procedures, it was determined that the activity-dependent cost for dismantling the ZGS can be approximated by calculating the total mass of the ZGS accelerator multiplied by 9 man-hours of labor per ton multiplied by $25 per man-hour, plus the cost of concrete decontamination or destruction. For ZGS, the activity costs associated with shielding concrete are small unless the ring tunnel building has to be demolished. The cost of dismantling other proton synchrotrons can be approximated in this fashion.

In the case of cyclotrons and synchrocyclotrons, attention must be given to the method of construction of the large magnet structure. If it were site-assembled by lamination of heavy plates, this would allow separation of the plates for disassembly and handling. If the magnet were constructed of large forgings, it might be necessary to cut the magnet into smaller pieces for removal.

Cyclotrons are generally old machines and have little value for reuse or salvage. Thus cataloging and other period-dependent costs, except for health physics, should be kept to a minimum. Residual activity will require that they be treated as radioactive assemblies. Health physics monitoring and documentation will comprise the major period-dependent cost.

Table 1

Estimated Period Dependent Costs for ZGS
(in chronological order of assignment)

Position	Pay Scale $/mo[a]	No.	Duration Months	Cost
General Manager	7,500	1	24	$180,000
Administrative Manager	6,500	1	30	195,000
Secretary	2,850	1 + 1[b]	30 + 24[b]	153,900
Property Manager	5,250	1	18	94,500
Clerk	2,500	1 + 3	18 + 12	135,000
Health Physics Coordinator	4,750	1	24	114,000
Health Physics Technician	3,500	2 + 4	12 + 6	168,000
Mechanical Equipment Manager	7,000	1	16	112,000
Electrical Equipment Manager	7,000	1	16	112,000
Facility Services Manager	7,000	1	18	126,000
Site Security Guards	3,500	4	9	126,000
Safety Engineer	5,250	1	9	47,250
Inspector	5,250	1	12	63,000
Junior Engineer	5,000	9	6	270,000
Engineer Associate	3,500	8	6	168,000
Average	4,478		461	$2,064,650[c]

[a]Includes general and administrative overhead and contract fees, at about 1.85 x direct salary.

[b]This notation indicates one secretary for 30 months and one for 24 months.

[c]The number of significant figures shown is for computational accuracy and does not imply precision to the nearest one hundred dollars.

Table 2

Estimated Activity Dependent Item Cost Summary for ZGS Dismantling (in $1,000's)

Item Description	Contract Costs	Supply Costs	Activities Performed by Outside Contractor	Total Activity Dependent Costs
Ring				
Magnet Steel	--	27.0	598.0	625.0
Magnet Coils	5.0	20.0	19.7	44.7
Vacuum Chamber	--	10.0	5.8	15.8
Ring Supports	--	1.0	13.5	14.5
Operating Floor	--	5.0	90.4	95.4
Cabling	1.0	--	15.7	16.7
Accessory Equipment	2.0	2.1	44.8	48.9
Straight Sections	--	1.0	31.0	32.0
Decontaminate Building	7.5	--	17.2	24.7
Ring Subtotal[a]	15.5	66.1	836.1	917.7
Injector-Linac				
High Voltage Terminal	--	--	2.0	2.0
High Voltage Power Supply	--	--	1.3	1.3
High Voltage Terminal Room	--	0.5	8.0	8.5
Beam Line-Source to Linac	--	0.2	2.3	2.5
Linac	--	1.0	31.5	32.5
Linac Accessories	--	--	34.2	34.2
Beam Line-Linac to Ring	--	--	9.8	9.8
Remove Remaining Equipment	--	--	4.5	4.5
Decontaminate and Final Clean-up	5.0	--	6.0	11.0
Injector-Linac Subtotal[a]	5.0	1.7	99.6	106.3
Center Building				
Remove All Equipment	--	--	355.0	355.0
Center Building Subtotal[a]	--	--	355.0	355.0
Control Room				
Equipment and Cabling	--	--	90.0	90.0
Clean-up and Repairs	5.0	--	14.0	19.0
Control Room Subtotal[a]	5.0	--	104.0	109.0
Power Building				
Motor Generator	--	--	71.3	71.3
Clear Generator Control Room	--	--	4.0	4.0
Exciter Room	--	--	1.4	1.4
Cabling	--	--	2.6	2.6
Switch Gear Room	--	--	10.1	10.1
Mechanical Equipment	--	--	4.0	4.0
Minor Building Decontamination	--	--	26.7	26.7
Power Building Subtotal[a]	--	--	120.1	120.1
ZGS Total[a]	25.5	67.8	1,514.8	1,608.1

[a]The number of significant figures shown is for computational accuracy and does not imply precision to the nearest one hundred dollars.

The cost estimate for dismantling the 60" cyclotron located at Argonne National Laboratory (ANL) has been based upon the assumption that the magnet components could be moved by riggers through the dee door and outer wall door to grade level outside the building for loading onto a heavy-duty, low-bed truck trailer. (This is not a physical possibility at ANL since the building was constructed around the massive cyclotron magnet with no portals for its exit. This is not an uncommon occurrence at research cyclotron facilities.)

The period-dependent cost for dismantling the 60" cyclotron is estimated to be \$160,000. In sharp contrast to the ZGS, the entire project will be of only about four months duration. In Table 3, an itemization of activity dependent costs is provided. As was the case in the ZGS large ring prototype, the cost of dismantling the 60" cyclotron can be approximated as the total mass multiplied by 9 man-hours per ton multiplied by \$25 per man-hour, plus the cost of concrete shielding removal. This same approximation could be applied to any cyclotron not requiring significant preparation of the magnet for rigging.

The 22-MeV electron linac at ANL consists of easily dismantled modules. Some of these modules such as the klystrons and vacuum pumps have reuse potential. The water-cooling modules can be retained intact for possible use in experimental set-ups (if storage space is available). The modulator, amplifier power supply, and auxiliary power-supply modules are considered surplus unless an immediate use is apparent. Activity-dependent costs are estimated for the dismantlement and removal of the linac and the removal of the modules and control console. Removal of electrical substation components is not included.

The dismantling of the electron linac would require only two months, at an estimated period-dependent cost of \$26,000. For low energy linacs, such as the one at ANL, there are essentially no health physics staff requirements, although 0.1% of the total mass of such a linac is estimated to be radioactive. Work would be performed almost entirely by an outside contractor. The activity-dependent costs are shown in Table 4. Since the modules of most linacs are not heavy, a cost of 16 man-hours per ton multiplied by \$25 per man-hour multiplied by the total tonnage of the linac facility gives a good approximation of the removal costs of this type of accelerator. In general, no work will be required on the concrete walls and floor surrounding smaller linacs.

Salvage value of the Tandem Van de Graaff at ANL is no more than that of scrap unless the machine can be used in its entirety elsewhere. However, the negative ion sources will have many components useful elsewhere in the laboratory and would be returned to stock. An allowance of five months of technician labor has been made as a period-dependent cost for in-house dismantling of

Table 3

Estimated Activity Dependent Item Cost Summary for 60" Cyclotron Dismantling (in $1,000's)

Item Description	Contract Costs	Supply Costs	Activities Performed by Outside Contractor	Total Activity Dependent Costs
Cyclotron				
Vacuum Tank, Dees	--	--	5.6	5.6
Coils and Tanks	--	--	5.8	5.8
Magnet	--	--	33.1	33.1
Cyclotron Subtotal[a]	--	--	44.5	44.5
Shielding				
Demolish Concrete Vault	--	--	557.8	557.8
Shielding Subtotal	--	--	557.8	557.8
Mechanical Equipment Room				
Mechanical Equipment	--	--	14.6	14.6
Mechanical Equipment Room Subtotal	--	--	14.6	14.6
Electrical Equipment Room				
Electrical Equipment	--	--	5.8	5.8
Electrical Equipment Room Subtotal	--	--	5.8	5.8
Dee Storage and Hot Lab				
Dee Storage	2.0	--	1.2	3.2
Hot Lab	--	--	--	--
Dee Storage and Hot Lab Subtotal	2.0	--	1.2	3.2
Control Room				
Equipment and Cabling	--	--	2.9	2.9
Control Room Subtotal	--	--	2.9	2.9
60" Cyclotron Total[a]	2.0	--	626.8	628.8
60" Cyclotron (excluding[a] shielding)	2.0	--	69.0	67.0

[a]The number of significant figures shown is for computational accuracy and does not imply precision to the nearest one hundred dollars.

these ion sources. There will be only a small contribution of effort for health physics monitoring and contractor services procurement. About 0.5% of the component mass of the Tandem Van de Graaff is estimated to be radioactive.

The dismantlement of the Tandem Van de Graaff can be completed in a three month period for an approximate period-dependent cost of $69,000. In Table 5 the activity-dependent costs associated with the dismantlement of the ANL Tandem Van de Graaff are presented. Since the Tandem Van de Graaff tank comprises the bulk of the mass of a Van de Graaff and since it can be classified as light structural steel, the cost of dismantling a Van de Graaff can be approxi-

Table 4

Estimated Activity Dependent Item Cost
Summary for Dismantling a 22-MeV Electron Linac
(in $1,000's)

Item Description	Contract Costs	Supply Costs	Activities Performed by Outside Contractor	Total Activity Dependent Costs
Linac Assembly	--	--	0.8	0.8
R. F. System	10.0	--	6.6	16.6
Cooling System	7.5	--	3.8	11.3
Control Console	5.0	--	0.9	5.9
22-MeV Electron Linac Total[a]	22.5	--	12.1	34.6

[a]The number of significant figures shown is for computational accuracy and does not imply precision to the nearest one hundred dollars.

Table 5

Estimated Activity Dependent Costs
for Dismantling a 9-MV Tandem Van de Graaff
(in $1,000's)

Item Description	Contract Costs	Supply Costs	Activities Performed by Outside Contractor	Total Activity Dependent Costs
Tandem Van de Graaff	10.0	--	20.1	30.1
Ion Source	--	--	--	--
Controls and Accessories	5.0	--	2.3	7.3
TOTAL[a]	15.0	--	22.4	37.4

[a]The number of significant figures shown is for computational accuracy and does not imply precision to the nearest one hundred dollars.

mated by taking the total mass multiplied by 16 man-hours per ton multiplied by $25 per man-hour. There are no major costs associated with shielding concrete removal or cleanup, since after shutdown the radiation levels outside the Van de Graaff tank are not considered problematic. For other types of electrostatic accelerators, also of light structural steel, the cost of dismantling can be approximated similarly.

Table 6 summarizes the comparison of estimated costs associated with dismantling and disposal of three reference fuel cycle facilities and the four prototypic accelerators located at ANL. The three reference fuel cycle facilities used in the comparison are a fuel reprocessing plant (FRP),[10] a pressurized water reactor (PWR)[11] and a mixed oxide fuel fabrication facility (MOX).[12] The decommissioning of an accelerator would be a much simpler task than either a fuel reprocessing plant or a power reactor, since the amount of radioactive contamination is much lower. The radioactive material at an accelerator is limited to that in the accelerator itself and that in the shielding. Because there are no radioactive liquids processed, as in a fuel reprocessing plant, there will be no piping which requires decontamination. The accelerator also has essentially no surface contamination of any of its components, as would the three fuel cycle facilities. Generally, the radioactivity within the accelerator components has been caused by activation. This is similar to the pressure vessel in a PWR, except that accelerators (with one exception - see Task 4 below) will have much lower levels of activation and no long-lived isotopes.

The cost of dismantling the ZGS is much lower than that for an FRP and PWR since much of the ZGS can be simply unbolted and dismantled. The cost of decommissioning the ZGS is comparable to that of dismantling a small mixed oxide fuel fabrication plant. This is merely a coincidence and is not due to any similarities between these two types of facilities.

<u>Task 4.</u> To survey the quantity of radioactivated material and the radiological health problems at operating particle accelerator facilities with emphasis on the highest energy machines. For additional information, the reader is referred to two previous papers.[13,14] In most cases, the induced radioactivity is confined to relatively few parts of the machine and the disposition of these is through the normal radioactive waste channels without complication. However, some machines have low-level induced radioactivity in massive components. Estimates of radioactive material at proton and electron machines can, in general, be estimated by using the techniques of Gollon[15] and Swanson,[16] respectively. Any accelerator with particle energy greater than 10-MeV will produce induced activity. This indicates that as the energy of accelerators used for medical and industrial application continues to be increased above the 10-MeV level, the problem of induced activity will need to be addressed as part of the eventual decommissioning of these machines.

Health physics aspects of the decommissioning plan vary widely with different generic accelerator categories. For electrostatic positive ion accelerators (van de Graaffs, Cockroft Waltons, dyna-

Table 6

Estimated Costs ($1978) of Immediate Dismantlement and Disposal of Four Prototypic Accelerators and Three Reference Fuel Cycle Facilities

Facility	Dismantling	Packaging	Transportation	Disposal	Total[1]
ZGS	3.7×10^6	4.1×10^4	1.0×10^6	2.9×10^5	6.3×10^6
60" cyclotron	7.9×10^5	3.6×10^4	1.8×10^5	6.2×10^4	1.3×10^6
35 MeV electron linac	6.1×10^4	5.0×10^2	2.1×10^3	7.3×10^2	8.0×10^4
Tandem Van de Graaff	1.1×10^5	5.0×10^2	2.1×10^3	7.3×10^2	1.4×10^5
Fuel Reprocessing Plant[2]	2.3×10^7	3.2×10^6	6.6×10^6	1.4×10^7	5.8×10^7
Pressurized Water Reactor[3]	2.3×10^7	3.7×10^6	5.2×10^6	2.2×10^6	4.2×10^7
Mixed Oxide Fuel Fabrication Facility[4]	5.6×10^6	7.4×10^4	1.1×10^5	3.9×10^5	7.7×10^6

[1]Sum of dismantling, packaging, transportation, and disposal, including a 25% contingency factor on all activities.

[2]In $1975, Reference 10, Tables 7.8-13, 7.8-14, and 7.8-24.

[3]Reference 11, Tables 10.1-1 and 10.1-4.

[4]Reference 12, Tables 10.1-1 and 10.1-6.

mitrons, pelletrons, and insulated core transformers) and small electron accelerators, health physics activities would generally be limited to radiation surveys of accelerator and beam line components for potential induced activity and segregation of activated items for proper disposal or storage to await decay. For cyclotrons, synchrocyclotrons, and synchrotrons, the levels of induced activity in certain components will be high initially and a delay to allow for radioactive decay will generally be advisable. Advance planning of the work sequence to minimize radiation exposure will be necessary. Maximum usage should be made of any available remote handling capabilities for removal of target probe, beam extraction components, and other highly activated materials. The remainder of the dismantlement could probably proceed using hands-on operation. For large electron accelerators (linacs and synchrotrons), health physics problems will be comparable to the situation at cyclotrons, except at target stations, beam switch yards, and beam dumps the requirements would be essentially identical to the synchrotrons and synchrocyclotrons. Building decontamination or selective demolition may be needed to remove areas of induced activity in the vicinity of the points of intense proton or electron beam interactions.

For the highest energy operating proton linac, the 800-MeV Los Alamos Meson Physics Facility (LAMPF), the levels of induced activity in the primary target areas are greater than 1000 R/hr. Remote dismantlement techniques will be necessary unless a very long delay time is allowed prior to the start of decommissioning. LAMPF is the only accelerator in the United States which presents radiation levels in major components of the same order of magnitude as those found in a reactor pressure vessel. Mothballing or entombing the target areas for a long period of time might be a desirable decommissioning option.

It is interesting to compare the total dose commitment estimated for the decommissioning of the ZGS facility to the estimates developed for reference FRP, PWR and MOX facilities. Table 7 shows this comparison. The values for the ZGS facility were calculated using the Sullivan and Overton decay curve[17] applied to the initial dose estimates. It should be noted that the FRP and PWR estimates are based on the use of remote-handling techniques while the ZGS work is entirely hands-on. The radiation doses associated with each dismantling activity for each prototypic accelerator were estimated in the same fashion as in Table 2 for cost. The details are omitted in this paper because of space limitations. The total radiation dose for the cyclotron, linac, and Tandem Van de Graaff dismantlements should not exceed 66 person-rem, 0.003 person-rem, and 0.01 person-rem, respectively. As expected, 80% of the person-rem involved in the ZGS dismantling occurs in the dismantling of the ring straight sections and pole pieces. This intensive dismantling effort occurs in the highest radiation environment. About 80% of the person-rem of the cyclotron dismantlement is associated with

Table 7

Estimated Radiation Dose Commitment Comparison
Accelerator and Fuel Cycle Facilities

	Occupational Man-Rem Dose Commitment		
	Immediate Dismantlement	10 Year Delay	30 Year Delay
ZGS	95-320	11-37	1-5
FRP[a]	532	445-453	312-333
PWR[b]	1300	770[d]	474[d]
MOX[c]	76	165	307

[a]Reference 10, Table 2.6-1.

[b]Reference 11, Table 2.10-1.

[c]Reference 12, Table 2.9-1.

[d]These estimates do not include doses to transportation workers.

---

the removal of the dees and vacuum tank, despite the fact that 90% of the hours spent are involved in concrete dismantlement at very low radiation levels.

In Table 8, the amounts of radioactive mass associated with the dismantlement of various generic types of accelerators are presented. The largest accelerators would produce almost 10^8 kg of activated material. By comparison, only 3 x 10^7 kg of radioactive waste is generated each year by nonfuel cycle activities.[18] Detailed mass estimates of the specific radioactive and nonradioactive components at the four prototypic accelerators were determined but are not presented in this paper. While small electrostatic accelerators in medicine and industry will generate only a small amount of activated material, there are currently about 1000 in operation.

Recommendations. Planning for the decommissioning of particle accelerators should begin during the design and construction phase. Problems such as inaccessible radioactive components could be avoided. Material selection and location of components can be chosen to minimize residual radioactivity. The commitment of scarce resources to an accelerator's radiation environment should be reviewed.

Entombment will not generally be a realistic option for decommissioning of particle accelerators, as the radiation levels are not

Table 8

Expected Radioactive Mass (Kg) From Immediate Dismantlement and Disposal of Four Prototypic Accelerators and Three Reference Fuel Cycle Facilities

Facility	Expected Average Mass Above Background	Radioactive Mass Between 2.5 mr/hr and 1 R/hr @ 1m	Radioactive Mass greater than 1 R/hr @ 1m
Electrostatic, small cyclotrons (< 50 cm. diam.)	1.5×10^2	1.5×10^2	0
Cyclotrons (50 cm. to 130 cm. diam.)	2.7×10^4	2.3×10^3	0
Cyclotrons (130 cm. to 250 cm. diam.)	2.9×10^5	4.4×10^4	8.2×10^1
Cyclotrons (> 250 cm. diam.)	2.6×10^6	1.2×10^5	2.0×10^4
Proton Synchrotrons (18 m. to 60 m. diam.)	1.4×10^7	9.5×10^6	2.3×10^4
Proton Synchrotrons (> 60 m. diam.), proton linac	9.5×10^7	2.0×10^7	3.3×10^6
Reference Fuel Reprocessing Plant[a,d]	5.4×10^6	---	---
Reference Pressurized Water Reactor[b,d]	1.9×10^7	---	---
Reference Mixed Oxide Fuel Fabrication Plant[c,d]	4.7×10^5	---	---

[a]Reference 10, Tables E.6-1 through E.6-5.

[b]Reference 11, Tables G.4-3 throught G.4-6.

[c]Reference 12, Table H.2-6.

[d]Total radioactive mass shown for comparative purposes.

generally high enough to justify entombment. Furthermore, surface contamination is generally limited and readily decontaminated.

The possibility of placing the decommissioning of accelerators under the federal government should be considered, as this would guarantee a consistent, uniform decommissioning policy. The number of accelerators to be decommissioned will increase, especially in the medical treatment and isotope production areas. Even though the decommissioning of an accelerator does not pose as serious a problem as other nuclear facilities, accelerators should be considered as an integral part of the entire nuclear decommissioning picture.

Federal de minimus standards for induced activity should be developed. Present practice is to send unclaimed accelerator components having low-level, short half-life activity to above ground storage areas or to radioactive waste burial grounds for disposal, which is a tremendous waste of natural resources.

References

1. Comptroller General Report to the Congress, "Cleaning Up the Remains of Nuclear Facilities--A Multi-Billion Dollar Problem," EMD-77-46 (June 16, 1977).
2. AEC contract file numbers AT (30-1)-975 and AT (30-1)-4246.
3. AEC contract file numbers AT (11-1)-3066 and AT (30-2)-72.
4. AEC contract file number AT (11-1)-3063.
5. AEC contract file number AT (11-1)-3076.
6. M. J. Kikta, "Accelerator Decommissioning Trip Report; June 21-June 23, 1978," Memo to file at ANL-EIS (28 June).
7. W. M. Brobeck and Associates, "Engineering/Cost Analysis of Particle Accelerator Decommissioning," Report No. 4500-39-20-R1 (Nov. 15, 1978).
8. W. M. Brobeck and Associates, "Cost Estimate for Decontamination and Decommissioning the Zero Gradient Synchrotron, Report No. 4500-39-20-R2 (Nov. 15, 1978).
9. W. M. Brobeck and Associates, "Engineering/Cost Analysis General Methodology and Applications to Decommissioning Certain ANL Accelerators," Report No. 4500-39-20-R3 (Nov. 15, 1978).
10. Battelle Pacific Northwest Laboratory, "Technology, Safety and Costs of Decommissioning a Reference Nuclear Fuel Reprocessing Plant, NUREG-0278 (Oct. 1977).
11. Battelle Pacific Northwest Laboratory, "Technology, Safety and Cost of Decommissioning a Reference Pressurized Water Reactor Power Station, NUREG/CR-0130 (June, 1978).
12. Battelle Pacific Northwest Laboratory, "Technology, Safety and Costs of Decommissioning a Reference Small Mixed Oxide Fuel Fabrication Plant," NUREG/CR-0129 (Feb., 1979).
13. R. L. Mundis, M. J. Kikta, G. J. Marmer, J. H. Opelka, J. M. Peterson, B. Siskind, "Health Physics Considerations in Accelerator Decommissioning and Disposal," in Low Level Radioactive Waste Management, EPA 520/3-79-002 (May, 1979).
14. J. M. Peterson, J. H. Opelka, R. L. Mundis, "Waste Management Considerations in Particle Accelerator Decommissioning," Transactions of the American Nuclear Society 25th Annual Meeting, Volume 32, TANSAO 32 1-832 (1979).
15. P. J. Gollon, "Production of Radioactivity by Particle Accelerators," _IEEE Trans. of Nuc. Sci._, Vol. NS-23, No. 4 (Aug., 1976).
16. W. P. Swanson, "Radiological Safety Aspects of the Operation of Electron Linear Accelerators, Ch. 2.6, IAEA, Vienna (1979).
17. A. H. Sullivan and T. R. Overton, "Time Variation of the Dose Rate for Radioactivity Induced in High Energy Particle Accelerators," _Health Physics_ 11: 1101 (1965).
18. Appendix K, "Report of Task Force for Review of Nuclear Waste Management (Draft)," DOE/ER-0004/D (Feb., 1978).

VOLUME REDUCTION OF RADIOACTIVE WASTE RESULTING FROM DECONTAMINATION OF SURPLUS NUCLEAR FACILITIES

Rich F. Vance, John N. McFee and John W. McConnell

Energy Incorporated

Idaho Falls, Idaho USA

INTRODUCTION

Numerous surplus nuclear facilities have been ear-marked for decontamination and decommissioning. The decontamination of these facilities will produce significant quantities of radioactively contaminated waste requiring disposal. These wastes will include wet ion exchange resins, liquid wastes, and dry combustible materials. A fluidized bed concept for calcining liquid wastes and incinerating combustible wastes in the same process vessel is being developed by Energy Incorporated in Idaho Falls, Idaho, in a partnership with Newport News Industrial Corporation of Newport News, Virginia. A licensing topical report has been filed with the NRC(1) and an application for patent has been submitted. The initial benefit derived from such a waste volume reduction system is a reduced space commitment to waste storage at the affected waste management facility. The expense of space at such a facility is increasing because the storage techniques are being upgraded to allow for ease of retrieval. Subsequent benefits of reduced waste volumes will include reduced retrieval costs, reduced shipping costs, and a reduced impact at the final permanent storage site.

Although a large number of nuclear facilities are scheduled for decontamination and decommissioning (D&D), they are at relatively few locations with a number of facilities being located in close proximity to one another. Examples include facilities at Hanford, Oak Ridge National Laboratory, the Idaho National Engineering Laboratory, and Los Alamos Scientific Laboratories. The installation of a single radioactive waste drying and volume reduction system could support all D&D projects at one site. The D&D projects are being scheduled for accomplishment over a period of several years. Thus, a waste volume reduction system would be used regularly to support a number of projects over a prolonged period. Further, when not dedicated to a D&D project such a system could be used to support continuing operations at the laboratory.

PROCESS CAPABILITIES

The system has been demonstrated as capable of calcining concentrated liquid chemical wastes and incinerating aqueous slurries of spent ion exchange resin beads and also dry combustible wastes. The resulting anhydrous solids are readily adaptable to encapsulation by solidification.

Tests have demonstrated that the product can be solidified with cement, bitumen, and polymer binding agents.[2] Vitrification with sealing glass has been found to be effective on the calcine product of boric acid solutions.

The original calcining design intent was to process the evaporator bottoms from operating BWR and PWR reactor plants. Therefore, the system has been extensively tested for the sodium sulfate and boric acid wastes typical of these reactor systems. Sodium sulfate solutions of 25 wt.% experience a volume reduction of about 4 to 1. Boric acid solutions of 12 wt.% experience a volume reduction of about 8 to 1. Although exhaustive testing on the multitude of available decontamination agents has not yet been completed, the volume reduction of a number of miscellaneous compounds has been successfully demonstrated in the EI laboratory. For the standard design developed for the utilities, the system processes aqueous solutions at a net rate of 132 l/h.

Resin and filter sludge slurries of 60 to 70 wt.% water are accepted by the system at a rate of about 68 kg/h. Although the degree of volume reduction is a function of resin loading, 18 to 1 is typical for spent, unregenerated resins. Filter sludges, which contain a significant quantity of non-combustible material such as diatomaceous earth, undergo a volume reduction on the order of 5 to 1, being largely dependent upon the actual fraction of non-combustibles present.

Dry combustible wastes include paper, cloth rags, anti-contamination clothing, and miscellaneous packaging materials. These materials can be fed to the process at a rate of up to 90 kg/h. Following incineration in the fluidized bed, the volume of the ash is approximately 1/80 the volume of the original compacted waste.

PROCESS DESCRIPTION

General

The application of fluidized bed technology permits the processing of a wide variety of feed types within a single chamber process vessel. The various waste feed types are applied individually to a fluidized bed being operated at appropriate conditions. Incineration or calcination occurs within the vessel with the ash or calcine product being carried from the vessel by the fluidizing gases to a cyclone where the dry solid product is separated from the off-gas stream. Before venting to the atmosphere the off-gases are processed through a wet scrubbing system and filter train designed to remove particulate and radioactive gases. The scrub solution is recycled through the liquid waste system allowing operation with zero liquid discharge. The entire system is operated at less than atmospheric pressure to assure that no out-leakage of the contaminated gases occurs prior to or during the off-gas cleanup process.

Feed Systems

General. The feed systems are depicted schematically on Figure 1.

Liquid Feed Subsystem. The liquid waste feed subsystem consists of a tank, a pump, a strainer, a meter and atomizing nozzles. The tank is maintained at an elevated temperature, usually about 55°C and stirred, to prevent precipitation of the dissolved solids. The metering device automatically regulates the liquid waste flow, and the nozzles atomize the waste as it is injected into the fluidized bed. Sodium sulfate solutions, typical waste of BWR demineralizer regeneration, can be fed in concentrations of up to 25% by weight with no treatment beyond temperature control. Boric acid solutions, typical waste of PWR systems, can be fed in concentrations of up to 12% by weight, but must be neutralized. This is necessary because unneutralized orthoboric acid, H_3BO_3, will bond with water vapor and be carried through the cyclone to the wet scrub system.

Experience to date suggests that any true solution can be fed to the process vessel via this subsystem. There is little doubt then that the subsystems could feed any standard decontamination solution which may be applied as part of D&D projects.

Resin/Sludge Feed Subsystem. The resins and sludges are accumulated in a conical mixing/dewatering tank. Water is removed to leave a slurry of 60% to 70% water by weight. The water that results from the dewatering process can be returned to the slurry pumping system as makeup water or pumped to an evaporator or the

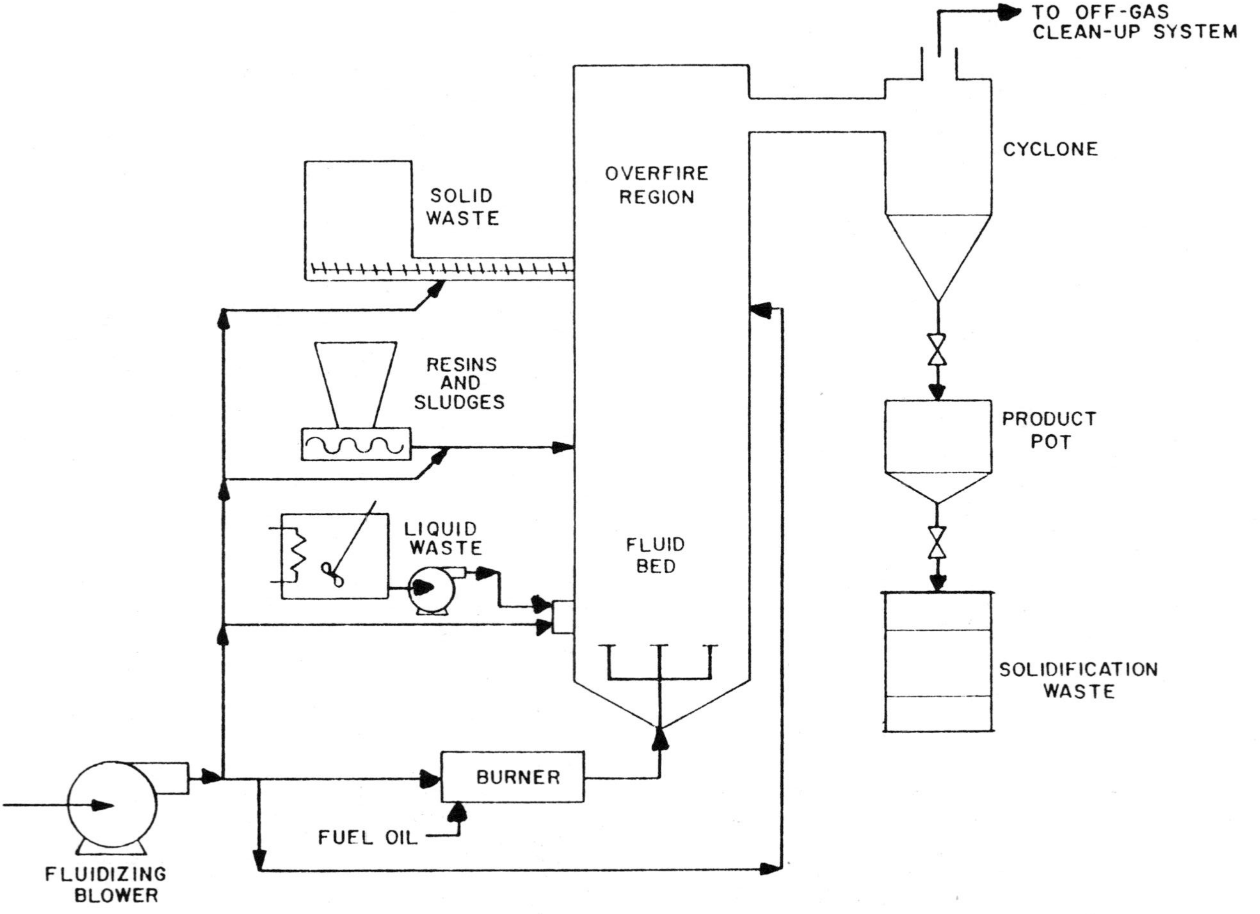

Figure 1. Process Vessel, Feed Subsystems and Product Removal.

liquid waste tank. Mechanical agitation within the tank prevents bridging, compaction or adhesion to the tank walls. Depending upon the filter aids used in a given liquid waste treatment system and the amount of inorganic materials attached to the resins, a melting point modification additive may be required when processing the resins and filter sludges. The conical tank directs the "dewatered" slurry into the inlet of a progressive cavity pump which is used as a metering device. Pressurized air is then used to pneumatically inject the slurry into the fluidized bed.

Dry Combustible Feed Subsystem. The dry combustible waste feed subsystem is comprised of a shredder, a storage hopper, a mechancial feed device and an isolation valve. The purpose of the shredder (not shown on Figure 1) is to reduce the size of the waste so that it can be transported efficiently to the process vessel. During incineration operation the shredder is not used and the hopper opening is sealed. A screw feeder system draws the waste from the bottom of the storage hopper at a uniform rate and delivers it to the isolation valve at the process vessel. Pressurized air strips the waste from the screw flights, carries it through the isolation valve and delivers it to the fluidized bed. The pressurized air keeps the valve seat clear of waste material and provides a very effective barrier against migration of the fire back into the screw feeder. The probability that a fire would migrate back is remote. However, should this occur, inert gases will automatically discharge into the hopper to extinguish the fire. When the dry waste feed system is not in use, the isolation valve is closed.

Process Vessel Operation

General. The process vessel is a metal alloy lined, air cooled right cylinder which houses the fluidized bed. A metal alloy was selected for the vessel interior due to concern that if the walls were fabricated from porous ceramic materials radioisotope migration into the walls would result in an ever increasing source term. The selected alloy is resistant to high temperatures, and will be air cooled to prolong its useful life. Cooling is by forced convection with the air being processed through an HVAC system and filters as a precaution prior to release from the building. The previously described feed subsystems, a fluidizing blower which delivers air through an array of fluidizing nozzles beneath the bed material, and a fuel oil burner which adds supplementary heat to the fluidizing gases when needed are manipulated to establish the proper operating parameters within the process vessel. The bed is composed of an unreactive high melting point material having the consistency of sand.

Liquid Calcination. The fluidizing air is preheated by the combustion of fuel oil in a commercially available burner for the calcination of liquid wastes. The liquid wastes coat the fluidized bed particles where drying occurs. A layer of dry solid covering the bed particles results. This dry solid is abraded by the scrubbing action of the fluidized bed and is carried from the vessel by the fluidizing gases. Calcination requires a bed temperature of 200°C to 400°C.

Resin Sludge Incineration. During resin or filter sludge incineration, supplemental heat is supplied by preheating the fluidizing air with the fuel oil burner. The supplemental heat is required to maintain adequate combustion conditions for the resin slurry. Resin incineration is accomplished at a bed temperature of 800°C to 950°C. The fluidizing air provides the oxygen required for combustion and the resulting gases carry the ash from the process vessel.

Dry Combustible Waste Incineration. During dry waste incineration, the fluidizing air enters the process vessel at near ambient temperature. The bed is kept hot (between 800°C and 1000°C) from the heat of combustion of the waste. Oxygen for initial combustion is provided by the fluidizing air. Air is also introduced to the process vessel above the fluidized bed to provide an afterburner affect, thereby assuring complete combustion of the waste. The resulting combustion gases carry the ash from the process vessel.

Off-Gas Cleanup

The off-gas cleanup equipment is depicted schematically on Figure 2.

The gases leaving the process vessel are first passed through a cyclone to remove dried calcination salts, or incineration ash, as the product of the system. The product is discharged to solidification equipment through a product pot which provides an air lock between the cyclone and that equipment. As the off-gas is essentially at the same temperature as the process vessel, the cyclone is fabricated from the same metal alloy and is also air cooled.

The gases are then passed through a wet scrub system comprised of a quench tank, venturi scrubber, wet cyclone, entrainment separator condenser, and mist eliminator.

The gases are cooled at the quench tank primarily by the heat required to vaporize the water in the aqueous scrub solution. Some decontamination also occurs here.

The venturi scrubber, wet cyclone and entrainment separator operate as a unit to wet the particulate and absorb radioactive gases, then separate the contaminated liquid from the scrubbed gas stream.

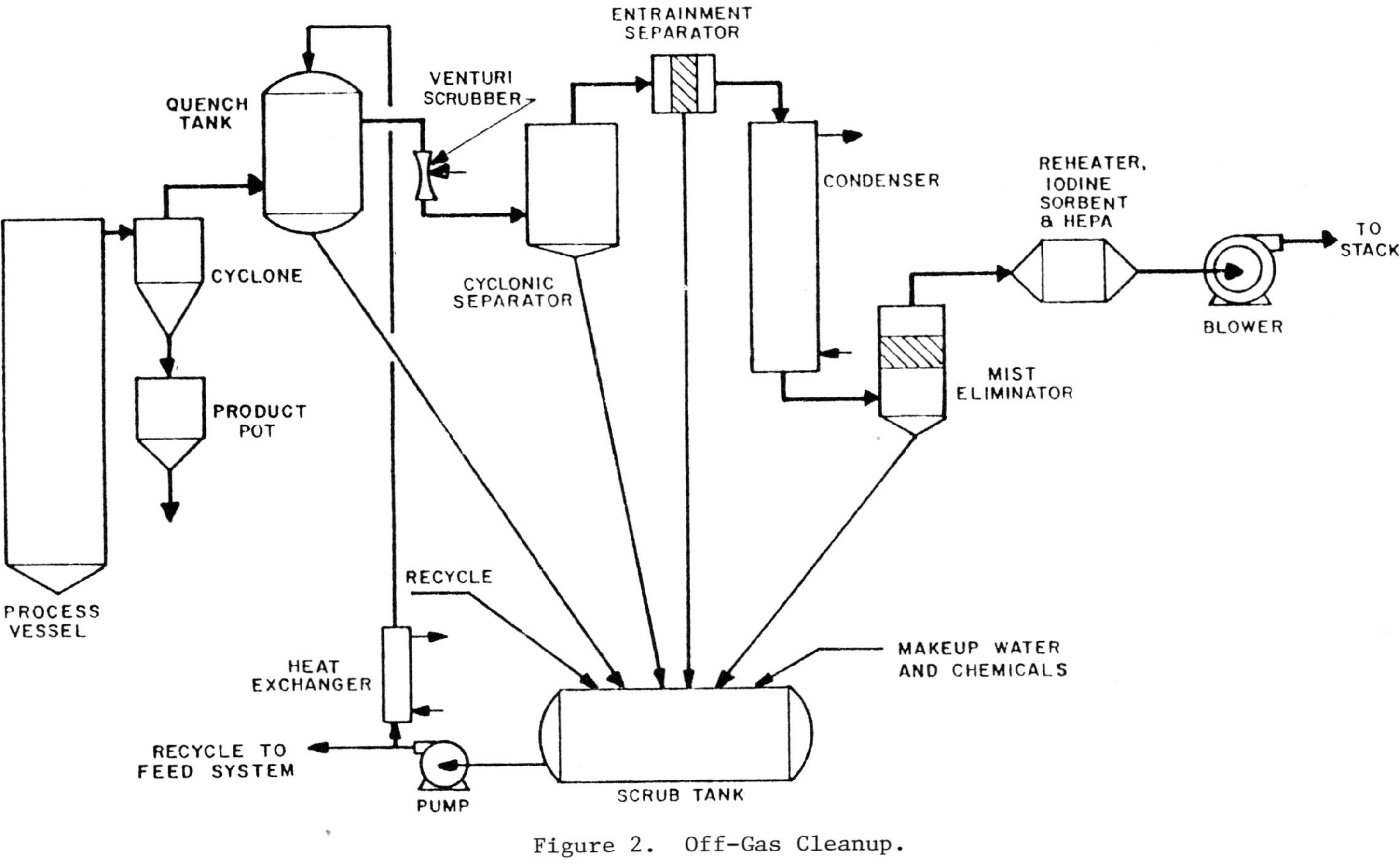

Figure 2. Off-Gas Cleanup.

The condenser and mist eliminator operate as a unit and have a dual purpose. First, the liquid inventory in the scrub is controlled by allowing a predetermined amount of water vapor to escape the system with the off-gases. Second, any unwetted particulate in the off-gas serve as nucleation sites for the condensing moisture providing additional decontamination when the condensate is separated from the gas stream by the mist eliminator. The liquids drawn from the quench tank bottom, wet cyclone, entrainment separator and mist eliminator are collected in the scrub tank where the solution inventory is monitored and the desired chemical characteristics, such as pH and chemical additive concentration, are checked and maintained. A small recycle stream is routed back to the liquid feed system to control the solids concentration. The remainder of the solution is subject to recycle back to the quench tank and venturi scrubber.

The scrubbed off-gases next pass through a filter train consisting of a reheater, solid iodine sorbent bed and high efficiency particulate air (HEPA) filters. The purpose of the reheater is to warm the gases to assure that the relative humidity is less than 100% prior to passing through the filters. The iodine sorbent bed would probably not be required for D&D projects as the presence of radioiodine should not be a problem due to the short half lives of I-131 and I-133. The final piece of equipment through which the gases pass is an off-gas blower. This blower maintains a negative pressure over the entire system, assuring that off-gas cannot leave the system except through the intended gas clean-up equipment. The off-gas clean-up system assures a clean gas effluent which can be safely vented to the atmosphere.

Instrumentation

The system's instrumentation and controls are designed to maintain process parameters within the limits which assure safe and efficient system operation. Safe operation is provided by control sequencing and interlocks which prevent improper operation and effect automatic system shutdown if system parameters are not maintained within prescribed limits. The instrumentation provides for two-level alarms and protective action. When a process parameter drifts outside the normal operating band, the first indication is an alarm which notifies the operator of the problem. If corrective action is not made, a second alarm, set slightly further outside the control band, is actuated. The second alarm is accompanied by automatic protective action. This two-level action calls developing problems to the attention of the operator allowing him to take corrective measures in time to prevent a possible automatic system shutdown.

CONCLUSION

Decontamination and decommissioning projects will generate substantial quantities of waste including wet ion exchange resins, liquid wastes, and dry combustible materials. Volume reduction of these wastes can reduce space requirements at interim waste storage sites, reduce retrieval efforts, provide fewer shipments to as yet undesignated permanent federal waste repositories, and reduce space required at those repositories. A fluidized bed waste volume reduction system is being developed which can handle all these waste types in a single system which provides for efficient off-gas cleanup and no liquid discharge. This system would be particularly advantageous where several facilities are to be decontaminated and/or decommissioned at the same site.

REFERENCES

1. Topical Report: RWR-1™ Radwaste Volume Reduction System, EI/NNI-77-7-P, Energy Incorporated, Idaho Falls, Idaho and Newport News Industrial Corporation, Newport News, Virginia, (June 24, 1977).
2. J. R. May, "Radioactive Waste Volume Reduction," Power Engineering, pp. 68-71, (October, 1978).

ENGINEERING ASPECTS OF DRESDEN UNIT 1 CHEMICAL CLEANING

W. S. Lange
Station Nuclear Engineering Department
Commonwealth Edison Company
Chicago, Illinois 60690

T. L. Snyder
Catalytic Inc.,
Philadelphia, Pa. 19102

Dresden Unit 1 is the first privately financed nuclear power reactor to be built in the United States. Dresden 1 began commercial power operation in April, 1960. It is a General Electric designed dual-cycle boiling water reactor with a thermal power of 700 MWt and a design electrical output of 207 MWe. The primary system consists almost entirely of 304 stainless steel.

In the Dresden 1 system, power is withdrawn from the cycle partly by boiling within the reactor, as is the case in a conventional boiling water reactor and partly by steam generation in secondary loops much like pressurized water reactors.

Besides the reactor itself, the steam supply comprises a steam separating drum, four secondary steam generators, four recirculating pumps, and unloading heat exchangers.

The primary steam drum is sized to provide approximately 11,000 gallons of water storage volume. The drum serves in the dual role of reservoir and separating device. The drum arrangement produces good natural circulation characteristics and increases the available net positive suction head on the pumps.

In the vertical type secondary steam generators, the primary fluid is pumped through inverted U-tubes. Each of the four separate secondary loops, consisting of the steam generator, recirculating pump and secondary valving, can be isolated from the system as a unit.

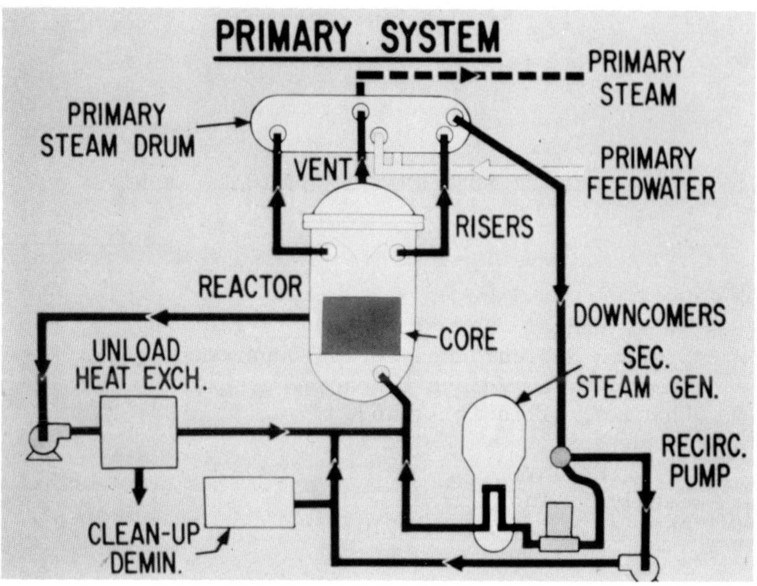

Figure 1 Dresden Unit I Simplified Flow Diagram

Corrosion problems in a nuclear unit are similar to those encountered in a fossil station. The important difference is that the oxides which form in the primary side of a nuclear unit also contain radioactive isotopes. During the 18 years of operation, such deposits have built up on the piping, on the bottom of the reactor vessel, on the tubes of the secondary steam generators, inside the primary system valves, and on the bottom of the steam drum.

While the most important result of a boiler cleaning is the improvement of heat transfer rate, the major concern in a nuclear unit cleaning is the removal of deposits containing radioactive isotopes, mostly Cobalt 60, to reduce the radiation exposure received by plant personnel.

Dresden Unit 1 has experienced increased radiation levels similar to other nuclear units. Radiation exposures as expressed in total annual man-rem has increased with cumulative operating time.

From 1960, when Unit 1 came on line, to 1970, the man-rem totals were between 100 and 400 man-rem per year. This compares favorably with the industry experience.

As the operating years continued to accumulate, the radiation fields became more troublesome. The dose rate buildups, of course, resulted in substantial increases in occupational radiation exposure. In 1973, Dresden 1 total was 900 man-rems. One example of a high radiation job took place in 1974. The replacement of a 20-foot section of 6" diameter pipe in the primary system required personnel to receive a total of 115 man-rem, took three months to complete, and used one month of critical path outage time. The job duration was not entirely due to high dose rates, since procedure and material problems developed as well. However high personnel exposure did delay the job completion.

Based on this type of experience, several trends can be observed. The more years that a unit operates, the greater man-rem dose received will be. Due to decreased flexibility in performing maintenance, it will be more difficult to maintain unit availability at acceptable levels. Reducing the radiation fields is necessary to combat the decrease in availability problems.

It was obvious that a method of reducing occupational exposure due to contaminated deposits at Edison's nuclear units had to be developed. Edison has a great deal of experience in cleaning fossil boilers, but cleaning a nuclear reactor and its associated components and piping was an entirely new concept.

There were two factors of importance to CECO in such a cleaning. One was the decontamination factor, which is the ratio of the amount of radioactive material originally in the system to the

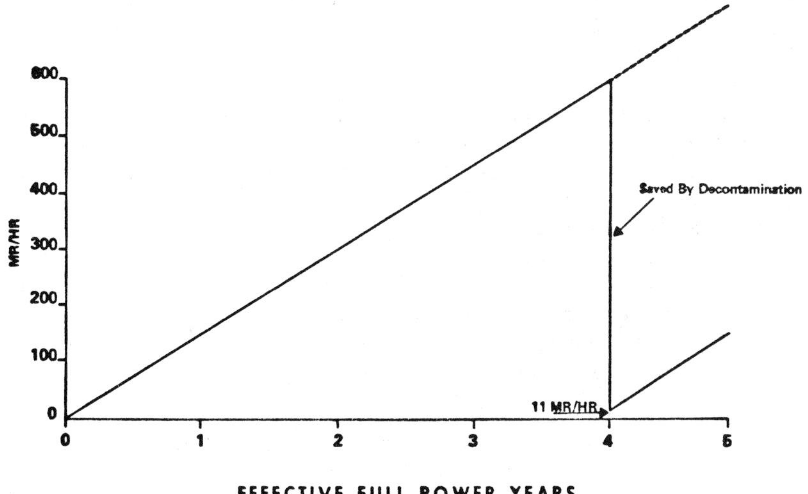

Figure 2 Typical Shutdown Radiation Levels on Primary Piping

amount left after the cleaning. The other was the corrosion rate of the cleaning process, which had to be necessarily low to assure that the primary system coolant boundary was not in any way degraded. The sawtooth curve (Figure 2) indicates the trend to be expected from a primary system decontamination of Dresden 1, assuming a decontamination factor of 50. However, the latest available lab data indicate that decontamination factors of 100 to 1000 can be expected.

In May 1972, the Dresden 1 Chemical Cleaning Project was formally begun, when Commonwealth Edison selected Dow Nuclear Services to perform the activities necessary to remove the radioactive oxide film from the primary side of the Dresden 1 steam supply system.

The project was divided into three phases. Phase I consisted of feasibility studies, development of methods, and corrosion investigations. Phase II consists of the modification of existing equipment and the construction of the facilities necessary to accomplish the cleaning as well as the actual operation itself. Phase III will address the recommissioning of Unit 1 after the chemical cleaning process is finished. The requirements of Phase III have not yet been fully developed. Plans for recommissioning will be formulated after an inspection is made to determine the condition of the system. At present, an internal reactor vessel inspection by ultrasonic testing is planned as well as an extensive ISI program for the entire unit.

Catalytic, Inc. was chosen as the Architect-Engineer to design the facilities exterior to the sphere, as well as the temporary modifications within the sphere.

The chemical cleaning of Unit 1 is expected to begin in early 1980. Dresden Unit 1 was shut down on November 1, 1978 because of licensing restrictions unrelated to the chemical cleaning. A one-year extension of the Operating License was requested to allow a shutdown closer to the chemical cleaning date, but the NRC would only approve a ten-month extension. This outage will be one of the most difficult and comprehensive ever attempted on a nuclear unit. In addition to the chemical cleaning, a high pressure coolant injection system (HPCI) and core spray system modification will be installed. Also, the reactor protection system will be upgraded to IEEE 279-1968 criteria, and two unloading heat exchangers will be replaced. A more extensive list of other Unit 1 projects will be discussed in the following paper. These projects are in addition to the normal refueling outage jobs.

The design process commenced in 1973 with extensive field investigation to review the as-built configuration of the existing systems. Considerable testing was undertaken to alleviate any of the aforementioned concerns and a field test operation was conducted using the proposed solvent in existing plant equipment.

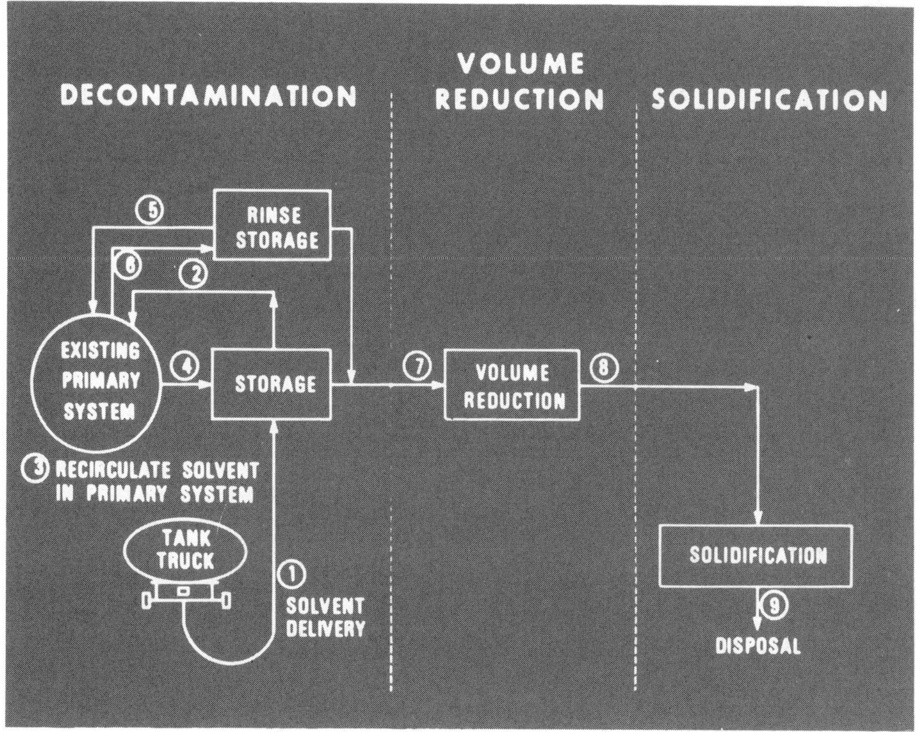

Figure 3 Decontamination Simplified Flow Diagram

The design vintage of Dresden 1 precluded the use of the existing radioactive waste treatment facilities. New equipment had to be added and a separate facility designed to accommodate this equipment.

The decontamination technique is shown by the simplified flow diagram in Figure 3. The solvent is delivered in common stainless steel tank trucks and transferred into one of the radwaste storage tanks. The solvent is mixed and diluted to "use" strength.

The solvent is then injected into the primary system by use of temporary pumps and piping.

The solvent is circulated in the primary system at 255°F for approximately 100 hours. At the end of this time, the solvent is spent and is transferred back into the radwaste storage tanks.

Then demineralized water is injected into the primary system for rinsing purposes. The first rinse is removed and stored in the shielded radwaste storage tanks. Subsequent rinses can be stored in the outside storage tanks.

The wastes are then processed through the evaporator or the demineralizer in the facility. The concentrated wastes are then solidified and shipped to a burial site.

The plot plan shows the new radwaste facility in relation to Dresden 1. The radwaste storage tanks are within the facility, and the rinse water storage tanks are in an area surrounded by a dike outside the facility. The new HPCI building can be seen east of the sphere. The existing Dresden 1 radwaste facility can be seen south of the new radwaste facility.

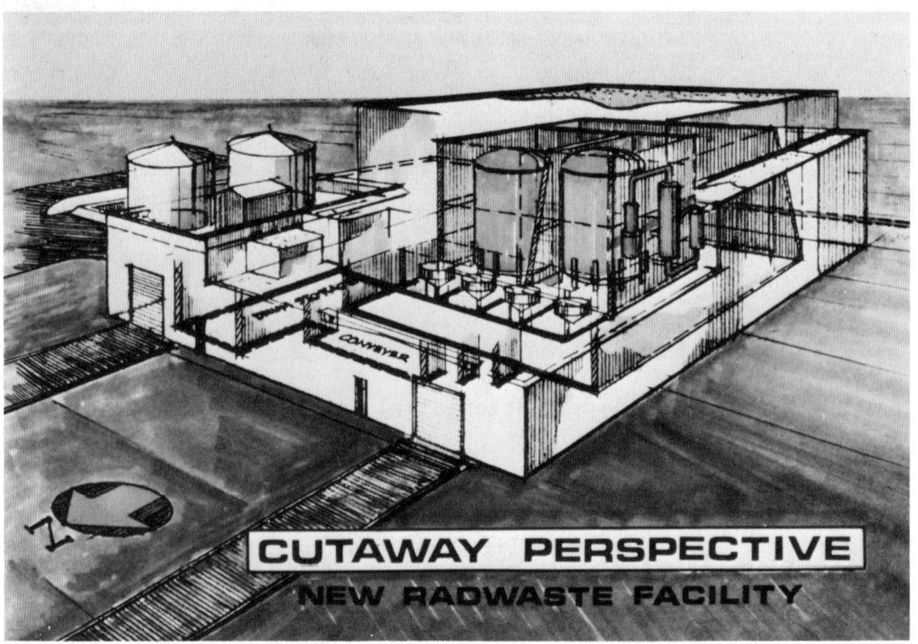

Figure 4 Cutaway Perspective

In this cutaway perspective, (Figure 4), some more detail can be seen. The evaporator is west of the radwaste receiving tanks. The evaporated concentrate storage tanks are located north of shielded storage tanks. The solidification conveyor can be seen in the foreground with the drum storage to the east.

This cross section of the site shows the transfer piping tunnel exiting from the sphere, crossing the yard into the new radwaste facility.

As mentioned earlier, a separate facility was determined necessary to receive, evaporate and solidify the wastes from the chemical cleaning.

To lessen the impact upon existing site services, the facility incorporates its own air compressor, electrode boiler and service water supply and power source. The location adjacent to the existing Dresden 1 radwaste hold tanks will allow for future conversion of the facility to augment Dresden 1 waste processing systems which do not have evaporation and solidification capabilities.

The building is reinforced concrete and is designed to meet the requirements for a seismic category I and tornado-resistant structure for those areas containing the highly radioactive liquid and concentrate. The remaining areas (auxiliaries, and control) are designed to meet the requirements of the Uniform Building Code with adequate provision for radiation shielding. To assist in the design and construction of the facility, an engineering design check model (Fig. 5) was prepared. A model of the facility was scaled at 3/4" to one foot. The model has been used by Commonwealth Edison operations people to generate comments on the design. It was later used by the Catalytic engineers to check for interferences. It is now being used by Dow engineers as an aid in writing operating procedures. A model of the existing spherical containment was also built.

The general scheme for treatment of the wastes is not new to radwaste processing. The water will be processed through the evaporator with the concentrate ultimately being solidified and the distillate being demineralized for reuse in the system.

Two 158,000 gallon radwaste hold tanks will receive the spent solvent and first rinse which contains the bulk of the 3,000 curies. The tanks are located inside separate reinforced concrete vaults. The rest of the rinse water will be of low activity and will be received in the two 100,000 gallon rinse and hold tanks. Enough tankage is provided to accommodate one complete primary system rinse beyond the three rinses anticipated. The wastes will be processed as a parallel activity to recommissioning and operating of the unit.

If needed, the distillate will be processed through a 100 gpm deep bed ion exchange system for return to the Dresden 1 contaminated condensate storage tank. If the rinse water activity is as low as anticipated, the capability exists to process the rinse/hold tanks directly through the demineralizer system and avoid the lengthy evaporation process. Spent resin will be discharged to a resin storage tank and later will be solidified.

Figure 5 Decontamination Facility Design Check Model

The evaporator we selected is an HPD forced circulation evaporator incorporating 2" tubes. These large tubes effectively eliminate the need for inlet filters and the waste disposal and maintenance associated with them. The liquid is circulated at 6,000 gpm through the heater and evaporator bodies. The concentrate can be withdrawn on a semi-batch basis every eight (8) hours or other specified intervals. Concentrate is withdrawn via a Strahman ram valve from the bottom of the evaporator and is pumped to the concentrate tank. After completion of the drawdown, the piping is automatically flushed.

The vapor leaving the evaporator passes to an external separator where it is scrubbed and then demisted to agglomerate water particles, dissolved and suspended solids. The high quality vapor leaving the separator then passes to the distillate and recycled from this point to the entrainment separator before transfer to the outside rinse water tank.

We have designed the evaporator layout to compartmentalize its components to provide the necessary shielding when maintenance is required. With this evaporator, we expect to be able to achieve 17 gpm with spent NS-1 and 20 gpm on normal radwaste liquids at 50% concentration.

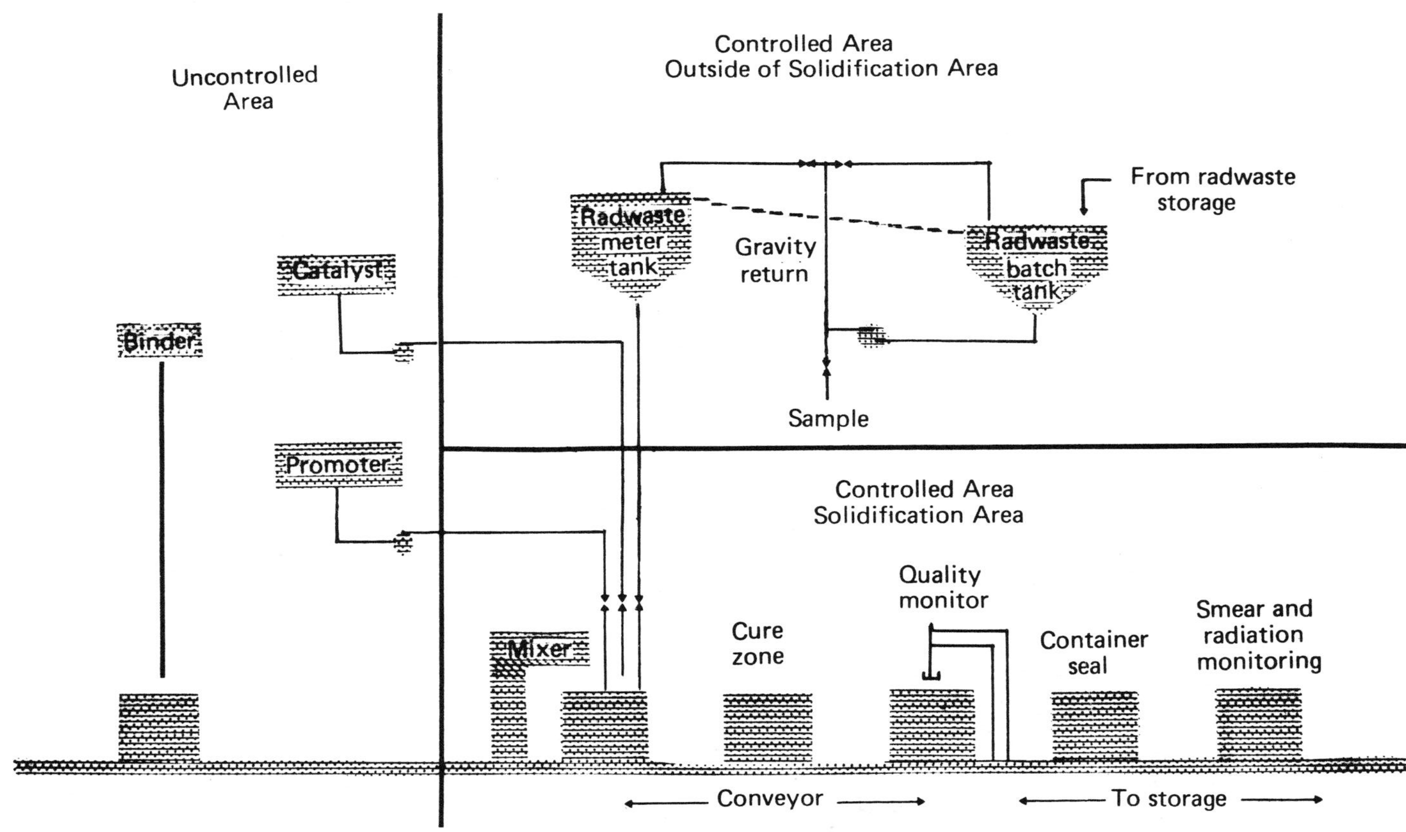

Fig. 6. Solidification System

The solidification of evaporator bottoms and the spent ion exchange resin is an intergral part of radwaste processing. Several criteria were established for the evaluation of new or existing systems, including ease of system operation, weight and volume of finished product per unit of waste, physical strength, heat and flame resistance, resistance to leaching, and reliability of process. Most important is the compatibility of the spent solvent with the solidification agent. The materials subjected to laboratory comparison were portland cement, ureaformaldehyde and a modified, commercially available vinyl ester resin (Dow process).

For the cleaning wastes in particular, the Dow process was a clear choice. This process was successfully field tested at Dresden to process NS-1 from cleaning of the corrosion fatigue loop between June 3 and June 16 in 1976. This loop cleaning verified the expected decontamination factors as well.

The solidification area in the facility includes a control room, the solidification process equipment, a drum storage area, and a shipping area for loading onto trucks.(see Figure 6).

A great deal of first-time knowledge has been gained during this project. The NS-1 solvent has been shown to result in little metallurgical effect as a result of extensive testing. The Dow solidification process appears to be the best available, particularly with respect to Appendix I consideration. The facility design that has been developed would be applicable, with minor modifications, in situations where large quantities of low-level radwaste are encountered.

We have received many inquiries from utilities and consultants about our facility. In a few cases, there were expressions of interest in our design for a specific application. In most cases, a nuclear power plant newer than Dresden 1 will have a better existing radwaste system. Considerable savings could result from incorporating the existing system in the facility design. Whatever the situation, early comprehensive planning is essential for a successful project.

RESULTS TO DATE OF THE DRESDEN-1 CHEMICAL CLEANING

David E. Harmer

The Dow Chemical Company
Dow Nuclear Services
47 Building
Midland, Michigan 48640

Jerry L. White

Commonwealth Edison Company
Station Nuclear Engineering Div.
One First National Plaza
P.O. Box 767
Chicago, IL 60690

SUMMARY

After more than 18 years of commercial operation, Dresden-1 has now been shut down for an outage to reduce radiation levels and to perform various ISI activities and retrofitting tasks. One of the first tasks will be the complete chemical cleaning of the internal primary system to remove radioactive deposits from pipe and equipment.

The original objectives of the program included:

1. Reduce radiation levels to improve plant accessibility.
2. Ensure continued safe and efficient operation.
3. Develop and prove techniques applicable to other reactors.
4. Encourage vendor, manufacturer, and consultant participation.

Laboratory tests of the NS-1 solvent on samples of actual Dresden-1 internal surfaces have given radiation reduction in excess of 99%. A field result close to this efficiency is anticipated.

Plans for this chemical cleaning have been progressing since 1973. The cleaning solvent has been selected and tested. Process details and procedures are essentially complete. Construction activities went forward in two areas; new piping and equipment connections in the containment sphere, and an entirely new building to process the spent cleaning solvent and rinses. Photographs and drawings of these facilities are included.

The actual chemical cleaning of Dresden-1 is scheduled during early 1980. This date represents a moderate amount of slippage from the earlier target date. The new schedule, and results of the cleaning operations to date, are included in this report.

INTRODUCTION AND HISTORY

This paper presents an update of the Dresden-1 Chemical Cleaning project. Because of slippage in the construction schedule many of the results are not yet at hand. However, it is of interest to indicate exactly where the project stands as well as to fill in a few more specific details of the actual operations to be carried out.

Dresden Unit 1, being the oldest fully commercial nuclear reactor, was shut down on October 31, 1978 after 18 years of continual operation. Construction activities pertaining to the chemical cleaning operation and the handling of wastes to be generated during the operation had already been in progress well in advance of the beginning of this outage. The chemical cleaning is now scheduled early in 1980. Following the cleaning operations numerous ISI activities and plant modifications will be possible in the reduced radiation localities throughout the plant. These are summarized in Table 1.

This project has been supported in part by the U.S. Department of Energy. Commonwealth Research Corporation (a wholly owned subsidiary of Commonwealth Edison Company) acts as the contract manager. The primary contractor for the operation of the project is the Dow Nuclear Services Department of The Dow Chemical Company. Catalytic, Inc. has been Dow's subcontractor for all of the engineering in connection with this project. The Nuclear

Table 1. Additional Plant Outage Activities

1. Install High Pressure Coolant Injection (HPCI).
2. Install New Core Spray System.
3. Upgrade Reactor Protection and Post Accident Systems.
4. Install System to Handle "Anticipated Transient Without Scram" (ATWS).
5. Replace Unloading Heat Exchangers.
6. Upgrade Fire Protection System.
7. Perform In-Service Inspection.

Division of the General Electric Company has also participated in the role of a consultant and project team member, since they were the original vendor of Dresden Unit 1. Additional consulting, particularly in the area of metallurgical testing, has been afforded by Professor Roger Staehle, Mr. Craig Cheng, and Mr. Tom Hendrickson.

The project has already completed many of its phases. Initial plant studies were carried out in 1973 and from this a study of metallurgical conditions was completed. Solvent development began even earlier but was then directed specifically to Dresden, in parallel with the plant studies and metallurgical work. Corrosion testing was based on the metallurgical conditions which were found in the plant and developed as a major program involving many thousands of individual tests and samples. The development of a cleaning process and the implementation of it as an engineering design was an intensive activity from early 1974 until early 1979 . Procurement and construction began formally in fall, 1976, and are still continuing as of this date. The actual cleaning, treatment of radwaste and activities remain to be accomplished in the next several months followed by plant recommissioning.

In contemplating the full scale cleaning of Dresden-1, decontamination factors (DF) for the inside surfaces of piping and equipment are expected to range from 100 to 1000. Experience has shown that the DF is highest in those areas containing the highest amount of activity. This estimate is based on laboratory work on samples from three different locations within the Dresden System. An in-place chemical cleaning of a small sub-system at the Dresden Unit 1 achieved decontamination factors which were essentially comparable to the laboratory results. (The details are outside the scope of this paper.) This experience, however, did support the expectation that the field DF's could be achieved in the same order of magnitude as the laboratory experiments.

OPERATIONAL DESCRIPTION

The previous paper[1] has described the items of equipment for the chemical cleaning as well as their function. For that reason this paper concentrates on the operation of the system as a whole, assuming knowledge of individual equipment components. Procedures for all of the activities are written and most of them have received their final review by the responsible parties. Before the cleaning itself is started certain tasks must be completed. Metallurgical specimens have already been placed in the turning vane and in a side loop in one of the steam generator recirculating systems. There has already been a complete baseline gamma scan of the piping itself in order to obtain an accurate before-and-after series of radiation readings. Temporary equipment is about 75% installed within the containment sphere. Pre-operational testing of all components and systems has started. Just before the beginning of the cleaning procedures, the primary steam lines, the secondary side of steam generators, and the main feedwater line will all be backfilled in a manner to prevent direct introduction of solvent into these parts of the system. A very important part of the operations will be the carrying out a mock run immediately prior to the actual cleaning with solvent. It is a key part of safety and quality which are an integral part of this project. This mock run will include all of the activities of the cleaning procedures but will employ deionized water rather than the actual solvent. It will serve as a final check for the functioning of all equipment and procedures.

The solvent for use in this cleaning is Dow Solvent NS-1, especially developed for BWR corrosion product removal. The actual volume of the primary system of Dresden-1, which will be cleaned, is 90,000 gallons. A twelve inch transfer line, which runs between the reactor containment vessel and the new radwaste processing building, contains another 5,000 gallons of solvent. There will be 125,000 gallons of solvent mixed on-site to allow extra in case of minor leaks or in case that sweetening (replenishment of solvent capacity) is required during the run. This sweetening process is only necessary if total amounts of corrosion products greatly exceed the best current estimates.

The cleaning will take place at a temperature of 255°F with an allowable range of ± 10°. It will last for a minimum of 100 hours. If corrosion product removal has not stabilized at the end of this time the run can be continued for a total of 200 hours. The system will be pressurized to 35 psig using nitrogen. Throughout this time a laboratory trailer will be at the site to perform analyses for solvent strength, iron, copper, and nickel content, dissolved radioactivity, and pH.

At the completion of the cleaning time a normal drain of the system will occur after the solvent has been pre-cooled below its boiling point. A direct drain is possible down to the level of handhole covers on the secondary steam generators. Below that point, solvent will be drained through the Reactor Enclosure Drain Tank system and transferred in a separate line to the radwaste handling facility building. In case of a major problem during the cleaning, an emergency dump can be effected in only 15 minutes. Since the solvent cannot be cooled in this time the emergency receiving tank will contain sufficient cold quench water to prevent flashing of the hot solvent. Any minor leaks which occur during the course of the cleaning are to be routed through the Reactor Enclosure Drain Tank system and its separate two-inch line to the radwaste facility. By this means small leaks can be segregated and repaired or simply allowed to continue without interruption of the ongoing recirculation for the chemical cleaning process.

The detection of leaks during progress of the cleaning is important in order to decide the safest course of action, that is, emergency dump, segregate and tighten, or allow to continue. Several features have been built into the cleaning system to facilitate leak detection and control. There are floor drain leak detectors which were especially built for this project and will be installed at 28 floor drain locations throughout the primary containment building. The level of solvent in the steam drum can be detected within 50 gallons variation while the liquid accumulating in the Reactor Enclosure Drain Tanks can be detected within 20 gallons. Humidity indicators are available in several locations throughout the primary containment and will give a signal if a leak were to become large enough to increase the water vapor content of the air in the vicinity. Continuous air monitors will also be employed, although it is known that small leaks do not increase the amount of airborne activity. Finally, areas which have a high potential for leakage will be equipped with protection in the form of waterproof sheeting which would direct a leak towards a floor drain and will be disposable after completion of the cleaning operations.

After the initial drain of the system additional solvent removal activities will complete the cleaning operation. All deadlegs within the system are to be either drained or backflushed. A series of rinses will take place. The first rinse will contain chemicals for removal of small amounts of metallic copper which may deposit during the cleaning. Additional rinses will continue until the final water quality is obtained with less than 30 micromhos per centimeter conductivity and less than 6 parts per million total organic carbon.

A few additional words about the copper rinse are instructive. Copper oxide in small quantities (estimate of 30 pounds total) was found throughout the Dresden primary system. This oxide will be dissolved by the solvent along with other corrosion products. During extended periods of cleaning the possibility exists that some of this copper may plate out onto metal surfaces. This is not a heavy copper plating such as would be found upon a bare carbon steel surface, but is rather a "flash" of copper across a metal surface. The purpose of the copper rinse is to remove such copper so that particles cannot break loose at a later date to plug orifices or lodge in pump bearings. The copper rinse conditions are covered by a Dow patent (U.S. 3,438,811). It will use up to 5% of residual solvent together with added hydrogen peroxide and ammonia at low concentrations. This first rinse, which will contain residual solvent whether or not the copper removal conditions are imposed, will be maintained at 120°F and atmospheric pressure until the copper content of the solution is constant by analysis. At that time the rinse is to be transferred to the spent solvent facility building and rinsing is continued. Table 2 shows expected rinse conditions.

Processing of the effluents from the chemical cleaning begins with the lowest concentration materials and progresses to the primary charge of spent solvent at the last stage of the operation. The low level rinses can be processed most effectively by simple demineralization over a mixed bed resin until a water quality less than 1 micromho per centimeter has been achieved. Schematically, this is represented in Figure 1 by pumping the lowest level of activity from the outdoor storage tank farm through the demineralizer beds to a holding area from which it can be returned to the inventory of water at the Dresden Station.

Processing of the intermediate concentration rinses takes place by use of the HPD evaporator, chemical treatment of overheads if organic carbon has been found, re-evaporation and finally demineralization of the clean condensate. Schematically, this is shown in Figure 2 as transferring water from the lower level of the two internal shielded tanks through the evaporator and back into storage in the low level tanks.

If chemical treatment to remove organic carbon is required in the condensate from the evaporation process, it is carried out by return of the water to the internal 155,000 gallon tank, addition of an oxidizer solution, re-evaporation, and demineralization to reach reactor quality water which can be returned to the normal station water inventory (Figure 3).

Table 2. Rinse Calculations

Fluid	K Gal	Curies	μ Ci/ml	Processing Classification
NS-1	95	2850	7.3	High
Rinse - 1	102	142	0.37	Intermediate
Rinse - 2	95	7.1	0.02	Low
Rinse - 3	95	0.38	0.0011	Low
		2999.48		

Basis: 3000 Curies in Sphere. 95% Drain Factor.

Figure 1

LOW LEVEL RINSE PROCESSING

155K
gal
155K
gal
100K
gal
100K
gal
To Station
High Level
Waste Storage
Low Level
Waste Storage
Evapo-
rator
Demineralizer

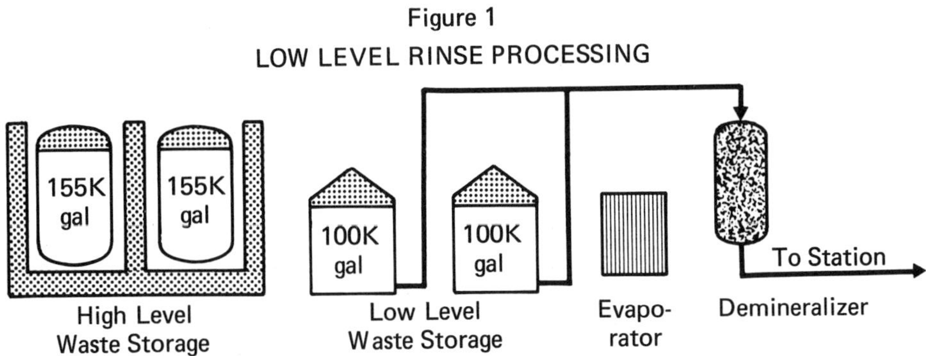

Figure 2

INTERMEDIATE RINSE AND SOLVENT PROCESSING

155K
gal
155K
gal
100K
gal
100K
gal
High Level
Waste Storage
Low Level
Waste Storage
Evapo-
rator
Demineralizer

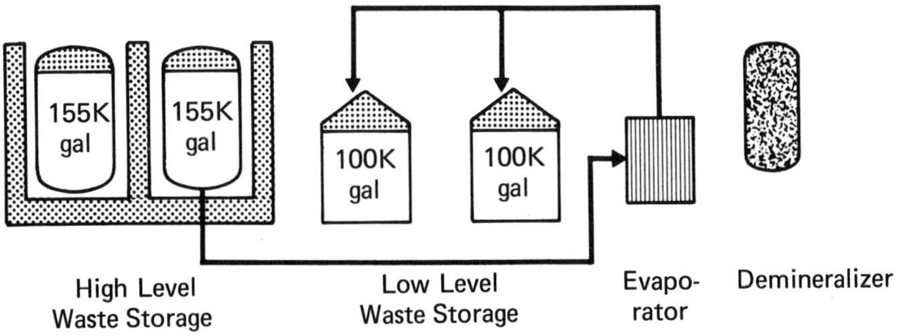

Figure 3

INTERMEDIATE AND SOLVENT PROCESSING

EVAPORATOR OVERHEADS TREATMENT

155K
gal
155K
gal
100K
gal
100K
gal
To Station
High Level
Waste Storage
Low Level
Waste Storage
Evapo-
rator
Demineralizer

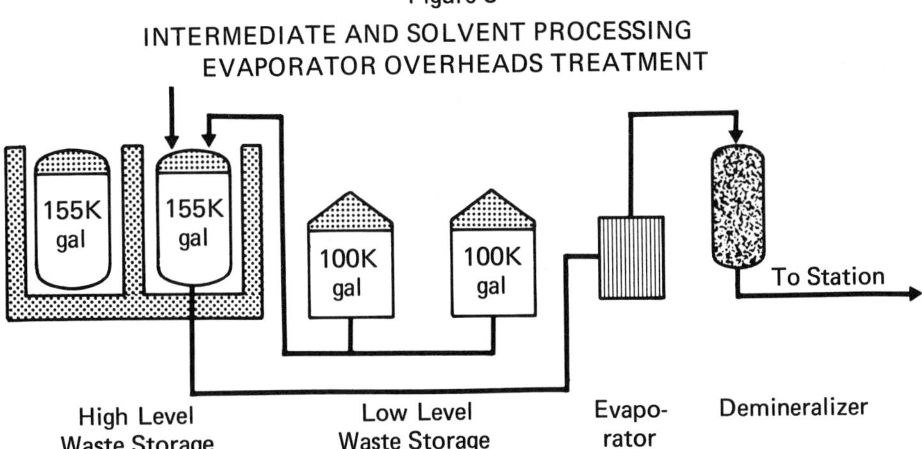

The primary charge of solvent is treated by the same sequence of events: evaporation, oxidation, re-evaporation, and demineralization. In this case it is known that the oxidation step must be carried out to achieve final water quality for return to normal plant inventory.

As the solids content of the evaporator rises to a predetermined level (about 50% solids) this material is transferred to special evaporator bottoms storage tanks. When each of these tanks is full it constitutes a radwaste batch which will be processed for solidification without addition of further new materials. This procedure was covered in the previous paper[1].

Meanwhile, final activities connected with the chemical cleaning will have moved forward in the primary containment building. The metallurgical samples are to be removed and tests performed on a portion of them. The remaining portion will be returned along with fresh samples for long term observation under the return-to-service reactor conditions. The same portions of reactor piping which have already been measured before the cleaning will be measured again in order to obtain accurate decontamination factors.

PROJECT STATUS

A few photographs of progress at the Dresden Site are now presented. This series of photographs was made during August of 1979 and give the best idea of the current state of the project. At this point it should be stated that considerable difficulty was experienced in maintaining a construction schedule for this project. While the engineering and technical inputs were capable of being carried out according to the original schedule, a problem was encountered in obtaining a sufficient number of the required construction craftsmen. Joliet and South Chicago are enjoying a surge of construction. For that reason delays have been continually experienced in completing the various work segments according to the intended schedule.

An idea of the type of temporary piping which was installed within the containment sphere is afforded by Figure 4 which shows piping and temporary equipment in the vicinity of the heat exchanger.

Figure 5 is a photograph taken in April of 1979 to show construction details at the top of one of the 255,000 gallon shielded holding tanks. It can be seen from the amount of reinforcing rod that this area is within the seismic bathtub containment. It should be noted that the original waste treatment facilities for Unit 1 were not sufficient to handle the cleaning solvent or the amount of radioactivity to be removed.

Figure 4. PHOTOGRAPH OF NEW PIPING IN SPHERE

Figure 5. DRESDEN STATION-DECON BUILDING- R/W RECEIVING TANK 102A (TOP), LOOKING NORTHWEST

Figure 6. DRESDEN STATION- RINSE WATER & CONDENSATE HOLD TANKS, LOOKING NORTHWEST

Figure 7. HPD EVAPORATOR HEATER AND VAPOR BODY BASE, IN-PLACE

Figure 8. SOLIDIFICATION LINE DURING CONSTRUCTION

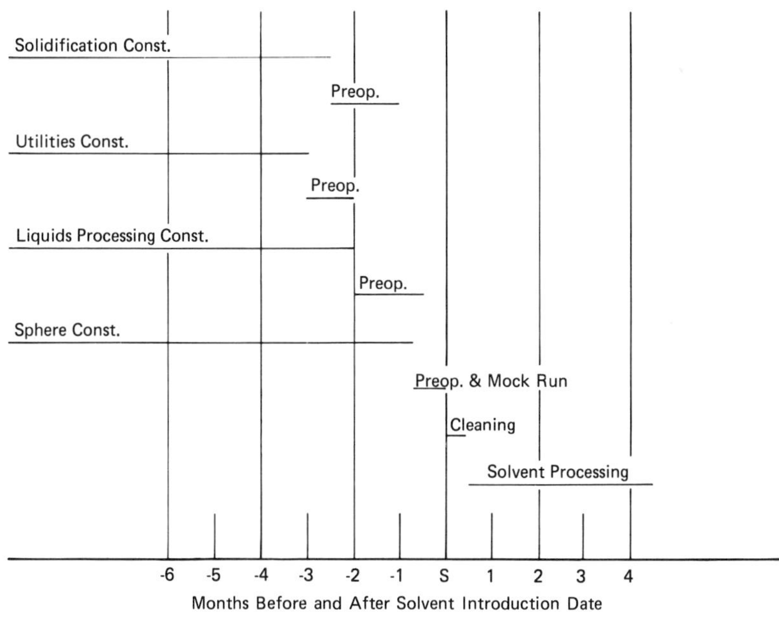

Figure 9. SCHEDULE OF CHEMICAL CLEANING ACTIVITIES

Figure 6 shows the two external storage tanks for make-up water and dilute rinses, within their diked conclosure. All water within these systems is controlled.

A portion of the HPD evaporator equipment is shown within its shielded room in Figure 7.

The solidification line is shown in Figure 8 as it existed in late August. Major pieces of equipment are all in place with final detailed connections still to be made. The adjacent drum storage area is large enough to contain 459 drums in a three layer pyramidal stack.

To complete the update on the Dresden-1 project the chart in Figure 9 shows the scheduling of activities directly associated with the chemical cleaning. The date for actual introduction of the solvent is the key date for all associated activities. This date is now expected in early 1980. The tasks and their duration before and after the cleaning can be observed. It should be noted that once the rinsing procedure has finished, the remaining tasks are not on the critical path for the ISI and modification work within the reactor containment building. Delays, as previously explained, have occurred in the construction schedule. However, every indication still remains that this job will be performed successfully and that high decontamination factors will be obtained for the Dresden-1 chemical cleaning.

REFERENCES

1. W. S. Lange and T. L. Snyder, "Engineering Aspects of Dresden Unit 1 Chemical Cleaning Project" ANS Topical Meeting on Decontamination and Decommissioning of Nuclear Facilities, September 16-19, 1979

This paper has been prepared under Commonwealth Research Corporation as a Subcontract effort under U.S. Department of Energy Contract EY-76-C-02-4014A-000. By acceptance of this article, the publisher and/or recipient acknowledges the U.S. Government's right to retain a nonexclusive, royalty-free license in and to any copyright covering this paper.

DISSOLUTION CHARACTERISTICS OF METAL OXIDES IN WATER COOLED REACTORS

J.L. Smee

London Nuclear Decontamination Limited
P.O. Box 1025, 4056 Dorchester Road
Niagara Falls, Ontario L2E 6V9

ABSTRACT

The CAN-DECON process is a dilute chemical decontamination process developed in Canada to decontaminate CANDU PHW reactors. The chemical reactions involved in dissolution of magnetite, the principal component of oxide films in CANDU PHW reactors, are described.

The compositions of oxide films in CANDU BLWs, BWRs, and PWRs are also described. The types of reagents needed to dissolve these films are discussed. Reducing type reagents are needed for PHWs, BLWs, and BWRs while an oxidizing type reagent is needed for PWRs. The reasons for this are discussed.

Recent loop decontamination tests have been performed in Canada with BWR and PWR specimens. Using the CAN-DECON process and a reducing type reagent, typical decontamination factors on BWR specimens are 10 to 20. With a new oxidizing reagent, decontamination factors of 5 to 15 are being obtained on Inconel and stainless steel PWR specimens.

CANDU	CANada Deuterium Uranium
PHW	Pressurized Heavy Water
BLW	Boiling Light Water
BWR	Boiling Water Reactor
PWR	Pressurized Water Reactor

1. INTRODUCTION

Chemical decontamination involves the dissolution of metal oxides. During development of the CAN-DECON process[1], a type of dilute chemical decontamination (DCD) process which was developed in Canada specifically for CANDU PHW reactors, considerable work was performed on the dissolution of magnetite in dilute reagents. This is the principal oxide in PHW reactors.[2] Since then, the work has been extended to include those oxides formed in the oxidizing environment of the BWR and the reducing environment of the PWR.

Before discussing this work, the main features of the CAN-DECON process will be reviewed, and it will be shown how an understanding of the chemical reactions involved in magnetite dissolution led to selection of a reagent for CANDU PHW decontamination.

2. CAN-DECON PROCESS

Figure 1 shows the main features of the CAN-DECON process. It is a DCD process that does not require draining or flushing of the system to be decontaminated. When the whole reactor system is decontaminated, it is not necessary to defuel. This simplifies operation and at the same time results in cleaning of the fuel. This is a great advantage to the overall decontamination process since it removes from the system a significant amount of activated corrosion products which, if allowed to remain, would eventually dissolve and redeposit out-core, thus partially negating the effects of the decontamination. Cation exchange resin is used for impurity removal and reagent regeneration, while anion exchange resin is used for reagent removal at the end of the process. Only solid wastes, consisting of ion exchange resins and filters, are produced, thus greatly simplifying disposal.

Figure 2 shows the steps that are involved in applying CAN-DECON. The reactor is shut down and the coolant is kept circulating at about 90°C. The coolant is purified with mixed bed ion exchange resin to remove additives such as Li, which is used for routine chemistry control, and to neutralize the coolant. At the same time, temporary equipment such as the reagent injection system is tied into the main circuit. Next, a small amount of reagent is added, typically 0.05 to 0.1 weight percent. In a CANDU reactor, it is added directly to the heavy water in the primary heat transport system; that is, the coolant itself becomes the decontaminating solution.

Figure 1

CAN-DECON CONCEPT

- o Low Reagent Concentration
  - minimizes corrosion
  - minimizes downgrading of heavy water
- o Reagent Added Directly to Reactor Coolant
- o System is neither Drained nor Flushed
  - no large tanks required
- o Reactor is not Defuelled
  - simplifies operation
  - fuel is also decontaminated
- o Regeneration of Reagent by Cation Exchange Resin
  - reagent is reused
  - effective concentration much higher than initial concentration
- o Removal of Reagent by Anion Exchange Resin
- o Solid Wastes Only
  - ion exchange resins
  - filters

The chemicals circulate through the system, attacking the deposits and releasing contaminants from the walls. Once the contaminants are in the liquid, they can be removed from the system by purification. Dissolved metals such as Fe and Co are removed by cation exchange resin, and particulate matter is removed by filtration. Large amounts of particulate matter are generated as the chemicals attack the deposits, and typically 30 to 50 percent of all the activity eventually removed is removed by the filters. An efficient submicron filtering system is essential.

Apart from removing dissolved metals, the cation resin performs another important function. It converts the spent contaminated solution into a clean reusable form. This is regeneration. The regenerated stream is recirculated to the reactor to be used over and over again as long as contaminants are still being removed. Regeneration is briefly described in Section 4 of this paper.

Figure 2

STEPS IN THE CAN-DECON
DECONTAMINATION PROCESS

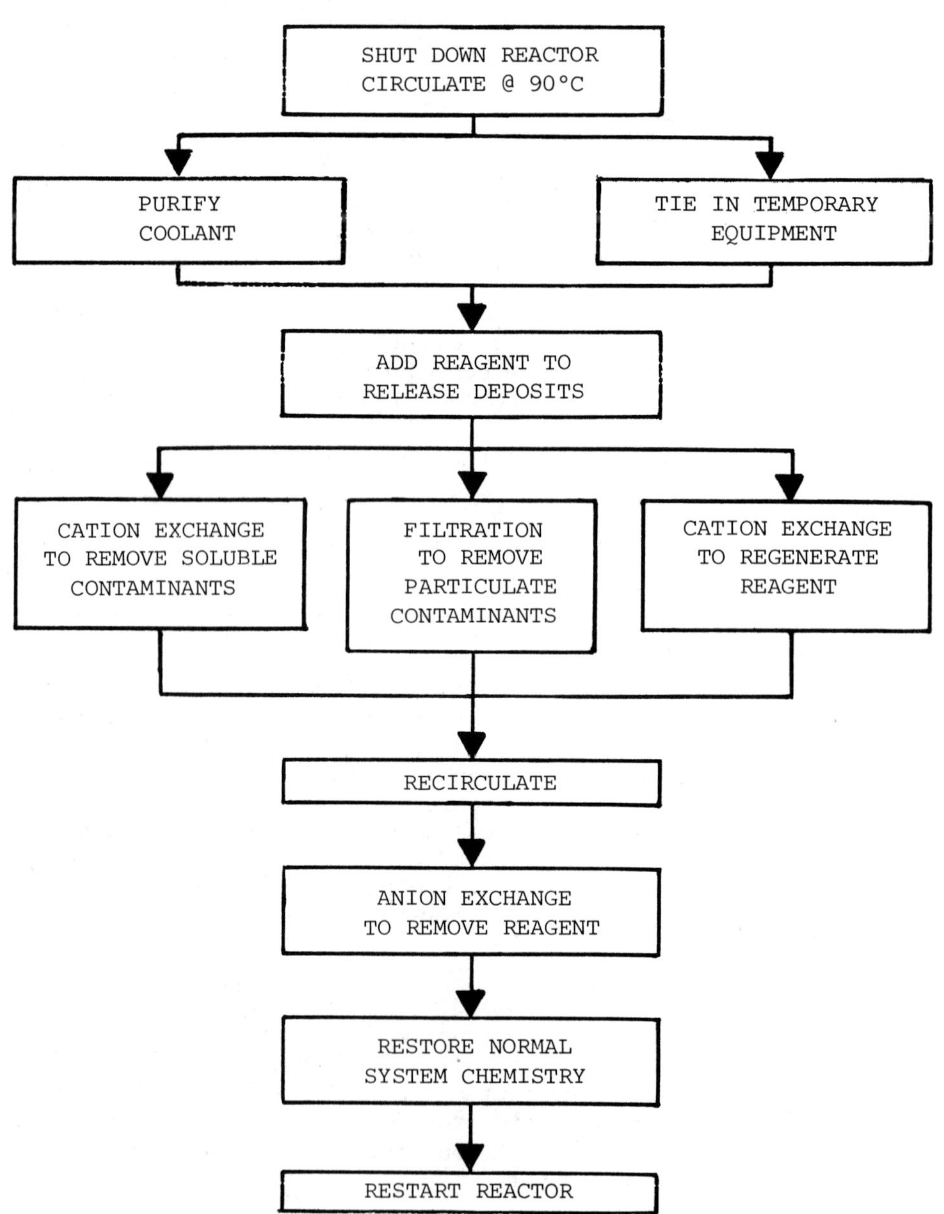

The decontamination is terminated by valving out the cation resin and valving in a mixed bed column. The anion part of the resin removes the chemical reagents themselves, and the cation part removes any remaining dissolved metals. Normal system chemistry is restored and the reactor restarted. During decontamination of the Douglas Point reactor, the entire process took 72 hours.[1]

These are the basic steps involved in CAN-DECON. Clearly, the heart of the process is in the chemical reagent and the purification-regeneration with ion exchange resin. The CAN-DECON process is independent of any particular chemical and will operate with several different chemicals or mixtures of chemicals. The optimum formula will depend on the composition of the deposit to be dissolved and this will obviously vary from one reactor type to another. In a CANDU PHW, the deposit is primarily magnetite, Fe_3O_4. The Co-60, which is the principal target of the decontamination, is embedded in the magnetite as cobalt ferrite and is released when the magnetite is dissolved. However, the actual weight of Co involved is very small, so for all intents and purposes, it can be assumed that the film is pure magnetite.

Figure 3 shows some of the chemical reactions involved in magnetite dissolution. This process was studied in detail by Atomic Energy of Canada Limited staff by making an oxide-covered disc the working electrode in an electrochemical cell and by measuring the electrode potential against a standard electrode.[3] It was found that simple oxide dissolution, reaction 1, is slow even in highly acidic solutions. However, once the solution penetrates through the oxide to the metal surface, base metal dissolution begins. The electrons produced in reaction 2 can increase oxide dissolution as shown by reaction 3, and for a short time the rate of oxide dissolution actually increases. However, the electrons produced by reaction 2 can also be consumed by reaction with H^+, as shown in reaction 4. Reactions 2 and 4 become predominant as substantial surface area of base metal is exposed. The result is a marked decrease in the availability of H^+ ions and electrons for magnetite reduction via reaction 3.

The obvious solution was to have an independent source of electrons to make reaction 3 dominant as soon as the reagent is added. This was done by incorporating a chemical reducing agent into the reagent. The reagent must also contain an acid. An organic chelating agent was also added to prevent reprecipitation and allow a much higher dissolved metal concentration than would be possible from pH effects alone. It is also the main agent transporting metal ions to the ion exchange columns.

Figure 3

CHEMICAL REACTIONS OF OXIDE DISSOLUTION

1. Simple Oxide Dissolution - Slow

$$Fe_3O_4 + 8H^+ \rightarrow 2Fe^{3+} + Fe^{2+} + 4H_2O$$

2. Base Metal Dissolution - Increases with Time

$$Fe \rightarrow Fe^{2+} + 2e^-$$

3. Electron Catalyzed Oxide Dissolution - Fast

$$Fe_3O_4 + 8H^+ + 2e^- \rightarrow 3Fe^{2+} + 4H_2O$$

4. Proton Reduction - Increases with Time

$$2H^+ + 2e^- \rightarrow H_2$$

Figure 4 shows how all these components work together to dissolve the deposits. The organic chelating agents are also weak acids and when dissolved, they ionize into H^+, which works with the other acids and reducing agent to dissolve the deposit, and the chelating part Y^-, which complexes the metal ions as shown in Figure 4.

Figure 4

DISSOLUTION OF MAGNETITE

$$HmY \rightleftarrows mH^+ + Y^{-m}$$

$$8H^+ + Fe_3O_4 + 2e^- \rightarrow 3Fe^{2+} + 4H_2O$$

$$Fe^{2+} + Y^{-m} \rightarrow Fe{:}Y^{2-m}$$

4. REGENERATION

Figure 5 shows how the cation resin, designated by HR, works to break the chemical bond between the metal and the organic chelating agent, capture and retain the metal, and release the acid and the chelating agent. The acid goes on to help dissolve more deposits and the chelating agent can be used over and over again. This is the regeneration step described earlier.

Figure 5

REGENERATION BY CATION RESIN

$$Fe{:}Y^{2-m} + 2H{:}R \rightarrow Fe{:}R_2 + 2H^+ + Y^{-m}$$

5. OXIDE DEPOSITS IN WATER COOLED REACTORS

Figure 6 shows the types of metallic oxide deposits found in different types of reactors. In a CANDU PHW with no in-core boiling, carbon steel piping, and Monel boiler tubes, the chemistry conditions are reducing and the stable form of Fe is magnetite, Fe_3O_4. This forms a dense black protective layer on the piping walls. The discussion so far has centred on dissolving this compound and releasing the entrained Co-60 which is held as Co ferrite. There is also some Ni ferrite. There are no copper deposits.

In advanced CANDUs, such as the reactors at the Bruce station, there is some in-core boiling and, under these conditions, Monel is not the most suitable boiler tube material, so Inconel has been substituted. The chemistry conditions will still be reducing, but the in-core boiling will lead to slightly higher dissolved O_2, especially at the core outlet, so the value has been increased to less than 10 ppb. All of the oxide deposits listed for the present CANDU reactor will be present, likely some hematite, and possibly some Cr deposits. If there is an appreciable amount of Cr in these deposits (say, 30 to 45 percent), it may mean the use of a different reagent since the original reducing reagent developed for present CANDU reactors does not dissolve iron chromite to any significant extent. If the Cr content is low (say, less than 10 percent), the deposit will probably dissolve and release the Cr compounds as particles.

Figure 6

OXIDE DEPOSITS IN
WATER COOLED REACTORS

Reactor Type	Chemistry	Deposits	Decontamination Reagent Required
present CANDU PHW No steam CS, Monel	Reducing O_2 5 ppb	$FeOFe_2O_3$ (Fe_3O_4) $NiOFe_2O_3$, $CoOFe_2O_3$ no Cu deposits	Reducing
advanced CANDU PHW Some steam CS, Inconel	Reducing O_2 10 ppb	Spinels as above Fe_2O_3 ? $FeOCr_2O_3$?	Reducing Oxidizing ?
CANDU BLW Steam CS, SS	Oxidizing O_2 10 ppb	$FeOFe_2O_3$ (50%) Fe_2O_3 (50%) no Cr deposits	Reducing
PWR No steam SS, Inconel	Reducing	$FeOFe_2O_3$, $NiOFe_2O_3$ $CoOFe_2O_3$, $FeOCr_2O_3$ $Ni_xFe_yCo_zOCr_aFe_bO_3$ ($x+y+z=1$) ($a+b=2$)	Oxidizing
BWR Steam SS	Oxidizing	$FeOFe_2O_3$, Fe_2O_3 NiO_aOH_b, $NiOFe_2O_3$ ($a+1/2b=1$) no Cr deposits	Reducing

The third type of CANDU is the BLW as exemplified by Gentilly-1.[4] It is similar to a direct cycle BWR in that there are no boiler tubes, but the piping is primarily carbon steel with some stainless steel. The dissolved O_2 is still low, but the conditions are certainly more oxidizing than in the previous two reactor types considered. In a BLW reactor, approximately 50 percent of the oxide film is hematite and the other 50 percent magnetite. When G-1 was decontaminated by the CAN-DECON method, the hematite dissolved as readily as the magnetite. Decontamination factors in the range 2 to 4 were obtained. This demonstrates that the reagent developed for PHW reactors can also be used successfully in CANDU BLW reactors. Independent work in the United Kingdom has confirmed that hematite solubility can be very high under reducing conditions.[5]

In PWRs, the primary circuit piping is stainless steel and

a different type of film is formed. Chromium, a major component of stainless steel, is usually a major component of the oxide film, typically 40 percent. Under reducing conditions, Cr exists as Cr^{+3} and readily enters into a spinel type structure such as $FeOCr_2O_3$. Oxides of this type containing Cr are very insoluble. To dissolve them and release the entrained activity, the Cr must first be oxidized to the +6 valence state where it readily dissolves. This is the basis of the AP-CITROX process where alkaline permanganate is used to oxidize and dissolve the Cr and a citric acid-oxalic acid mixture is used to dissolve the remaining magnetite and Ni ferrite, etc.

Note that strong oxidizing conditions are required. A reagent that operates under reducing conditions, such as the reagent used in CANDU PHW decontaminations, will have little effect since there is nothing to reduce. If the major component of the oxide film is iron chromite, $FeOCr_2O_3$, both Fe and Cr are already in their lowest oxidation states. This is why a different reagent will have to be used in the advanced CANDUs if they have a significant amount of Cr in their oxide film. A program to develop an oxidizing reagent that operates within the CAN-DECON framework is under way in Canada.

In BWRs chemical conditions are considerbly more oxidiz than in any of the other reactor types considered, and the most stable form of Cr in solution is +6, not +3. In the +6 state, Cr does not form spinels or any other type of deposit, but rather stays in solution until removed by ion exchange resin in the reactor water cleanup system. Since there is no Cr, the deposits will be very similar to the deposits found in a CANDU BLW. The only difference may be the presence of more Ni in the form Ni ferrite because of the use of stainless steel, which contains about 8 percent Ni. Previous workers in BWR decontamination studies stressed the great difficulty in removing the film from BWR surfaces because of the extreme stability of the deposits produced in the oxidizing environment of the BWR.[6] Canadian decontamination experience does not support this statement. While it is true that the less soluble hematite will be formed in addition to magnetite, it does not form a tenacious protective coating as does magnetite, and so is more susceptible to flaking and erosion once the chemical reagents start to attack the deposit and undermine it. The decontamination of G-1 has shown that even with relatively mild reagent conditions, hematite can be successfully removed. But more important than this argument is the absence of Cr, which actually makes the deposits easier to dissolve than PWR deposits, but different conditions are required. Different means a reducing reagent instead of an oxidizing reagent. If Ni ferrite is the dominant species in the deposit, there is nothing that can be touched by an oxidizing reagent. Fe is already in its highest valence state, and tests by AECL have shown that even persulfate will not oxidize Ni to the +3 state when it is incorporated into a spinel structure.[3] A reducing reagent will dissolve Ni ferrite by the same reaction described earlier for magnetite, but the rate may be somewhat slower since Ni ferrite is thermodynamically more stable than magnetite. Both reactions are shown in Figure 7, as well as the reaction for the dissolution of hematite under reducing conditions.

Figure 7

DISSOLUTION OF DEPOSITS

Magnetite

$$Fe_3O_4 + 8H^+ + 3Y^{4-} + 2e^- \rightarrow 3Fe{:}Y^{2-} + 4H_2O$$

Nickel Ferrite

$$NiOFe_2O_3 + 8H^+ + 3Y^{4-} + 2e^- \rightarrow 2Fe{:}Y^{2-} + Ni{:}Y^{2-} + 4H_2O$$

Hematite

$$Fe_2O_3 + 6H^+ + 2Y^{4-} + 2e^- \rightarrow 2Fe{:}Y^{2-} + 3H_2O$$

---

6. DISSOLUTION OF DEPOSITS

In summary, a reducing reagent is required for deposits not containing Cr, while an oxidizing reagent is required for those containing Cr, as shown in Figure 8. The statements in this figure are not meant to be absolute, but rather to indicate the relative effectiveness of the two types of reagents with various kinds of deposits.

Recent work in Canada has substantiated these conclusions. Actual BWR specimens have been decontaminated in test loops, using the CAN-DECON process with a reducing reagent. Decontamination factors as high as 80 have been obtained. More typical values are 10 to 20, as shown in Figure 9. PWR specimens decontaminated under the same conditions gave DFs of 1.1 to 1.2.

However, when a new, experimental oxidizing reagent was used, DFs as high as 20 were obtained with more typical values being 5 to 10 on Inconel and 10 to 15 on stainless steel. A patent application on this new reagent is pending, and additional details will be released at an appropriate time.

Figure 8

SUMMARY OF DISSOLUTION CHARACTERISTICS OF VARIOUS METAL OXIDES FOUND IN WATER COOLED REACTORS

Fe_2O_3 Fe_3O_4 $NiOFe_2O_3$ $CoOFe_2O_3$	Reducing Reagent - will dissolve Oxidizing Reagent - will not dissolve
Cr_2O_3 $FeOCr_2O_3$ $NiOCr_2O_3$	Reducing Reagent - will not dissolve Oxidizing Reagent - will dissolve

Figure 9

DECONTAMINATION FACTORS OBTAINED ON LWR SPECIMENS USING THE CAN-DECON PROCESS

Specimens	Reducing Reagent	Oxidizing Reagent
BWR SS	10-20	-
PWR SS	1.1-1.2	10-15
PWR Inconel	1.0-1.1	5-10

References

1. P.J. Pettit, J.E. LeSurf, W.B. Stewart, R.J. Strickert, S.B. Vaughan, "Decontamination Of The Douglas Point Reactor By The CAN-DECON Process", Paper No. 39, CORROSION/78, Houston, Texas (Mar. 78)

2. B. Montford, T.E. Rummery, "Properties of Douglas Point Generating Station Heat Transport Corrosion Products", Report No. AECL-4444, Atomic Energy of Canada Limited, Pinawa, Manitoba (Sept. 75)

3. T.E. Rummery, D.W. Shoesmith, Private Communication, Atomic Energy of Canada Limited, Pinawa, Manitoba (Nov. 76)

4. Nuclear Engineering International (Nov. 72)

5. D.J. Turner, "Some Thermodynamic Aspects Of Decontamination in SGHWR And Other BWRs, Part II: Decontamination With Citric Acid And Other Polybasic Acids", CEGB Report, RD/L/N213/75 (Feb. 77)

6. D.E. Harmer, O.U. Anders, J.J. Holloway, "Developing A Solvent, Process, Equipment, And Procedures For Decontaminating The Dresden-1 Reactor", presented at the 37th Annual Meeting of the American Power Conference, Chicago, Illinois (21-23 Apr. 75)

TEPCO BWR DECONTAMINATION EXPERIENCE

Shiro Sasaki, 1; Hiroaki Koyama, 2; Jiro Kani, 3; Tadao Yoshida, 4

1. Tokyo Electric Power Co., Inc., Tokyo, JAPAN
2. Central Research Institute of Electric Power Industry
3. Toshiba Corporation, Toshiba, JAPAN
4. Toyo Engineering Corporation

INTRODUCTION

The status of our program for research and development on BWR decontamination was described in the previous paper[1]. The major items conducted recently are as follows:

(1) Identification of the major radionuclides contributing to the radiation dose rates, characterization of the crud deposits on pipings and fuel surfaces, and prediction of future radiation levels, are carried out.

(2) Decontamination factors and corrosiveness were examined with several domestic and foreign reagents which were screened in non-radioactive condition and the specimens taken from the reactor primary system at the Fukushima Dai-ichi plant.

(3) Test for evaluation of the effects of residual reagents on the surface of the structural materials after decontamination and test for treatment of decontamination wastes are under progress.

(4) The evaluation method is beeing studied as a step to evaluate various kinds of decontamination process.

EXPERIMENTAL RESULTS

1. Analysis of crud deposits

The chemical and radiochemical analysis of the crud deposits on the specimens taken from reactor water cleanup system and feedwater sparger at Fukushima Dai-ichi plant were carried out.

Crud deposits were separated as the soft and the hard crud by ultrasonic and cathodic cleaning, respectively. Table 1 shows the elemental compositions of crud deposits on these specimens.

There are not so much differences in the composition between units or locations.

Fig. 1 shows the one example of the radioactivity distribution in these crud deposits. Radionuclides such as Co-60, Mn-54 and Zn-65 were detected in the crud on the specimen from feedwater sparger of Unit 2. These radionuclides were predominantly included in the hard crud as well as the cleanup system pipings, on the other hand, the base metal included about 20% of the total radioactivity. This value is much higher than that of the clean-up system piping which is 2%.

Table 1. Elemental composition of crud at Fukushima Dai-ichi plant

Element	Unit 1 (Cleanup System)				Unit 2 (Feedwater sparger)			
	Main pipings		Bypass pipings		Reactor water side		Feedwater side	
	mg/cm^2	%	mg/cm^2	%	mg/cm^2	%	mg/cm^2	%
Fe	0.625	90.7	0.300	86.0	0.637	80.8	0.409	88.4
Ni	0.053	7.6	0.025	7.2	0.066	8.4	0.040	8.7
Cr	0.010	1.5	0.022	6.3	0.083	10.6	0.012	2.5
Co	0.002	0.2	0.002	0.5	0.002	0.2	0.002	0.4
Total	0.690	100.0	0.349	100.0	0.788	100.0	0.463	100.0

2. Evaluation of decontamination reagents

Several reagents, which were screened by the preliminary test with non-radioactive simulated crud, were tested with those specimens described in previous section in static and dynamic conditions to evaluate the decontamination factor and corrosiveness. The test equipments is shown in Fig. 3.

The several screend domestic and foreign reagents are the similar type for Dresden-I and CAN-DECON.

The decontamination factors of Co-60 is shown in Table-2 and the general corrosion of carbon steel is shown in Table 3. Very little amount of general corrosion to stainless steel was measured. Furthermore, SCC acceleration test using creviced bent beam (CBB) specimens shown in Fig. 2 to evaluate the effect of the reagents is progressing at present, no indication of SCC was found under decontamination conditions.

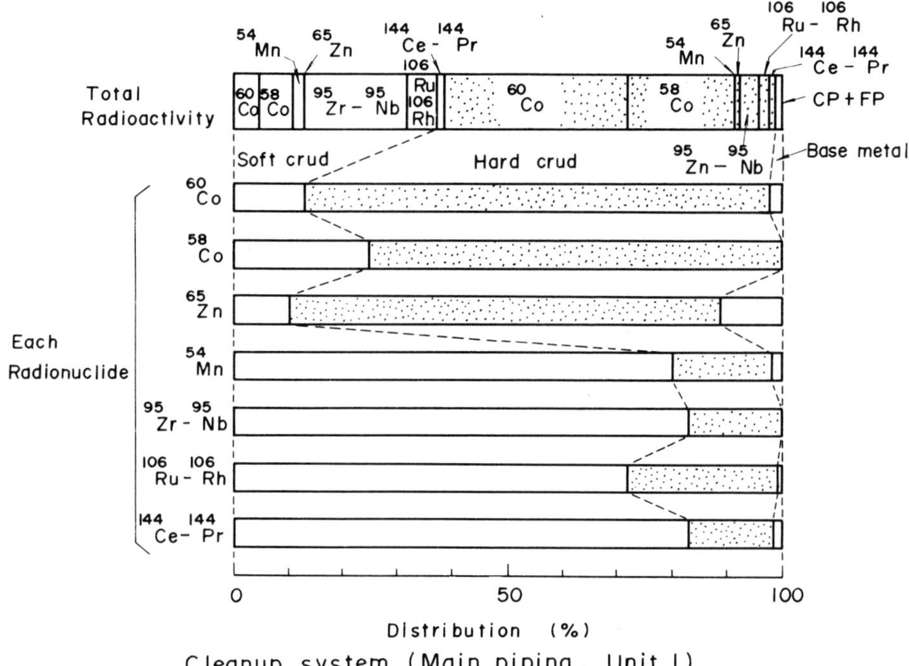

Cleanup system (Main piping, Unit I)

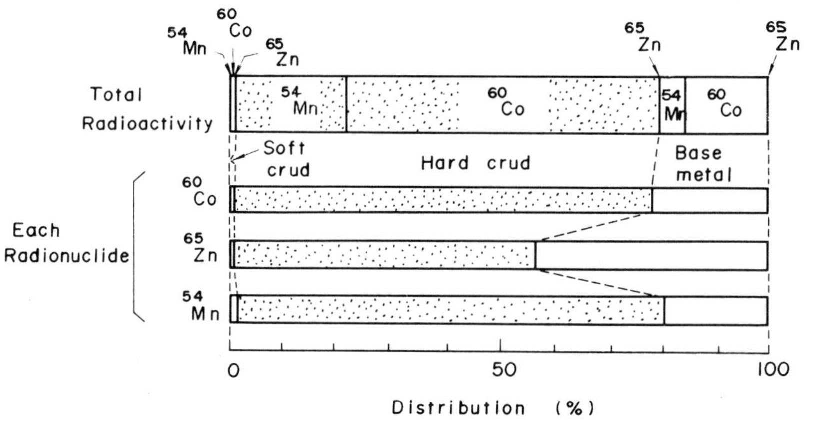

Feedwater sparger (Reactor water side, Unit 2)

Fig. 1 Radioactivity distribution in crud deposits

Table 2. The Co-60 decontamination factors of reagents

Test type	Specimen	Reagents		
		A	B	C
static test	feedwater sparger	~40	~25	~20
	cleanup system piping	~200	~2.5	~1.5
dynamic test	feedwater sparger	~200	~30	~15
	cleanup system piping	~300	~30	~2

Table 3 The general corrosion of carbon steel with reagents

Reagent	A	B	C
Time (hrs)	100	24	24
Corrosion loss (mg/cm^2)	5~20	0.5~3	25~35

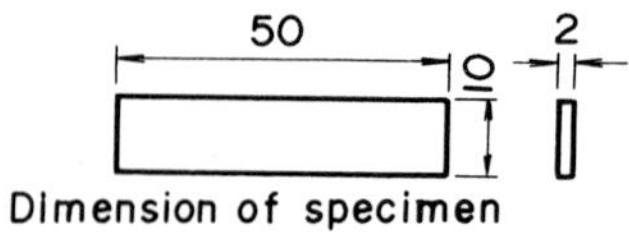

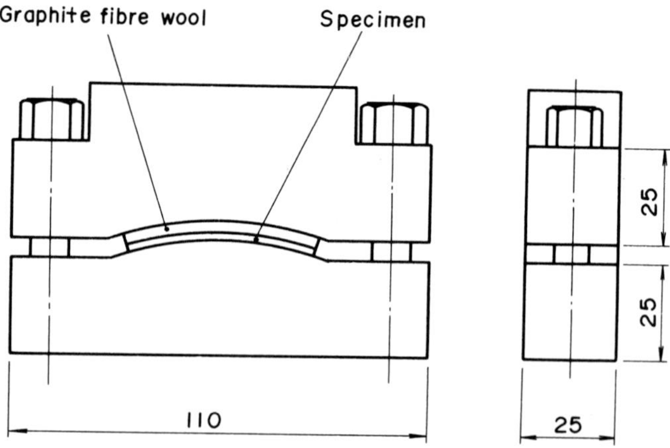

Fig. 2 Dimension of CBB test specimen and jig

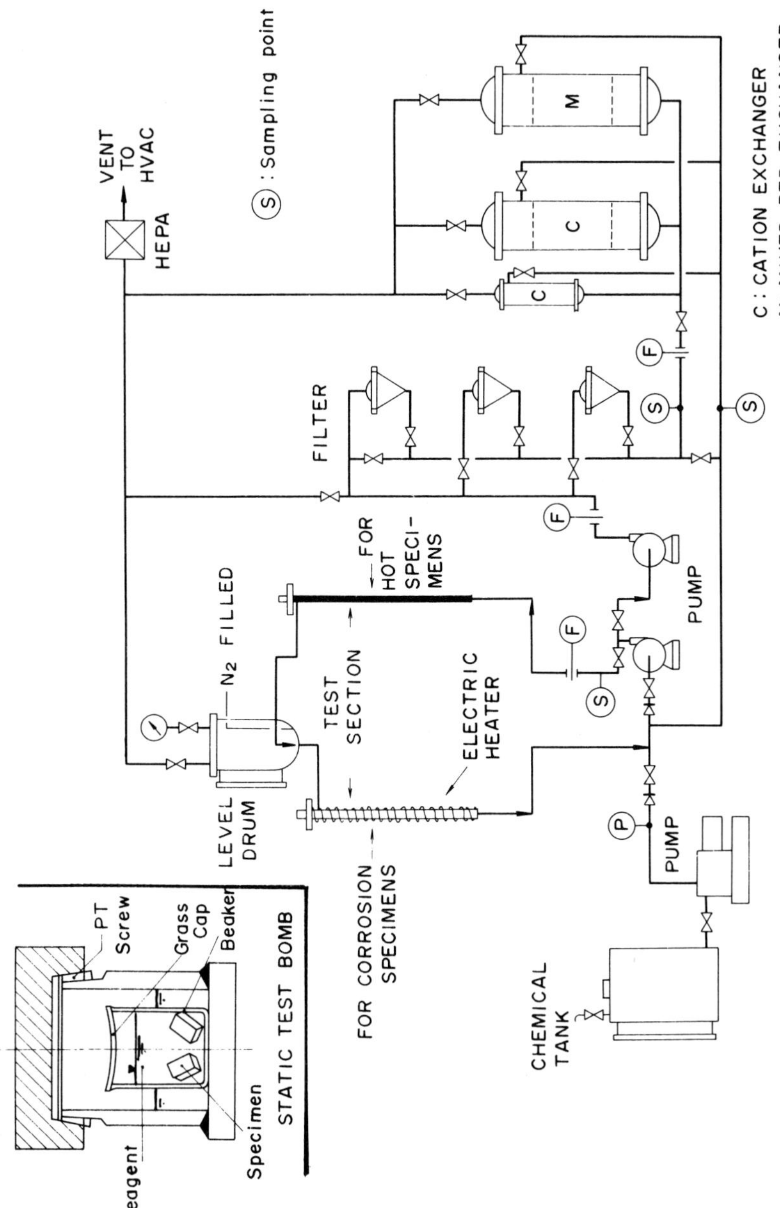

Fig. 3 Decontamination and corrosion test equipments

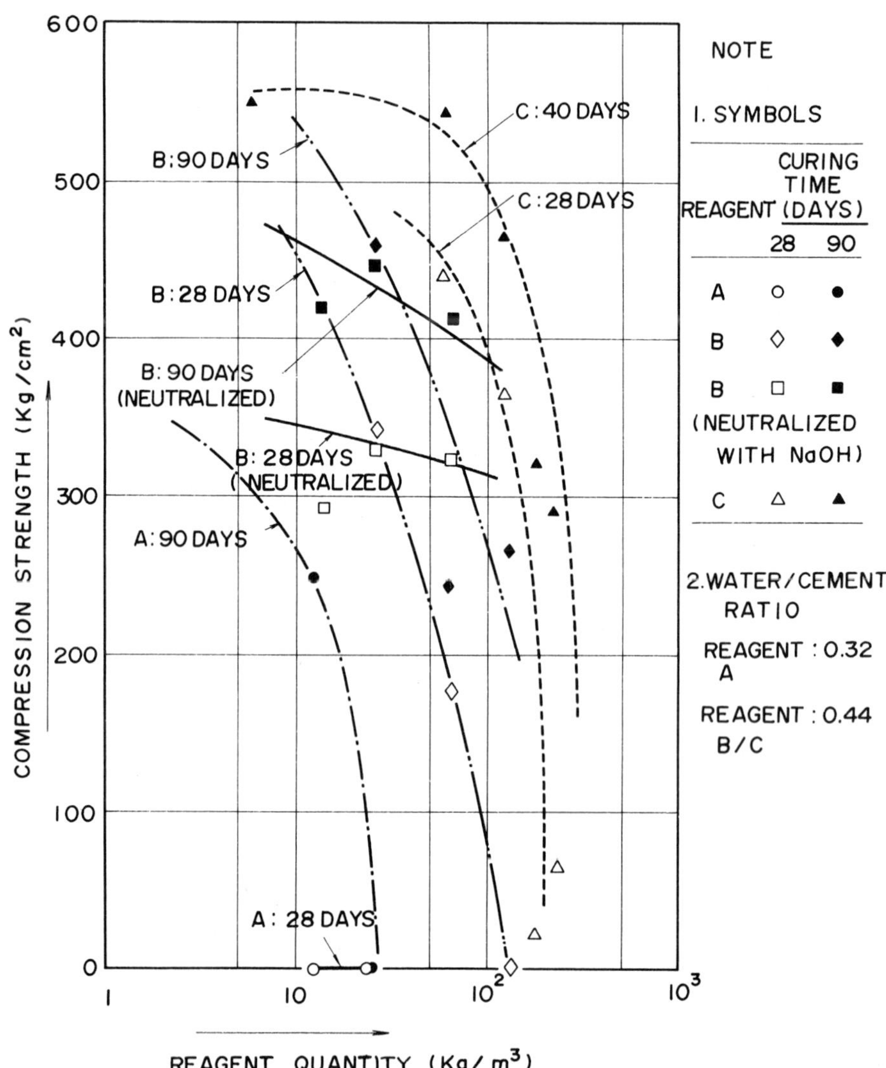

Fig. 4 Compression strength of solidified reagent with portland cement

3. Fundamental test for radwaste treatment

A series of the test is conducted to obtain the basic data for the design of the decontamination waste treatment system. These tests include concentration, drying and solidification of the wastes.

From the concentration and drying test, the thermal decomposition characteristics of these wastes were identified, and it was found that the reagents in these wastes were decomposed above around 150°C.

In solidification test, these wastes were solidified with portland cement, bitumen and plastics, and the tests on the strength and the decomposition characteristics were performed. The test of leachability of solidified wastes is now continuing.

Fig. 4 shows the strength of solidified wastes with portland cement. From this figure, maximum amount of a reagent in the cement to get specified strength is known.

4. Evaluation of the effects on the materials

In order to evaluate the effects of resudual reagents on the surface of the structural materials after decontamination, a long term test under BWR operating conditions is progressing using a high temperature and high pressure non-radioactive loop shown in Fig. 5.
At the loop, the test section No. 1 is a autocrave where the general corrosion test is carried out, the test section No. 2 is a welded pipe stand where the effect on welded pipe is tested and at the test section No. 3, the SCC acceleration test with CBB and double U-bend specimens is carried out.

EVALUATION OF DECONTAMINATION PROCESSES

In this study, we have been evaluating various decontamination process using a flow diagram as shown in Fig. 6.

As a first step, the following items should be considered.

(1) The decontamination operation should be performed safely (i.e.; the occupational radiation exposure should be minimized, gas evolved with the decontamination operation should be treated safely etc.)

(2) The decontamination should not have the serious effect on the system (i.e.; corrosion of the material, malfunction of the equipment etc.)

(3) Generated radwastes should be treated and stored safely.

In evaluating the effectiveness and cost of each decontamination process, following method as shown in Fig. 7 are considered.

Since the principal object of decontaminations is to reduce the occupational exposure, the effectiveness will be expressed by a value of total Man-Rem saving through the plant life-time. And the cost will be consist of the investment, operation, R & D

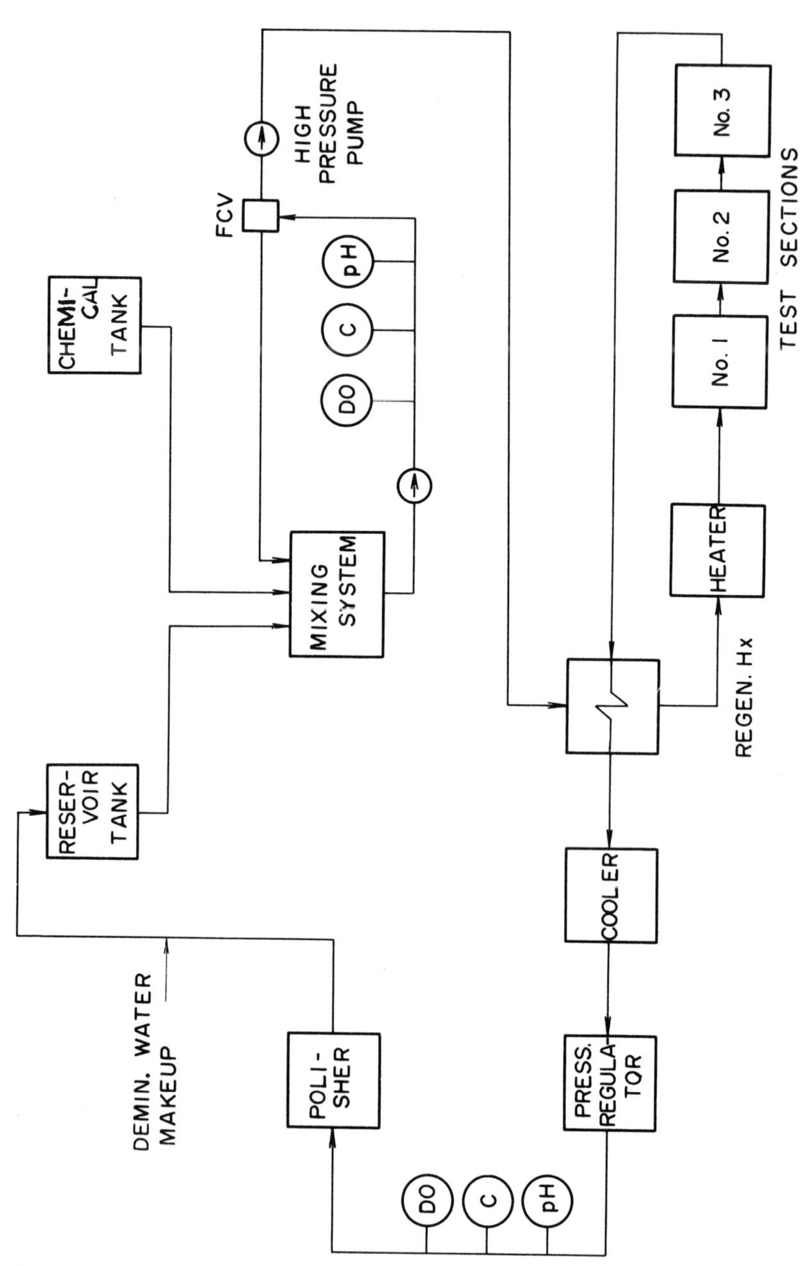

Fig. 5 High temperature and high pressure test loop

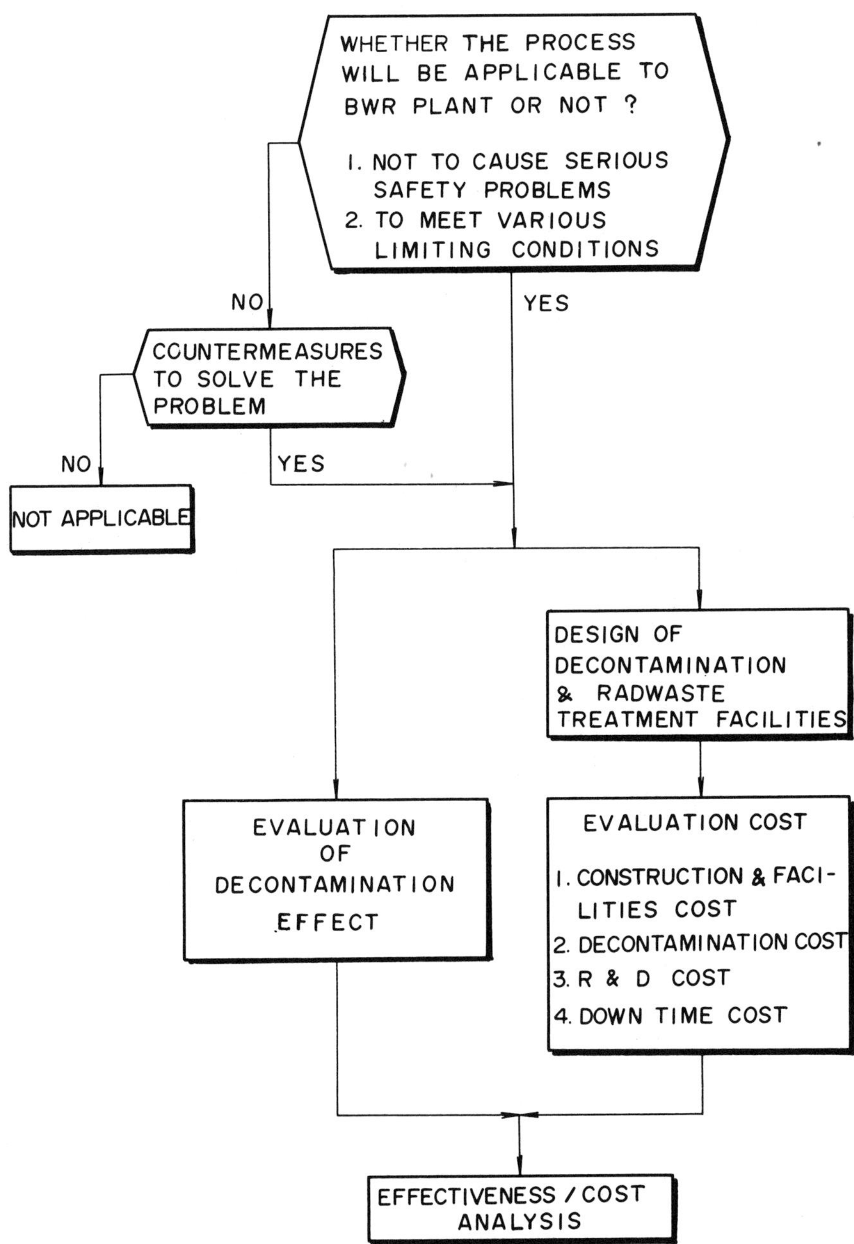

Fig. 6 Simplified flow diagram for evaluation of decontamination process

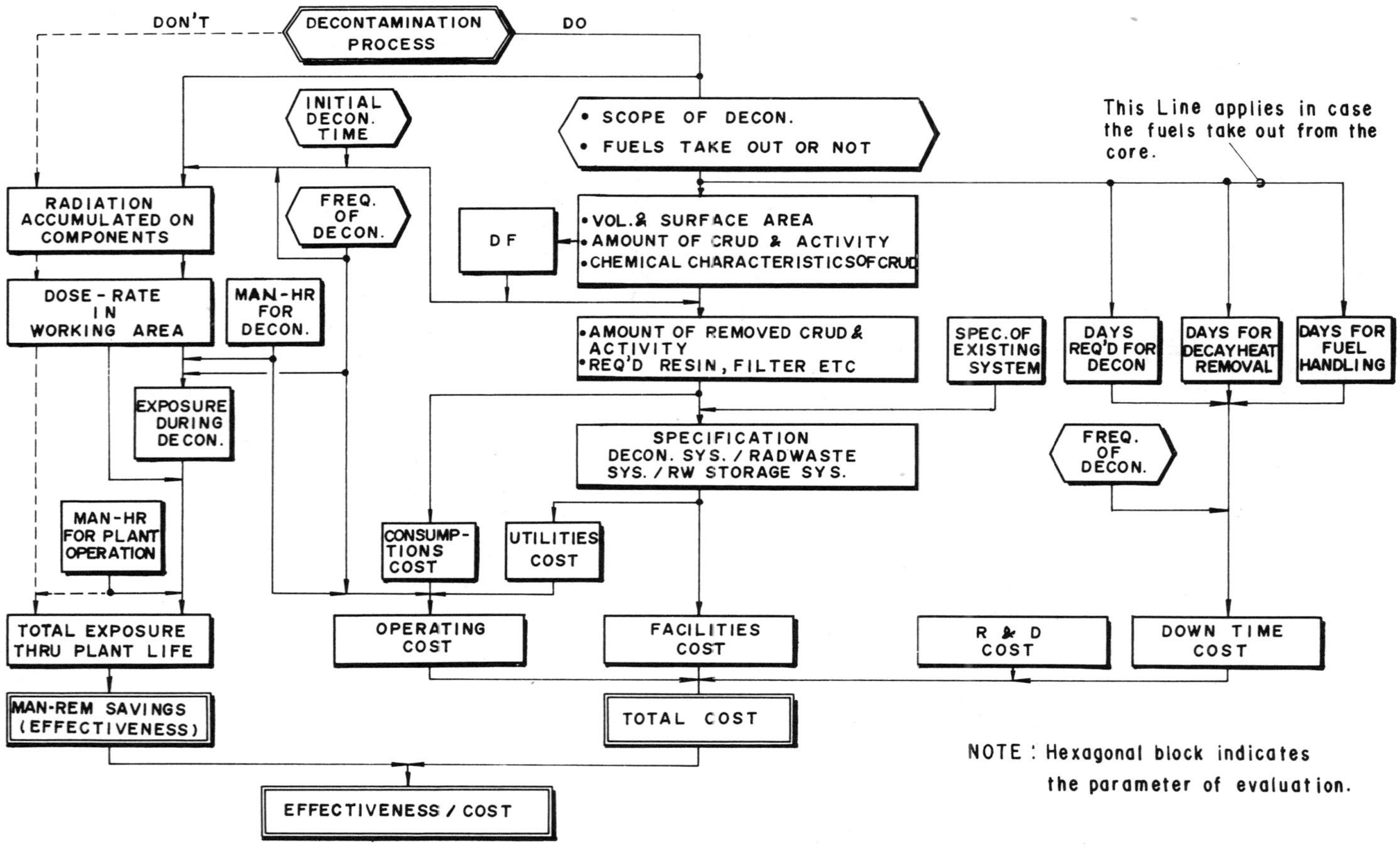

Fig. 7 Flow diagram for effectiveness-cost evaluation

costs and lost revenue for a down time.

With the above methods we are presently investigating following decontamination processes with or without fuel decontamination.

. Dresden-1 process
. Can-Decon process applied to BWR
. Dilute chemical process
. Mechanical decontamination
. Reduction of the crud into the reactor primary system

REFERENCE

1. S. Sasaki (TEPCO), H. Kōyama (CRIEPI) and A. Tani (TOSHIBA), Research and development on BWR decontamination at the Tokyo Electric Power Company, ANS 24th Annual Meeting, San Diego (June 1978)

DILUTE CHEMICAL DECONTAMINATION PROGRAM REVIEW

L. D. Anstine, J. C. Blomgren, and P. J. Pettit

General Electric Company, Commonwealth Edison Company
and U.S. Department of Energy*

INTRODUCTION

It is desirable to reduce radiation exposure to workers in nuclear power plants to values that are as low as reasonably achievable. One method of reducing occupational exposure is to reduce the source of the exposure. Decontamination of the primary coolant system would reduce shutdown radiation levels and occupational radiation exposure in nuclear power plants.

One promising method is the use of dilute organic chemicals for rapid decontamination of the primary coolant systems as has been demonstrated in Canadian reactors.[1] A similar approach could be used for operational decontamination of nuclear power plants in the United States. In 1977, the U.S. Department of Energy contracted with the Commonwealth Research Corporation (CRC)** to study the technical feasibility of this approach for boiling water reactors (BWRs). In turn, CRC subcontracted the experimental work to General Electric.

The objective of the Dilute Chemical Decontamination Program is to develop and evaluate a process which utilizes reagents in dilute concentrations for the decontamination of BWR primary systems and for the maintenance of dose rates on the out-of-core surfaces at acceptable levels.

---

*Present address: The Atomic Industrial Forum
Washington, D.C. 20014

**CRC is a wholly-owned subsidiary of Commonwealth Edison Company.

PROCESS CONCEPT

The process concept on which the feasibility study was based consists of the injection of concentrated organic reagents into the primary coolant until the desired concentration (∿0.01M) is reached. Decontaminations would be performed during a reactor shutdown with the fuel remaining in the core. The solvent would be recirculated throughout the primary system and would dissolve most of the out-of-core and fuel deposits. Some corrosion products may be released as particles. A side stream would be processed continuously through a filtration system to control the particles and through an ion-exchange system to regenerate the reagent by removing the dissolved corrosion products. The decontamination process would be terminated by a purification cycle during which the solvent would be circulated through mixed-bed ion-exchange resins until the coolant is returned to BWR specifications. It is envisioned that the entire process would be completed in less than 3 days.

SOLVENT DEVELOPMENT

In early screening tests 0.01M solutions of oxalic acid, nitrilotriacetic acid (NTA), and ethylenediaminetetraacetic acid (EDTA) were identified as being capable of dissolving 80% of the Co-60 in BWR out-of-core films in less than 72 hours. Exploratory tests were performed to evaluate the properties of these three reagents relative to:

1. dissolution of the out-of-core films and fuel deposits,
2. regeneration by ion-exchange,
3. purification by mixed-bed ion exchange,
4. radiolytic stability,
5. thermal stability,
6. precipitation and redeposition of dissolved crud, and
7. general corrosion of low alloy steels.

Based upon the results of these tests, oxalic acid was selected to be the principal component in the solvent formulation. A summary of the advantages and disadvantage of each reagent is given in Table 1. The exploratory tests demonstrated that at conditions considered acceptable for a dilute chemical decontamination, 0.01M solutions of oxalic acid are capable of dissolving 80% of the Co-60 in BWR out-of-core oxide films in less than 12 hours.

The major process disadvantage identified for oxalate base solvents was precipitation of ferrous oxalate. Several approaches to eliminate, minimize, or control the precipitation of ferrous oxalate were evaluated. These approaches included the addition of corrosion inhibitors to reduce the ferrous iron generation rate, oxygenation of the solvent to oxidize the ferrous iron to

Table 1. Advantages and Disadvantages of Reagent Systems

Oxalic Acid

Advantages

1. Rapid dissolution of out-of-core oxide films
2. Innocuous decomposition byproducts
3. Can be regenerated
4. Easily injectable
5. Low general corrosion rates
6. Does not require a corrosion inhibitor
7. Previously used in reactor decontaminations

Disadvantages

1. Precipitation of ferrous oxalate
2. Requires additives to control precipitation
3. Potential slow fuel deposit dissolution rate

NTA

Advantages

1. Rapid dissolution of fuel deposits
2. Can be regenerated
3. Chemically stable - no precipitation
4. Thermally stable at process conditions

Disadvantages

1. Erratic dissolution results with out-of-core oxide films
2. Potential depletion of reagent due to unknown mechanism
3. High general corrosion rates
4. Decomposes to intermediate degraded nonvolatile forms of NTA
5. Requires addition of corrosion inhibitor and probably a reducing agent

EDTA

Advantages

1. Rapid dissolution of fuel deposits
2. Chemically stable - no precipitation
3. Thermally stable at process conditions

Disadvantages

1. Potential Fe^{+3} retardation of dissolution rate
2. Cannot be regenerated
3. High general corrosion rates
4. Decomposes to intermediate degraded nonvolatile forms of EDTA
5. Requires addition of corrosion inhibitor and reducing agent
6. Difficult to inject

ferric iron, rapid regeneration to maintain the ferrous iron concentration below the solubility limit, and addition of complexing agents to increase the solubility of ferrous iron. A combination of oxygenation, regeneration, and addition of citric acid is capable of preventing precipitation of ferrous oxalate in bulk solution and of minimizing the formation of ferrous oxalate films on carbon steel surfaces.

A second concern with the use of oxalic acid is the potential for a lower dissolution rate of fuel deposits. The few scoping tests performed with fuel-deposit scrapings removed from fuel rods with an abrasive stone indicated the fuel-deposit dissolution rate might be only 50 to 70% in 12 hours. Additional experimentation may be required in future programs.

VNC LOOP TESTS

A test loop (Figure 1) was constructed at the General Electric Vallecitos Nuclear Center (VNC) to provide a test vehicle for evaluating solvent systems in integrated decontamination process tests. The loop contains the necessary subsystems, such as ion exchange and filtration, as well as process and instrumentation controls. A 30-in. section of 6-in.-diameter Type-304 stainless steel piping was decontaminated in each test. This piping was removed from the Nine Mile Point (NMP) clean-up system and was provided by Niagara Mohawk Power Corporation. A summary description of the corrosion product films[2] on the NMP piping is given in Table 2. The basic characteristics, including dissolution rate in oxalic acid, of the NMP films are similar to those observed for the films on the piping of other BWRs.[2-4]

Table 2. NMP Out-Of-Core Corrosion Products

	Crystal Structure	Elemental Composition (%) Fe	Cr	Ni	Co	Film Thickness (mg/dm^2)	Co-60 Concen. ($\mu Ci/cm^2$)
Loosely Adherent	αFe_2O_3	95	2	3	0.1	50	1.4
Tightly Adherent							
Species 1	$NiFe_2O_4$	83	1	15	0.5	36	8.1
Species 2	$NiFe_2O_4$	70	9	20	0.6	13	1.8

The VNC loop tests were designed to model most of the chemical and hydraulic conditions expected during a dilute chemical decontamination of BWR primary system piping. The normal operating procedures and conditions for the VNC loop tests are given in Table 3, while the specific conditions and results of VNC loop tests are given in Table 4.

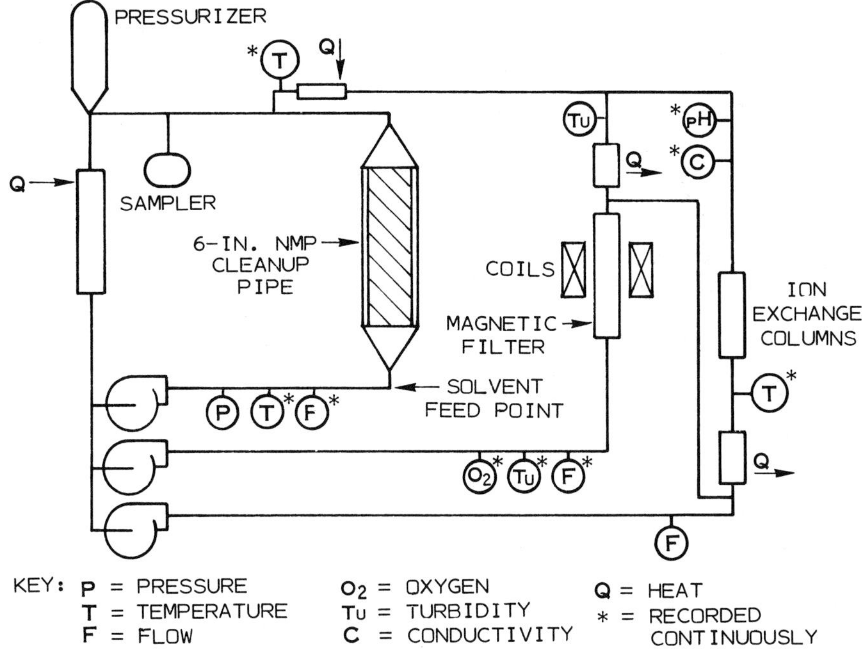

Fig. 1. VNC Decontamination Test Loop (Simplified)

Tests A, B, and C were exploratory and were used to define process conditions; Tests D and E were conducted as process demonstrations. The latter two tests were considered highly successful, since the desired decontamination rate was obtained while all the process parameters were maintained within specification. The results of the VNC loop tests indicated the most reasonable chemical system was 0.01M oxalic acid and 0.005M citric acid at pH 3.0 with a controlled level of dissolved oxygen.

Typically, with constant chemical conditions, the Co-60 dissolution rate, after a short induction period, could be reasonably expressed by a first order rate equation with a rate constant, k, expressed as:

$$k = \ln(1-x)/t \tag{1}$$

where x is the fraction of Co-60 dissolved and t is time. Early tests with a small glass loop and the later VNC loop tests indicated the value of k is dependent upon the concentrations of oxalic acid, iron, and dissolved O_2 as well as the pH and temperature. The disso-

Table 3. VNC Loop Procedures and Conditions

1. A new section of contaminated pipe was loaded into the test section;

2. The ion-exchange columns were filled with the appropriate resins;

3. The loop was filled with demineralized water ($\sim 5\ell$);

4. The loop was heated to 90°C and the dissolved O_2 was adjusted to the desired specifications by sparging with a mixture of N_2 and air;

5. Concentrated solutions of the reagents were injected into the demineralized water;

6. The Co-60 activity on the pipe and in several of the tubing lines was monitored continuously with NaI detectors;

7. The loop pH, dissolved O_2, conductivity, turbidity, temperature, and pressure were monitored continuously;

8. The solution was continuously filtered with either a high intensity magnetic filter or a 0.22 μm filter paper;

9. The reagents were continuously regenerated by ion exchange at a rate of 1.4 system volumes per hour;

10. The pH was maintained at 3 by passing the solution through a H^+-form cation-exchange resin bed;

11. 10 ml samples of the solution were collected periodically and analyzed;

12. The solution was recirculated until the desired level of decontamination was obtained; and

13. The test was terminated by passing the solution through a mixed-bed of ion-exchange resin.

Table 4. VNC Loop Tests

Test	Reagent*	Dissolved O_2 (ppm)	Co-60 Dissolution Percent	Co-60 Dissolution Time (h)	Comments
A	0.01M NTA	<0.2	80	60	
B	0.02M OA	<0.2	95	15	FeC_2O_4 precipitation
C	0.01M OA 0.005M CA	0.2 to 20	82	30	Multiple dissolution rates due to varying O_2 level
D	0.01M OA 0.005M CA	0.75	85	12	
E	0.01M OA 0.005M CA	0.75	85	15	60% regeneration rate - 0.85 system volumes per hour
F	0.01M OA 0.005M CA	0.75	-	-	Corrosion test No contaminated coupons.

*OA = Oxalic Acid
CA = Citric Acid

lution curves for VNC Loop Tests D and E are given in Figure 2. In Test E the regeneration rate was 60% of the rate in Test D. As expected, the lower regeneration rate resulted in a lower dissolution rate. Also, the induction period at the beginning of the Test E was approximately twice as long. These results are considered caused by an increased iron concentration.

In VNC Loop Test B, the dissolved O_2 was maintained below 0.2 ppm and approximately 20% of the total iron removed from the pipe section precipitated as ferrous oxalate; an equivalent percentage of Co-60 also precipitated. Even though essentially all the ferrous oxalate and Co-60 redissolved during the purification cycle, it is still considered highly undesirable to allow bulk precipitation during a dilute chemical decontamination. As indicated earlier, appropriate levels of dissolved oxygen will prevent bulk precipitation of ferrous oxalate by oxidizing the ferrous iron to ferric iron. However, higher levels of dissolved oxygen suppress the dissolution rate of BWR oxide films in solutions of oxalic acid. In VNC Loop

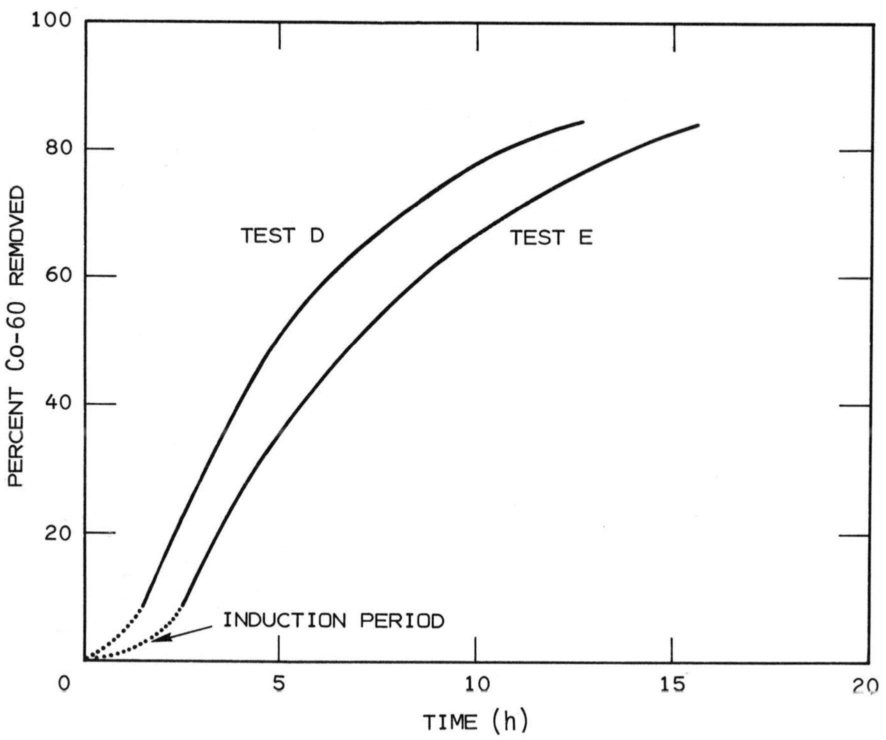

Fig. 2. Co-60 Dissolution Curves

Test C, the dissolved oxygen concentration was varied from 0.2 to 20 ppm to assess the optimum level of dissolved oxygen. The results of this test indicate 0.5 to 1.0 ppm dissolved O_2 is an acceptable range.

The Co-60 activity in solution increased to a maximum during the first few hours of each VNC loop test. The Co-60 activity in solution then decreased, as the oxide film dissolution rate decreased and was exceeded by the solvent regeneration rate. The regeneration resins are essentially 100% efficient for removing dissolved metals. The Co-60 activities in solution as a function of time for Tests D are shown in Figure 3.

At the conclusion of the decontamination cycle, the solvent is passed through a mixed-bed of ion-exchange resin to remove the reagents and residual dissolved corrosion products, and return the coolant to shutdown BWR water conditions. Typically, in the VNC loop tests a conductivity of less than 2 μmhos was obtained within 8 hours.

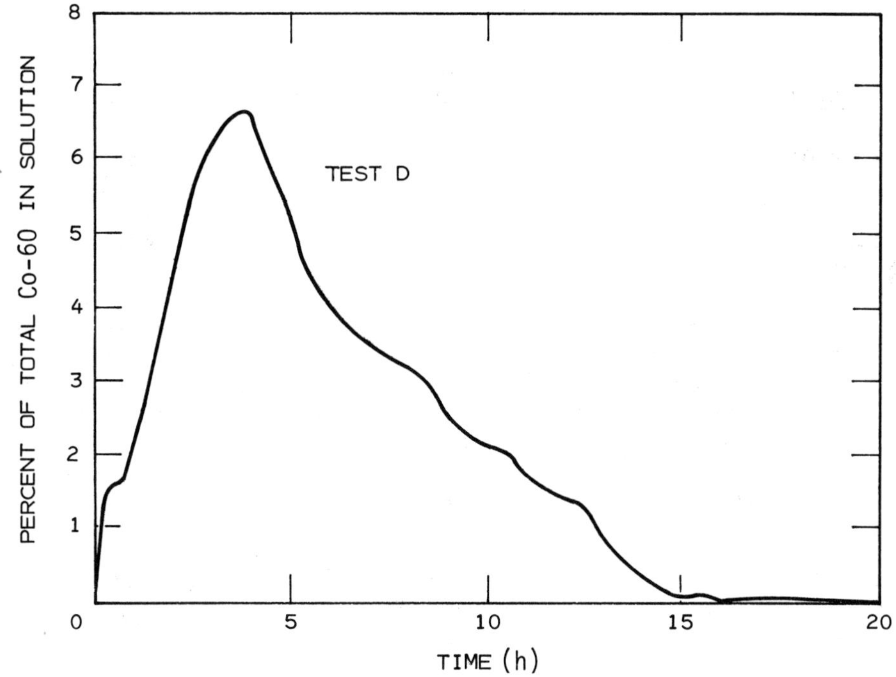

Fig. 3. Co-60 Activity in Solution

In the VNC loop tests most of the radioactivity and corrosion products were removed by the regeneration resins. Only small quantities of filterable contaminants were released from the test specimens. A summary of the final distribution of the corrosion products and Co-60 activity for Test D is given in Table 5.

VNC Loop Test F was conducted to evaluate the corrosion properties of generic BWR materials in a simulated dilute chemical decontamination. A 12-liter vessel was utilized in the VNC loop in place of the contaminated pipe section. Three, creviced, constant-radius, bent-beam specimens from each of 3 heats of 16 different alloys/ conditions were exposed to the decontamination and purification cycles to evaluate crevice, galvanic, general, and pitting corrosion; intergranular attack; and stress corrosion cracking. Uncreviced general corrosion coupons of 5 alloys with potentially high general corrosion rates were also included. In addition, creviced, constant-radius, bent-beam, and uniaxial tensile specimens of alloys potentially susceptible to stress corrosion cracking were included and subsequently exposed to oxygenated, high-temperature, low-conductivity water.

Table 5. Final Corrosion Product and Co-60 Distribution for VNC Test D

System	Total Corrosion Products (mg)	Total Co-60 (mCi)
Regeneration Resins	3500	13
Purification Resins	400	0.2
0.22 μm Filter	2	<0.1
Solution	1	<0.1
NMP Pipe	-	2.3

Typical weight losses for all 16 materials are given in Table 6. The weight losses are expressed in milligrams per square centimeter per decontamination. The decontamination cycle was 12 hours and the purification cycle was 8 hours. Metallographic examination of the corrosion specimens is in progress.

Table 6. Weight Loss During Decontamination

Material	Weight Loss Per Decontamination (mg/cm^2)*	
	Creviced Bent Beams	General Corrosion Coupons
Austenitic Stainless Steels		
Type-304, Furnace Sensitized	0.1	-
Type-304, Welded and Low Temperature Sensitized	0.4	-
CF-8 (Cast Type-304)	0.1	-
Type-316	0.1	-
Nitronic-50	<0.05	-
Martensitic Stainless Steels		
Type-420	1.0	3.6
Nickel Base Alloys		
Inconel-600	1.4	-
Inconel-X750	0.8	-
Low Alloy Steels		
A508	2.2	5.8
A302B	1.9	4.2
Carbon Steel		
A106A	1.2	4.5
Zircaloys		
Zirc-2	<0.05	-
Zirc-4	<0.05	-
Hard Facing Alloys		
Stellite	<0.05	-
Colmonoy	0.3	-
Copper-Nickel Alloys		
90-10 Copper Nickel	0.1	0.4

*1 mg/cm^2 ≅ 5 x 10^{-5} in. = 0.05 mils

CONCLUSIONS

Based on the work completed to date, the following conclusions have been reached:

1. Rapid decontamination of BWRs using dilute reagents is feasible.

2. Reasonable reagent conditions for rapid chemical decontamination are: 0.01$\underline{M}$ oxalic acid + 0.005$\underline{M}$ citric acid, pH3.0, 90°C, 0.5 to 1.0 ppm dissolved oxygen.

3. Control of dissolved oxygen concentration is important, since high levels suppress the rate of decontamination and low levels allow precipitation of ferrous oxalate.

4. Demineralization is capable of removing both contaminants and residual reagents from the reactor coolant.

5. The release rate of particulates from the out-of-core surfaces during a dilute chemical decontamination is low and the filtration requirements may be minimal.

REFERENCES

1. P. J. Pettit, J. E. LeSurf, W. B. Stewart, R. J. Strickert, and S. B. Vaughan, "Decontamination of the Douglas Point Reactor by the Can-Decon Process," presented at National Association of Corrosion Engineers Conference, Corrosion/78, Houston, Texas, March 1978.
2. L. D. Anstine, "BWR Decontamination and Corrosion Product Characterization," General Electric Report NEDE-12665 (1977).
3. G. Romeo, "Characterization of Corrosion Product on Recirculation and Bypass Lines at Millstone-1," General Electric Report NEDO-12680 (1978); EPRI NP-949 (1978).
4. L. D. Anstine and G. E. Von Nieda, "Characterization of the Corrosion Product Films on the Dresden-1 Decontamination Pilot Loop After Re-Exposure to Reactor Water," General Electric Report NEDC-12691 (1977).

RESEARCH AND DEVELOPMENT ON BWR-SYSTEM DECONTAMINATION IN SWEDEN

Jan Arvesen, Hans-Peter Hermansson and
Ramond Gustafsson

Studsvik Energiteknik AB
Nyköping, Sweden

INTRODUCTION

The following is a description of the studies of the decontamination of BWR systems and components which have been carried out during the last three years at Studsvik Energiteknik AB. The work has been carried out in collaboration between the Swedish nuclear utilities and Studsvik Energiteknik AB.

OUT-OF-PILE DECONTAMINATION TESTS

The first tests were carried out in a high pressure out-of-pile loop in the Heat Transfer Laboratory in Studsvik. They were aimed at testing a decontamination method using low concentrations of the chemicals involved, and also its efficiency, and the corrosion resistance of reactor component materials to the decontaminating environment. The solution is composed of citric and oxalic acids and EDTA.

This solution was circulated at 85°C. During decontamination the solution passed through a cation exchanger which trapped the radioactive corrosion products and regenerated it. The advantage with this method, by now well known, is the low concentration of chemicals and that the radioactivity is trapped in the ion exchanger.

The test was also intended to improve the routine of the decontamination of reactor systems.

The high pressure loop used had the following basic data:

Total volume	~ 200 l
Pressure	max 12.0 MPa
Temperature	max 324 ^{o}C (saturation temperature)
Circulation rate	max 3 kg·sec^{-1}
Effect	max 80 kW
Pressure maintained by	steam
Cleansing rate	~ 30 l·h^{-1}

The high temperature part of the loop was made in stainless steel, mainly AISI 316; the housing and wheels in the circulating and spray pumps are manufactured in AISI 304, as is the pump spindle, whereas the spindle casing is chromium plated. Metallic gaskets were used.

The cleaning units in the loop during decontamination comprised a mechanical filter, AKA-Cuno Micro-Clean 5 µm mesh, and two ion exchangers each of 9 l capacity. Duolite ARC 351 resin was used in the cation exchanger and Duolite ARA 366 resin in the anion exchanger. During decontamination the solution passed through a cation bed and for the subsequent cleaning it passed through a combined anion/cation bed (65 %:35 %). The cleaning constant was 0.6 h^{-1}.

Water and chemicals for decontamination were dosed from three 25 liter containers. Nitrogen was used to de-aerate the water.

Several objects for decontamination and corrosion testing were placed in the test section: a regulating valve, a check valve, a radioactive pipe and coupon specimens. The radioactive pipe had been taken from the primary circuit cleaning system of the BWR Oskarshamn 1.

During the test the temperature of the water phase was measured before the test section and after the heater. The water flow rate was measured after the circulation pump. The flow rate in the different parts of the test section varied between 0.3 and 3 m·s^{-1}. Instruments for measuring the conductivity and pH of the water in the loop were positioned at a sampling point following the circulation pump. A Tallium-cell was used to measure the oxygen concentration in the loop during operation under simulated BWR conditions. Spectrophotometric and titration apparatus were also available for analysis of the decontamination chemicals.

The surface dose rate on the radioactive pipe was measured with a NaI detector connected to a single channel recorder adjusted to register γ-radiation between 0.6 and 1.6 MeV. This covers the three isotopes Mn 54, Zn 65 and Co 60.

The coupon specimens for corrosion tests were mostly made in the following material:

Stellite No 6, Inconel 600, Inconel 182, Inconel X 750 and Zircaloy-2.

There were also some coupons of AISI 316 and AISI 304. Some of the coupons had been heat treated such as to simulate the normal welding procedures in nuclear power stations. Their susceptibility to crevice corrosion was studied during the tests.

Ten cycles were carried out during the test period. Each cycle comprised operation under BWR conditions followed by decontamination.

Whilst BWR operating conditions prevailed the oxygen content was maintained at 200±100 ppb. For the first 24 hours after each decontamination the oxygen consumption was high, and consequently the level of 200 ppb could not be maintained. The pH was about 6.5 and the conductivity between 0.5 and 1 $\mu s \cdot cm^{-1}$ and the pH dropped to ~ 4.5. This was partly due to residual chemicals in branch pipes and thermocouple fixtures.

The chemicals decomposed rapidly to CO_2 and water at the temperature in question, and the pH and conductivity returned to the normal values for operation. When the loop had been run under BWR conditions for the desired length of time the power supplied was reduced and the loop cooled down to 85°C. Then a slurry containing the decontamination chemicals, mixed in about 15 l of the de-areated water, was added. After dissolution at 85°C the cation exchanger was connected into the loop, and the decontamination was continued under controlled temperature and chemical conditions. Analysis indicated that additional dosing, of in particular EDTA, was necessary after a few hours decontamination. For the last decontamination EDTA was overdosed by 30 %. The concentration fell during the day to about 90 % of what it should have been. No additional dosing occurred.

When decontamination ceased, or in the later cases after eight hours decontamination, the ion exchanger was reconnected to the mixed bed to permit cleaning of the chemicals. Cleaning continued over night and the following morning the loop was returned to BWR conditions. Normally a decontamination, from stopping to restarting the loop, took 24 hours.

During the test only unimportant stoppage was caused by leaks. It is very likely that most of these would have occurred even without exposure to the decontamination solution.

Leaks caused by the decontamination solution occurred in two thermocouple fixtures when the soldered joint failed. They were made using a zinc-cadmium solder, (Castolin).

In the first run nearly four hours elapsed after chemical dosing before any decontamination could be noted. After that removal was rapid until it levelled off when the activity had decreased by 67 %.

The coupon specimens were examined first after three cycles. They were still in the same positions, and were covered by a thin, greyish brown even oxide film.

After five cycles the coupon specimens were examined again. On this occasion some of them were removed for descaling and weighing.

The coupons were returned to the loop after weighing. All the coupons were removed from the loop after ten cycles. Some underwent metallographic examination and the remainder were descaled and weighed. Loss of material on the coupons was slight.

The resin in the ion exchangers was changed between each decontamination. Following the first cycle the radioactivity of both ion exchangers was measured. The upper portion of the cation exchanger was strongly radioactive. The mixed bed also showed an increase, although to a lesser extent, in radioactivity. After the final decontamination the cation exchanger was treated with 50 1 NaOH to measure the amount of EDTA.

Analysis of the loop water during decontamination and eluation after the final cycle showed that the cation exchanger traps some of the EDTA. This could be due partly to the low pH in the cation exchanger, which would result in a lower solubility of EDTA, and partly to the formation of EDTA-ion complexes containing positive charges. It is therefore recommendable to overdose with EDTA.

IN-PILE TESTS

The primary aim of these tests was to develope routines for the dosing of the chemicals, managing the filters and ion exchangers, sampling, and tracing any operational disturbances after starting-up a full scale loop following decontamination. A number of water and radiochemistry parameters have also been studied. Based on earlier experience no decontamination effects worthy to note were anticipated.

The tests were carried out in an in-pile PWR loop of the materials testing reactor in Studsvik. In order not to disturb the normal operation of the loop, an outer system was erected outside the machine room for the loop. This outer system included a circulation pump, a system for dosing and mixing the chemicals, a 10 kW electric heater and a filter bank, cation exchanger and mixed bed ion exchanger with radiation shielding.

The in-pile loop had the following basic data:

Total volume	~ 400 l
Temperature during decontamination	$85^{o}C$
Pressure during decontamination	0.3 - 1.5 MPa
Pressure maintained by Helium	Helium
Circulation rate	$1.5\ kg \cdot s^{-1}$
Cleansing rate	$0.2 - 0.4\ kg \cdot s^{-1}$
Cleaning constant	$\sim 2.7\ h^{-1}$
Filter porosity	1 µm

The filter bank contained 12 248 mm AKA-CUNO Micro-Wynd cotton filter cartridges of mesh 1 µm. Manometers were positioned before and after the filters to measure the pressure drop across them.

The cation resin Duolite ARC 351 was used and for the anion resin Duolite ARA 366. For the cation resin a 20 l container was used, for the mixed bed, 60 % anion and 40 % cation, two 10 l containers. The decontamination solution circulated through the ion exchangers without prior cooling, i.e. at $85^{o}C$.

The chemicals were directly dosed into the outer system using a pneumatic pump. They were dissolved or emulsified in five litres of water. Only a small amount of the EDTA was dissolved in that volume. The emulsion was however soon dissolved once diluted in the warm ($85^{o}C$) loop water. During the dosing the flow to the ion exchangers was shut-off to prevent preferential trapping of EDTA.

Radioactivity was measured using a 4 000 channel device (ND600) and Ge-Li detectors positioned in the machine room on one of the main pipes.

The same chemistry as previously described for the out-of-pile tests was used with the exception of an overdosing of EDTA, as discussed above.

For the greater part of the decontamination process the circulation pumps of the loop stood still. On two occasions, each of one hour duration, the circulation pumps were however exposed to decontamination solution.

When the decontamination was finished the cation exchanger was isolated and the mixed bed ion exchanger was connected. One of them started to leak after a couple of hours and therefore it was shut-off. The flow through the remaining exchanger then increased. When the conductivity had sunk to 100 $\mu S \cdot cm^{-1}$ the cleaning was stopped even though the ion exchanger was still functioning well. Chemicals were by then leaking into the loop from dead volumes so fast as to almost balance cleaning in the ion exchanger.

After tapping-up and restarting the loop its own ion exchanger was reconnected and the conductivity sank to 5 $\mu S \cdot cm^{-1}$. The loop temperature was then raised to 150°C to decompose the residual chemicals. During restarting the conductivity increased somewhat. When the pressure holding tank was reconnected it rose to 45 $\mu S \cdot cm^{-1}$, but returned quickly.

The normal procedure for restarting was followed and operational conditions were established. The conductivity in the circuit was then 2 $\mu S \cdot cm^{-1}$.

The decontamination loop was then cleaned from chemicals and its conductivity was reduced from 120 to 10 $\mu S \cdot cm^{-1}$ in five hours.

The activity was monitored on a pipe in the machine room continuously throughout the test. It was also measured using a hand monitor on both horizontal and vertical pipes in the machine room, as well as on the pipes and radiation shielding of the decontamination loop.

No significant reduction in the activity levels, either with regard to the general level or to the particular measurement positions, could be detected.

The measurements indicated that about 90 % of the activity was due to Co 60, about 10 % to Co 58 and about 1 % to Fe 59. The ND 600 measuring device could not record any reduction in the activity at the measurement position.

Water samples taken during the test were analysed by chemical and gammaspectrometric methods.

Solution samples for fractionated filtering were taken from points before and after the particle filter at the beginning and the end of the decontamination period. Each sample, volume one litre, was filtered through a filter packet containing filters with porosities of 0.45, 3, 20 and 100 μm. The filtrate and residue were analyzed with gammaspectrometry which showed that the radioactivity is present in a particulate form, < 0.45 μm, or as ionic species. The dominant isotopes were Cr 51 and Co 60.

Chemical analysis of the water samples showed that the release of iron reached a maximum a few hours after decontamination began; it then decreased. The amount of activity in the water samples followed the same pattern.

After completion of the test, the activity gradients in the ion exchangers were measured. In the case of the mixed bed exchangers this was done directly on their outsides; but in the case of the cation exchanger a detector was positioned 15 cm from its surface.

The search area of the detector was restricted to 10 cm along the length of the ion exchanger, which was therefore moved 10 cm between each measurement.

Samples from the mixed bed and the particle filter were also analyzed. No samples could be taken from the cation bed because the dose rate was too high. The analyses showed that the ratio of nuclides was the same in the mixed bed substances as it was in the water: in both cases Cr 51 and Co 60 nuclides were completely dominant.

Extra time was necessary to re-establish the water chemistry. The cleaning operation continued normally for the first few hours. Thereafter chemicals started to leak out of the dead volumes in the system and to affect the cleaning process. Comparison with the corresponding process in the out-of-pile test showed that the water chemistry could be re-established much more rapidly in that test.

This indicates that the time required to re-establish the water chemistry depends greatly upon the specific loop, where dead volumes are particularly important. It is also important to ascertain prior to decontamination whether or not the dead volumes or parts of the system can be drained whilst the water chemistry is re-established.

From the measurement of the activity gradients it was evident that the ion exchangers had not been fully exploited either in respect to activity or dosed chemicals. Despite the fact that only one of the mixed beds was used for the greater part of the cleaning process, it could still be used for the final cleaning of the decontamination loop. When the cleaning from the chemicals was stopped only small amounts of them could have remained in the loop since the conductivity had sunk from 1 800 to 100 $\mu S \cdot cm^{-1}$, and the pH risen from 2.8 to 4.0.

DECONTAMINATION TESTS ON COMPONENTS FROM A PWR LOOP

In an attempt to determine what effect a decontamination solution, containing the same chemicals in the same concentrations as in the tests described above, would have on oxides formed under PWR conditions, tests were made on a steam separator and on one of the main circulating pumps from a PWR loop. The decontamination was performed in a specially built loop, volume ~ 200 l. Only one cation exchanger, with about two litres of resin, was used. The system was drained after each test.

First a steam separator, which had been running under PWR conditions with hydrogen dosing, was decontaminated.

The test was only aimed at determining the decontamination factor. Activity was monitored partly using hand monitors and partly with TLD monitors.

Decontamination was started in the morning and continued for 48 hours.

A main circulation pump was decontaminated next. It took 24 hours. After decontamination had been under way for six hours, one litre samples were taken before and after the ion exchanger; they were subjected to fractionated filtering. The filter porosities were 100, 20, 30 and 0.45 μm.

These tests showed that the oxide in the steam separator was affected very little and that the mean value for the decontamination factor was very low.

The surface dose rate of the pump on the other hand was reduced by ~ 30 %, which is also very low.

Analysis of the ion exchanger and the filters demonstrated that most of the corrosion products were present in the form of ionic species.

DECONTAMINATION TESTS WITH FUEL RODS

These tests were carried out twice on segmented fuel rods from the BWR Oskarshamn I. Four rods each about 500 mm, were used in both tests.

A loop was built specifically for these tests and had the following basic data:

Total volume	15 1
Pressure	atmospheric
Circulation rate	1.6 $kg \cdot s^{-1}$
Effect	2.0 kW
Cleansing rate	15 $1 \cdot h^{-1}$
Cation exchanger resin	2 1
Mixed bed ion exchanger	2 1

The housing and wheel in the circulation pump were made of AISI 316 stainless steel. The container for the ion exchanger resin was made of plexiglass pipes. Duolite ARC 351 was used as the cation resin and Duolite ARA 366 as the anion resin.

The decontamination solution was the same as for the tests described above: citric and oxalic acids and EDTA, at the same concentrations. The fuel rods had been in the reactor for between two and five years.

Low decontamination factors were determined throughout the tests.

Towards the end of the test the EDTA concentration had become very low and additional dosing was necessary.

DEVELOPMENT OF A NEW CHEMISTRY

The chemistry as used up to this point had shown that it contained at least the following weaknesses:

- moderate or poor capability of dissolving the older oxide layers in a BWR system;

- low capacity as a solvent of oxides.

It was therefore considered to be motivated to try to improve the existing chemistry, or if possible develope an entirely new one.

The sparse literature information available, and thermodynamic consideration imply that a decontamination solution, which contains both a complex former and a strongly reducing agent, is particularly suitable for dissolving BWR oxides.

The solubility of $CoFe_2O_4$ (the most stable Co compound formed under BWR conditions) increased the more reducing the environment is. The chances of the Co metal precipitating are considered to be small under reducing conditions.

A series of tests was performed using several possible reducing agents. The decontamination solution was the same as before but the oxalic acid concentration was considerably higher. This increase was aimed at increasing the solvent capacity.

The most important results of the tests are:

- The original chemistry, without the addition of reducing agents, gave a decontamination factor of 4 - 5 on a specimen from the Oskarshamn II BWR reactor after ~ 23 000 EFPH. Similar specimens from the Ringhals-1 BWR reactor gave, with the same chemistry, decontamination factors of 2.2 - 2.6, after ~ 14 000 EFPH. The authors consider these values to be on the low side.

- When considering the periodic decontamination of a relatively newly commissioned reactor however, such a low decontamination factor as two is deemed acceptable.

- When the first decontamination of a system which has been operational for several years is carried out, the activity levels are often of such a magnitude that they should be reduced by ~ 75 %, which corresponds to a decontamination factor of 4.

Decontamination of the more difficult specimens from the Ringhals-1 reactor could be achieved using the appropriate reducing agent. Decontamination factors of 4.4 - 5.2 were obtained. This chemistry will be used in future decontamination tests.

REGENERATION TESTS

When decontamination of an entire core or individual system in the primary circuit of a reactor is to be carried out the question arises of continual regeneration in ion exchangers combined with the filtering of particles.

For the regeneration tests contaminated solutions from the tests on the new chemistry and an anion resin loaded with the same chemistry were used. The isotope trapping thus achieved was acceptable.

DECONTAMINATION OF SPACERS FROM THE OSKARSHAMN I REACTOR

As a part of the work to study the decontamination of reactors and reactor systems four spacers from the Oskarshamn I reactor have been decontaminated. The fuel rods from which the spacers were dismantled had been in operation for about 14 000 EFPH. The surfaces of the spacers were covered by a greyish black deposit.

The spacers were decontaminated one at a time because of the high surface dose rates. Decontamination was performed at 90°C using the new chemistry. Each test was started using fresh decontamination solution. After decontamination the spacers were rinsed in the de-ionized water.

During one of the tests a sample of the solution was taken and filtered. The filter and filtrate were analyzed chemically and by gammaspectrometry; the filter was only examined qualitatively.

A further study of the decontaminated spacers will be carried out after they have been in position in the core for about a year.

DECONTAMINATION TESTS ON FUEL CLADDING

Decontamination tests were carried out using the new chemistry on two pieces of cladding material taken from the central part of a fuel pin which originated from the Oskarshamn I reactor. The fuel rods had been exposed for about 27 000 EFPH. The tests were perforemed at 90°C and with stirring.

The remaining activity on the pieces and the total concentration of radionuclides in the solution were measured after the treatment. Acceptable decontamination factors for all the nuclides present were achieved in all the tests.

DECONTAMINATION TESTING WHILST ELECTROPOLISHING

A test series with several different electrolytes has been performed. Electrolytes containing 30 - 65 % ortho-phosphoric acid had high decontamination factors when tested on components which had been deemed difficult to decontaminate using conventional methods. It is not considered likely that development of this will be continued.

The problems appear to lie in the field of waste disposal. Expended solutions can be decontaminated or regenerated with ion exchangers or filters. Sooner or later however the electrolyte can no longer be cleaned and then dumping can become necessary. When considering electrolytes containing high concentration of ortho-phosphoric acid, it is the opinion of the authors that they cannot be dumped without contravening the strict regulations against the discharge of chemicals.

Tests with other electrolytes have shown that, for example, very weak solutions of organic and inorganic acids in combination with ultrasonic cleaning have proved to be successful. The results of these tests are promising and worth further work. The advantage with this electrolyte is that it can, after use and dilution, be reprocessed in the plant for cleaning radioactive water wastes, which already exist in nuclear power stations.

CONCLUDING WORDS

This conference contribution provides a number of examples of various efforts in Studsvik, both recent and current, in the field of system and component decontamination. All of these are aimed at, amongst other things, finding a method of decontaminating BWR-systems which has acceptable decontamination factors, in combination with little or no corrosion attack.

Research is currently under way to determine the corrosion resistance of the most sensitive structural steels in the reactor to the new decontamination chemistry.

Work is also under way to try to define the design parameters and dimensions for ion exchangers and particle filters for use in regenerating solutions with the new chemistry for system decontamination.

Tests are planned with the new decontamination chemistry in combination with a dispersing agent. The aim of this is to decrease the deposition of oxide particles from the core in the extremes of the primary circuit.

The studies described here are in the near future approaching use in decontamination of a real system. The authors are today hopeful that their new chemistry will fulfill the demands upon it.

IN SITU DECONTAMINATION OF PARTS OF THE PRIMARY LOOPS AT THE GUNDREMMINGEN NUCLEAR POWER PLANT

Norbert Eickelpasch, Manfried Lasch

Kernkraftwerk RWE-Bayernwerk GmbH
Postfach
8871 Gundremmingen, W.-Germany

INTRODUCTION

The Gundremmingen nuclear power plant, installed in 1966, is a 250 MW_{el} boiling water reactor. The nuclear steam supply system (GE type) includes three external loops each with a secondary steam generator. In 1977, after more than ten years of operation, extensive testing and welding works at the steam generators and the primary loops due to the investigation of material problems had to be planned. Activated corrosion products, mainly Co-isotopes, which were deposited on the system surfaces during the long operation time, caused increasing dose rates at the system as shown in fig. 1.

THE NEED OF DECONTAMINATION

The main works to be performed were cutting off and welding totally 6 safe ends and dye penetrant test of weldings and inner surfaces at the recirculation system (see fig. 2).

For reducing the personal exposure and to prepare the dye penetrant tests a decontamination became necessary.

The welding was expected to require more than 800 h at a radiation level of 800 mR/h at the working place; that means a total dose of ~ 650 manrem for high qualified welders. Besides of health physics demand, it seemed to be impossible to provide several hundreds of these persons.

Experiences showed that reliable dye penetrant testing needs a carefully cleaned metal surface. We found out that chemical cleaning, i.e. decontamination is the most effective method

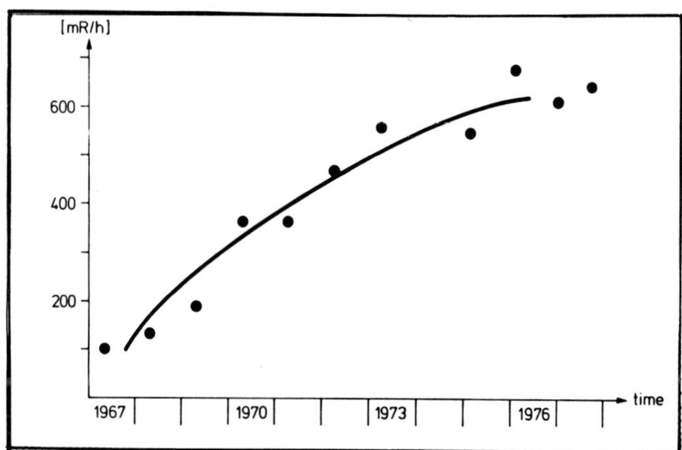

Fig. 1 Dose Rate at the Primary Loop

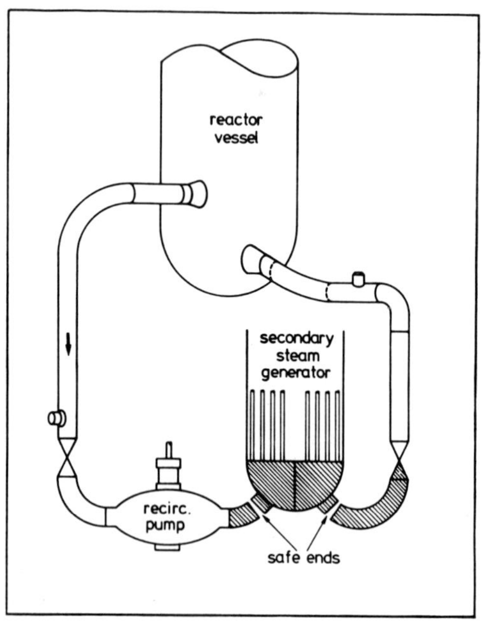

Fig. 2 Primary Loop at KRB

to get well prepared metal surfaces for this test. The cleaning only by abrasive method means long time working in a high radiation field.

THE DECONTAMINATION METHOD

The deposited corrosion products form an arrangement in two layers:

- An upper red layer, which is easily removable by mechanical means, e.g. by extensive brushing.
- A lower grey film, which is very solid and hard to remove.

The chemical analysis of the crud is shown in fig. 3. The main components are iron- and nickel-oxide. Stoichiometric considerations and X-ray examination indicate magnetite, a spinell type.

The results of radiochemical analysis are presented in fig. 4. The total activity amounts ~ 5 $\mu Ci/cm^2$. As function of decay time, Co-60 and Mn-54 become predominant.

We tested the various decontamination methods published in literature on dismantled parts of the primary system in our laboratory. We did not find any really good working chemical method for the old BWR-oxide-layer. Only those methods, called APAC (Alcaline Permanganate Ammonium Citrate) [1, 2] yielded decont factors up to 4.

For solving our actual problem we finally choosed the method offered by KWU [3, 4], a modified APAC method. For preparing this work KWU performed extensive corrosion tests to prove this method as harmless even to the non stabilized austenite steel SS 304 regarded here.

To get experience with the limits of decontamination treatment time with respect to material damage we decontaminated a piece of a safe end until the first failures could be detected. The selected piece of unstabilized sensitized material represented the most critical sample for corrosion risk.

By intensive non destructive material tests (i.e. dye penetrant-, ultrasonic-, X-ray-, eddy current test and metallography) it was ensured that the sample showed no cracks at all.

After 8 decontamination treatments, round about 64 hours, the first indications for beginning crack-corrosion were revealed. The dye penetrant test showed a very slight reddish colour. The metallography showed intercristalline attacks less than 130 µm maximum depth of penetration.

Confirmed by these tests we choosed a decontamination method with 16 hours of application.

Fe_3O_4	67 %
NiO	29 %
Cr, Cu, Zn, Co	< 4 %

Fig. 3 Chemical Analysis of the Oxides

	HLP	shut down	7 months later
Cr 51	27 d	~ 5%	—
Mn 54	300 d	8%	18%
Fe 59	44 d	~ 5%	—
Co 58	70 d	50%	7%
Co 60	5 a	30%	75%
total activity [γ]		~5-10 μCi/cm²	~4-5 μCi/cm²

Fig. 4 Radiochemical Analysis of the Crud

To minimize corrosion risk of primary pump materials, we decided to avoid penetration of decontamination solution into the pump.

DECONTAMINATION ENGINEERING

The decontamination method [5, 6] exists of:

- Treatment with the oxidizing solution (alkali permanganate) for 2 h at 90 - 100 ^{o}C;
- rinsing with cold water;
- treatment with the decont-solution (ammoniated citric acid-oxalic acid-edta-solution) for 6 h at 90 - 100 ^{o}C;
- rinsing with water.
- Repeating these steps.

The engineering provisions to be foreseen are shown in fig. 5:

- Tanks with heating appliance for the preparation of the solutions;
- a continious flow heater (18 kW);
- a circulation pump;
- tanks for intermediate storage solution to reuse them in the next bottom shells;
- waste treatment facility.

The volume of the decontamination solution to be handled was ~ 2 m^3.

To succeed in an effective decontamination the following measures were proved to be necessary:

- The solution had to be stirred violently.
- The removal of the oxide layer had to be supported by mechanical aids. We put taprogge balls with corundum coating into the solution. If possible, a mechanical cleaning with a high pressure water jet is preferable.

Special engineering and radioprotection problems arose because the work had to be performed in situ:

1) The parts to be decontaminated had to be sealed effectively. The sealing equipment must be extremely easy to handle, because it had to be installed in a high radiation field. Its installation had to be trained perfectly by a model before the first "hot" installation.
2) The whole procedure had to be surveyed by telecontrol.
3) Secondary radiation sources - mainly the recirculation pump and the steam generator tubes - had to be shielded (see fig. 6).

WASTE TREATMENT

The oxidizing solutions were chemically spent after six times of application. The acitivity concentration amounted $2 \cdot 10^{-3}$ Ci/m^3. This solution could be treated without problems in our chemical waste water facility, that works with precipitation-flocculation methods.

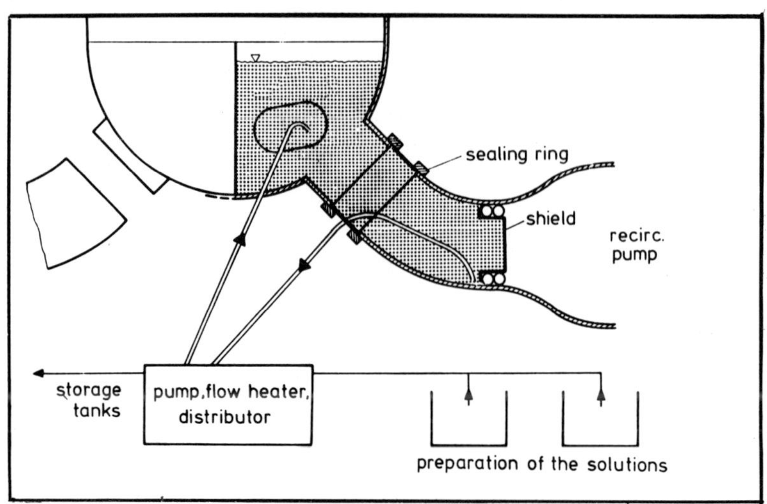

Fig. 5 Chemical Engineering

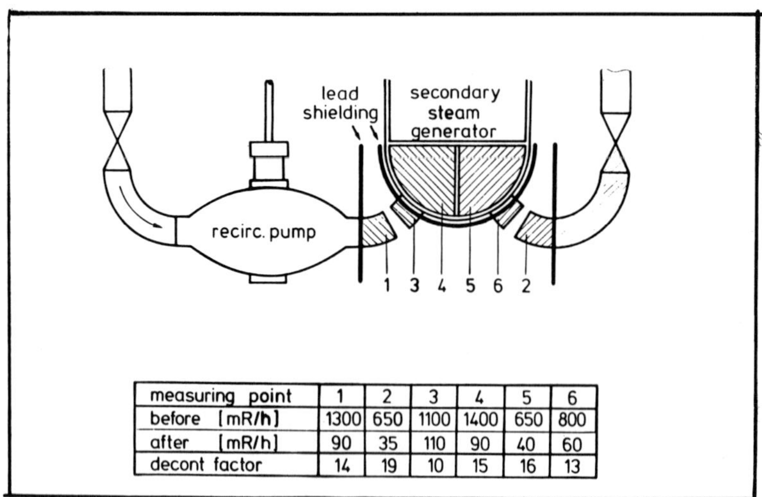

measuring point	1	2	3	4	5	6
before [mR/h]	1300	650	1100	1400	650	800
after [mR/h]	90	35	110	90	40	60
decont factor	14	19	10	15	16	13

Fig. 6 Dose Rates Before and After Decontamination

The decontamination solution (citric acid etc.) was spent after two times of application. This solution contained about $3 \cdot 10^{-3}$ Ci/m^3 total activity. The relation of Co-60 to Mn-54 was nearly the same as found in the oxide layer. Because this solution contained a high concentration of complexing agents, it could not be treated by precipitation methods. For that reason the solutions were neutralized and stored in a 20 m^3 tank. Later on they were solidified in concrete without further concentration. In total 16 m^3 of decont solution with a total activity of 2,8 Ci had to be disposed like that.

RADIATION EXPOSURE

The decontamination work and shielding measures resulted in totally 50 manrem (see fig. 7). Especially the installation of sealings required about 30 %.

	[man rem]
installation of the decont equipment	3
installation of the sealing	16
decont. operation	3
mechanical cleaning	6
waste treatment	1
	Σ 29
installation of the lead shielding	18 rem

Fig. 7 Radiation Exposure During Decontamination

RESULTS

The chemical decontamination combined with additional mechanical aids yielded a very effective cleaning of the metal surface. The dose rates measured with a shielded detector before and after decontamination are shown in fig. 6. Decontamination factors of more than 10 were reached.

The radiation level in the area of welding was reduced from 800 to less than 50 mR/h. So the planned extensive welding can be performed.

After decontamination dye penetrant testing could be performed without further cleaning.

CONCLUSIONS

It has been demonstrated that not dismantled parts of the primary system of an old BWR could be effectively decontaminated by the KWU-APAC-method.

The following provisions have to be considerated:
- high temperature and mechanical aids,
- previous material testing,
- easy handling of the sealing equipment
- previous personal training,
- previous test of the whole proceedings,
- telecontrol device,
- additional shielding,
- waste treatment facilities.

REFERENCES

[1] J.A. AYRES, Decontamination of Nuclear Reactors and Equipment, the Ronald Press Company, New York 1970.

[2] N. EICKELPASCH, J. RENNER, KRB, Aktennotiz C 52/76.

[3] R. RIESS, H. BERTHOLD, BMFT-FB K 77-01 (1977).

[4] R. RIESS, H. BERTHOLD, Reaktortagung Hannover 1978, Tagungsbericht p. 962.

[5] N. EICKELPASCH, H. BERTHOLD, M. LASCH, Reaktortagung Hannover 1978, Tagungsbericht p. 959.

[6] N. EICKELPASCH, M. LASCH, Atomwirtschaft 5/79, p. 247-251.

U. S. LICENSED REACTOR DECOMMISSIONING EXPERIENCE

Peter B. Erickson

Division of Operating Reactors

U. S. Nuclear Regulatory Commission

INTRODUCTION

To date, 64 reactors licensed under 10 CFR Part 50 have been decommissioned. In addition, four demonstration nuclear power plants have been decommissioned. These demonstration plants were government owned but operated by utilities under special government authorization. The 64 licensed reactors consist of five power reactors, six test reactors, one nuclear ship and 52 research reactors and critical facilities. Of the decommissioned research reactors and critical facilities, 42 have been dismantled and the remaining ten are either in the process of being dismantled or plans for early dismantlement are being developed. Tables I through V list each decommissioned reactor, the reactor type, the power level, the location and date the license was terminated or the present status if a license is still in effect.

SAFE STORAGE

Licensees with facilities in safe storage are required to control access to radiation areas, perform periodic radiation surveys both inside and outside of the facility, inspect the facility and report findings of surveys and inspections to the NRC. NRC Regulatory Guide 1.86 "Termination of Operating Licenses for Nuclear Reactors" describes acceptable ways for maintaining a facility in a safe status.

Access control at a decommissioned facility in safe storage has usually involved upgrading or minor modification of existing fences, radiation signs, containment buildings, steel doors, and concrete shielding structures and the use of security personnel from adjacent company facilities. Where security personnel are not available from adjacent facilities, such as Saxton and SEFOR, intrusion alarm systems, which are continuously monitored, have been installed to detect unauthorized entry. When continuously manned security coverage is not maintained, the NRC has required that access to high radiation areas be made very difficult. We have accepted the use of combinations of heavy shielding blocks and welded entry portals for the high radiation areas in combination with the intrusion alarms. Since all fuel, liquids and easily movable radiation sources have been removed from the site, access control is used primarily for protecting an intruder from serious overexposure.

Annual reports received by the NRC for facilities in safe storage state that there has been no evidence of release of radioactivity to the environment or any unauthorized entry into high radiation areas. The Office of Inspection and Enforcement of the NRC audits the containment of radioactivity with independent radiation surveys and measurements both inside and outside of the facilities.

The NRC has uncovered no material migrating to clean areas in a facility or outside the controlled areas. Some facilities do, however, show some evidence of rusting of carbon steel structures such as water tanks and carbon steel containment buildings. To date, this deterioration has not affected the integrity of the retention of radioactive material which is largely confined to the activated pressure vessel, pressure vessel internals and the primary system. Also, since the primary systems have all been drained and are essentially at atmospheric pressure a release of radioactive liquid is not likely to occur. The licensee is responsible for maintenance of the facility in a manner to assure that structures are adequate for access control and retention of radioactivity.

All five power reactors, six test reactors and the Nuclear Ship Savannah (N. S. Savannah) have been placed in safe storage with future dismantling delayed. Discussions with licensees for these facilities indicates that while no definite date for dismantlement has been selected, most intend to remove residual radioactivity within approximately 50 years after reactor shutdown. Two facilities in safe storage, Plum Brook and the N. S. Savannah are discussed to illustrate the range of differences in facility conditions.

TABLE 1

DECOMMISSIONED POWER, TEST AND NUCLEAR SHIP REACTORS IN SAFE STORAGE/MOTHBALLED WITH CONTINUED LICENSE

DOCKET NO. REACTOR	THERMAL POWER	LOCATION	PRESENT STATUS
50-16 Fermi 1 Fast Breeder Power Reactor	200 MW	Monroe Co. Mich.	Possession Only Lic.
50-18 GE VBWR BWR Power Reactor	50 MW	Alameda Co. Calif.	Possession Only Lic.
50-114 CVTR Presssure Tube, Heavywater, Power Reactor	65 MW	Parr S.C.	Byproduct Lic. (St.)
50-130 Pathfinder Nuclear Superheat, Power Reactor	190 MW	Sioux Falls S. Dak.	Byproduct Lic. (NRC)
50-171 Peach Bottom 1 HTGR Power Reactor	115 MW	York Co. Pa.	Possession Only Lic.
50-22 Westinghouse Test Reactor	60 MW	Waltz Mill Pa.	Possession Only Lic.
50-30 NASA Plum Brook Test Reactor	60 MW	Sandusky Ohio	Dismantling Plans Being Developed
50-146 Saxton PWR Test Reactor	28 MW	Saxton Pa.	Possession Only Lic.
50-183 GE EVESR Exp. Superheat Test Reactor	17 MW	Alameda Co. Calif.	Possession Only Lic.
50-200 B&W BAWTR Test Reactor (Pool Type)	6 MW	Lynchburg Va.	Byproduct Lic. (NRC)
50-231 SEFOR Sodium Cooled Test Reactor	20 MW	Strickler Ark.	Byproduct Lic. (St.)
50-238 NS Savannah PWR	80 MW	Charleston S.C.	Possession Only Lic.

TABLE II

DECOMMISSIONED RESEARCH REACTORS AND CRITICAL FACILITIES IN SAFE STORAGE WITH CONTINUED POSSESSION ONLY LICENSE

DOCKET NO. REACTOR	THERMAL POWER	LOCATION	PRESENT STATUS
50-6 Battelle Memorial Institute Pool Type Research Reactor	2 MW	Columbus Ohio	Dismantling Plans Being Developed
50-47 Watertown Arsenal U. S. Army Pool Type Research Reactor	5 MW	Watertown Mass.	Dismantling Plans Being Developed
50-94 Rockwell Inter. Corp. L-77 Research Reactor	10 W	Canoga Park Calif.	Dismantling Authorized
50-106 Oregon State Univ. AGN-201 Research Reactor	0.1 W	Corvallis Oregon	Dismantling Authorized
50-111 North Carolina State Pool Type Research Reactor	10 KW	Raleigh N.C.	Dismantling Plans Being Developed
50-129 West Virginia Univ. AGN-211P Research Reactor	75 W	Morgantown W. Va.	Dismantling Plans Being Developed
50-141 Stanford Univ. Pool Type Research Reactor	10 KW	Stanford Calif.	Dismantling Authorized
50-147 Rockwell Int. Corp. FCEL Split Table Critical Facility	200 W	Canoga Park Calif.	Dismantling Authorized
50-185 NASA MOCKUP Pool Type Research Reactor	100 KW	Sandusky Ohio	Dismantling Plans Being Developed
50-384 Calif. Polytechnic State Univ. AGN-201 Research Reactor	0.1 W	San Luis Obispo Calif.	Dismantling Plans Being Developed

Plum Brook Facility

The Plum Brook Facility in Sandusky, Ohio is owned by NASA and consists of the 60 MWt Plum Brook Test Reactor and the 100 KWt Plum Brook Mockup Reactor. Both reactors have been shutdown since January 1973 and all fuel has been removed from the site.

The Plum Brook Test Reactor is a heterogeneous light water cooled and moderated reactor that used MTR type fuel. Since 1973 the reactor has been maintained in safe storage. In addition to removing all fuel from the site, all resins were removed, the reactor vessel and all piping systems were drained, and areas with high radiation were shielded and sealed. Fuel storage canals have been cleaned and drained and hot drain systems and sumps have been flushed and kept dry.

Access control has primarily involved the use of existing doors, fences, shielding, intrusion alarms and security personnel. For instance, doors to the containment building, subpile room and hot cells are locked and the keys administratively controlled. Radiation surveys and sampling is performed quarterly to verify retention of radioactive material in constrolled areas. The integrity of physical barriers is verified by routine security guard checks and monthly inspections.

In 1977 NASA considered a plan for entombing the Plum Brook Test Reactor with monitoring for a limited period of time to assure that entombment structures were adequately retaining the radioactivity. This plan was not pursued, however, in view of the possibility that the license would remain in effect and some monitoring would be required as long as any radioactive material, above levels acceptable for release to unrestricted access, remained on site.

The Mockup Reactor is a pool type reactor that duplicated the Plum Brook Test Reactor in core characteristics but operated at a maximum power level of only 100 KWt. The Mockup Reactor was used for verifying nuclear characteristics of in-core experiments before they were placed in the test reactor. In addition to removing the fuel, all water has been drained from the reactor pool. The radiation level near the remaining Mockup Reactor core components is approximately 100 mr/hr. Access to pool area is controlled by locked doors and radiation signs.

NASA is now developing plans for dismantlement of both reactors at the facility. Buildings and structures will be retained to the extent allowable but all radioactive material will be removed from the site. The major residual activity is in the

TABLE III

DISMANTLED RESEARCH REACTORS (LICENSE TERMINATED)

DOCKET NO. REACTOR	THERMAL POWER	LOCATION	DATE LIC. TERMINATED
50-1 Illinois Inst. of Technology (Water Boiler Research)	100 KW	Chicago Ill.	04-28-72
50-4 USN Research Lab (Pool Type)	1 MW	Washington D.C.	03-18-71
50-8 N.C. State (Aqueous Homogeneous)	100 W	Raleigh N.C.	09-07-66
50-17 Industrial Reactor Labs. (Pool Type)	5 MW	Plainsboro N.J.	11-04-77
50-43 U.S. Naval Post-graduate School (AGN-201)	0.1 W	Montery Calif.	10-11-72
50-50 North American Aviation (L-47 homogeneous)	5 W	Canoga Park Calif.	06-30-58
50-58 Oklahoma State University (AGN-201)	0.1 W	Stillwater Okla.	03-19-74
50-60 U. S. Navy Hospital (AGN-201M)	5 W	Bethesda Md.	06-24-65
50-64 University of Akron (AGN-201)	0.1 W	Akron Ohio	10-09-67
50-84 University of Calif. (AGN-201)	0.1 W	Berkeley Calif.	08-23-66
50-98 University of Delaware (AGN-201)	0.1 W	Newark Del.	02-26-79
50-101 Gulf United Nuclear (Pawling Lattice Test Rig)	100 W	Pawling N.Y.	06-25-74

TABLE III (CONT'D)

DISMANTLED RESEARCH REACTORS (LICENSED TERMINATED)

DOCKET NO. REACTOR	THERMAL POWER	LOCATION	DATE LIC. TERMINATED
50-114 William March Rice University (AGN-211)	15 W	Houston Texas	09-26-67
50-122 University of Wyoming (L-77)	10 W	Laramie Wyoming	12-05-75
50-135 Walter Reed Medical Center (L-54, Homogeneous Solution)	50 KW	Washington D.C.	07-26-72
50-167 Lockheed (Pool Type)	10 W	Dawson Co. Georgia	09-01-60
50-172 Lockheed (Radiation Effects Reactor)	3 MW	Dawson Co. Georgia	08-31-71
50-202 University of Nevada (L-77)	10 W	Reno Nevada	02-24-75
50-212 General Dynamics Fast Critical Assembly	500 W	San Diego Calif.	03-05-65
50-216 Polytechnic Inst. N.Y. (AGN-201M)	0.1 W	Bronx N.Y.	12-21-77
50-227 General Atomic Co. (TRIGA Mark III)	1.5 MW	San Diego Calif.	12-10-75
50-235 Gulf General Atomic (APFA)	500 W	San Diego Calif.	10-22-69
50-240 Gulf General Atomic (HTGR)	100 W	San Diego Calif.	04-02-73
50-253 Gulf Oil Corp. (APFA III)	500 W	San Diego Calif.	08-10-73
50-310 NUMEC and Commonwealth of Pa. (Pool)	1 MW	Quehanna Pa.	12-02-66

TABLE IV

DISMANTLED CRITICAL FACILITIES (LICENSE TERMINATED)

DOCKET NO. REACTOR	MAX. POWER	LOCATION	DATE LIC. TERMINATED
50-13 Babcock & Wilcox (Split Table)	1 KW	Lynchburg Virginia	06-01-73
50-14 Battelle Memorial Platics Moderated Critical Assembly	200 W	W. Jefferson Ohio	05-11-70
50-23 Nuclear Development Corp. of America (Crit. Ex.)	100 W	Pawling N.Y.	06-22-61
50-24 General Electric (BWR Crit. Ex.)	200 W	Alameda Co. Calif.	12-01-69
50-34 Westinghouse Electric Corp. (PWR Crit. Ex.)	1 KW	Waltz Mill Pa.	12-08-69
50-37 Gen. Dynamics (CIRGA Zirconium Hydride Mod.)	25 W	San Diego Calif.	03-15-60
50-75 NASA (ZPR-1, Solution Type Crit. Fac.)	100 W	Cleveland Ohio	10-13-73
50-87 Westinghouse Electric Corp. (Crit. Ex. Station)	100 W	Waltz Mill Pa.	01-26-72
50-108 Allis Chalmers (Crit. Ex. Fac.)	100 W	Greendale Wis.	01-20-67
50-153 Westinghouse Electric Corp. (CVTR MOCKUP, Heavy Water)	3 KW	Waltz Mill Pa.	04-24-63
50-154 Martin Marietta Corp. (Liquid Fluidized Bed Crit. Ex.)	10 W	Middle River Md.	02-07-66
50-191 Babcock & Wilcox (Plutonium Recycle Crit. Ex.)	-	Lynchburg Virginia	06-01-73

TABLE IV (CONT'D)

DISMANTLED CRITICAL FACILITIES (LICENSE TERMINATED)

DOCKET NO. REACTOR	MAX. POWER	LOCATION	DATE LIC. TERMINATED
50-197 NASA (ZPR-2 Solution Type Crit. Fac.)	100 W	Cleveland Ohio	10-13-73
50-203 GE (Mixed Spectrum Crit. Assembly)	400 W	Alameda Co. Calif.	03-11-68
50-234 Gulf Oil Corp. (Thermionic Crit. Fac.)	200 W	San Diego Calif.	08-10-73
50-246 General Dynamics Corp. ACRE	10 KW	San Diego Calif.	12-30-66
50-290 Gulf United Nuclear (Water Mod. Proof Test Fac.)	100 W	Pawling N.Y.	06-25-74

TABLE V

DECOMMISSIONED DEMONSTRATION NUCLEAR POWER PLANTS

DOCKET NO. REACTOR	THERMAL POWER	LOCATION	PRESENT STATUS
115-1 Elk River BWR	58.2 MW	Elk River Minn.	Dismantled Federal Control Terminated
115-2 Piqua Organic Cooled	45.5 MW	Piqua Ohio	Entombed DOE Monitoring
115-3 Hallam Sodium Cooled	256 MW	Hallam Nebr.	Entombed Doe Monitoring
115-4 Bonus BWR Nuclear Superheat	50 MW	Rincon Puerto Rico	Entombed DOE Monitoring

reactor vessel and reactor vessel internals of the 60 MW test reactor. NASA estimates that this inventory consists primarily of 156,000 curies of tritium in the beryllium reflector segments, 2,640 curies of cobalt-60 in the reactor vessel and internals and 7,340 curies of Iron-55 in the reactor vessel and internals. The reflector segments and other reactor internals will be detached, removed and disposed of prior to remotely cutting up the reactor vessel. Dismantling of the 100 KW Mockup Reactor will involve disposal of much smaller amounts of induced activity in the reactor internals and the reactor concrete pool walls.

N. S. Savannah

The N. S. Savannah is the first and only nuclear powered cargo and passenger ship in the United States. The 80 MWt Savannah reactor has been shutdown since August 1970. All fuel, radioactive resins, primary and secondary system water and loose radioactive material has been removed and the ship is in safe storage at the U. S. Army Depot Berth in North Charleston, South Carolina. All radiation areas are controlled through the use of the shielding barriers of the reactor containment structure and locked hatches in other parts of the ship. Also, the U. S. Army provides access control with their existing guard force. The U. S. Maritime Administration, the licensee, continues to maintain cathodic protection for the ship's hull. Authorization has been given to move the Savannah to the James River Reserve Fleet to be retained there in safe storage until the reactor vessel radiation levels have decayed to reduce exposures for eventual dismantling.

The N. S. Savannah may, however, be refurbished and put on display at the Naval and Maritime Museum at Patriots Point, Charleston, South Carolina. The Patriots Point Development Authority has proposed to lease the ship from the Maritime Administration for that purpose. The Patriots Point Authority was established and funded by the State of South Carolina and now has the U. S. S. Yorktown, an aircraft carrier, on display at their Maritime Museum.

Prior to transfer of the Savannah to the Maritime Museum, all highly radioactive areas such as the pressure vessel and surrounding containment vessel would of course have to be secured adequately to prevent unauthorized entrance. Also, further decontamination of other areas may be required. The Maritime Administration would retain ownership of the vessel with radiation control and monitoring accomplished by the State Public Health Service. The NRC will evaluate the adequacy of any proposed license changes to allow the use of the Savannah for this

Maritime Museum. The Savannah will remain at its berth in North Charleston until a decision is made on moving it to the Maritime Museum or the James River Reserve Fleet.

DISMANTLING

Experience in dismantling has involved 42 research reactors and critical facilities and the Elk River Demonstration Power Plant. The major effort in dismantlement to date has involved the Elk River Reactor and larger research reactors in which considerable quantities of activated and contaminated concrete, steel and soil have had to be removed. Regulatory Guide 1.86 has been used for guidance on surface contamination with activation and soil contamination limits evaluated on a case basis. The licensee has been required to show through analysis that radiation exposures to any member of the public would be a small fraction of 10 CFR 20.105 limits (500 mr/yr) for activated materials and soil contamination. We have, also, required that activated material be removed such that the radiation level three feet from the surface of the activated material would be less than 50 microrem/hr. The licensee has also been required to demonstrate with cost benefit analysis that the residual radioactivity was as low as reasonably achievable.

ENTOMBMENT

Three Demonstration Nuclear Power Plants have been entombed (Table V). These are government owned reactors which were operated by private utilities. Radiation surveys and sampling is accomplished by local agencies for the U. S. Department of Energy (DOE). There has been no evidence of deterioration of the entombment structures or release of radioactivity from these entombed facilities per discussions with DOE personnel. Two of the entombed facilities were used for other purposes following entombment. The Piqua Containment Building is still used by the City of Piqua, Ohio as a warehouse and the Bonus Facility in Puerto Rico was used as a museum following entombment. At Hallam, all above ground structures were removed prior to entombment.

REGULATIONS AND GUIDES ON DECOMMISSIONING

We are now involved in reviewing our regulations and guidance on reactor decommissioning in light of our experience and studies that we have funded.

A study by Battelle PNL "Technology, Safety and Costs of Decommissioning a Reference Pressurized Water Reactor Power Station" (NUREG-CR 0130) was published in June 1978 and a similar study by Battelle for boiling water reactors (NUREG-CR 0672) is near completion. In addition Battelle is performing an environmental study for the NRC on decommissioning (NUREG-0586). An Oak Ridge study "Potential Radiation Dose to Man from Recycle of Metals Reclaimed from a Decommissioned Nuclear Power Plant" NUREG-CR 0134) was published in December 1978.

We anticipate that the result of our reviews will be to develop more specific rules and guidance for each decommissioning alternative and more specific facility radiation release criteria.

CONCLUSION

We believe that the experience to date has demonstrated that there are very viable options for decommissioning both smaller nuclear reactors and today's larger commercial facilities. A single decommissioning route is not appropriate for all facilities. Methods and time of decommissioning must be tailored to accommodate the specific site characteristics including future use of the site, cost and potential exposure to workers without, of course, any compromise to the health and safety of the public.

EXPERIENCE AND PLANS FOR THE DECOMMISSIONING OF NUCLEAR REACTORS IN OECD - NUCLEAR ENERGY AGENCY COUNTRIES

Eli Maestas

OECD Nuclear Energy Agency
38 Boulevard Suchet
75016 Paris, France

Decommissioning refers to the orderly disposition of nuclear facilities after their retirement from service, taking account of environmental, radiation protection, waste management and safety considerations. The activities involved in a decommissioning exercise can vary from simple closure of the facility with a minimum removal of radioactive materials, to complete plant disassembly and the restoration of the site for unrestricted release for other uses. As the nuclear industry reaches a mature state, decommissioning becomes an important consideration with significant activity foreseen in the not too distant future.

It is not the intention of this paper to justify or suggest a particular alternative to decommission a nuclear power reactor or nuclear facility, but rather to describe the experience and plans that have been made in the Member countries of the Nuclear Energy Agency of the Organisation for Economic Co-operation and Development. For those who are not familiar with the OECD-NEA, it is an inter-governmental organisation based in Paris comprising 23 Member countries, principally the western European nations plus Australia, Canada, Japan and the United States. The primary objective of NEA is to promote co-operation between its Member governments on various aspects of nuclear energy development as a contribution to economic progress, as well as health, safety and regulatory considerations. The NEA has recently undertaken, within the framework of a decommissioning programme, the organisation of a

TABLE I: NUCLEAR POWER PLANTS OPERATING AND UNDER CONSTRUCTION IN OECD/NUCLEAR ENERGY AGENCY MEMBER COUNTRIES*

Country	Operating**		Under Construction	
	Number	MWe	Number	MWe
Austria			1	692
Belgium	4	1,676	2	1,799
Federal Republic of Germany	16	9,038	10	10,638
Finland	2	1,080	2	1,080
France	15	7,258	22	21,195
Italy	4	1,382	2	1,960
Japan	22	13,399	6	5,114
Netherlands	2	499		
Spain	3	1,073	7	6,302
Sweden	6	3,700	5	4,682
Switzerland	4	1,926	1	942
United Kingdom	33	6,980	6	3,714
TOTAL	111	48,011	64	58,118

*Excludes United States and Canada **As of 1 May 1979

TABLE II: NUCLEAR POWER PLANTS
POSSIBILITIES FOR DECOMMISSIONING*

Probable period of shutdown 1981-1990

Station Name	Country	Type**	Capacity MWe	Commercial Operation
Marcoule G2,G3	France	GCR	2 x 39	1959-1960
BR-3	Belgium	PWR	10.5	1962
VAK KAHL	FRG	BWR	15	1962
Windscale AGR	UK	AGR	32	1963
JPDR-II	Japan	BWR	10.3	1963
MZFR	FRG	PHWR	52	1966
EL-4	France	HWGCR	70	1968
Calder Hall	UK	GCR	4 x 50	1956-1959
Chapelcross	UK	GCR	4 x 50	1959-1960

* Excludes the United States and Canada

** PWR Pressurised light water moderated and cooled reactor
BWR Boiling light water cooled and moderated reactor
GCR Gas cooled, graphite moderated reactor
AGR Advanced gas cooled, graphite moderated reactor
PHWR Pressurised heavy water moderated and cooled reactor
HWGCR Heavy water moderated, gas cooled reactor
SGHWR Steam generating heavy water reactor
HTGR High temperature gas cooled, graphite moderated reactor
FBR Fast breeder reactor

TABLE II (continued): NUCLEAR POWER PLANTS POSSIBILITIES FOR DECOMMISSIONING*

Probable period of shutdown 1991-2000

Station Name	Country	Type	Capacity MWe	Commercial Operation
Berkeley	UK	GCR	2 x 143	1962
Bradwell	UK	GCR	2 x 125	1962
Latina	Italy	GCR	150	1964
Hunterston A	UK	GCR	2 x 150	1964
Garigliano	Italy	BWR	150	1964
Trino Vercellese	Italy	PWR	242	1965
Hinkley Point A	UK	GCR	2 x 230	1965
Trawsfynydd	UK	GCR	2 x 195	1965
Dungeness A	UK	GCR	2 x 205	1965
Chinon-2	France	GCR	200	1966
Sizewell A	UK	GCR	2 x 210	1966
Tokai I	Japan	GCR	158	1966
KRB Gundremmingen	FRG	BWR	237	1967
Chinon-3	France	GCR	320	1967
Ardennes	France	PWR	270	1967
Oldbury A	UK	GCR	2 x 205	1968
KWL Lingen	FRG	BWR	256	1968
Winfrith	UK	SGHWR	92	1968
KWO Obrigheim	FRG	PWR	328	1969
AVR Juelich	FRG	HTGR	13.5	1969
Dodewaard	Netherlands	BWR	51.5	1969
St. Laurent A1	France	GCR	480	1969
Jose Cabera 1	Spain	PWR	153	1969
Beznau 1	Switzerland	PWR	350	1969

*Excludes the United States and Canada

specialist meeting on decommissioning requirements in the design of nuclear facilities and the setting up of expert groups to assess the state-of-the-art in remote handling and cutting technology and in decontamination methods. Because there are many papers presented at this Conference from the United States and Canada, I shall confine my remarks to the European countries and to Japan.

TERMINOLOGY

After a nuclear power plant has been retired from service, the nuclear fuel, radioactive process materials and wastes produced in normal operations should be first removed by routine procedures. The remaining course for the plant may be postulated to fall generally into three options (or stages), each of which can be defined in terms of the physical state of the plant and its equipment including the type and degree of surveillance which will be required. These three options have been defined by the International Atomic Energy Agency and will be briefly stated with their most commonly quoted equivalent as used here in the United States [(1)].

Option I Decommissioning: Lock-Up With Surveillance

This stage is commonly called "mothballing" and leaves the facility virtually intact. The mechanical opening systems in the primary containment barrier are blocked and sealed. The plant is maintained under continuous surveillance and security.

Option II Decommissioning: Restricted Site Release

This stage is commonly called "entombment". The primary containment barrier is reduced to minimum size and sealed by physical means. The biological shield is extended so that it completely surrounds the barrier. The containment building and other parts of the plant can be dismantled after decontamination to acceptable levels. Surveillance continues,but at a reduced scale compared to option I.

Option III Decommissioning: Unrestricted Site Release

The plant, including active materials which remain despite decontamination, are removed and the site is released for other uses without restrictions. In many instances in which the future use of the site has not been determined, unrestricted site release will be the most desirable result to a decommissioning exercise, and is possible with today's technology.

Using these options, a comparison can be made of the various experiences which will be discussed later in the paper.

REASONS FOR DECOMMISSIONING

A nuclear power plant will eventually be decommissioned to one of the above options at the end of its service life. The decision to decommission to a particular option is dependent on many factors, primarily economics, radiation protection and safety. The time when service ends is one of those grey areas which is determined by many variables. Based on past thinking, it has been generally accepted that reactors would have a productive lifetime of 20 - 40 years; however, the present generation of nuclear power plants could have longer lives in the future due to the total energy demand requirements, economies of scale and plant performance. If we consider the "design" life of a reactor to be 30 years, this is the same period which is usually chosen as the capital repayment period; there is some economic incentive to operate the plant beyond that period. However, the time must come when reasons dictate that the nuclear power plant be decommissioned. This could be due to uneconomical operation as a result of increased maintenance and repair requirements or decreased plant availability. New or supplementary government regulations may require additional expenditures for the plant to meet revised conditions. Similarly, a design may be rendered obsolete by new technical developments. Finally, the occurrence of a major accident, where subsequent repairs cannot be accomplished due to economic, technical or safety factors, may be another reason for decommissioning.

The decommissioning operations which have been undertaken so far have been characterised by many parameters, such as the type and size of reactor, the operating history, levels of radiation, and contamination,

TABLE III: NUCLEAR POWER PLANTS RETIRED FROM SERVICE IN OECD/NEA MEMBER COUNTRIES*

Station Name	Country	Type	Capacity MWe	Operating Period	Status
Chinon 1	France	GCR	70	1964-1973	Option I
KKN Niederaichbach	FRG	HWGCR	100	1974-1974	Option I
HDR Grosswelzheim	FRG	BWR	23	1970-1971	Option I
Dounreay FR	UK	FBR	14	1962-1978	Option I
Agesta	Sweden	PHWR	10	1964-1974	Option I

* Excludes United States and Canada

etc., so that each project has been a special case. Similarly, the next generation of nuclear power plants decommissioned will continue to be special cases which should contribute to a broad base of new experience and information.

EXPERIENCE IN DECOMMISSIONING

From very early in nuclear power development, some forms of decommissioning activities have been undertaken. To date, numerous demonstration, test or research reactors plus various supporting facilities in the nuclear fuel cycle have been decommissioned. However, the number of nuclear power plants which have been retired in Europe and Japan is small. Presently there are 111 operating nuclear power stations in Europe and Japan producing about 48,000 megawatts of electrical energy. Another 64 nuclear stations are under construction and are planned for completion by the middle of the coming decade (2). Table I provides a summary by country of nuclear power plants operating and under construction. Of the operating plants, about 40 will be at least 30 and some over 40 years old by the year 2000. To present these data in a more comprehensive form, Table II identifies nuclear power plants which may be expected to be decommissioned through the year 2000. It is premature to make firm estimates on the schedule of retirement of nuclear power plants from service because their operational life is uncertain within a wide range. However, the table is based on the ages of the plants and should only be used for comparative purposes. For the expected decommissioning period, a 30-year life was used, which may be conservative as experience dictates that nuclear power plants may be operated beyond this period. On the other hand, there may be instances where unexpected shutdown may occur, or where certain prototype plants are conceived to have shorter planned lives.

With regard to actual European experience in the decommissioning of nuclear power plants, the number of plants thus far disassembled is small; however, some valuable experience does exist with regard to demonstration, as well as test and research reactors (3). Table III summarises the most notable projects undertaken concerning decommissioning of demonstration power plants. I will go through some of the more significant accomplishments on a country-by-country basis as well as presenting a brief description of development work and plans.

France

France has a very aggressive nuclear programme with 15 operating stations producing more than 7,200 megawatts plus 22 additional power stations expected to come on line between now and 1982. A total of 8 experimental, test and demonstration reactors have been decommissioned to various stages of option I to option III. The largest of these is the Chinon 1, an 80MWe prototype graphite moderated, gas-cooled power reactor belonging to the Electricité de France. This plant operated for about 10 years producing about 3 billion KWh and was shut down for economic reasons in 1973. A detailed study was performed in which each decommissioning option was evaluated. Eventually, the plant was decommissioned to option I in a time of six months. It is presently converted into a nuclear museum. Final disassembly is foreseen to occur after 30 years' duration, by which time the technologies of cutting reinforced concrete and the remote cutting of steel vessels, plus the burning of graphite, should be well advanced.

A great deal of work is going on in planning for the future decommissioning of French nuclear reactors. In addition to studies on waste management and safety, theoretical studies have been performed on the disposition of a reference 1,000 MWe PWR nuclear power plant, and the 1,200 MWe fast breeder Super Phénix now being built. In support of a dismantling project, extensive R&D is being done on various telemanipulators applicable to remote operations and on self-propelled manipulators which can travel on vertical walls and on ceilings. These prototype devices are capable of carrying miscellaneous tools, cameras, inspection devices, etc.

Federal Republic of Germany

The Federal Republic of Germany has 16 nuclear stations operating which produce slightly over 9,000 megawatts. Another 10 plants are under construction to be operational by 1983. There are two significant nuclear plants which have been shut down. The Kernkraftwerk Niederaichbach, heavy water moderated, gas cooled reactor generated 100 MWe and operated a few years. It was shut down in 1974 and is now in an option I state of protective storage. Its complete removal is being considered and planning for this is now under way. The smaller HDR Grosswelzheim BWR rated at 23MWe, operated commercially for less than a year and was shut down in 1971. It is also in option I protective storage.

In regard to planning for decommissioning, the Federal Republic of Germany requires in its licensing procedure for nuclear power plants that an analysis be performed showing that the plants can be decommissioned at the end of their operating lives. As a result of this, extensive engineering and theoretical studies have been performed concerning the conceptual decommissioning of nuclear power reactors, such as a 800 MWe BWR and 1,200 MWe PWR reference plant. In addition, studies on estimating the waste arisings and the distribution of radioactivity present in nuclear plants after they have been shut down have been prepared in this connection.

United Kingdom

In the United Kingdom, an early start in nuclear power plant construction has resulted in 33 reactors now operating, generating about 7,000 MWe. Another 6 reactors in 3 stations are under construction and will provide an additional 3,700 MWe by 1981. The Dounreay Fast Reactor is the UK's most significant power reactor recently shut down in the spring of 1978. This fast breeder reactor of 14 MWe began commercial operation in 1962. It was retired from service because it had reached the end of its planned experimental life. Presently, it is in an option I condition with evaluations underway for its future disposition. It has been suggested that perhaps the facility would be converted into a technical museum.

Another reactor which may likely be shutdown in the near future is the 32 MWe Windscale Advanced Gas-Cooled Reactor (WAGR). A detailed evaluation on the decommissioning alternatives available for this plant has been completed and is based on carrying out the decommissioning work progressively through the complete plant removal and restoration of the site to a green field condition. It is likely that the methodology used in this project would ultimately be applicable to the Magnox reactors, many of which are beyond the halfway point in their planned operating economic lives. These gas-cooled reactors will probably form the bulk of the European plants which will become redundant by the end of the century. An important difference which exists between the Magnox reactors compared to light water reactors is the lower specific activities in the radioactive inventory in the reactor; however, this is offset by a larger waste volume. In addition, there is a preponderance of mild steel over stainless steel and large amounts of graphite will need to be recycled or disposed.

Sweden

Sweden has 6 nuclear power plants producing 3,700 MWe of electricity and 5 under construction which will produce 4,682 MWe of which 2 are about to go into commercial service. Sweden is among the European countries deriving a large percent of its electrical energy load from nuclear power plants at this time. There is one shutdown reactor in Sweden which will probably be dismantled in due course. This is the 10 MWe Agesta nuclear power plant, a demonstration heavy pressurised water reactor built in a rock cavity. It was shut down in 1974 and is now in an option I condition. The owners are now planning and evaluating future decommissioning alternatives, with some consideration being given that parts of the project may be offered as a joint international co-operative undertaking.

TEST AND RESEARCH REACTORS

I have briefly reviewed the status of shutdown nuclear power plants in several European countries. Though emphasis has been given to the decommissioning of nuclear power plants, much additional experience exists from the disassembly of test and research reactors as well as supporting nuclear fuel cycle facilities.

In Switzerland, the 30 MWth Lucens heavy water moderated, gas-cooled test reactor was dismantled during the 5 years subsequent to its shutdown in 1969 due to a fuel element which failed and caught fire. The plant, at the time of its shutdown, had operated a few months at full power. The majority of the plant has been disassembled and much of the resulting radioactive waste is being stored, pending disposal, in the plant underground rock cavity. Since the decommissioning effort has begun, costs amounting to about 4.1% of the original construction cost have been expended to reach a level somewhat short of the total plant removal.

The Italian experience on decommissioning is limited to the small research reactor Avogadro RS-1, a MTR pool type reactor rated at 7 MWth and shut down in 1972. It is interesting to note that the pool of the reactor has been converted to a spent fuel storage facility.

In Japan, a country with an impressive nuclear power programme now operating 21 nuclear power plants producing over 13,000 MWe, experience in decommissioning has been limited to nuclear research facilities, for example the decommissioning of JRR-1 and the Homogeneous Test Reactor. In addition, partial dismantling of the JDPR II, a 10 MWe BWR, was made for modification purposes. This plant can be expected to terminate operation in several years. In this connection, several committees have been formed to investigate decommissioning activities of nuclear facilities in other countries and to carry out feasibility studies on decommissioning of some Japanese nuclear facilities.

One last nuclear power plant which is faced with an unusual situation is Austria's new 692 MWe BWR, Tullnerfeld station at Zwentendorf. This plant was in the final assembly state when a national referendum was held to decide whether it should be completed. Last November 1978, Austrians voted against nuclear power by a margin of 1 per cent. Now the owners and the government are trying to decide what to do next with the nuclear power plant.

CONCLUSIONS

1. There exists, as a result of decommissioning activities of retired demonstration nuclear power plants plus test and research reactors, a broad base of experience in Europe and Japan covering a variety of different types of facilities. This experience supports the generally accepted consensus that nuclear power plants can be decommissioned with today's disassembly technologies and better dismantling efficiencies in the future can be expected with improvements in methodology.

2. A number of plants can be expected to be shut down in the coming decade. To provide for the orderly disposition of these plants, studies evaluating the technical, economic, radionuclide inventory, safety, waste management and radiation protection and public health aspects have been prepared or are underway in many countries. To broaden the use of the still

limited number of cases, it would be worthwhile to place some of these undertakings within the framework of an international co-operative exercise in decommissioning.

3. The materials which arise from the dismantling of a nuclear power plant could pose a volume burden to waste treatment and disposal facilities. Adequate planning to manage radioactive wastes from decommissioning should be considered a long-term task.

4. The legal aspects of decommissioning operations are subject to general nuclear regulations. However, the specific detailed guidance pertaining to the decommissioning of nuclear facilities in NEA countries with few exceptions, is rather limited. As more experience with larger plants becomes available in the next decade, special decommissioning regulations and criteria can be expected to be formulated. Areas where the situation needs to be clarified, for example, are the criteria for the release of equipment and sites either with or without restrictions, and the financial responsibility for the dismantling of a plant many years after shutdown. This last point is important in that it appears that the preferred option in many countries may be to defer the final dismantling of the reactor except in special case situations.

REFERENCES

(1) Decommissioning of Nuclear Facilities, 1975 Report of a Technical Committee Meeting on the Decommissioning of Nuclear Facilities, IAEA 179.

(2) Power Reactors in Member States, 1979 Edition, IAEA Vienna, June 1979

(3) K. J. Schneider, "Decommissioning of Nuclear Facilities, Report on the International Symposium held in Vienna, 13-17 November 1978", Atomic Energy Review, IAEA, Vol. 17, No. 1, March 1979.

DECOMMISSIONING SITUATION IN THE FEDERAL REPUBLIC OF GERMANY

Dr. G. Watzel, Rheinisch-Westfälisches-Elektrizitätswerk AG, Essen
Dipl. Phys. J. Essmann, Preußische Elektrizitätswerke AG, Hannover
Dipl. Ing. G. Lukacs, NIS Nuklear-Ingenieur-Service GmbH, Frankfurt

CONSIDERATION OF DECOMMISSIONING IN THE LICENSING PROCEDURE

As the number of nuclear power plants in the Federal Republic of Germany increases, the discussion of whether and how these nuclear power plants can be decommissioned later will also become more intense. It is indisputable that as a result of the activity inventory present during the decommissioning of nuclear power plants -- in contrast to the decommissioning of conventional power plants-- special conditions prevail, so detailed considerations are required of how these decommissioning operations can be carried out without endangering the population and the decommissioning personnel.

Accordingly, the question of decommissioning has been included in the licensing procedure under the Atomic Energy Act of the Federal Republic of Germany. In the "Safety Criteria" promulgated by the Federal Ministry of the Interior in 1974 / 1 /, therefore, the following requirement is specified under Section 2.10, "Decommissioning and Removal of Nuclear Power Plants":

> "Nuclear power plants shall be in such a condition that they can be decommissioned in compliance with the Radiation Protection Regulations. A concept for the removal of the plant after its final decommissioning in compliance with the Radiation Protection

Regulations shall be provided".

Correspondingly, in the revised edition of the "Atomgesetz", i.e., the Atomic Energy Act of the Federal Republic of Germany, dated 31 October 1976 / 2/, in Article 7, the following Paragraph No. 3 has been added

> "The decommissioning of a plant ... as well as the protective storage of the plant finally shutdown, or the dismantling of the plant or of plant components are subject to approval ...".

As a result of these requirements, presentation of a corresponding document has been required with all nuclear power plants for which licensing is pending in the Federal Republic of Germany. The statements required for decommissioning were defined in the check list for a standard safety report issued by the Federal Ministry of the Interior and in the Guidelines of the Reactor Safety Commission.

The German operators of nuclear power plants, acting within the framework of their "Vereinigung der Deutschen Elektrizitätswerke" ("VDEW") (Association of German Utilities), have formed a working group to coordinate the handling of decommissioning questions, to arrange for the exchange of information and standardization of procedures followed by the individual utilities, and to serve as a point of contact on basic problems to be discussed with authorities and experts. Independent of this working group, however, each utility must handle the complex of decommissioning questions in the licensing procedure for its own plant with the particular responsible officials and reporting personnel.

The situation discussed below is described in detail in / 3 /.

From the beginning, the German utilities have represented the opinion that it is not reasonable to carry out a specific decommissioning study for every nuclear power plant with a light-water-reactor, but instead it is only necessary to prove the general feasibility of the decommissioning using techniques available today. This is especially true since it must be assumed that at the time of the decommissioning -- which generally will take place 40 years after commissioning of the plant at the earliest-- techniques required for decommissioning will have been developed further. Consequently, a study prepared at the present time under this premise -- general proof that the decommissioning can be carried out-- not only should be valid for all nuclear power plants with light-water reactors, but also should eliminate the necessity for preparation of separate studies for specific plants. The technology and organization of a nuclear power plant have already led to essential features at the present time which facilitate a later decommissioning. This is all the more true because relevant experience gained from plants already in operation is always considered when planning a nuclear power plant. The emphasis here is placed on suitable layout of the plant components, on good possibility for inspection, maintenance, and repair, and on the feasibility of in-service inspections, as well as on the possibility of disassembly and reassembly of large plant components.

The ease of decommissioning certain plant components can only be increased if the safety standard achieved with the plants, which alone guarantees operational safety and thus the protection of the general public, is lowered. As a result,existing requirements for greater ease of decommissioning must be subordinated to the requirements --which must indisputably be evaluated more highly-- for operational safety.

The German utilities are also of the opinion that, in conjunction with the decommissioning of nuclear power plants, economic viewpoints must also not be disregarded, i.e., the financing of the measures required must be ensured. Since the German utilities which build and operate nuclear power plants are corporations (i.e., organized on the basis of private law), the corresponding financial reserves must be created for decommissioning. An agreement must be reached with fiscal authorities on the amount of the yearly reserves. In every case, however, the reserves must be sufficient so that the costs of the decommissioning can be covered at the time when they are incurred.

For these reasons, the general considerations devoted to this complex of problems in 1974 / 4 / were intensified, and detailed studies were called for in order to examine the questions of whether, from today's point of view, additional measures are required to decommission nuclear power plants in conformity with legal regualtions at the end of their operating time. In particular, the question should also be studied whether changes must be made in

the technical concept of the nuclear power plants with light-water reactors constructed in the Federal Republic of Germany for reasons of the decommissioning.
Finally there is the additional objective of determining the costs for the decommissioning so that the amount of the reserve funds which the utilities must build up can also be substantiated.

The German utilities have therefore issued a contract through the VDEW to NIS, Nuklear-Ingenieur-Service GmbH to make a detailed decommissioning study of two reference plants; this study is to be conducted in a number of phases, in the last of which subcontractors will provide assistance. The objectives of this study are to determine the technical status of the decommissioning which has been reached today both domestically and internationally, to point out the decommissioning techniques and other technical measures required to decommission a nuclear power plant with a light-water reactor, and to make a cost calculation. The results are to be presented in such a way that application to other plants with a light-water reactor requires only minor supplementation for matching purposes.

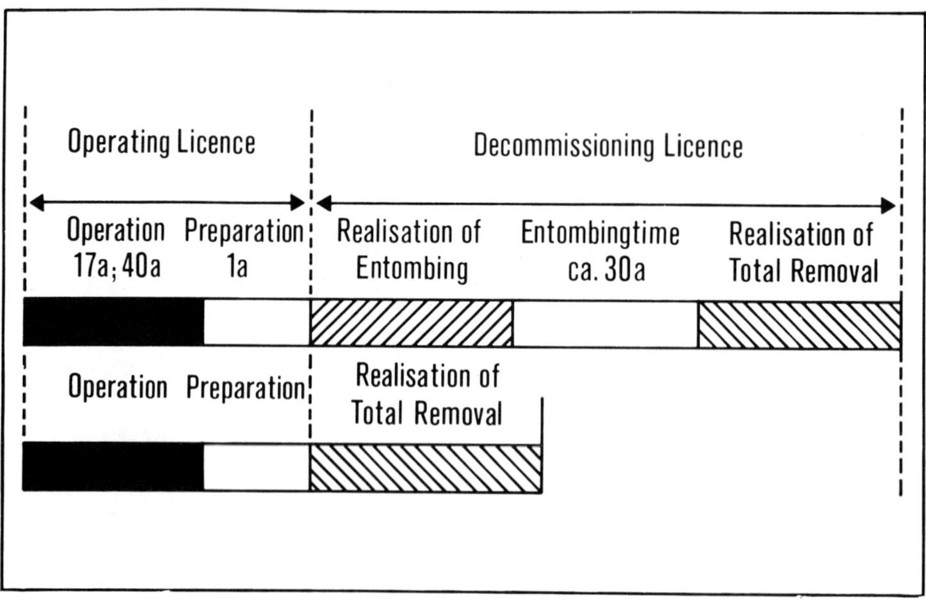

Fig. 1 Variants for Decommissioning

DECOMMISSIONING STUDIES

The study to be made under the contract has the ultimate goal of determining, in as exact a form as possible, the operational expenditures to be expected in terms of both technology and personnel, the radiation exposure that will be accumulated during this work, and finally to calculate the total costs. For the two reference plants selected -- Biblis A (1,204 MW, PWR) and Brunsbüttel (805 MW, BWR)-- the study considers, in the first place, immediate complete removal (after about 1 year of waiting time following final shutdown) and, in the second place, protective storage for a duration of 30 years followed by complete removal (Fig.1). As a result, with respect to the calculation of costs, the most unfavorable case (immediate complete removal) is considered on the one hand, while on the other hand, as a result of the section-by-section procedure, a parametric analysis is developed, by means of which cases other than the ones treated here can be dealt with later.

As the initial state, the activity inventory of the reference plants after 17 and 40 years of operating is determined, whereby it is found that the operation beyond 12 to 15 years contributes only insignificantly to further increase in the activity inventory. The inventory is composed, in the first place, of activated materials, mainly concentrated in the reactor pressure vessel and its internals as well as in the biological shield, and in the second place of contaminated material.

The calculation of the activated material is based in part on manufacturers' data on the neutron flux (design values), and in part the flux between these points is extrapolated considering the material values. For purposes of the activation calculation, three energy ranges with the boundaries of 0 eV, 1.855 eV, 0.821 MeV, and 10 MeV are distinguished. The functions for the cross sections are taken from / 5 / to / 7 /. For the activation calculation thick-walled components were subdivided into zones, for each of which an average value of the neutron flux is assumed; as a result of this subdividing the value is averaged over not more than two orders of magnitude. Using Biblis as an example, the result is given in Fig. 2, whereby the activation is added for all isotopes considered (data for 1 year and 30 years after shutdown).

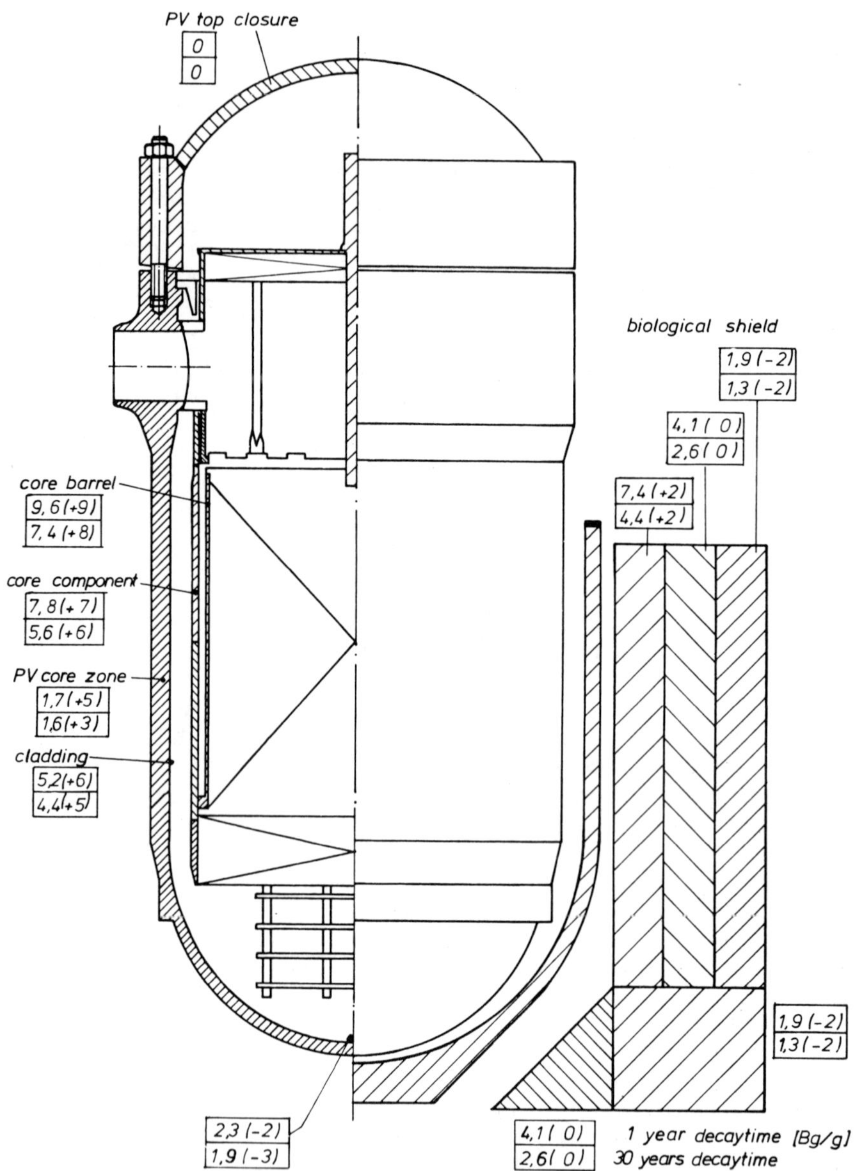

Fig. 2 Specific Activity of Several Components (Biblis A) After 40 Years Operation

Since the reference plants have been in operation for only relatively short periods of time to date, recourse was made to reports in the literature on other plants with respect to experience with contamination values /6/ to /11/. The isotope Co 60 is of dominant importance here. The magnitude of the contamination depends mainly on the activity content of the medium (primary coolant) and on the surfaces contacted in the systems by this medium. The following average values are based on the evaluation of the literature cited above / Table 1 /.

Table 1 Contamination of Systems and Building (One Year Decaytime)

System, Area	Contamination / Bq / cm² / PWR	BWR
Reactor Cooling system		
primary side, water	$3{,}7 \cdot 10^5$	$3{,}7 \cdot 10^5$
primary side, steam	-	$3{,}7 \cdot 10^2$
feedwater system	-	$3{,}7 \cdot 10^2$
Auxiliary systems	$3{,}7 \cdot 10^4$	$3{,}7 \cdot 10^4$
Nuclear ventilation system	37	37
Surface of building in nuclear area	3,7	3,7

In order to calculate the internal surfaces, specific values based on the mass of the systems were used, while the masses were derived in part directly from design data, in part by a system survey made during personal inspections of the rooms, and in part by estimates based on characteristic values of the plant.

Using the activity data calculated in this way, the dose rate values were calculated for the various system components and room areas in the reference plants. Radiation protection measures, including decontamination procedures, have not yet been considered here.
This takes place with the analysis of the individual phases of the decommissioning work. An example for the reactor pressure vessel is shown in Fig. 3.

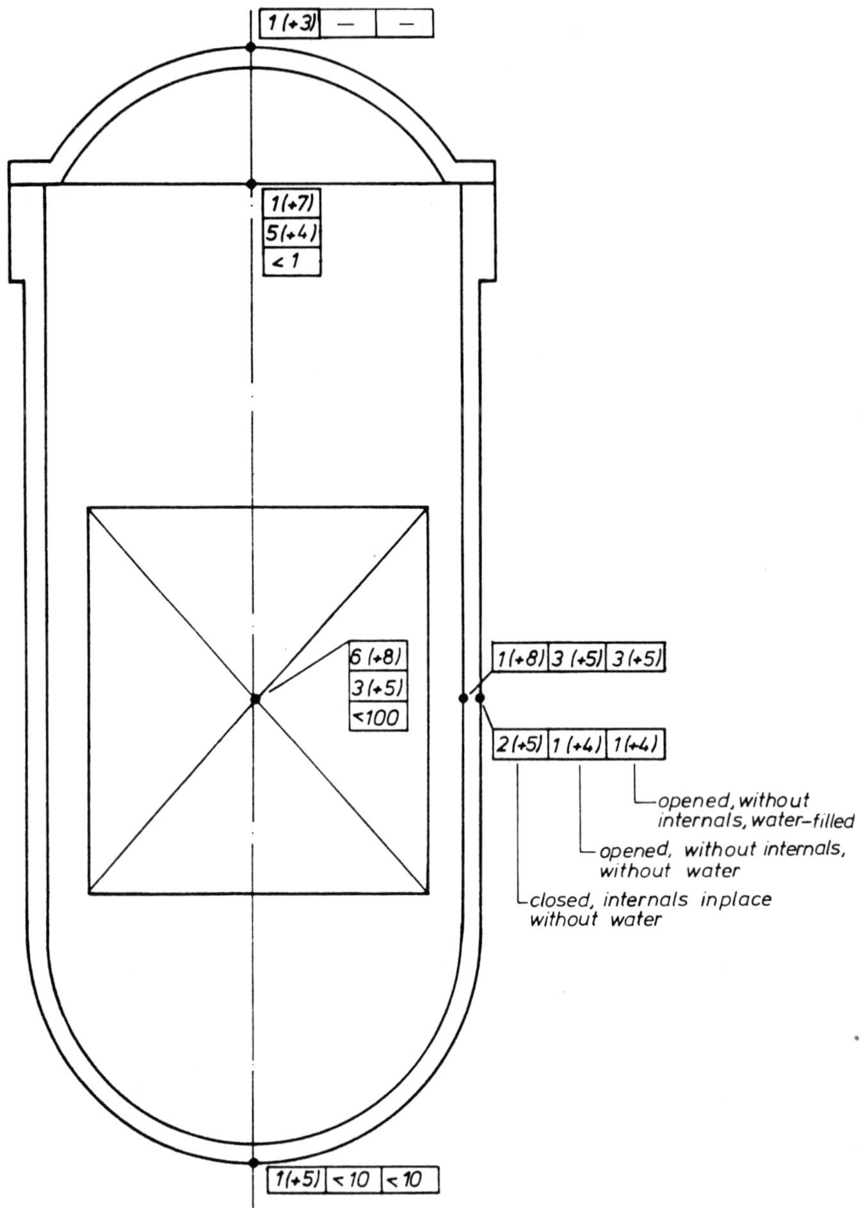

Fig. 3 Gamma Dose Rate On the Surface of Components (Biblis A) After 40 Years Operation and One Year Decaytime (mrem/h)

With the contaminated systems a differentiation is made of whether or not they are freely decontaminable in a simple manner. The former applies for a large number of the components which have been contacted by active air or steam. Such components can be pipe supports, steel structures (stairs, grates, platforms, etc.), ventilation systems, hoisting equipment, motors, building structures, and the containment.
The total activity inventory is shown in Fig. 4

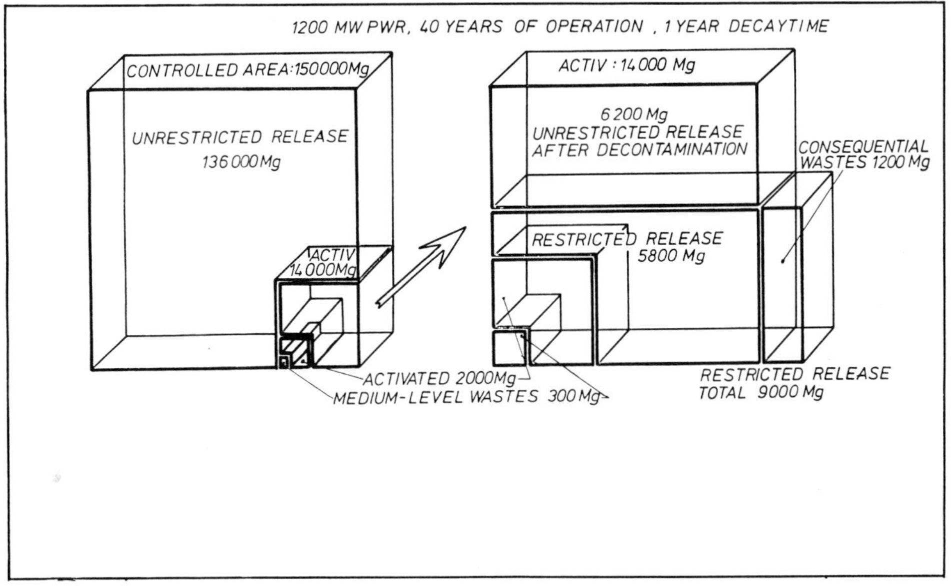

Fig. 4 Partition of the Decommissioning Wastes

For Biblis the total mass is about 7,900 Mg, while for Brunsbüttel it is about 7,300 Mg. The corresponding activity inventories are $2 \cdot 10^{+17}$ and $6 \cdot 10^{+16}$ Bq respectively. The greatest part of the activity is concentrated in just a few components (reactor pressure vessel, internals). On the basis of consideration relating to the cutting and the stacking in waste containers (with 21.6 m^3 usable volume), the following transportation units are found for the most important radioactive parts / Table 2 /.

Table 2 Number of Containers for Packing of Radioactive Waste (Biblis A)

	Weight / Mg /	Volume / m³ /	Radioactivity / Bq /	Volume after treatment / m³ /	Number of Containers
Pressure vessel	580	74	$1{,}6 \cdot 10^{14}$	296	14
Pressure vessel	164	21	$2 \cdot 10^{17}$	124	6
internals shield	1312	570	$4 \cdot 10^{11}$	1140	53
Primary system					
-Steam generator	1780	226		3767	175
-Pressurizer	170	22	$7 \cdot 10^{13}$	88	4
-Coolant pumps	200	25		500	23
-Piping	150	19		127	6
Auxiliary-system					
Piping	625	80		533	25
Vessels	510	65		260	12
Heat exchanger	1190	152		2533	117
Fittings	135	17	$1{,}4 \cdot 10^{13}$	212	10
Pumps	575	73		1460	68
Storage racks	175	22		69	3
Linar	55	7		11	1
	7621	1375		11120	517

Transportation and, most important, storage in large vessels (containers) are assumed to be a reasonable procedure, even though corresponding regulations for this procedure do not yet exist in the Federal Republic of Germany. By way of comparison, to pack the wastes according to the specifications of the experimental ultimate storage facility know as "Asse II", 32,455 low-active and 820 medium-active drums, each with a capacity of 200 litres, would be required. The Asse II experimental ultimate storage facility is a former salt mine and for years the conditioned operating wastes from German nuclear power plants have been sent there. Assuming a conditioning of wastes ready to be packed in 200 litres drums then a classification based on an IAEA proposal / 12 / can be made for these wastes regarding their surface dose rates / Table 3 /.

Table 3 Classification of Radioactive Waste (One Year Decaytime)

Class	Field of dose rates / R/h /	Weight of activated waste / Mg /		Weight of contaminated waste / Mg /	
		Biblis	KKB	Biblis	KKB
1	0,0002 ÷ 0,2	464	528	3550	5375
2	0,2 ÷ 2	607	975	2300	-
3	2	328	238		-
		1399	1741	5850	5375

In one section, the study compares in detail the various working methods and processes available which are considered with a decommissioning. Here, the particular advantages and disadvantages, as well as the possible areas of application, are explained. Differentiation is made between the working areas of decontamination, cutting, handling and conditioning, as well as packing and transportation.

On the basis of the fundamental considerations described up to this point, roughly subdivided decommissioning schedules have already been developed, one example of which is shown in Fig. 5; more detailed descriptions are given in / 13 / to / 16 /. A much more detailed analysis of the sequence of decommissioning operations will be made in the last phase of the study. On the basis of the radioactive components of the nuclear power plant determined, the available decommissioning techniques, and the estimated schedules, working packages that are precisely defined and separated from each other by clear boundaries will now be established, the elaboration and thorough planning of which will be carried out by those companies which could carry out these phases of the work in the actual case.

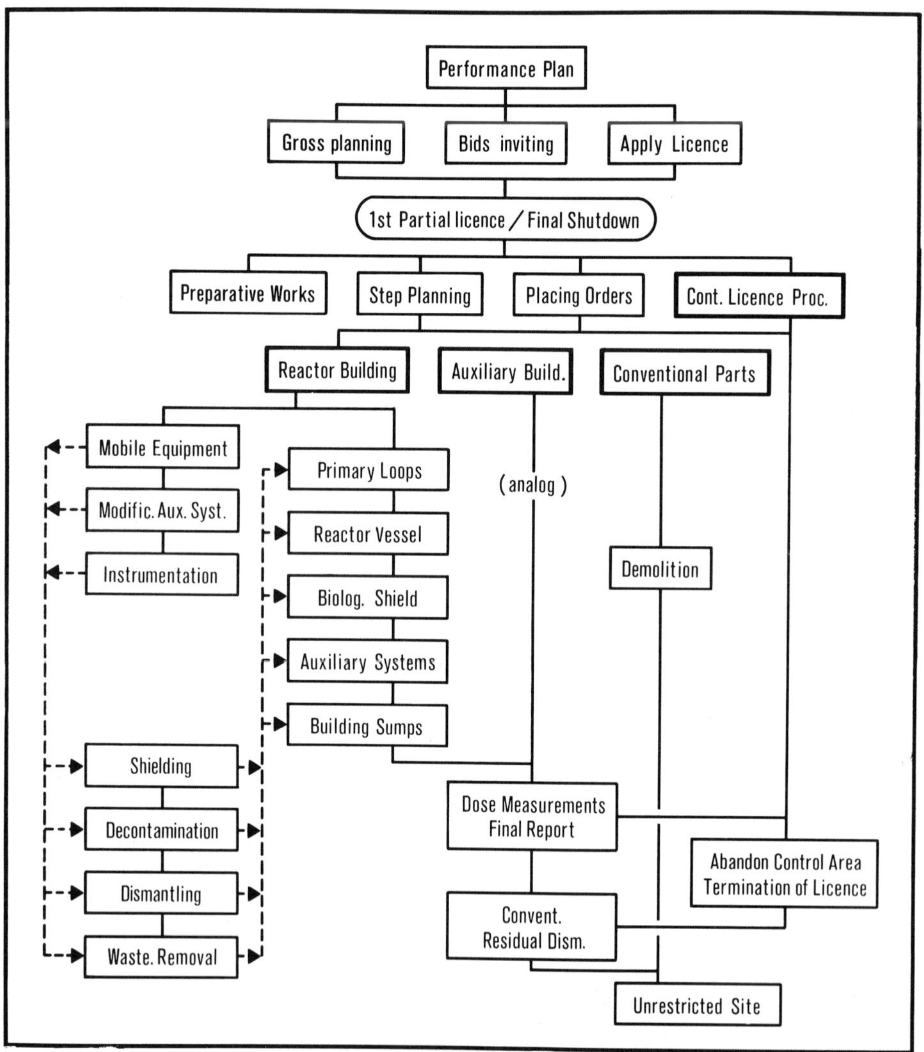

Fig. 5 Decommissioning Flow Diagram- Total Dismantling

This elaboration of the working packages takes place on a subcontractual basis through the coordinating engineering company (NIS). The working packages are shown in Table 4, whereby in order to attain results which are as well-founded as possible, great importance is attached to having several companies prepare these working packages independently of each other. In all, 11 companies are assisting in this work, some of which are preparing several working packages. Each company receives detailed specification for the working package(s) to be prepared, containing the exact scope of the work and the boundary conditions with respect to dose rate, system dimensions, accessibility, and auxiliary materials available. The individual results are reworked accordingly and are combined into an overall result.

Table 4 Working Packages

No.	Working packages	Nuclear of cooperative companies	Necessary for		
			Protective Storage	Total removal (immediately)	Total removal (after 30 years)
1	Project management	1	X	X	X
2	Health physics	2	X	X	X
3	Licensing	1	X	X	X
4	Preparation of the plant	1	X	X	X
5	Decontamination	4	X	X	X
6	Plant operating	1	X	-	-
7	Measures for protective storage	1	X	X	X
8	Modification of excisting equipment	1	-	X	X
9	Dismantling reactor pressure vessel	3	-	X	X
10	Dismantling radioactive components	3	-	X	X
11	Dismantling radioactive concrete	3	X	X	X
12	Treatment and packup of radioactive Waste	2	X	X	X
13	Removal of radioactive Waste	2	-	X	X
14	Cannibalizing of free parts	5	-	X	X
15	Dismantling of buildings and foundation	5	-	X	X
16	Ground works	5	-	X	X

The individual results should be given as far as possible in the form of cost values related to characteristic quantities in order to guarantee easy transferability to other decommissioning parameters or other plants. These parameters differ from one working package to another and, for example, characteristic quantities such as masses, surfaces, volumes, specific activities, surface dose rates, etc. can occur; the cost values are then calculated in DM per characteristic quantity considered. Other values can be the personnel requirement, personnel qualification (costs for working time), implementation time, etc. In addition, fixed costs independent of the magnitude of the characteristic quantities can also occur (for example, planning work, equipping the decommissioning site, development of special tools, etc.).

Fig. 6 shows the flow chart for the method of calculating the cost incurred with a decommissioning project for any desired plant with a light-water reactor, to the extent that it is similar in terms of its quality for this purpose, although differing characteristic quantities can arise in its quantitative value, and the decommissioning process can be carried out according to the same division into working packages as this was studied for the reference plants.

This study is expected to be finished by the middle of 1980, and will then provide not only the quantitative values for the decommissioning quantities mentioned above relative to the reference plants, but also a procedure for simplified calculation of these quantities for other LWR systems and other decommissioning variants.

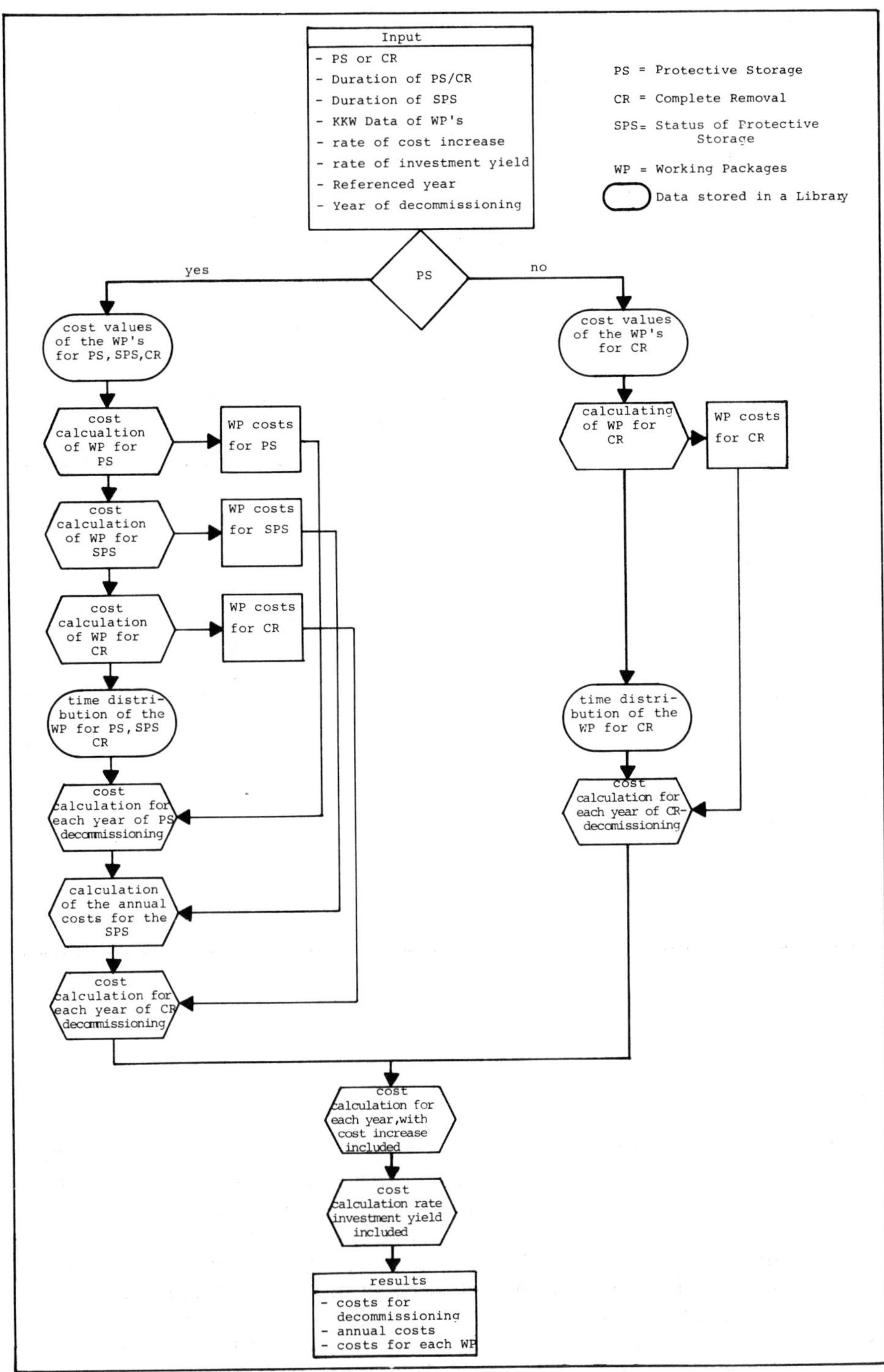

Fig. 6 Computer Program Flow Diagram

CURRENT DECOMMISSIONING WORK IN GERMANY

Although no nuclear power plant has been decommissioned by complete removal to date in the Federal Republic of Germany, various phases of decommissioning work have been conducted or are currently in the planning stage with experimental and pilot nuclear power plants, and research reactors. The status of the various projects is shown by Table 5.

Table 5 Decommissioning in Germany (Present and Projected)

Plant	Reactor	Output MWe	First criticality	Shutdown	Situation	
					present	projected
HDR	BWR+nucl. superheating	25	1969	1971	Research	Complete removal
KKN	Pressure tube	100	1972	1974	Protective storage	Complete removal
NS "Otto Hahn"	PWR	38 (therm)	1968	1979	out of action	Intermediate storage of waste
KWL	BWR+fossile fired superheating	160	1968	1977	out of action	Protective storage
FR 2	HPWR	44 (therm)	1962	1981	in operating	Removal of nuclear part
VAK	BWR	15	1961	ca.1985	in operating	
MZFR	HPWR	57	1966	ca.1985	in operating	

HDR

The Superheated Steam Reactor (HDR) in Großwelzheim was finally shut down in 1971. This is a boiling-water reactor with nuclear superheating, with a total of 25 MW electrical power. The plant was in operation for only 1 year and during this time developed an average power factor of hardly 5%. The reasons for the final shutdown were, in addition to defective components, primarily damage to the fuel assemblies an a lack of interest in nuclear-superheating with LWR nuclear power plants.

The activity inventory in the plant 2 1/2 years after shutdown was about 3.7 · 10^{+14} Bq. In the plant there were various forms of equipment and components with a total weight of 2,500 Mg, of which 740 Mg were contaminated or activated, and the rest were free of radioactivities. The concrete structural parts of the plant weigh a total of 75,000 Mg, but only about 100 Mg of this amount must be removed as radioactive waste. After prolonged discussions the decision was made to convert the superheated steam reactor to a nonnuclear experimental facility for research relating to technical safety. Preparations for this research continued until 1977, experimental operation was started then, and it is to continue until 1983. After that date the entire plant is to be removed. In order to eliminate the problem of radiation exposure and thus to permit unrestricted experimental operation, those components which have contributed to the radiation exposure were dismantled, cut, and removed before the experiments were started. These components were mainly the core internals, the saturated steam converter, and the superheated steam converter, and they contained more than 99% of the activity inventory. After these components were dismantled they were cut in accordance with the conditions of the Asse II experimental storage facility, packed in 200-litre drums, and shipped to Asse II. All of this work was carried out largely with conventional techniques and the personnel were subjected to only slight radiation exposure.

KKN

The Niederaichbach Nuclear Power Plant (KKN) was finally shut down in 1974, also after only a short period of operation. This is a pressure tube reactor with an electrical power of 100 MW, with CO_2 gas cooling and heavy water moderation. The reasons for the decommissioning were several defective components which would have had to be replaced;apart from the fact there also have been requirements of back fitting. However, since the construction of this type of reactor had been stopped even before the power plant was put into operation the additional investments which were necessary for the prototype plant were no longer reasonable. The activity inventory in the plant 1 year after shutdown was about 5.6 · 10^{+14} Bq. During the years following shutdown the plant has been converted to the condition of protective storage. The contaminated components were removed from the reactor auxiliary building and from the fuel assembly storage building and were placed in the reactor building.

At the present time the entire plant, with the exception of the reactor building, is free of activity. A decision has been made recently to remove the entire plant except for the foundations. The contract for detailed planning of a complete removal of the KKN was issued just a few days ago. This work will probably last 5 years. It is assumed that the radioactive wastes will be cut, packed in drums and ultimately stored in Asse II. About 3,500 drums with low-active wastes and 850 drums with medium-active waste are expected. The latter type of waste is composed of the pressure tubes.

NS "OTTO HAHN"

The last nuclear powered merchant ship still operating in the world, the NS "Otto Hahn", was also taken out of operation in the spring of this year (1979). The "Otto Hahn" had a drive rating of 10,000 shaft horse power. Since October 1968 this ship travelled more than 600,000 nautical miles on over 120 research and cargo-carrying trips. Because it is not possible to meet costs during the operation of this ship due to its small size, decommissioning became necessary. The total activity inventory of the "Otto Hahn" will amount to about $1.5 \cdot 10^{+16}$ Bq after one year of decay time. The total mass of the "Otto Hahn" is about 14,000 Mg. About 3,000 Mg of this mass are located inside the controlled area, and of this amount about 300 Mg must be considered active.The decommissioning is to be carried out as follows:

The active components will be dismantled in the largest possible units and will be transported to the Geesthacht Research Center (GKSS) in containers built specially for them. These radioactive wastes will be stored intermediately in a special hall on the grounds of the GKSS. Ultimate storage of these wastes will take place when a special ultimate storage facility for bulky radioactive wastes is established. Work on the NS "Otto Hahn" is to begin toward the end of 1979 after the fuel assemblies have been removed.

KWL

At the beginning of 1977 the Lingen Nuclear Power Plant (KWL) was shut down after 9 years of operation. This plant was shut down in order to carry out necessary repairs (replacement of the steam converter). Since the licensing authorities took the occasion of the repairs to demand extensive backfitting measures, whereby future operation was nevertheless not guaranteed, the operator decided to decommission the nuclear section of the plant. The secondary section of the plant -- a boiling-water reactor with fossile-fired superheating -- is to be operated further in conjunction with a gas turbine system to be constructed and a new boiler. The following procedure is proposed for decommissioning of the nuclear section of the plant:

- by the end of 1983: establishment of protective storage
- 1984 - 2010 : condition of protective storage
- 2011 - 2015 : complete removal

FR 2

The FR 2 Research Reactor in the Karlsruhe Nuclear Research Center has been in operation since 1962, and is used for radiation experiments. Plans have been made to take this plant out of operation by the end of 1981 and then to remove the nuclear section completely. Even though an experimental reactor is installed in this plant, extensive experience is expected to be acquired from the decommissioning work because the activity inventory in the plant is in the same order of magnitude as in a commercial nuclear power plant ($3.7 \cdot 10^{+16}$ Bq). The extensive and very complicated concrete construction should place demanding requirements on the cutting technology. Additional difficulties will arise from the fact that it is intended to use the building for other purposes later on. The contamination in the building must therefore be eliminated completely.

VAK

The first nuclear power plant in Germany, the Experimental Atomic Power Plant in Kahl (VAK, 15 MW) is to be shut down and decommissioned at the end of 1985 after 25 years of operation. This is a boiling-water reactor of the type constructed by GE. NIS has received a contract to ascertain the boundary conditions for the decommissioning (calculation of the activity inventory, determining the condition of the plant, etc.) and to study the possible decommissioning variants. The following alternatives are being considered:

- immediate complete removal of the plant shortly after final shutdown, and

- complete removal of the plant after a preceding period of protective storage.

The degree of difficulty involved in decommissioning this plant should be in the same order of magnitude as is the case with commercial nuclear power plants.

MZFR

The Multi-purpose Research Reactor in Karlsruhe (MZFR, 57 MWe) is also to be decommissioned during the 1980's. This should be the first decommissioning of a pressurized water reactor in Germany. Since this plant is located on the grounds of the Nuclear Research Center, it is especially well suited for research purposes and for testing new decommissioning techniques.

SUMMARY

Knowledge acquired to date, as of the current status of the study, can be summarized in the following points:

1. Nuclear power plants can be removed with the means known today, including on a short-term basis; however, a preceding limited period of protective storage is clearly more expedient if only for radiological reasons, i.e., radiation exposure of the personnel, as well as for cost reasons.

2. Nuclear power plants which meet today's requirements for easy maintenance, inspection, and repair, and for the possibility of replacements of large components in accordance with the radiation protection regulations do not require any special conceptual change for easier decommissioning. In general, if such conceptual changes should be considered, they should not detract from the safety of the plant for operation and they should not interfere with easy management of the plant.

3. Detailed studies are expedient for reference plants, but not for every individual plant at the present time. For purposes of the study, companies which can carry out the work studied in the actual case of the decommissioning must be asked for assistance.

4. The study of a decommissioning after a possible accident is -- because of the wide spectrum which is conceivable here -- neither reasonable nor necessary because the technical safety equipment installed limits the effects of an accident on the plant, on a long-term basis as well, and prevents such an accident from affecting the environment. This means, however, that time is available a posteriori for an analysis of the specific case and to take adequate technical measures.

5. Research work for further development of techniques which, although well known on principle and already employed operationally in modified form today, appears reasonable to us in view of the special aspects of the decommissioning.

6. Transportation and the ultimate storage of large low-active parts must be optimized. Moreover, there is a need to establish permissible radiological limits for the reuse of valuable waste.

7. According to today's point of view, even the variant with which higher decommissioning costs must be expected ultimately imposes only a relatively slight burden on power generating costs.

REFERENCES

/ 1/ Sicherheitskriterien für Kernkraftwerke Bundesanzeiger Nr. 206 (03.11.1977)

/ 2/ Bekanntmachung der Neufassung des Atomgesetztes v. 31.10.76, BGBl 131 p.3053 (1976)

/ 3/ J.Essmann et al., "Provision of Decommissioning of the german utilities for LWR Power Plant" IAEA-Symposium of the Decommissioning of Nuclear Facilities, SM-234/2, Vienna (1978)

/ 4/ D.Brosche, J.Essmann, "Zur Stillegung von Kern-kraftwerken", Atom und Strom Vol. 22 No. 3, (1976)

/ 5/ C.M.Lederer et al., "Table of Isotopes", Siseth Edition, John Wiley u.Sons Inc. (1967)

/ 6/ "Neutron Cross Section", BNL-325, Sec.Ed.; (1958-66)

/ 7/ M.K.Drake, "A Complication of Resonance Integrals" Nucleonics, Vol.24 No 8, (Aug. 1966)

/ 8/ G.Frejaville, "Contamination radioactive des circuits primaires des réacteurs á eau sous pression", B.I.S.T/CEA, N, 2120 (1976)

/ 9/ H.John, J.Vehlow, "Messung zur Verteilung aktivierter Korrosionsprodukte im Primärkreislauf des Kernkraftwerkes Obrigheim", Karlsruhe Nuclear Research Center, KfK1983 (1974)

/10/ W.Ahlfänger et al., "Dekontamination nuklearer Anlagenteile" VGB-Speisewassertagung pp.41-46 (1971)

/11/ D.Eickelpasch, "Untersuchung zur Reduktion der Strahlenbelastung das in Kernkraftwerken eingesetzten Personals", Dissertation (1978)

/12/ Standardization of Radioactive Waste Categories Technical Report Series No 101, IAEA, Vienna (1970)

/13/ G.V.P.Watzel, G.Thalmann, J.Vollradt, "Stillegung von Kernkraftwerken mit Leichtwasserreaktoren" DAtF, Reaktortagung (1978)

/14/ I.Auler, G.Lukacs, B.Bröcker, "Radioaktivitäten und Abfallmengen bei der Stillegung von Kernkraftwerken mit Leichtwasserreaktoren", DAtF, Reaktortagung (1978)

/15/ G.V.P. Watzel, J.Essmann, G.Lukacs, G.Thalmann "Decommissioning Study for Nuclear Power Plants with Light-Water Reactors", International Meeting on Nuclear Power Safety, Brüssel (1978)

/16/ I.Auler et al., "Decay Behavior Radioactive Inventory during Decommissioning of Nuclear Power Plants" IAEA-Symposium of the Decommissioning of Nuclear Facilities, SM-234/1, Vienna (1978)

/17/ A.Gasch et al.,"Results of an Analysis of the Quantities of Radioactive Wastes which Develop during the Decommissioning of Nuclear Power Plants" IAEA-Symposium of the Decommissioning of Nuclear Power Facilities,SM-234/3, Vienna, (1978)

DECOMMISSIONING PEACH BOTTOM UNIT 1

William C. Birely

Philadelphia Electric Co.
Philadelphia, PA 19101

John C. Schmidt

Catalytic, Inc.
Philadelphia, PA 19102

INTRODUCTION

Peach Bottom Unit No. 1, owned and operated by the Philadelphia Electric Company, was a 40-MW(e) High Temperature Gas Cooled Reactor demonstration plant situated 80 miles southwest of Philadelphia on the Susquehanna River. The plant was operated successfully for over seven years until October 31, 1974, at which time the plant was shutdown for decommissioning.

The Peach Bottom nuclear steam supply system was a helium-cooled, graphite moderated 115 MW(t) reactor operating at high temperature on a thorium-uranium fuel cycle. The NSSS produced superheated steam at 1000°F and 1450 psig, with a gross thermal efficiency of 39%.

The decision to shutdown and decommission was based on several factors. First and foremost was the fact that the program for which the plant had been originally designed was complete; this is, the objective of demonstrating the technical feasibility and commercial operation of an HTGR had been accomplished with outstanding success. Second, the size of the Peach Bottom prototype plant 40 MW(e) made it uneconomical in terms of operating costs or manpower compared to the large nuclear plants recently placed in operation. Finally, the changes that had been incorporated in the NRC safety and licensing requirements since initial operation in 1966 would necessitate major retrofitting for continued operation. A study indicated that the benefits to be derived from continued operation were not sufficient to justify the large expenditures which would result from plant modifications incurred in satisfying the new requirements.

PLANNING

Philadelphia Electric Company authorized the evaluation of three options for decommissioning. The options considered were:

1. Total removal of the facility
2. In-place entombment
3. Mothballing

Based on technical, economic, and safety evaluation of these options, mothballing of the facility under a Part 50 Possession Only license was selected. This option resulted in the least personnel exposure during decommissioning and the least hazard to the public as a result of high level radioactive waste shipping. On site entombment, involving the sealing of all residual activity within a massive concrete structure, presented the objectionable feature of having to remove the concrete in the event future regulations required total removal of the facility.

The decision to request final licensing status under a Part 50 Possession Only license offered definite advantages over a Part 30 By-Product license. Being in a Part 50 status means that the license is issued for an indefinite period of time and is under the control of NRC Reactor Licensing. In the Part 30 status, the plant would be under Materials Licensing and would be transferred to the state in the event that Pennsylvania became an agreement state. Under Part 30, the license is issued for a definite period (usually 5 years). A Part 50 license appeared to be the most advantageous because of its continuity.

All amendments to the Peach Bottom operating license were reviewed by Philadelphia Electric's on-site and off-site safety review committees prior to their submittal to the Nuclear Regulatory Commission for approval. The Technical Specifications were revised so that appropriate operating, testing, and staffing requirements were automatically deleted in steps related to changing plant conditions. The key decommissioning milestones permitting relaxation of the technical specifications without further approval of the Commission were:

1. Removal of all fuel from the reactor.
2. Degassing of the purification system, reactor temperature monitoring test program, and the disposal of the primary coolant gas (helium).
3. Shipment of all spent fuel from the Peach Bottom site.
4. Final lay up of containment and removal of radioactive components as specified in the Decommissioning Plan.

This method of phasing out the technical specification requirements proved to be very successful in minimizing testing and staffing requirements, simplifying the scheduling of decommissioning activities, minimizing the time spent obtaining approval for license revisions, and avoided unnecessary delays in the decommissioning program. Administrative procedures and controls were also relaxed in stages consistent with changing plant conditions.

Engineering of the Decommissioning Plan and Safety Analysis Report was performed by Catalytic, Inc. in conjunction with a Philadelphia Electric Co. engineering task force that met periodically to review all proposals. The Plan was reviewed and approved by both the Peach Bottom on-site and off-site safety review committee prior to its submittal to the NRC on August 29, 1974, and subsequently accepted with one amendment.

To properly implement the decommissioning according to the approved plan and engineering specifications, 71 Control Work Packages (CWP) were prepared by Catalytic, Inc. A CWP is a step-by-step procedure that contains the specific work steps, documentation, and signoffs to complete a particular task.

DECOMMISSION ACTIVITIES

The objectives of decommissioning were the removal of radioactive material that might have a potential of migrating beyond the plant boundaries, the construction of appropriate barriers, and the reduction of combustibles; so that, all active monitoring equipment could be retired, and the facility could be left in an unmanned status except for a semi-annual inspection.

The decommissioning plan called for work to progress in four phases. Phase I of the plan included unloading, canning and transfer of the fuel to the spent fuel pool, and degassing of the helium purification system delay beds. During this period, preliminary work that did not affect nuclear safety was undertaken. Phase II of the plan included shipping fuel to the reprocessing facilities in Idaho. Substantial mechanical decommissioning work was also performed during this period. However, all fuel handling and necessary nuclear safety systems were maintained intact. Once all fuel was accepted at the reprocessing facility, Phase III was initiated. These tasks included removal and decommissioning of all fuel handling systems and the radwaste facilities, as well as final decommissioning of all systems. Phase IV concerns the mothballed status and involves continued surveillance of the facility on a periodic basis.

During the planning for decommission, three areas of the plant were designated, each with its own unique decommission criteria. These areas are (1) unrestricted area, (2) controlled area within the exclusion area, and (3) inspection area within the exclusion area.

A chain link fence was installed around the containment and spent fuel pool building to establish the exclusion area. Access to the exclusion area requires entry through both a locked gate in the fence, and locked doors in the containment or spent fuel pool building. Barricades were installed within the exclusion area to segregate the inspection area from the controlled area.

A summary of decommission activities follows:

1. The removal and shipment of all nuclear fuel to the NRC fuel depository in Idaho.
2. All liquids, and pressurized gases such as the primary coolant helium, were removed and where necessary, processed as radioactive waste.
3. All flammable material other than electrical cable and graphite components within the reactor vessel were removed.
4. All mechanical penetrations into the containment were cut and sealed.
5. The containment was vented via a small breather pipe and absolute filter assembly to the atmosphere in order to prevent a temperature induced pressure buildup. The filter will be changed and surveyed as part of the surveillance program.
6. All vessels containing charcoal in the helium purification system were removed. This involved 18 vessels. A major consideration for removal was their location below the postulated flood elevation, and potential flammability of the material.
7. The Liquid Radioactive Waste Processing System, and several contaminated cooling systems, were completely removed and shipped offsite for their burial.
8. The primary coolant system was sealed.
9. The reactor vessel was sealed and left in place within an inaccessible concrete enclosure. The vessel internals, including the non-fuel graphite reflectors and dummy elements, were also left in place following defueling.
10. The fuel handling equipment was deactivated and secured in place. The spent fuel pool was drained and decontaminated. The fuel transfer chute was removed and sealed.
11. A portion of the exclusion area is inspected every six months to verify that radioactive material is not escaping or migrating through the containment barriers in the facility. The inspection area includes the refueling floor, and a passageway leading to the sump pit situated in the lower basement.

12. All exposed surfaces within the inspection area were decontaminated below the allowable limits of 5,000 dpm β-γ/100 cm^2 fixed contamination, and 1,000 dpm β-γ/100 cm^2 removable contamination. On contact, radiation levels were reduced below the allowable level of 1 mR/hr in the inspection area.
13. Appropriate barricades were established to restrict access to the controlled area within the exclusion area. While the radiological limits were not applicable to the controlled area, the levels of contamination and radiation were for the most part minimal.
14. All utility services to the containment, with the exception of lighting service, was permanently secured by disconnecting or locking out the power supplies. All monitoring and alarm systems were also secured.

All contamination systems outside the exclusion area were removed or decontaminated to levels less than those specified in Regulatory Guide 1.86 (5,000 dpm/100 cm^2 fixed contamination, 1,000 dpm/100 cm^2 removable). This area outside the exclusion zone, including the offices, laboratories, control room, and Turbine and Auxiliary Building, was released for unrestricted use. Radiation levels in these areas are less than 0.01 mR/hr.

RADIOLOGICAL ASPECTS OF DECOMMISSION

Almost all of the radioactive material remaining within the decommissioned facility is fixed as neutron activation products contained in the reactor vessel. The principle nuclide is ^{55}Fe in the steel pressure vessel. This activity will be reduced substantially by radioactive decay while the plant is in the mothball status. It has been calculated that the vessel surface dose rate will decay from the present level of 100 R/hr to 50 mR/hr in 40 years. The vessel access ports were sealed and concrete beams were positioned above the ports. The reactor vessel remains as originally designed within an inaccessible concrete enclosure.

The remaining radioactivity was ^{134}Cs and ^{137}Cs as contaminants in the primary piping. These nuclides presented the principal contamination source and the source of most exposure encountered during decommissioning. Nearly 100 curies of ^{137}Cs remain in the primary system.

The safety concern associated with the removal and transportation to a burial site, favored the decision to mothball the reactor vessel in place. The shielding, seismic design, and containment features, which are inherently satisfactory for operating plants, are likewise satisfactory for plants in a mothball status.

Health physics procedures were applied to limit personnel

exposure and to satisfy the ALARA concept. In addition to traditional radiation protection training and exposure control procedures, extensive planning and dry running of radioactive waste removal and decontamination activities were implemented, resulting in improved job efficiency and low personnel exposure levels. One of the activities performed involved removing and shipping a purification system vessel (Condensible Trap) reading 30 R/hr on contact.

Before the condensibles trap was removed, a planning meeting was held to develop exact procedures for rigging out and moving the vessel to the shipping container. Subjects dicsussed included: assignment of specific personnel to specific tasks; monorail installation; temporary shielding of the vessel during transport; rigging and transport methods to be used; timing; staging areas for workers while waiting to perform tasks; and health physics control.

The vessel was lifted with electric hoists with the special monorail, moved to a transport dolly and shielded, moved out of containment to the auxiliary building, and lifted into the shipping container. Then the container was moved to a waiting concrete truck and filled with concrete.

The total procedure was accomplished in less than three hours, with 16 workers involved in the operation. The maximum radiation exposure received by any of the workers was 180 mrem and the total exposure to all workers was 1040 milli-rem.

Approximately 400 gallons of contaminated oil and tritiated water were solidified prior to shipment. Several solidification techniques were tested using uncontaminated fluid in the interests of developing an effective method and reducing radiation exposure. Effective cement solidification methods were developed for both systems.

Various methods of decontamination were used, depending on surface porosity, type of contamination, level of contamination, and accessibility. Removable contamination was disposed of with an absolute filtered vacuum cleaner or by washing with solvents and rags. Fixed contamination on metal surfaces was usually cleaned with foaming bathroom cleaner and rags. Painted walls in the spent fuel pool responded best to the use of a commercial liquid cleaner. Some stubborn fixed contamination on concrete required abrasives (sandpaper and wire brushes), and when this method was ineffective, a jackhammer was used.

The most difficult area to decontaminate involved the 82 radwaste system drain pipes embedded within the concrete floors and walls of the facility. Acid backing and flushing using sulfuric acid to remove contamination on the internal surface of the buried pipes was only partially successful. Hydrolaser employing a 6500 psi water nozzle proved to be the only effective decontamination technique for the pipes.

The highest contamination levels experienced were 10^7 dpm/100 cm^2 within the primary coolant system. The spent fuel pool equipment was contaminated on the exterior surface to 10^6 dpm/100 cm^2. Most cutting on contaminated equipment was done with either reciprocating saws or band saws. This method of cutting produces relatively large chips or material which fall in the vicinity of the cut and are too large to become airborne. Flame cutting and arc-gouging were allowed only when contamination levels were less than 10,000 dpm/100 cm^2. With the exception of cutting out the dust collector on the primary coolant system and for some work in the spent fuel pool, all airborne activity was less than $1x10^{-9}\mu Ci/ml$. Airborne activity during cutting of the dust collector was $2x10^{-8}\mu Ci/ml$.

The only areas of the decommissioned facility with radiation levels over 100 mR/hr (whole body) are in the steam generator and reactor vessel rooms. Other areas within the plant are below 5 mR/hr and the dose rate within the inspection area is less than 1 mR/hr. The total exposure to all personnel during the decommission operation was 8.95 man-rem. The maximum exposure received by any individual was 750 mrem, and all exposures were less than 200 mrem/week during all phases of decommissioning.

A total of 490 containers of solid radioactive waste with a volume of 14,000 cubic feet were packaged and shipped during decommissioning. The total radioactivity of the solid waste from the entire decommissioning program was only 380 curies. Of the approximately 20,000 ft^3 of clean waste removed, 14,000 ft^3 of metal scrap was salvaged and 6,000 ft^3 of debris was disposed as trash. During the decommission phase, gaseous and liquid radioactive releases totaled only 3.75 curies and 0.3 curies respectively.

ADMINISTRATION

The Peach Bottom decommission program was accomplished by Philadelphia Electric Company with assistance from Catalytic, Inc. The utility, as the licensee, retained the responsibility for ensuring that all activities were performed in accordance with the NRC operating license and regulations. Administrative controls, procedures, audits, review committee, etc. were established to ensure compliance by Catalytic with all rules and regulations. Frequent detailed planning meetings were conducted prior and during decommission activities between utility and vendor personnel to establish areas of responsibility, personnel contacts, rules and regulations, resolve potential problems, and maximize utilization of available expertise. As a result, both organizations worked in an efficient and cooperative manner ensuring that the objectives of decommission were safely and economically accomplished. All phases of the decommission program and the final mothball status of the facility were monitored and reviewed by the Nuclear Regulatory Commission through on-site inspections. A summary of Philadelphia Electric and Catalytic responsibilities follows.

DECOMMISSION RESPONSIBILITIES

PHILADELPHIA ELECTRIC	CATALYTIC, INC.
1. Provided personnel for: a. Operate and maintain active equipment, and block out equipment for decommission. b. Construction crafts & foreman. c. Auditors of H.P. activities. d. Licensed engineer-ensure compliance with Operating License and direct operating personnel. e. Security.	1. Provided personnel for: a. Engineering activities. b. Project supervision. c. Health physics activities. d. Laborers and carpenters.
2. Performed activities associated with defueling, loading shipping cask, and degassing purification system.	2. Performed activities associated with removal and shipment of plant equipment, erection of barriers, decontamination. Provided subcontract administration and material control.
3. Administration: a. Task Force reviewed Decommission Plan. b. Onsite and offsite safety committee reviewed Decommission Plan, controlling procedures and reports. c. Utility engineer approval required prior to start of each decommission activity.	3. Developed Decommission Plan, controlling procedures and final report. Provided documentation associated with physical removal of equipment and decontamination.
4. Provided some consumable items, protective clothing and masks.	4. Provided most H.P. instrumentation, survey meters shipping containers.

5. Station management approved allowable exposure limits.	5. Controlled exposure within allowable limits.
6. Performed general site indoctrination and respiratory equipment training.	6. Provided specialized training.
7. Provided dosimetry and bioassay services.	7. Provided backup dosimetry.
8. Maintained formal communication with NRC.	8. Assisted in NRC audits.

Decommissioning activities involving the physical removal of equipment and decontamination required a total of forty-seven weeks. The work force peaked at 39 men during these operations. This force was composed mainly of Philadelphia Electric construction mechanics, contract carpenters and laborers. A distribution of maximum and average work force personnel is shown in the following table. These figures do not include manpower associated with reactor defueling and loading of the shipping cask.

PERSONNEL	MAXIMUM	AVERAGE
Construction Foreman	2	1
Construction Crafts	20	10
Carpenter - Laborers	8	4
Health Physics technicians	3	2
Field Staff (engineers, health physicist)	6	4

The total cost of decommissioning was \$3,524,000 of which \$1,517,000 was spent for the physical removal of components. Fuel shipping costs are not included in these figures. Approximately 40,000 manhours were required to complete the total project.

The gradual reduction in manpower needs at Peach Bottom Unit No. 1 coincided with the startup of two new BWR units at the site. As a result, most of the staff, crafts, and operating personnel were reassigned to the new units. Manpower relocation planning well in advance of a scheduled plant decommission should be given a high management priority so as to minimize personnel problems.

CONCLUSION

The application of advance planning and effective health physics techniques used during the Peach Bottom decommission program demonstrated the feasibility of decommissioning a nuclear facility economically at low personnel exposure levels and with a negligible environmental impact.

REFERENCES

1. "Decommissioning Peach Bottom Unit 1", Philadelphia Electric Company, July 1978.

2. "Decommissioning Plan and Safety Analysis Report", Peach Bottom Atomic Power Station Unit 1, May 1975.

3. W. C. Birely, "Operating Experience of the Peach Bottom Atomic Power Station", ANS Topical Meeting; Gas-Cooled Reactors: HTGR and GCFBR, May 7-10, 1974.

4. R. S. Stoucky, E. J. Kohler, "Planned Decommissioning of the Peach Bottom Unit No. 1 High Temperature Gas Cooled Reactor", ASME Publication 73 - WA/NE-7.

5. J. L. Everett, III, E. J. Kohler; "Peach Bottom Unit No. 1, A High Performance Helium - Cooled Nuclear Power Plant", ANS Annual Meeting, June 1978.

6. "Peach Bottom Unit No. 1 Decommission and Evaluation of 3 Options", Suntac Nuclear Corporation, December 1972.

THE USE OF ENGINEERING SCALE MODELS IN DECOMMISSIONING NUCLEAR FACILITIES

Chester S. Ehrman, P.E. and C. David Lilly, P.E.

Burns and Roe, Inc.
Oradell, New Jersey 07649

ABSTRACT

The use of engineering scale models as design tools for chemical process plants has been well developed for over 25 years. Within the past ten years models have been similarly well developed for application to commercial nuclear power plant design. As the nuclear industry prepares to decommission the first commercial nuclear power reactor, this paper presents the applications of this valuable tool to the decommissioning process.

INTRODUCTION

Engineering scale models have been used in the shipbuilding industry as a design and construction tool since the 17th century[1]. Over 25 years ago the chemical process industry recognized the application of scale modeling as a design process which simplified pipe routings, promoted maintainability of components, and served as a valuable communications tool between designer, constructor, and operator. The Navy Nuclear Propulsion Program began using full size wooden mock-ups of lead ship submarine design in the early 1950's in order to verify constructability, operability, and maintainability features within the limited confines of a submarine pressure hull. Radiation workers have learned to limit personnel exposure in high radiation areas by rehearsing complex tasks in full or partial mock-ups located in low radiation areas. Finally, about ten years ago, the engineering scale model was introduced into the commercial nuclear power plant design process as a means of eliminating expensive construction problems resulting from improper physical interfaces. Since that time, growing sophistication in the use of scale models and increasing precision in their fabrication have made the scale model an integral part of the design process rather than just a checking tool. Not only has modeling become an integral part of the modern efficient design process, it has allowed early incorporation of a set of complex and overlapping design requirements such as reliability, maintainability, equipment removability, accessability, operability, and availability. At present these models are highly developed tools of the engineering design profession[2]. A modern nuclear plant designed with a model is far more likely than a plant of an earlier generation to have essential functional requirements incorporated in its design which promote ease of decommissioning and/or decontamination. Burns and Roe has recently constructed detailed scale models for several nuclear power plants, such as Three Mile Island Unit No. 2, Washington Public Power Supply System Unit No. 2, Forked River, and the Clinch River Breeder Reactor Plant.

Description of a Decommissioning Model

With this historical perspective in mind, let's turn to the next step in this natural progression of the use of modeling by the nuclear industry, namely, the decommissioning process. A model used for decommissioning could be the plant's original design model if one exists or a new model constructed expressly for the purpose of decommissioning. Since the latter is the expected case for the generation of commercial reactors about to be decommissioned, the following points address the features of a model constructed for use in decommissioning.

Areas to be Modeled. The principal benefits perceived for the use of a model during the planning process are reduction of planned radiation exposure and cost reduction by the selection of an optimum sequence of equipment removal containing as many non-interfering paths as possible. Most of these benefits can be achieved by modeling only those portions of the plant in which conditions inhibit access for taking on-site measurements such as high radiation areas or areas containing substantial quantities of airborne or surface contamination and which also present complex spatial relationships between components, piping, ducts, and access openings.

Scale. The scale selected for decommissioning models should be chosen so that the model is large enough to allow extensive handling of components and piping in order that they may be manipulated through the process of removal including temporary shielding, staging, bagging, rigging, laydown, and transport. The choice of scale will ultimately depend upon the extent of the high radiation areas and the complexity of the equipment arrangements in these areas.

As-Built Conditions. In some plants, the record drawings have not been maintained in accord with changes in plant design or changes have not been consistently recorded. For reasons of expense and practicality as well as for maintaining radiation exposure As Low As Reasonably Achievable we would not expect that detailed checking of as-built drawings would be necessary, rather the model should be built to the best documentation which is available and then compared by measurement and photograph to the plant. Of particular concern in this area is the small diameter piping which is typically field routed such as sampling and instrument lines. These need to be accurately located on the model where they form radiation hot spots, as is frequently the case, or where they intrude into equipment removal envelopes. Where tight areas or high risk sequencing are encountered during the planning process, then and only then should detailed dimensional checks be considered. Of course, where tight areas are encountered, the design modeling practice of considering the construction of a larger scale model of the design area to study the problem in detail should be followed. In many situations involving decommissioning, full scale mock-ups for worker rehearsals may be planned. Depending upon how extensive these models are to be, they may be constructed early and used to work out tight area problems.

Precision. The degree of detail and accuracy provided for those areas which are modeled will vary depending upon the potential interference in an area. Equipment offering straightforward removal paths may be lumped together with its connecting pipe and supporting steel and shown simply as a block of styrofoam until it is discovered that the removal path of some other piece of equipment enters this area at which time additional details will be made in the presentation of the first equipment envelope. In highly complex areas containing high radiation areas, extremely precise positioning of all equipment, piping, conduit, cable trays, and ducting may be necessary.

Modeling Details. Several details of the decommissioning model would be of great importance to the project. All padeyes and structural steel capable of being used for rigging, staging, or temporary shielding support should be shown on the model together with the load rating of the fixture. General area radiation levels, component and piping contact readings, and all hot spots identified during radiation surveys should be displayed on the model. It may be desirable to select a color code for the piping based upon average contact radiation level rather than system identity. Where the decommissioning procedures call for them, temporary shielding, staging or scaffolding, temporary filtration exhaust systems, and surface contamination containment bags or tents should also be shown so that the interference that these items present to component removability can be easily recognized. Components should be fixed in place on the model in such a way that they are removable for routing studies. The points selected for pipe cutting should be distinctively marked. Other details which should be displayed for each area of the model are the health physics special clothing requirements for entry to each area and the locations of all emergency equipment such as emergency breathing apparatus, fire extinguishers, first aid kits, eye washes, deluge showers, and spill kits. During the recent emergency recovery actions at Three Mile Island Unit 2, the coordination between radiation survey maps and the model was essential to the process of selecting tap-in points and routings for standby emergency systems so as to reduce radiation exposure associated with conducting these operations.

The decommissioning model would most efficiently be constructed at the plant site so that model makers would have ready access to the plant to confirm needed details with photographs, measurement, and closed circuit TV. The input of the plant operators would be solicited during the model construction and pertinent information added to the model display. When the model is completed, its use for the decommissioning planning process could begin as described below.

Use of a Model in Decommissioning Planning

The model would be used to evaluate the basic feasibility of the major activity sequences which have been developed as the model is being built. With this completed, the model is used to develop and evaluate the detailed work procedures for equipment removal and decontamination. The following are examples of the model's use during the planning phase.

It is desirable to allow for as many concurrent activities as possible consistant with ALARA considerations to shorten the duration of the schedule and reduce schedular risk. Where procedures call for the erection of temporary shielding and scaffolding, it will be placed on the model to determine if interferences exist with other concurrent sequences. We feel that the model will be invaluable in making the classic temporary shielding judgment, namely, "will I use more exposure in shield installation and removal than I save in the operation which is shielded?" One of the values of the model is simply the verification that dismantled pieces can be properly rigged from a contamination or high radiation area to a staging area and from there to a waste packaging area. This is the reason that components on the model should not be permanently fixed in place. Reductions in radiation exposure will result if contaminated piping is removed to the staging area in the largest easily rigged pieces. This reduces the number of thick-walled piping cuts that are required and therefore the total personnel radiation exposure. On our Clinch River and Washington Public Power Supply System projects, a model was constructed to verify that the large primary coolant piping was installable in as large sized spools as possible in order to reduce the number of field welds. The decommissioning model would be used to perform this process in reverse. Since the model would be reasonably simple to dismantle and reassemble, it would be possible to "let your fingers do the walking" through the selected decommissioning sequences to verify that no interferences exist between plant features or with temporary rigging, staging, or shielding associated with other concurrent sequences. The model would be used to evaluate alternative methods for cutting piping based upon access requirements for machinery and operators. Shielding casks would also be modeled in order to examine relevant handling and contamination control steps. Evaluation of rigging sequences may change the order of equipment removal and of steel structures removal in order to minimize radiation exposure and optimize the project schedule. Another valuable benefit of using a decommissioning model for planning is that the needs for full scale mock-up of certain areas will become identified early and coordinated with the rest of the project plan.

Use of a Decommissioning Model in the Field

As valuable a tool as the decommissioning model would be during the planning process, it would be even more valuable during performance of the decommissioning itself. During this phase of the work, the model would be used as a training and familiarization tool for craft labor. Identification of emergency equipment on the model would be an aid to job site safety. The model would be available at the beginning of each shift for the shift briefing. The model could then be split into sections for each work team to familiarize themselves with the shift's assignment. When difficulties are encountered in a high radiation area, such as greater than anticipated dismantling operations or rigging problems, work teams will leave the area to identify and evaluate alternative problem solutions on the model or in a mock-up rather than in a radiation area. Radiation survey data including hot spot locations would be maintained up-to-date on the model. This would assist craft labor in planning their work and reducing personnel radiation exposure. The model would be a very accurate status keeping tool for a decommissioning project if it were dismantled at the same rate as the facility. This is best accomplished by having model makers updating the models on backshifts by adding staging, temporary shielding and rigging equipment as well as by removing components, using progress photographs. By maintaining job site status on the model and having plentiful progress photographs taken from vantage points identified on the model, exposure to supervisors will be reduced.

As the field work proceeds, the model is dismantled at the same rate as the plant, including backfilling with concrete, and finally may serve as a permanent record of the configuration of selected substructures or entombments which are left on the site.

CONCLUSION

Engineering scale models have shown themselves to be invaluable aids to the design and construction of nuclear facilities of all kinds. Application of this tool to the decommissioning process promises to reduce radiation exposure and costs through sequence optimization and better control of the craft labor force.

REFERENCES

1. Sumrall, Robert F., Curator of Ship Models, U.S. Naval Academy Museum, Annapolis, MD. Personal Communication. September 1979.
2. Ehrman, C.S., Edelman, R.A., Garvin, T.G., "Models - Their Use and Effectiveness." Proceedings of ANS Topical Meeting. The Construction, Licensing, and Startup of Nuclear Power Plants. 1977.

PLANNING FOR DECOMMISSIONING OF THE SHIPPINGPORT ATOMIC POWER STATION

R. L. Miller

Office of Surplus Facilities Management
UNC Nuclear Industries
Richland, Washington 99352

J. L. Landon

U. S. Department of Energy
Richland Operations Office
Richland, Washington 99352

INTRODUCTION AND OBJECTIVES

The Shippingport Atomic Power Station (Shippingport) was constructed in the mid-fifties as a joint project of the Government and Duquesne Light Company. The nuclear portion of Shippingport has been in operation since that time under supervision of the Division of Naval Reactors (NR) of the Department of Energy.

Upon completion of defueling, Shippingport becomes the responsibility of the Office of Nuclear Waste Management (ONWM) to decommission. Initial decommissioning activities will be concurrent with end-of-life testing and defueling of the reactor.

The overall objective of the project is to place Shippingport in a long term radiologically safe condition following termination of operations and to perform the decommissioning in such a manner so as to provide engineering, technology and cost information for future projects. To meet the overall objective, the following activities are required:

- Decommissioning Assessment
- Environmental Assessment

- Engineering for the Chosen Decommissioning Alternative(s)
- Preparation of an Environmental Impact Statement (EIS)
- Selection of a Decommissioning Contractor
- Decommissioning Operations.

The Decommissioning Assessment, Environmental Assessment, and draft outline for the EIS have been completed. The results of the assessments are summarized in this paper. In addition, the management of the Shippingport decommissioning project is discussed.

DECOMMISSIONING ALTERNATIVES

At the present time, the U. S. Department of Energy is considering three decommissioning alternatives for Shippingport. These alternatives are:

1. Prompt Removal/Dismantling
2. Safe Storage/Delayed Dismantling
3. Entombment (with vessel internals removed)

Following is a description of the alternative decommissioning methods.

Prompt Removal/Dismantling

Prompt Removal/Dismantling requires removal from the site of all radioactive fluids and wastes, and other materials having radioactivity and surface contamination. The U. S. NRC Regulatory Guide 1.86, Table 1, provides guidance for the release of materials for unrestricted use that are contaminated with surface contamination. Materials which contain induced radioactivity would be evaluated on a case-by-case basis. The Decommissioning Assessment assumed that materials with an exposure rate exceeding 0.1 mR/hr at 1-inch could not be released for unrestricted access. This exposure rate would be for surface contamination as well as induced radioactivity.

For the prompt dismantling alternative, a minimal decontamination effort may be undertaken primarily to remove loose contamination from plant areas. This would lower the occupational radiation exposure to workers in the plant. Decontamination of piping internals could be accomplished with clean water or acid flushing.

The systems normally exposed to primary coolant water or fuel pool water are generally considered contaminated systems. For Shippingport, examples include the coolant purification system and core cooling removal system, and canal cleanup system. Such systems or portions of them would be decontaminated (or removed) as required.

Safe Storage/Delayed Dismantling

This alternative consists of removing all radioactive fluids and wastes, and selected components external to the reactor vessel to be shipped off-site for disposal. All remaining highly radioactive or contaminated components (e.g., reactor pressure vessel and materials) would be sealed within an entombment structure barrier. For this mode, an appropriate and continuing surveillance program would be required to assure public health and safety until dismantling.

After a dormancy period to allow the decay of radioactive material to reduced levels (although not necessarily unrestricted access levels) the plant would be dismantled as in the prompt dismantling alternative. Delayed dismantling after a dormancy period of about 100 years would not require decontamination since radiation levels will have decayed sufficiently to permit occupational access. In general, this method permits dismantling to be accomplished with less sophisticated tooling and lower occupational exposure.

The Safe Storage structure is discussed in the next decommissioning alternative.

Entombment

The entombment mode is similar to Safe Storage with the exception that certain radioactive components removed at plant shutdown so that the entombed radioactivity required life does not exceed the entombment structure design life.

Typically, this requires removal of materials containing long-lived isotopes from the reactor vessel. This alternative also requires an appropriate and continuing surveillance program to assure public health and safety throughout the entombment lifetime.

The feasibility of the Entombment alternative is dependent on the entombment structure design life and the radioactive inventory required life. The inventory required life is the period for the long-lived radioactive isotopes to decay to non-hazardous levels. The entombment structure must be demonstrated to maintain its integrity during the required entombment period. The evaluation of structure design life and inventory required life are discussed in the following paragraphs.

The entombment structure design life is estimated to be about 200 years, and is based on the strength of the concrete entombment barriers and their resistance to environmental factors. The steel containment chambers provide a second entombment barrier. The following environmental factors will be considered in design of the entombment structure:

- Floods
- Hydrology
- Precipitation
- Temperature Variation
- Weathering/Corrosion
- Tornado/Wind
- Earthquake

The entombment boundary would consist primarily of the existing reinforced concrete and steel containment structures (i.e., Reactor Chamber, Boiler Chambers, and Auxiliary Chambers in the case of Shippingport). Openings in the concrete boundary, such as topside hatches, most passageways, and the entire fuel canal would be covered with 2 ft. thick reinforced concrete slabs. At least one passageway to an airlock would be locked closed, rather than concrete sealed, to permit access to the entombment for periodic inspections and maintenance.

Temperature variations, especially freeze/thaw cycles, would pose the most severe environmental phenomenom for the concrete entombment structure. Shippingport concrete structures above the frostline were constructed with air entraining concrete. This concrete will withstand through-cracking due to freeze/thaw cycles for at least 200 years, provided the structure is protected with a waterproof coating and periodic inspections and repairs are made as necessary.

Chemical attack from groundwater, heavy precipitation, weathering and corrosion, and tornado and wind damage are all expected to have a minimal impact on the concrete entombment structure. A schedule of routine maintenance and repair work would be employed to correct any minor defects as necessary.

Comparisons of the Three Decommissioning Alternatives

The determination of feasibility of the decommissioning alternatives is dependent upon the radioactive inventory, the entombment required life, the entombment structure design life, and the available methods for decontamination, removal, shipping and disposal. Each of these factors will be reviewed relative to the Shippingport decommissioning program. In order to properly evaluate the feasibility of a decommissioning alternative, the cost schedule, occupational exposure, manpower requirement, waste volumes, environmental impact, and any maintenance/surveillance requirements must be considered.

REMOVAL OF STRUCTURE AND COMPONENTS

The techniques available for removal of system piping, components and structures have been demonstrated as feasible during prior decommissioning programs. It is not anticipated that new or unique technology will be required to accomplish the removal of contaminated or clean equipment or structures.

There are several techniques available for mechanical cutting and removal of piping, pumps, heat exchangers and other contaminated or clean components. Pneumatically-operated power hack saws can be strapped to a pipe and a high pressure air supply drives the power hack saw until the section is cut. They are capable of unattended operation which permits the work crew to cut additional pipe simultaneously, or to interpose radiation shields between the workpiece and themselves. Another technique for pipe cutting is placing strap-life explosives on the pipe to be cut and detonating the explosive. The technique permits unattended operation and is well suited for remote cuts in inaccessible areas. More conventional techniques such as oxyacetylene torch, carbon arc and reverse polarity metallic arc cutting methods could be used.

Structures, floors and walls could be mechanically decontaminated by scarfing (removing 1- or 2-inches of surface) with surface grinders, scabblers, pneumatic drills and chippers or saw-cutting sections of floors or walls.

Massive, heavily-reinforced concrete structures (greater than 2 ft. thick) can be removed by controlled blasting. The exposed concrete surfaces would be water sprayed to reduce dust and contamination spread. After detonation, exposed reinforcing bar could be cut with an oxyacetylene torch to remove sections of concrete. Thinner concrete sections could be demolished with a conventional wrecking ball, or a ram hoe (a back hoe fitted with a large pneumatic hammer).

The arc saw is the reference method for prompt dismantling of the reactor vessel. It is a circular saw blade approximately 36 inches in diameter and 1/4-inch thick. The blade cuts by passing a high amperage, low voltage current (7500 amps, 80 volts) between the saw and workpiece. The heat generated in the workpiece cut melts the metal which is blown away by the blade. Another technique for cutting thick vessel sections is the plasma arc torch. This tool uses a high temperature plasma gas (40,000°F) to melt the metal.

DISPOSAL OF RADIOACTIVE WASTE

The radioactive materials generated during decontamination or dismantling can be sent to controlled licensed burial grounds for disposal. Most of the radioactive waste shipments (system piping and components, and structures) must be classified as Low Specific Activity (LSA) shipments and would be buried in the shipping containers (strong, tight packages - steel or wooden boxes). Shipments of the vessel and internals removed during the prompt dismantling mode are generally classified as Large Quantity Radioactive Material and will usually require special shielding for transportation. Since the shielding is reusable, inner liners would be used for burial.

Liquid wastes from decontaminated systems could be filtered evaporated, and demineralized. The resulting solids would be processed as solid waste and the liquid disposed of in accordance with standard acceptance practice.

Contaminated piping and components removed during Safe Storage/ Entombment or Prompt Dismantling will be "nested" in each other, where feasible, for reduction in burial volume and loaded into LSA containers. Large diameter pipe, only contaminated internally, may be sealed on the ends and buried as its own container. Where possible, contaminated piping and components will be nested inside the large diameter pipes before the pipes are sealed. This is an effort to reduce solid waste volume as much as possible.

Vessel and internal segments will be placed in disposable steel liners, and the liners loaded in reusable shielded container(s). The steel liner will provide a barrier against contamination leaching into the burial soil.

The transportation of radioactive material to controlled burial grounds can be accomplished via truck, rail, or barge. Truck shipments appear to be the most practical at this time since commercial controlled burial grounds in current use do not have rail spurs and barge shipping has not been used extensively for waste shipments due to the long turnaround time for shipping casks. However, specific applications of rail or barge shipping may offer advantages over truck shipments. Regardless of the type of shipping used, all radioactive waste shipments will comply with applicable U. S. Department of Transportation regulations.

ENVIRONMENTAL ASSESSMENT

The major environmental impacts, which will be evaluated for the decommissioning of Shippingport, include: radioactive and nonradioactive gaseous effluents; radioactive and nonradioactive liquid effluents; solid wastes; resources committed; transportation of radioactive materials; and an analysis of the credible significant accidents involved with the Shippingport decommissioning program. A preliminary assessment of these impacts is discussed in the following subsections.

Gaseous Effluents

Of the three decommissioning alternatives evaluated, Prompt Removal/Dismantling would create the highest concentrations of airborne radioactivity. In-air cutting of the high specific activity reactor vessel internals with an arc saw will release vaporized particulates from the kerf. Controlled blasting of concrete structures with surface contamination will also cause radioactive particulates to become airborne. The use of a contamination control envelope (CCE) and high efficiency particulate air (HEPA) filters in the CCE exhaust system and the building ventilation system will reduce the particulate release to the environment.

An NRC study (NUREG/CR-0130) of the decommissioning of an 1175 MWe pressurized water reactor provided estimates of vessel internals cutting releases. In the NRC study, vessel and internals were estimated to contain approximately 4.8×10^6 curies which is a factor of 17 greater than the estimated 286,000 curies in the Shippingport Station. The NRC estimated dose to the lung of the maximum exposed individual during vessel internals cutting was estimated to be 2.5×10^{-7} rem for the first year dose. This is less than the exposure experienced for an operating reactor of this size. Preliminary assessments show that the radioactive inventory would not exceed the NRC estimated exposure and the gaseous effluent environmental impact during Prompt Removal/Dismantling should be negligible.

The Safe Storage with Delayed Dismantling and Entombment alternatives will not create any greater environmental releases of radioactive gaseous effluents than Prompt Removal/Dismantling.

The largest concentration of nonradioactive gaseous releases would occur during the controlled blasting of the concrete structures for dismantling. Water sprays may be used to hold down the dust within the blast area. It is assumed that the suspended particulate concentration immediately after a blast will be 1000 mg/m^3 (conservatively estimated at 2000 times the occupational limit). The atmosphere is then passed through the building HEPA filters with a resulting concentration at the vent outlet of 11 μg/m^3. This is well within the suspended particulate annual average (continuous) ambient air standard of 60 μg/m^3 for the public as established in 40 CFR 50.7.

Liquid Effluents

Radioactive liquid effluents from decommissioning Shippingport will consist of two major components. The disposal of all existing liquids in piping, components, and the fuel handling canal will be the largest consitituent of liquid volume to be processed. Additionally, certain components may be water flushed to reduce radiation levels for access.

The former of these two liquid waste sources would be encountered with all decommissioning methods. The latter flushing waste source would be specific to each decommissioning method. All radioactive (or potentially radioactive) liquids would be filtered, evaporated, demineralized or solidified using the existing Shippingport liquid waste treatment system.

It is anticipated that liquid released from the Shippingport site will be the water-based distillate from drained and flushed systems. Title 40 of the Code of Federal Regulation Part 423 allows a daily liquid effluent release of 1.0 mg/l of total iron and 1.0 mg/l of total copper. Since iron and its oxides are not soluble in water, and since copper will not be present in typical solutions, the distillate will contain only trace amounts of iron. Therefore, the criteria of 40 CFR 423 will be met.

COMMITMENT OF RESOURCES

The two major resources considered important in this assessment were site land and licensed burial ground land. All other resources such as concrete for entombment structures, electricity for lighting and power fuel oil for heating and liquid waste evaporation,

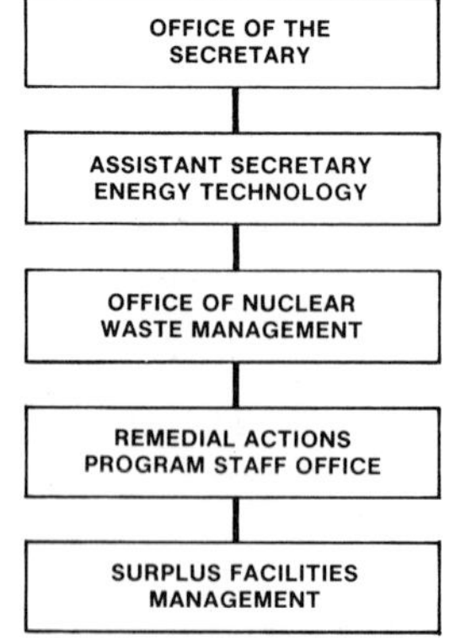

DOE ORGANIZATION
FOR SURPLUS FACILITIES MANAGEMENT PROGRAM

Figure 1

RICHLAND OPERATIONS OFFICE

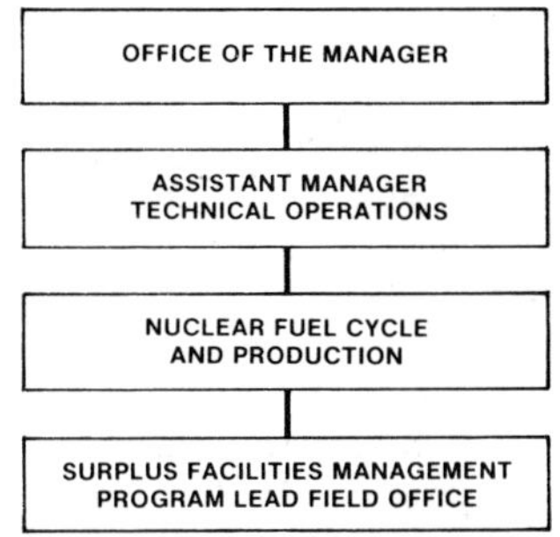

DOE-RL ORGANIZATION
FOR SURPLUS FACILITIES MANAGEMENT PROGRAM

Figure 2

UNC NUCLEAR INDUSTRIES
OPERATIONS DIVISION

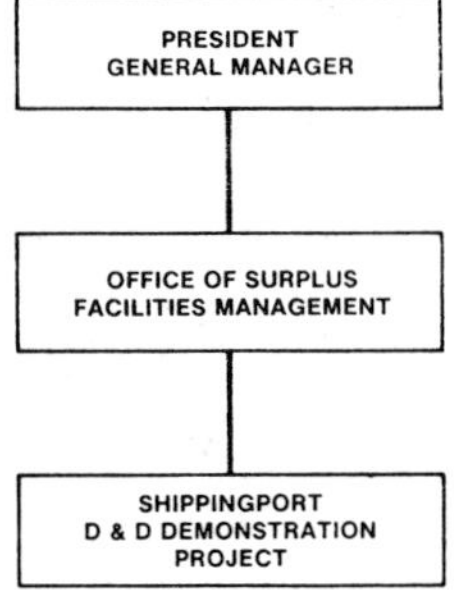

Figure 3

gasoline for equipment, etc., were not considered major environmental factors.

The Shippingport site land resource that would be committed for the Safe Storage/Delayed Dismantling or Entombment alternatives would consist of approximately seven acres of land adjacent to the Beaver Valley access. The land would not be available for power generation use for about 100 years of dormancy. Prompt Removal/Dismantling will allow unrestricted use and therefore, restore the entire DOE building-occupied area of seven acres. The Prompt Removal/Dismantling alternative will utilize clean rubble on-site for fill.

PROJECT MANAGEMENT

The decommissioning of Shippingport is managed by the Office of Assistant Secretary of Energy Technology (ASET) through the Office of Nuclear Waste Management (ONWM) and under the direction of the Remedial Actions Program Staff Office (RAPSO). The responsibility is delegated to the lead field office for decommissioning DOE surplus facilities, namely the Richland Operations Office (DOE-RL). Within DOE-RL, the Surplus Facilities Management Program (SFMP) has responsibility for the project management for Shippingport decommissioning. Technical assistance is provided to SFMP by UNC Nuclear Industries through the Office of Surplus Facilities Management (OSFM). Figure I shows the organization within DOE-Headquarters, Figure II shows the organization within DOE-RL, and Figure III shows the UNC Nuclear Industries organization. Overall project management responsibility remains at headquarters. ONWM provides guidance to DOE-Richland, approves project plans, provides project funding, and handles matters related to state governments and officials, and the public and media.

The lead organization (DOE-RL) has the responsibility to plan, engineer, and prepare environmental and safety documents for decommissioning activities; to complete decommissioning activities consistent with the provisions of DOE/Duquesne Light contract and lease; and to document decommissioning work on archives.

The lead contractor (UNC) is responsible for technical assistance to DOE-RL on all matters related to Shippingport decommissioning.

RESIDUAL ACTIVITY LIMITS FOR DECOMMISSIONING

Enrico F. Conti

Office of Standards Development

U.S. Nuclear Regulatory Commission

Introduction

The Plan for the Reevaluation of NRC Policy on Decommissioning of Nuclear Facilities, NUREG-0436[1], identified the need for specifying acceptable criteria for residual radioactivity following decommissioning for both surface and volumetric facility component contamination and for residual radioactivity in soil. The plan also identified the need for an NRC staff report on specific criteria for residual radioactivity levels following decommissioning. This paper presents the approach and considerations of an NRC staff technical task group, with particular focus on reactor decommissioning. These considerations have not been completed and have not been reviewed by the NRC. A technical report on residual activity limits for decommissioning various facilities operated under NRC license as well as a Regulatory Guide on this subject are scheduled to be issued by calendar year 1980.

Background

Guidance on residual activity limits for surface contamination has been available for reactors in Regulatory Guide 1.86[1] and for byproduct, source, or special nuclear material licensees in an NRC Staff Technical Position[1]. Similar guidance for surface contamination has recently been issued in a draft ANSI standard[2]. Analyses of exposures from recycle of metals with induced radioactivity and/or radioactive contamination were performed by F. R. O'Donnell[3].

Historically, the question of acceptable residual radioactivity levels in soil has arisen in cases which required some type of restoration to conditions approximating initial background. Specific cases include the Palomares accident, Grand Junction mill tailings, and the Florida Phosphate tailings. The recent focus on surveys of excess sites by ERDA (DOE) and NRC, as well as the need for criteria usable for making licensing decisions on facility decommissioning, necessitates the development of a practical method for specifying residual activity levels for purposes of regulation.

Residual contamination may be in or on structures, equipment, and soils. The acceptable residual level of any form of contamination is not a simply set, predetermined value similar to the allowable concentration values from the tables in 10 CFR Part 20. It is a matter of assessing the radiological impact of the residual contamination and the cost and advantages of removing it. Constructing and using models to determine this radiological impact can be difficult.

Models have been used for some time to simulate the release of radionuclides from operating facilities, the environmental transport of the radionuclides, and the exposure or ingestion by man of these radionuclides, which lead to estimates of the radiological impact (presented as a dose) to a hypothetical individual. This dose assessment methodology has been utilized in making regulatory decisions for some time in individual reactor licensing actions and was used for the generic analysis for the use of recycle plutonium in mixed oxide fuel (GESMO)[4]. The methodology used by the NRC staff is described in guides developed to implement Appendix I of 10 CFR Part 50. These are Regulatory Guides 1.109[5], 1.110[6], 1.111[7], 1.112[8], and 1.113[9].

Studies conducted at Battelle (PNL)[10, 11, 12] on generic facility decommissioning have demonstrated the utility of using the predictive methodology in a partial inverse manner, i.e., back-calculating the concentration and/or areal radionuclide levels equivalent to a unit radiation dose to a hypothetical individual. The exposure to a mixture of radionuclides with the involvement of different critical organs makes it very difficult to perform a summation of impacts equivalent to a whole body dose equivalent limit.

The lack of any authoritative definition of a "de minimis" dose (i.e., a dose-equivalent corresponding to a risk that is comparable to the risk from other activities that are generally accepted without special concern) also meant that a proposed acceptable level of residual radioactivity in soil was certain to be in jeopardy. The recommendation of the International Commission on Radiological Protection adopted on January 17, 1977 (ICRP 26)[13], provides the conceptual basis for constructing a methodology which can provide practical performance objectives for stipulating acceptable surface contamination and residual radioactivity levels in soil for purposes of regulation.

The conceptual approach in ICRP 26 makes it possible to sum impacts of groups of nuclides in terms of equivalent risk and to delineate surface contamination levels and soil concentrations which are correlated with a specified risk to an individual from unrestricted use.

The decommissioning of a facility usually has as its goal the return of the site to the public for unrestricted use after cessation of operations. Some of the site buildings could remain if they meet the regulatory standards for residual radioactivity and be used for other purposes, most likely of an industrial nature. The retrieved land is to be suitable for farming or any other unrestricted use.

There are many different types of nuclear fuel cycle facilities, each having different potential residual radionuclides and different exposure pathways. This means that there should not be a single criterion based on residual activity since different radionuclides carry differing stochastic risks per unit of activity. Each type of facility to be decommissioned may require separate consideration with a set of radioactivity levels specified both to protect the health and safety of the public and to permit unrestricted use of the facility and site. These criteria should be capable of implementation and amenable to verification by the Commission's inspection programs to ensure compliance. This paper focuses on the radionuclides and exposure conditions to be characteristic of Light Water Reactor facilities.

Approach to Problem

The NRC technical staff effort on developing residual activity limits for decommissioning has been predicated on the assumption that the following objectives would be met: residual activity limits should present a small risk from exposure, and the limits should allow conducting an effective measurement program to demonstrate compliance, and they should be consistent with existing guidance.

Existing Guidance and Recommendations

The U.S. EPA has the statutory authority for establishing generally applicable environmental standards for radioactivity. In the absence of such a standard for residual activity, the NRC staff has considered existing applicable guidance and recommendations in the literature as a benchmark of comparison for the residual activity limits under development. Close coordination with EPA will minimize the possibility that residual activity limits issued by the NRC would be in conflict with an EPA standard issued at a later date.

The standards and guidelines pertinent to the development of residual activity limits have been reviewed in reports by Dickson[14] and Schilling, et al.[15]. In addition, it should be noted that the radiation dose objective for the ambient waters of the Great Lakes Water Quality Agreement is one mrem per year[16]. The EPA standard for radioactivity in processed drinking water has been specified as 4 mrem per year.[17] Adler and Weinberg[18] have recommended setting a de minimis level at one standard deviation of natural background (22 mrem per year).

Consideration of risk of exposure as a basis for setting residual activity limits suggests using an annualized risk to an individual of about one chance in a million of a health effect from the radiation exposure resultant from the planned activity. This level of risk was identified in the proposed EPA FRC guidance for Plutonium in Soil for accident situations[19] and was also identified as an acceptable risk in ICRP 26. For purposes of this study, the NRC staff group has used the risk-dose conversion of 10^{-4} per rem from the BEIR Committee.[20]

Methodology

To place in perspective the residual radioactivity levels for the spectra of radionuclides associated with the operations of the various facilities, numerical estimates of radiation dose to man were developed. These estimates provide insight into a) what residual radioactivity levels would be related to a particular dose level, b) which of the various exposure pathways are significant, and c) the nuclides within the spectra of nuclides associated with the facility which are significant dose contributors. The important nuclides and pathways of exposure for reactor situations are identified below. As discussed previously, more detail on the methodology as well as application to other nuclear facilities will be included in the staff technical report to be issued at a late date.

Given the information on the particular radionuclides and their quantities, the computational model estimates:

a) the relative quantities of various spectra at various times following the decommissioning of the facility, and

b) the quantities that could exist at the time the facility was released for unrestricted use such that the dose at a future date does not exceed a particular numerical value.

This modeling approach assumes that the residual radioactivity is associated with soil or contained on or within the structure of the facility. The residual levels in these somewhat 'fixed' media serve as the model 'initial conditions' and the driving function for the various pathway models.

The mathematical modeling used drew heavily from past staff efforts, e.g., GESMO[4] and Regulatory Guide 1.109[5]. However, considerable effort was devoted to updating the methodology, in particular, adoption of the recently published ICRP-26 dosimetry methods.[13]

The environmental exposure modes considered are irradiation from surface deposits and resuspended radionuclides, inhalation of resuspended radionuclides, and the ingestion of food products contaminated by plant uptake of material in the soil and the foliar deposition of resuspended radionuclides. For reactor sites, the predominant exposure pathway is direct irradiation from surface deposits.

Over the past decade, information has become available to permit a prudent quantification of the relationship between absorbed dose and the risk of biological effects. This information permitted the International Commission on Radiological Protection (ICRP)[13] to consider two categories of radiation effects, namely, stochastic and non-stochastic effects. Stochastic effects are those disorders, e.g., cancer, for which the probability of an effect occurring, rather than its severity, is a function of the absorbed dose.

For stochastic effects, the ICRP recommended a dosimetric system that permits the total stochastic risk incurred from the irradiation of all body tissues to be calculated. This concept thus permits the summing of the absorbed doses to the various organs or tissues of the body. This methodology was used for the modeling done by the NRC technical staff.

The ICRP has not published, at the time of this writing, the report of Committee 2 setting forth the detailed dosimetric and metabolic models and the results of calculations by its task group. However, the National Radiation Protection Board has published a report [21] employing the Committee 2 methods in conjunction with the approved dosimetric and metabolic models of the ICRP Committee. The dose conversion factors for ingestion and inhalation exposures used in this study were obtained from this report. We note here, however, that:

a) A quality factor of Q of 20 for alpha particles and recoil particles has been employed

b) the modifying factor, n, has been taken as unity,

c) the dose is based on a fifty-year commitment period,

d) for inhalation calculations, particles of 1 μm AMAD (Activity Median Aerodynamic Diameter) were assumed,

e) the lung model of the ICRP Task Group on Lung Dynamics[22] was used and,

f) The GI tract model of Eve[23] was employed.

For external dosimetry, the work of Kocher[24] was used. In the case of external dosimetry, the body organs are exposed to much the same dose and thus the stochastic dose equivalent is equal to the total body dose.

Monitoring for Compliance

The NRC has a contract with Oak Ridge National Laboratory (ORNL) to develop monitoring programs for decommissioned sites to insure compliance with residual radioactivity limits. This monitoring program is being designed in a manner to assure that the survey of a site will have a high probability of detecting any areas of excess radioactivity or hot-spots. Hot-spots which are detected will need to be removed or decontaminated to meet the residual activity limits. Additional monitoring will be performed by an NRC inspector to verify compliance. Reports on surveys of several excess DOE sites by ORNL have indicated the need for post decommissioning surveys to be based upon statistical considerations.[25]

In cases where the radionuclide spectrum is not sufficiently established on site, it will be necessary to take samples of soil or other media from the site and perform laboratory analyses in order to determine the radionuclide composition of contaminants.

A great deal of the actual monitoring necessary in a termination survey will have to be conducted with portable intrumentation rather than laboratory analyses. It is expected that considerations of cost and time will indicate the need for conducting these surveys with portable instrumentation. The detection limits presented in Table I are derived from experience with the Department of Energy's Formerly Utilized Sites-Remedial Action Program. The detection limits listed in this table are based on practical considerations. Scanning speed has been considered as one of the limiting factors.[2] For example, the sensitivity of an alpha survey meter may be an order of magnitude better if the meter could be held in a fixed position for a minute or more. This is impractical in actual survey work since excessively long periods of time would be required to survey any large area or major piece of equipment. For reactor situations, this emphasis on field surveys is expected to be appropriate.

Table I Detection Limits for Direct Surveys with Portable Instruments

Radiation Detected	Monitoring Technique	Detection Limit
Gamma	NaI(Tl) scintillator with count-rate meter	2.5 - 5 μR/hr[a]
Beta	Thin window GM probe w/count-rate meter	2000 - 20,000 cpm/100 cm^2[b]
Alpha	ZnS scintillator with count-rate meter	200 dpm/100 cm^2

[a]Dependent upon background radiation levels which typically range from 5 to 10 μR/hr. This detection limit represents a 50% increase over the background. For fallout radionuclides, an exposure rate of 5 μR/hr corresponds to 0.1 to 20 $\mu Ci/m^2$ (e.g., 0.1 $\mu Ci/m^2$ for ^{60}Co and 0.5 $\mu Ci/m^2$ for ^{137}Cs). For a wide range of soil concentrations (e.g., 5 μR/hr corresponds to approximate 2.5 pCi/g of ^{226}Ra in the soil).

[b]Highly dependent on beta energy; low energy betas (<150 keV) may require special monitoring techniques. The stated sensitivity may not be achieved where background exceeds 100 cpm.

Conclusion

The staff analysis conducted to date indicates that residual activity levels which would be expected to result in exposures of 5 mrem per year to an individual from realistic exposure pathway conditions would both be consistent with existing guidance and result in activity levels which can be effectively monitored for enforcement. The radionuclides which are of particular importance to light water reactor decommissioning are Co-60, Cs-137, and Cs-134. The exposure pathways which are most significant for reactor sites are external irradiation from deposited radionuclides, with ingestion of contaminated foods and inhalation of resuspended activity of much smaller magnitude.

REFERENCES

1. U.S. NRC, Plan for Reevaluation of NRC Policy on Decommissioning of Nuclear Facilities, NUREG-0436, Office of Standards Development, December 1978.

2. Draft American National Standard N13.12, Control of Radioactive Surface Contamination on Materials, Equipment, and Facilities to be Released for Uncontrolled Use, American National Standards Institute, August 1978.

3. O'Donnell, F. R., et al., Potential Radiation Dose to Man from Recycle of Metals Reclaimed from a Decommissioned Nuclear Power Plant, NUREG/CR-0134, ORNL/NUREG/TM-215, December 1978.

4. U.S. NRC, Final Generic Environmental Statement on the Use of Recycled Plutonium in Mixed Oxide Fuel in Light Water Cooled Reactors, NUREG-0002, Volume 3, August 1976

5. U.S. NRC, Calculation of Annual Doses to Man from Routine Releases of Reactor Effluents for the Purpose of Evaluating Compliance with 10 CFR Part 50, Appendix I. Regulatory Guide 1.109, October 1977

6. U.S. NRC, Cost-Benefit Analysis for Radwaste Systems for Light-Water-Cooled Nuclear Power Reactors. Regulatory Guide 1.110, March 1976.

7. U.S. NRC, Methods for Estimating Atmospheric Transport and Dispersion of Gaseous Effluent in Routine Releases from Light-Water-Cooled Reactors. Regulatory Guide 1.111, July 1977

8. U.S. NRC, Calculation of Releases of Radioactive Materials in Gaseous and Liquid Effluents from Light-Water-Cooled Power Reactors. Regulatory Guide 1.112, April 1976.

9. U.S. NRC, Estimating Aquatic Dispersion of Effluents from Accidental and Routine Reactor Releases for the Purpose of Implementing Appendix I. Regulatory Guide 1.113, April 1977.

10. K. J. Schneider and C. E. Jenkins, Technology, Safety, and Costs of Decommissioning a Reference Nuclear Fuel Reprocessing Plant, NUREG 0278, Vol. I, October 1977.

11. Smith, R. I., Konzek, G. J. and Kennedy, W. E.m Jr., Technology, Safety and Costs of Decommissioning a Reference Pressurized Water Reactor Power Station. NUREG/CR-0130, Pacific Northwest Laboratory for U.S. Nuclear Regulatory Commission, June 1978.

12. Jenkins, C. E., Murphy, E. S., and Schneider, K. J., Technology, Safety and Costs of Decommissioning a Reference Small Mixed Oxide Fuel Fabrication Plant. NUREG/CE-0219 Pacific Northwest Laboratory for U.S. Nuclear Regulatory Commission, June 1978.

13. Recommendations of the International Commission on Radiological Protection, ICRP Publication 26, Annals of the ICRP, Vol. 1, No. 3, 1977.

14. H. W. Dickson, Standards and Guidelines Pertinent to the Development of Decommissioning Criteria for Sites Contaminated with Radioactive Material, ORNL/OEPA-4, August 1978.

15. Schilling, A. H., Lippek, H. E., Tedger, P. D., Easterling, J. D., Decommissioning Commercial Nuclear Facilities: A Review and Analysis of Current Regulations, NUREG/CR-0671, August 1979

16. "Refined Radioactivity Objective for the Great Lakes Water Quality Agreement", Federal Register, 42 (65), 18171, April 5, 1977.

17. EPA Drinking Water Regulations for Radionuclides, Federal Register, 41, (133), 28402, July 9, 1976.

18. Adler, Howard I., and Weinberg, Alvin M., "An Approach to Setting Radiation Standards", Health Physics, Vol. 34, pp. 719-720, Pergamon Press Ltd. Great Britain, June 1978.

19. Proposed Guidance on Dose Limits for Persons Exposed to Transuranium Elements in the General Environment, U.S. Environmental Protection Agency, September 1977.

20. National Academy of Sciences/National Research Council, The Effects on Populations of Exposure to Low Levels of Ionizing Radiation. Report of the Advisory Committee on the Biological Effects of Ionizing Radiations, NAS/NRC, Washington, D.C., 1972.

21. Adams, N., et al., 'Annual Limits of Intake of Radionuclides for Workers', National Radiological Protection Board, NRPB-R82, Oct. (1978).

22. International Commission on Radiological Protection, Task Group on Lung Dynamics, Deposition and Retention Models for Internal Dosimetry of the Human Respiratory Tract', Health Physics, 12, 173 (1966)

23. Eve, I.S., 'A Review of the Physiology of the Gastro-Intestinal Tract in Relation to Radiation Doses from Radioactive Materials', Health Physics, 12, 131 (1966)

24. D. C. Kocher, Nuclear Decay Data for Radionuclides Occurring in Routine Releases from Nuclear Fuel Cycle Facilities, ORNL/NUREG/TM-102, August 1977.

25. Leggett, R. W., Dickson, H. W., Haywood, F. F., "A Statistical Methodology for Radiological Surveying," Symposium on Advances in Radiation Protection Monitoring, Stockholm, Sweden, June 26 - 30, 1978

DESIGN CONSIDERATIONS FOR FACILITY DECOMMISSIONING

B. F. Ureda and G. W. Meyers

Rockwell International Corporation
Atomics International Division
Canoga Park, California 91304

The decommissioning of any nuclear facility is technically feasible. Unlike the disposal of most industrial plants which are routinely removed by demolition and salvage contractors, nuclear facility disposal, because of the radiation hazards, requires a more specialized controlled effort subject to ever increasing regulatory restrictions. Needed is a recognition of the decommissioning problem at the time of facility design so that design features which would simplify decommissioning can be incorporated. Though the addition of decommissioning aiding features will make the design more complex, such an approach during plant design will result in major savings. Government sponsored studies are currently being conducted which will lead to the establishment of design criteria for decommissioning nuclear facilities.

Atomics International has been active in decommissioning of nuclear facilities since 1966 and has developed a full appreciation of the difficulties associated with decommissioning facilities designed during the industry's infancy. We have examined the possibilities for simplifying the decommissioning tasks by a review of problems and their resolution encountered during the disposition of eight nuclear facilities at our Field Laboratories site.

DECOMMISSIONING EXPERIENCES

Decommissioning of the eight test nuclear facilities began in 1974 and work on four has been completed. The eight facilities are surplus facilities remaining from the reactors in space programs, nuclear systems support laboratories, a waste handling facility and the Sodium Reactor Experiment (SRE). These facilities were radiologically contaminated by neutron activation of structural materials, accidents involving radioactive materials, system failures, and the transport of radioactive material in the reactor coolant system. Decommissioning began several years after operations ceased, in which period the radioactivity from short-lived radioactive isotopes decreased significantly.

Design and construction of these facilities is representative of nuclear facilities in the late 1950's and early 1960's. Concrete was the primary material for structure, shielding, and containment, and stainless steel the structural material for vessels and process systems.

The principal objective for decommissioning of these facilities is to reduce the radioactive contamination to a level which will permit a return of the site to an unrestricted future use status. The objective is to be accomplished safely and economically with a minimum potential for radiation exposure to operational personnel and the environment.

SPECIFIC AI DECOMMISSIONING PROBLEMS

Achieving the decommissioning objectives required solving many problems. The more important ones are described below and facility design recommendations applicable to the specific problem follow the description.

1. Decontamination of Concrete Surfaces - This is particularly troublesome. Various techniques were employed ranging from washing surfaces with solvents to direct removal of the surfaces. A scabbling tool was most frequently used to spall concrete surfaces at the SRE. None of the techniques are wholly satisfactory since they involve intensive manual labor. Surface cracks, concrete expansion joints, and the porous concrete are very resistant to being cleaned.

Concrete surfaces in areas where contamination may occur, such as in vaults, trenches, pits, building columns and walls need to be protected. Metal liners, plastic covers, paint or hard, smooth coatings should be used to cover the concrete. Selection of the covering material should consider, in addition to permeability characteristics, the frequency of maintenance and traffic in the area which could increase the risk of damage to the material. If metal liners are used, they should be installed for easy removal. The use of expansion joints or discontinuities of any kind should be avoided. Surface geometrics should be optimized for cleaning. For example, rounded corners and edges should be used where possible.

2. Disposal of Massive Concrete Structures - Handling or demolition of contaminated massive concrete structures, such as found in reactor biological shields, can be difficult and costly. These operations also have a high potential for radiation and radioactive material exposures to operating personnel. Extensive control measures are consequently required. At the SRE, the demolition of the biological shield was accomplished using a hydraulically powered ram. The upper portion of the shield was not significantly contaminated by induced radioactivity. However, the section at and near the reactor core line was activated requiring packaging and shipment of the ram-generated rubble to burial. Separation of the clean concrete from the activated concrete was not attempted in the disposal of the activated shield center section.

The use of massive monolithic concrete designs should be reconsidered. Other shielding materials, for example lead shot or interlocking concrete blocks, could be used. Where integrated concrete structures are necessary for containment, means for easily removing or spalling the concrete surfaces should be included in the designs. A two-phase biological shield structure could be used. An outer physically independent layer to provide containment and some shielding and an inner layer constructed to a thickness calculated to provide the major shielding requirements and to contain all the significant induced activation expected during the reactor operating lifetime.

Strategically located heating elements could be embedded in the concrete for later use as a means for dehydrating and pulverizing the surface material. Concrete reinforcing steel specially designed could become a resistance heater element. Explosive charge holes or tubular reinforcing steel to provide for the insertion of explosives or heating elements could be specified for the concrete. Reinforcing steel should be positioned in the shield structure as far away as possible from the activating radiation. Concrete aggregates and reinforcing material should be selected to minimize radioactive activation products. Replaceable concrete plugs inserted in the shield should be used to assist in assessing irradiation effects.

3. Disposal of Reactor Vessels - Disposal of the highly activated reactor vessels required remotely operated plasma torch cutting equipment. The cutting equipment was designed to fit the SRE vessels.

The location and attachment of the piping, grid plates, and other reactor vessel internals should consider possible interference with the installation and operating requirements of the vessel cutting equipment. Cutting equipment installation support structure could be included in the vessel design. At some time in the vessel construction and installation, the compatibility of the cutting equipment with the vessels should be tested. Vessel cooling coils should not be embedded in concrete unless they are required for concrete cooling. Material selection for vessels, pipe, and support structure should consider irradiation effects.

4. Disposal of Waste Holdup Tanks - Excavation and removal of waste holdup tanks buried in soil was relatively easy. However, leaky valves, tank fittings and piping contaminated the nearby soil.

Radioactive waste tanks should not be buried but should be contained in isolated vaults with provisions for containing leaks. Tracks or guides for remotely operated vessel cutting equipment should be included in the tank design.

5. Removal of Process Equipment - Process system equipment such as heat exchangers, pumps, sodium purification, waste handling and their support structures were removed without major difficulty.

Dismantling of process equipment would be simplified if the large system components were designed for ease of disassembly and for isolation of sections which are likely to be contaminated. Process system equipment installations should include adequate access for removal operations.

6. Disposal of Handling Machines - The 50- and 60-ton fuel and moderator handling machines were stripped of clean reusable exterior equipment and the machines were shipped, intact, to the burial site.

Handling machine designs, which would permit easy disassembly and isolation of contaminated sections, would allow a greater salvage of material and a decrease in waste volume for burial. Handling machines are particularly vulnerable to airborne contamination. A smooth shroud over the exterior wiring, controls and instrumentation would facilitate cleaning.

7. Sodium Coolant Disposal - Disposal of the bulk sodium and sodium residues from the reactor vessel and from the process system equipment was a chemical and radiological hazard.

Better drainage for the system and components would be helpful during decommissioning. Inclusion of facilities for reacting the sodium with alcohol, steam, or water in the coolant system design would also be helpful.

8. Disposal of Insulation - Removal of asbestos from pipe and coolant system components required protection for personnel against airborne asbestos fibers and radioactivity, and required special packaging.

Asbestos in any form should not be used. Similar problem-causing materials, such as flammables, toxic substances, caustics and acids, if used, should be contained or installed for easy removal.

9. Decontamination of Facility Structures - Decontamination of facility structures, such as open beams and columns, electrical wire trays, conduits, exhaust ducts, etc., may be necessary to save the facility for future use. However, the cost of cleanup in some cases can exceed replacement costs.

Facility designs should recognize that radioactive spills will occur and that plant interiors and especially horizontal surfaces will become contaminated. Easily cleaned containment or covers should be provided and a re-routing of services through areas where spills are less likely to occur should be considered.

10. Contamination Assessment - The radiological assessment of the plant contamination at the time of decommissioning is an important factor for planning.

A well kept and complete record of radioactive spills and other plant operating history is necessary. Access for instrumentation in shielding materials, soil and concrete would be helpful.

GENERAL RECOMMENDATIONS

Considerations for simplifying decommissioning not directly related to problems experienced at AI are listed below.

1. Plant operating equipment, such as fuel handling machines, should be designed to handle highly radioactive products of reactor vessel cutup.

2. Plant air exhaust systems should have the capability to handle airborne contamination from decommissioning systems.

3. Waste reduction capability should be provided on site. Installation of saws, shears, arc cutters, electropolishers and compacters should be considered.

4. Contaminated water handling facilities such as evaporators or deionizers should be provided.

5. Design of plant maintenance facilities should include features which will support decommissioning tasks. Areas should be designated and be sufficiently large to accommodate dismantling of components, packaging, shipment or storage of waste.

Work Performed Under DOE Contract DE-AI-03-76SF75008

DECOMMISSIONING METHODS AND EQUIPMENT

Thomas S. LaGuardia, P. E.

Nuclear Energy Services, Inc.
Danbury, Connecticut 06810

INTRODUCTION

The U.S. Department of Energy (DOE) Division of Environmental Control Technology identified the need for a handbook for the decommissioning of government-owned and commercially-owned radioactive facilities.
The objectives were: (1) to assemble, in one reference, a state-of-the-art guide for the nuclear industry on methods and equipment available for decommissioning; (2) to provide necessary information for estimating costs and environmental impact. Nuclear Energy Services (NES) was requested to prepare the handbook for DOE.

The main topics covered in the handbook include the following:

Description of Decommissioning Alternatives
Selection of Decommissioning Alternatives
Estimation of Radioactive Inventory
Decontamination
Removal of Radioactive Metals
Removal of Radioactive Concrete and Structures
Removal of Contaminated Systems
Disposition of Waste
Environmental Impact Assessments
Cost Estimating Procedure

NES expects to complete the handbook this fall and to have it ready for publication by the end of the year.

This paper will briefly describe the primary methods and equipment that have been developed and used in decommissioning.

The major areas include: decontamination; removal of contaminated piping and components; vessel and internals cutting methods; and removal of radioactive concrete. The equipment required for these major activities is available and has been satisfactorily demonstrated for these applications.

DECONTAMINATION

Decontamination of radioactive surfaces may be accomplished using chemical or mechanical decontamination techniques.

Chemical Decontamination

Non-destructive chemical techniques are used when the piping, component or structure is intended for reuse and thus, there can be no detriment to the integrity of the item. The methods and procedures for non-destructive decontamination are aimed at protecting the integrity of the system or connected systems, whereby chemical attack is preferential to the contaminating constituents rather than to the base metal.

The cost of such a program is quite high. For example, the DOW Chemical/Commonwealth Edison program to chemically decontaminate Dresden 1 for continued operation, is currently estimated to cost in excess of $36 million. What's more, high decontamination factors (D.F. = 1000) are claimed to be possible.

By comparison, destructive chemical decontamination, by using acids such as hydrochloric or phosphoric for removal of up to 3 mils of internal pipe surface (including the contamination layer) on the same size reactor, can be accomplished for approximately $2.5 million. Paradoxically, the decontamination processes often result in higher occupational exposure than decommissioning without decontamination. Obviously, waste disposal volume and overall costs are increased when extensive decontamination programs are employed. So in general, it is concluded that it is best to minimize decontamination in decommissioning, whenever possible.

Electropolishing is a new application of a chemical technique used in the steel-making industry to clean finished products. Thin layers (2 mils thickness) of contaminated metals are removed using an electrical potential difference between the workpiece and the cathode in a phosphoric acid electrolyte. Figure 1 shows a schematic of the system. Battelle Pacific Northwest Laboratories is developing the technique with an aim toward reducing the volume of material to be buried and increasing the volume of salvageable metals.

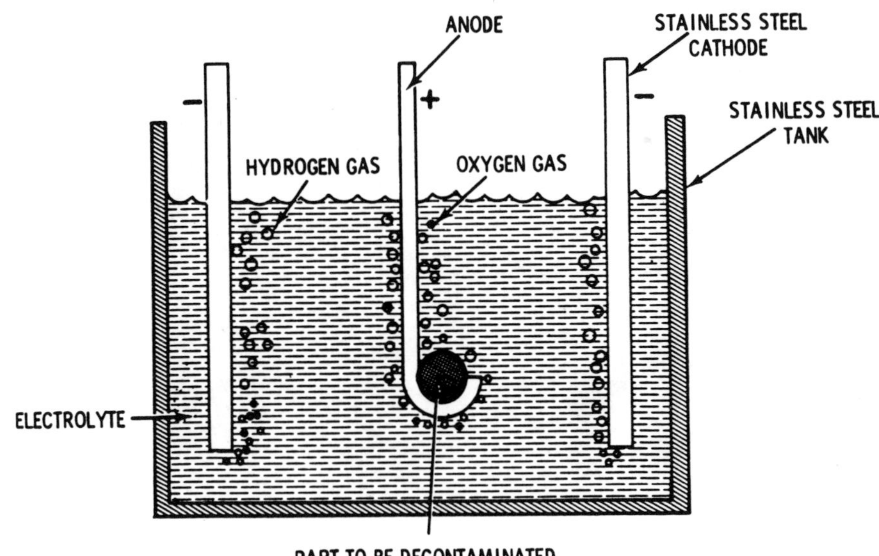

FIGURE 1. SCHEMATIC DRAWING OF THE TYPE OF ELECTROPOLISHING CELL USED TO DECONTAMINATE METAL SURFACES

Mechanical Decontamination

Non-destructive mechanical decontamination may be accomplished with ultrasonic methods. Here high-frequency sound generators, in a fluid couplant, agitate the adherent surface contamination and lift it from the surface. Ultrasonics are best suited for loose contamination on small parts.

Destructive mechanical decontamination methods remove the surface using grinders, scrapers or machine cutters. A relatively new method of concrete surface decontamination is available using a scarifying tool called a Scabbler, as shown in Figure 2. The Scabbler tool is pneumatically operated and can remove up to 1/4" thickness of concrete per pass; approximately 5 square yards per hour per piston. For the 7-piston tool, shown in Figure 2, the removal rate would be 30-35 square yards per hour. The tool was used at the Sodium Reactor Experiment (SRE) decommissioning for concrete decontamination, but was modified to include a HEPA filtered vacuum exhaust system.

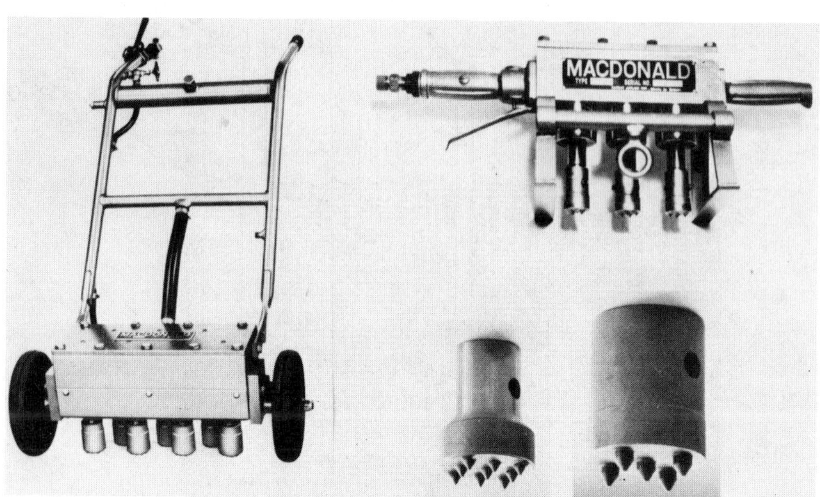

FIGURE 2 FLOOR AND WALL SCABBLER TOOLS
(Courtesy of Macdonald Air Tool Corp.)

REMOVAL OF PIPING AND COMPONENTS

There is not a great deal of new technology available for the removal of contaminated piping and components. The existing techniques, such as sawing, machining, abrasive cutting and oxyacetylene torching, are well developed and the considerable amount of nuclear and non-nuclear experience available allows reliable predictions of cutting speeds, consumables' costs and safety precautions.

Explosive Cutting

One technique adopted from the oil industry is explosive cutting. It was used successfully in the SRE decommissioning to segment and remove inaccessible piping from the SRE reactor vessel. Typical piping cutters are shown in Figure 3 and a mock-up of the underwater cut made at SRE is shown in Fig. 4.

Explosive cutters typically use RDX explosive and must be especially sized and designed for each cut. They can be installed remotely or directly with low radiation exposure. It is reported that explosive cutters have been used for materials greater than six inches in thickness, both in air and underwater. There are obvious limitations with respect to noise, vibration and structural integrity of surrounding structures which must be factored into the cutter design.

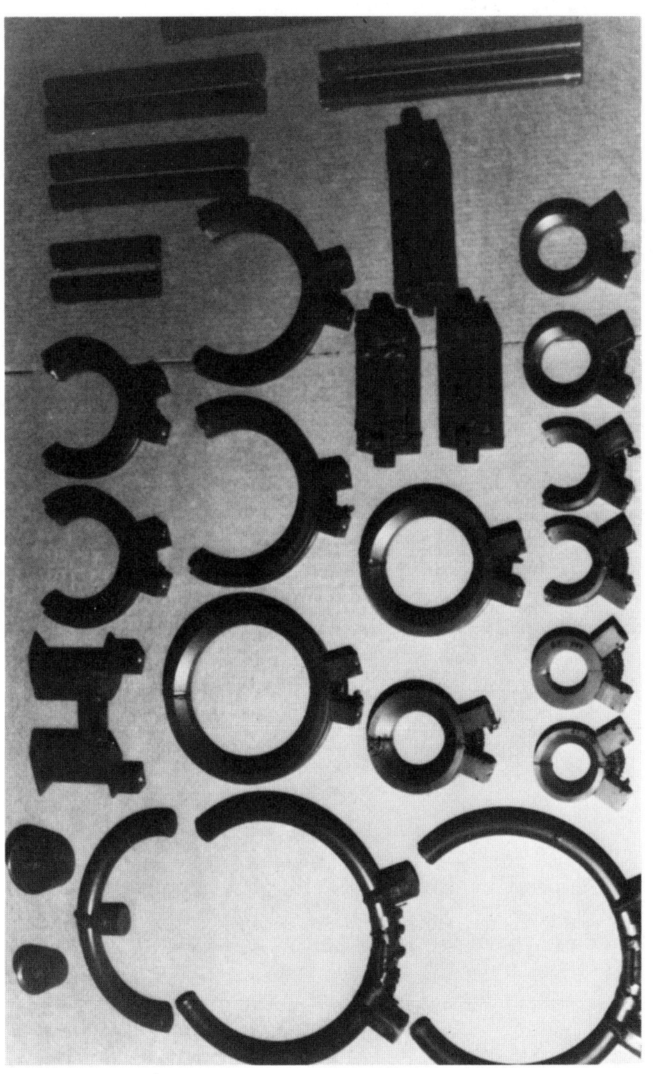

FIGURE 3. EXPLOSIVE CUTTERS

SRE PIPING MOCK-UP BEFORE DETONATION

SRE PIPING MOCK-UP AFTER DETONATION

FIGURE 4

REMOVAL OF VESSEL AND INTERNALS

The contact radiation exposure levels of reactor vessels and internals are very high and require the use of remotely operated cutting devices for segmentation. The vessel and internals must be segmented, either underwater or with shielding, to package the segments into shippable sizes and curie levels. The two processes most suitable for this application are the plasma arc cutting torch and the arc saw.

Plasma Arc Torch

The plasma arc system was successfully developed and used for segmenting the Elk River Reactor Vessel and internals in 1973. Subsequently, the same technique was successfully applied to the Sodium Reactor Experiment. The plasma arc torch cuts carbon and stainless steel using an extremely hot (approximately 24,000^{o}K) high energy plasma. To create this plasma, a direct current arc in a suitable gas such as argon is transferred from a tungsten electrode through a constricting orifice to the workpiece. The orifice forces the gas and the arc into a small diameter stream, creating high current densities and extremely high temperatures. The high temperatures cause ionization of the gas stream and formation of a plasma consisting of positively charged ions and free electrons. The hot plasma is ejected from the torch at very high velocity, which melts the metal and produces a kerf in the workpiece. Figure 5 shows a photograph of the plasma arc torch and Figure 6 a photograph of the plasma torch manipulator used at the Elk River Reactor dismantling. The state-of-the-art indicates that the plasma arc can segment 4" thick carbon steel while under water and 7" thick carbon steel in air. Further development of this technique is considered feasible.

Arc Saw

The arc saw is a patented development of the Retech Company in Ukiah, California. The process is effective for any electrical conducting material. The cut is made by a rotating circular, toothless saw blade through which a high amperage direct current is passed. The saw is operated in a constant voltage mode using a highly regulated D.C. power supply. The current of the arc between the saw and the workpiece is controlled by a fast and accurate servo-mechanism which receives its feedback input from the arc current. Hydraulic servo-motors position the blade relative to the workpiece to maintain constant amperage. A photograph of a stationary version of the arc saw is shown in Figure 7.

FIGURE 5 PLASMA ARC TORCH

FIGURE 6 PLASMA TORCH MANIPULATOR

The arc saw can cut underwater or in air, with or without a water spray. Underwater cutting is preferred for the following reasons: (1) large reduction in noise level; (2) cutting debris is retained in water; (3) saw cooling is more effective and blade life enhanced; (4) the arc is shorter, kerf is narrower, and quality of cut improved. In-air cutting is quite noisy and does produce a somewhat rougher cut.

Saw blade life depends upon the operating power level, cooling, blade material, thickness and diameter. Typically, blade loss is equal to about two percent of the material removed from the kerf.

The arc saw has been used to cut metals up to twelve inches in thickness. The depth of cut is limited to about 30 to 40 percent of the blade diameter, due to mechanical interference of the saw blade drive unit and support collar. Deep cuts may be accomplished using a thick blade for the initial passes in order to create a wider kerf and completing the cut with a thin blade in order to preclude mechanical interference. Blades as large as ten feet in diameter are considered technically feasible for this process.

Table I summarizes the physical, operational and cost data of the basic arc saw system as presently developed by Retech. Figure 8 shows the arc saw concept adapted to segmenting a large light water reactor pressure vessel.

The arc system, which was privately developed by the Retech Corporation, is going to be tested at the Hanford facility of the Department of Energy to develop its operational characteristics with respect to thick irradiated materials. The test program will include both in-air and underwater cutting with metal thicknesses of up to eighteen inches.

TABLE I

ARC SAW PHYSICAL, OPERATIONAL AND COST DATA

	Head Size		
Parameter	7 Inch	12 Inch	16 Inch
Max Blade Diam, in.	30	50	72
Max Depth of Cut, in.	10	19	30
Current Capacity of Head, A	6,000	15,000	25,000
Operating Voltage Differential, V	25	25	25
Weight of Head and Motor, lb	400	1,170	2,730
Cost of Basic System, $	110,000	185,000	270,000

FIGURE 7 STATIONARY ARC SAW

There is no physical contact between the blade and the workpiece. The arc causes melting in the workpiece and the rotation of the blade literally blows the molten metal out of the kerf. The rotation also provides uniform heat generation and dissipation in the saw. The process has some obvious advantages:

1. Since there is no blade contact there can be no binding in the workpiece.
2. Irregular geometries are easily cut since the process is self-starting. That is, the arc will re-establish itself after passing through a non-conducting medium.
3. The depth of cut is only limited by the physical diameter of the saw blade.
4. The workpiece need not have a void area behind it as does the plasma arc because, the slag is removed from the front face of the kerf and not blown through it.

Although blade rotation is necessary to the process, speed is not critical and 300 - 1800 rev/min is an acceptable range. Cutting rates of 300 square inches per minute have been achieved. The blade can be made of any conducting material such as copper, tool steel and carbon steel. Thin blades, i.e. with a thickness-to-diameter (t/d) ratio of about 0.001, are more efficient and allow greater cutting speeds, whereas thick blades (t/d ratio of about 0.01) can withstand large mechanical forces.

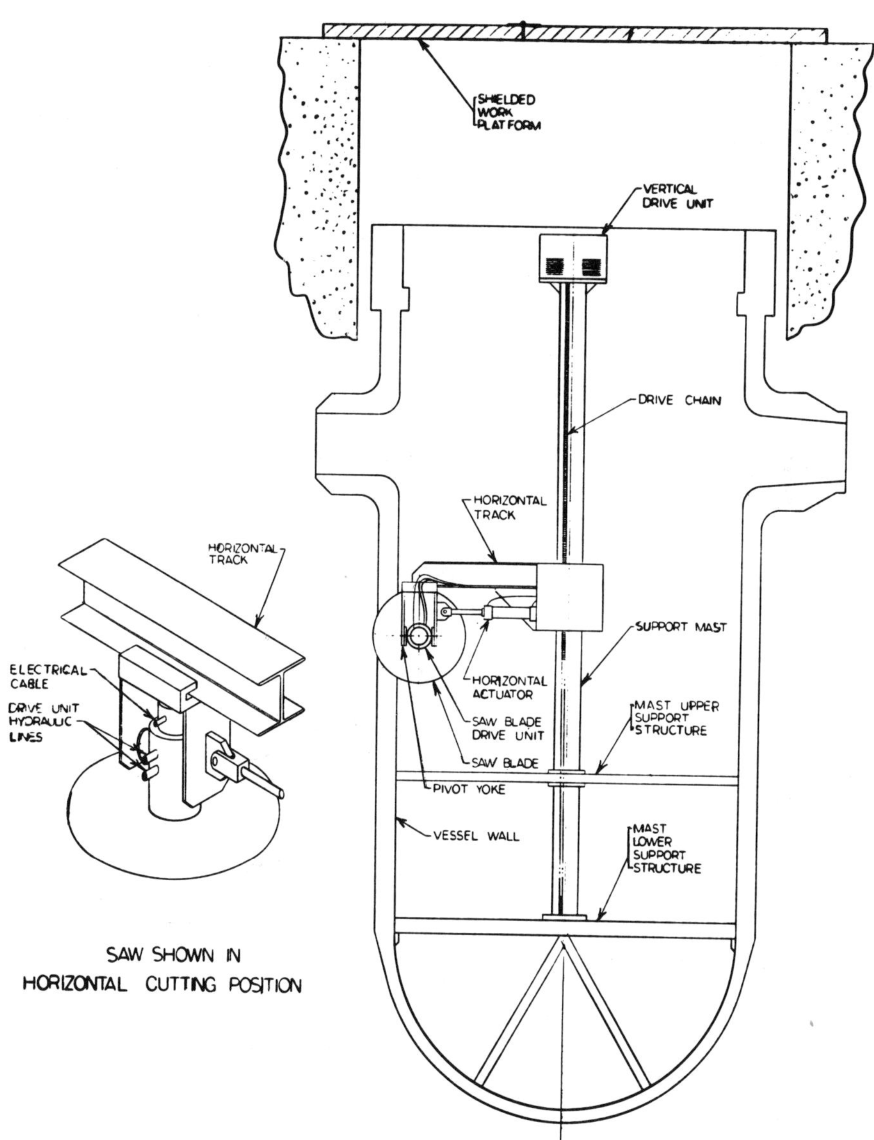

FIGURE 8 ARC SAW CONCEPT

CONCRETE DEMOLITION

There are many concrete removal techniques which have been adopted from the conventional demolition industry and, with minor modifications, successfully applied to nuclear facility decommissioning. The selection of a specific process should be based on the knowledge gained from the conventional demolition industry and the applicable (although somewhat limited) experience from actual decommissioning programs. Table II lists the concrete removal methods, typical applications, feasibility and relative equipment costs.

Of greatest concern is the demolition of massive, heavily reinforced concrete structures. The conventional wrecking ball or slab used for thin-section concrete is much too slow a method for massive concrete demolition. Controlled sequence explosives have been demonstrated as the most reliable and fastest technique for dismantling such structures.

The technique consists of drilling holes in the concrete at predetermined spacings, depths and burdens (distance from the free face), and loading each hole with a predetermined explosive charge. The firing is delayed in sequence such that subsequent burdens (rows) are fired 1 to 3 milliseconds after the first row. This delay permits the muckpile at the free face to move perpendicular to the boreholes, allowing subsequent muckpiles room to move perpendicular to the boreholes in a similar manner. This delayed ignition technique increases fragmentation and improves control of debris missiles. A sketch of a typical blastround in massive concrete is shown in Figure 9; the poor directional control from simultaneous (non-delayed) firing is shown Figure 10.

Contamination control is maintained by using a 3,000 lb. blasting mat and spraying the concrete with a fog spray before, during and after the blast. The fog spray is very effective in holding down dust levels (contaminated and nuisance dust). Reinforcing rods must be cut after each blast to free the muckpile for removal.

Other techniques listed in Table II may be substituted when the site work must be performed with minimum noise or vibration, or where the volume of concrete to be removed does not warrant the use of explosives.

CONCLUDING REMARKS

The methods and equipment identified for use in decommissioning are based on demonstrated techniques and proven equipment. The arc saw is recommended for use in cutting thick-section vessels. Although the application of the saw to vessel cutting has not

TABLE II

CONCRETE REMOVAL METHODS
SUMMARY OF APPLICATIONS

Process	Application	Feasibility	Relative Equipment Cost
Controlled Blasting	All Concrete ≥ 2ft	Excellent	High
Wrecking Ball	All Concrete ≤ 3ft	Excellent	Low
Ram Hoe (Hydraulic Ram)	Concrete ≤ 2ft	Good	Low
Hobgobbler (Air Ram)	Concrete ≤ 2ft	Good	Low
Flame Cutting	Concrete ≤ 5ft	Fair	Low
Thermic Lance	Concrete ≤ 3ft	Poor	Low
Rock Splitter	Concrete ≤ 12ft	Good	Low
Bristar (Demo Compound)	All Concrete ≥ 1ft	Fair	Low
Wall and Floor Sawing	All Concrete ≤ 3ft	Good	Low
Core Stitch Drilling	Concrete ≥ 2ft	Poor	High
Paving Breaker (pneumatic)	Concrete ≤ 1ft	Poor	Low
Air Hammer and Chisel	Concrete ≤ 3in	Poor	Low
Drill and Spall	Concrete Surface ≤ 2in	Excellent	Low
Scarifier	Concrete Surface ≤ 1in	Excellent	Low
Water Cannon	Concrete Surface ≤ 2in	Fair	High
Needle Scalers	Concrete Surface ≤ ½in	Poor	Low
Grinding and Sanding	Concrete Surface ≤ ¼in	Poor	Low

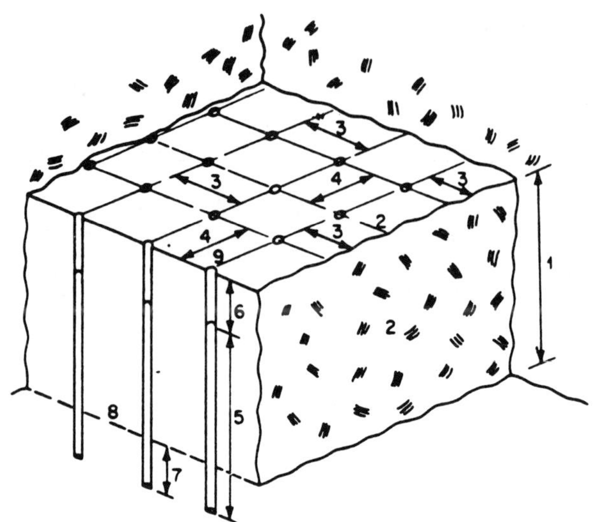

FIGURE 9 BLASTING ROUND

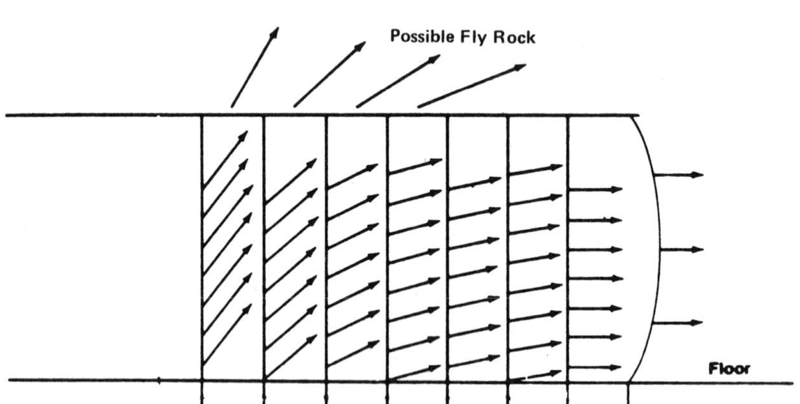

FIGURE 10 SIMULTANEOUS NON-DELAYED DETONATION

been demonstrated, a high probability of success is indicated based on the successful application of the saw in other configurations. A test program is in progress at Atomics International to demonstrate arc saw cutting of metal sections up to 18 inches in thickness. Similarly, the decontamination program at Dresden using Dow's NS-1 solvent is based on an extensive experimental program to demonstrate its effectiveness.

The availability of these techniques confirm the feasibility of decommissioning large light water power reactors and research facilities.

EQUIPMENT FOR REMOTE DISMANTLING

R. G. Brengle and E. L. Babcock

Rockwell International
Energy Systems Group

Dismantling of the Sodium Reactor Experiment at Rockwell International's Nuclear Development Field Laboratory in Chatsworth, California required the development of remotely operated tooling to dissect the reactor vessels and internal piping.

The SRE was a graphite moderated, sodium cooled, thermal reactor that operated from 1957 to 1964, producing more than 37 million kilowatt hours of electrical energy in over 27,300 reactor operating hours. The reactor cross section is shown in Figure 1. The 1-1/2" thick stainless steel reactor core tank was 132" in dia. x 226" high with the bottom of the tank 27' below grade. A 127" dia. x 179" high x 1/4" thick stainless steel liner was centrally located inside the core tank. The carbon steel outer tank, which provided outer atmospheric containment, was 150" in dia. x 226" high with a 1/4" thick wall and a 2-1/2" thick bottom.

The initial design and development of underwater plasma cutting and of a remotely operated rotating mast manipulator to guide the plasma torch have been reported by others [1].

The manipulator and plasma torch are operated from a remotely located cutting control console. The console allows programming direction, speeds, speed transitions, gas flows and start delay times to control the underwater plasma cutting process. Figure 2 shows the manipulator and console installed in the SRE mockup facility.

This paper covers the experience in using the equipment to completely dismantle the SRE vessels and the improvements incorporated

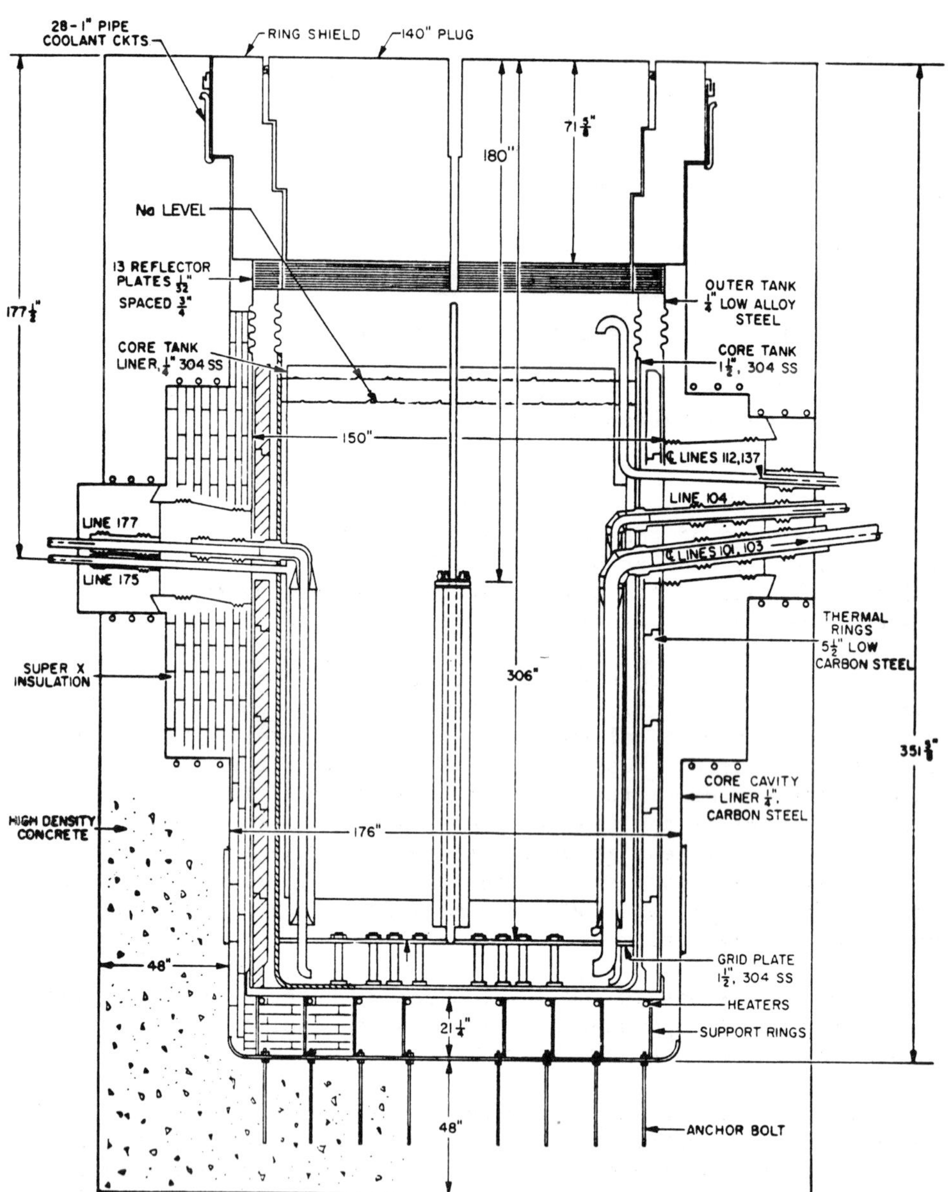

Figure 1. Sodium Reactor Experiment cross section.

Figure 2. SRE mockup facility showing the plasma torch manipulator and control console.

to make the system perform under field service conditions. Improvements to the plasma torch manipulator system include a radial drive system, a torch touch system for remotely setting the plasma torch standoff distance, and a minipulator arm-mounted TV camera which is useful for inspecting cuts and placement of remote grapples.

A radial drive system was required for SRE D&D to dissect the gridplate (1-1/2" stainless steel), core tank bottom (1-1/2" stainless steel), and outer tank bottom (2-1/2" carbon steel). In addition, special tooling was required to cut through the radius section that joined the core tank sidewalls to the bottom. When the need for a radial drive and radius cutter was identified, various system possibilities were considered and evaluated.

After evaluating electric, pneumatic and hydraulic systems, a drive system powered by a hydraulic cylinder was selected on the basis of minimum design and fabrication complexity, operational reliability, maintainability and initial cost. This system uses a double-acting hydraulic cylinder connected to a pump with 5000 psi hydraulic hose. The positive displacement pump is driven by an

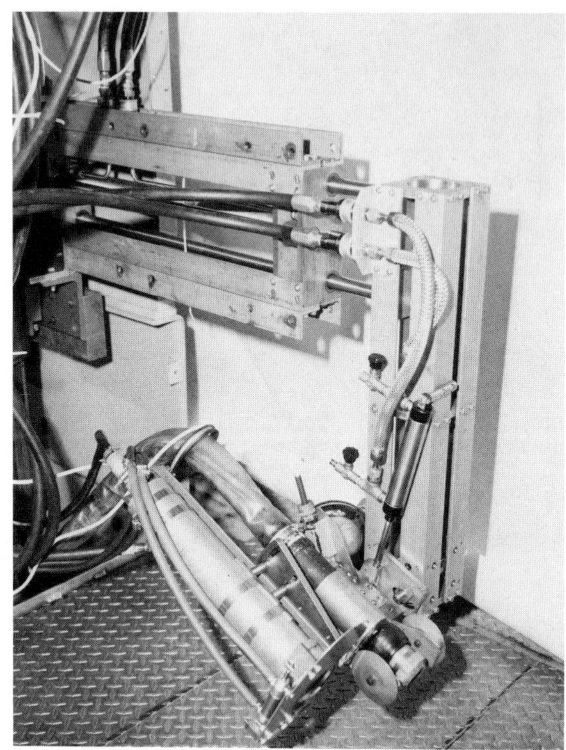

Figure 3. Side view of plasma torch manipulator radial drive arm with radius cutter.

electric motor, whose speed is determined by an SCR-type controller. A curve of motor speed vs. rate of linear displacement could then be determined. The position of the torch was sensed by a rack-driven potentiometer located on the radial arm and coupled to the console mounted position electronics system. The position and speed control electronics are virtually the same for the radial, azimuth and vertical directions. Since a hydraulic system was selected for the radial drive, it was also used to power the radius cutting attachment. The radius cutting attahcment was not coupled to the position electronics, but rather used a less complicated limit switch system to determine the zero, 30 and 90 degree positions. Figures 3 and 4 are front and side views of the system showing the radial arm, radius cutter, television camera and plasma torch.

The radial drive and position system described were used to segment the gridplate and bottoms of the SRE vessels. The system was also used to cut core tank access slots through 3/4" thick and 1-1/2" thick stainless steel rings that were welded to the inside surface of the core tank. The 3/4" thick ring provided the spacing and attachment point for the core tank liner and the 1-1/2" thick

Figure 4. Front view of plasma torch radial drive arm showing the radius cutter hydraulic cylinder, plasma torch and television camera

ring was the gridplate support ledge. Cutting slots in the rings flush with the core tank was necessary to provide torch clearance for making vertical cuts through these areas. Both series of slot cuts were started at the inside edge travelling radially outward toward the core tank cutting in the 1G or downhand position. Reliable positioning was of utmost importance in this application since a malfunction would cause the arm and torch to impact the core tank wall causing damage to the plasma torch and manipulator. Both series of slot cuts were performed successfully without incident. The hydraulically powered radial drive system proved to be both accurate and reliable.

The torch-to-workpiece standoff distance is one of the most critical parameters in plasma cutting and also one of the most difficult to control. The standoff distance is normally set by adjusting the distance from the torch to the workpiece with the manipulator guidewheels in contact with the workpiece. However, when cutting over obstructions, frequent adjustments must be made to maintain proper standoff distance since the guidewheels are not tracking on the actual surface that is to be cut. With guidewheels there is also difficulty in maintaining consistent speed and standoff distance when slag or other debris is present on the surface where the guidewheel must roll.

Increased rolling resistance causes additional deflection in the manipulator arm and when the arm unloads the speed increases. The net effect is jerky type of motion. The use of guidewheels also increases the overall size of the plasma cutting head which reduces the system effectiveness in cutting around and between obstructions.

To alleviate some of the problems associated with the use of guidewheels, a simple system was installed on the SRE manipulator to determine when electrical continuity occurred between the plasma torch nozzle and the workpiece. Using this position electronics the torch could then be retracted the required distance from the workpiece to set the standoff. When using this technique, the torch was operated without guidewheels. The torch touch technique for determining standoff distance requires a fair amount of operator skill to prevent interpreting arm deflection as standoff distance and the possibility of overloading the arm if contact point is overshot. The advantage of producing reliable cuts when the arm speed is not being influenced by debris proved to be the deciding factor and a major portion of the SRE vessels dissection was performed without the use of guidewheels.

Early in the development of the plasma manipulator system for SRE, it was recognized that a television system would be a tremendous asset during the remote dismantling of the vessels. A television system was procured complete with cameras with radiation resistant lenses and remote focus. Brackets were designed and fabricated, and the camera was mounted on the manipulator arm. A video tape recorder was used to document significant events in the dismantling process. Under conditions of poor shielding water clarity the TV system allowed placement of segment grapples from distances of up to 26 feet. Cutting patterns were scanned for unknown obstructions prior to actual cutting effectiveness was verified.

Further improvements to the system would be an arc voltage control, torch position feedback system to automatically maintain the torch standoff distance and solid state console electronics to improve the reliability and maintainability of the position control system.

The SRE vessels removal task has been completed and all the removed material has been shipped for land burial. Plasma cutting was performed under water on 1/4" and 1-1/2" thick stainless steel and in air on 1/4", 2" and 2-1/2" thick carbon steel. The actual cutting time is insignificant in the total time required to complete the task. Over 1500 ft. of lineal cutting was performed by the manipulator system during the dissection of the SRE vessels.

References

1. Graves, Streechon and Phillips, "Submerged Cutting of Reactor Components," ANS Transactions Vol 26, 1977.

THE DECONTAMINATION OF CONCRETE

R. G. Brengle

Rockwell International
Energy Systems Group

In power reactor construction, concrete comprises a major portion of the structural material. During the lifetime of the plant, the concrete may become contaminated from maintenance activities or failures of plant components.

The ability to decontaminate concrete efficiently and effectively is a major challenge for the decontamination and decommissioning of reactor plants.

The Sodium Reactor Experiment Facility (SRE) has many below-grade, contaminated concrete vaults. These vaults were unlined except for floor pans which terminated approximately one foot above the floor in some of the vaults. The remainder of the vaults was bare or painted concrete. Figure 1 shows a typical concrete piping trench after decontamination.

Mixed fission product contamination had penetrated the concrete approximately 1/8" to 3/8" on the average and up to the full thickness of the walls and floors where cracks and joints existed. The penetration is greater where contaminated liquid was in contact with the structure.

Various concrete removal techniques were investigated and used under D&D operating conditions at SRE; included were tractor-mounted percussion tools, conventional paving breakers, heavy chipping hammers equipped with chisels or bushing tools, multiple head pneumatic hammers with carbide bits, rotary flapper with carbide buttons, needle scalers, grinders and wet sandblasting.

Fig. 1. SRE kerosene piping trench after sandblasting and chipping.

The primary factors for selecting concrete removal tools are the speed of material removal at the depth desired, the ability to maintain radiological control of the operation and the time required for cleaning up removed debris.

For removing up to approximately 3/8" of contaminated concrete on floors, the multiple head pneumatic hammer with pointed carbide bits equipped with an absolute vacuum cleaner shown in Fig. 2 has proved to be the fastest method. When removing concrete to a depth of 1/8" to 1/4" with the three-headed type, rates of approximately 0.5 sq. ft./min. are achieved. The final surface is already vacuumed and no additional cleanup is required. The surface is slightly rough and readily accepts a concrete plaster refinishing coat. Larger floor model units are available with up to 11 heads per machine. These would be effective for large unobstructed floor areas. Bit life is approximately 500 sq. ft.

Modifications to the commercially available unit include fabrication of a shroud assembly to attach a vacuum and rerouting the piston air manifold to exhaust outside the enclosure on the larger models.

Fig. 2. Multiple head pneumatic hammer with carbide bits.

This tool is simple to use and requires a minimum of operator training. Since airborne activity is controlled and cleanup is continuous, the operation does not require respirators or local enclosure and has a minimum effect on other operations.

By using a spring balancer, it is expected that this type of tool would be effective in removing contamination from vertical surfaces, although some concrete would not be captured by the vacuum action.

For removal of contaminated concrete up to approximately 2" deep, and for chasing contamination in cracks, a pneumatic chipping hammer equipped with a chisel has proven to be an effective removal method. The chipping hammer removes material fairly selectively so the risk of structural damage is not nearly as great as would be for larger demolition type tools or explosives.

Operations using the chipping hammer are usually performed under temporary protective enclosures that are connected to an absolute filter exhaust system. The protective enclosure prevents chips of radioactive material from being scattered out of the controlled

Fig. 3. Concrete surface showing a bushed area to remove contamination from a local area that was not removed by sandblasting.

area and the exhaust system minimizes the amount of dust within the enclosure. Pre-filters in the exhaust system near the enclosure protect the absolute filters. Supplied air respirators are normally used since the concrete dust tends to rapidly plug canister time respirators. Cleanup after chipping is time consuming since the rubble must be shoveled into appropriate waste containers and the surface vacuumed prior to radiological survey.

Wet sandblasting has also been used successfully on the SRE D&D program. Water is injected into the abrasive stream at the nozzle. The amount of water used should be limited to that necessary to control dust since greater amounts reduce the velocity of the abrasive stream and hence the effectiveness of the process. Sandblasting is useful for removing contamination that has been fixed by painting or if the contamination in the concrete is less than approximately 1/16". To remove contamination to a depth of 1/16" requires approximately one pound of abrasive per square foot which adds to the overall volume of radioactive waste. Sandblasting is relatively fast at approximately 100 sq. ft./hr. and will remove material from irregular surfaces including shallow holes. The surface of concrete is composed of fine aggregate and cement with the larger aggregate located below the surface. Sandblasting removes the surface layer easily; however, the removal rate decreases once the larger aggregate is exposed.

Figure 1 illustrates the combination of techniques that were used to decontaminate this piping trench. The trench was initially sandblasted to remove contamination that had been fixed by painting. Concrete was then removed from the floor to a depth of approximately 3/8" using the multiple head chipper. The center floor joint was exposed to a depth from 3 to 8" using a chipping hammer with a chisel. A heavy chipping hammer equipped with a bushing tool was used to decontaminate areas that were still contaminated after sandblasting. The contrast between the sandblasted and the bushed surface is shown in Figure 3. The trench as shown is ready for demolition with heavy equipment.

The selection of a technique can only be as good as the information on which the selection was based. Concrete structures contain joints, cracks, pores, holes, piping penetrations and fastening anchors. With these conditions present it is extremely difficult to reliably determine the radiological condition of the structure. Because of these factors, decontamination tends to be an iterative process of materials removal, and radiological survey, etc., until the desired degree of decontamination is achieved.

All the techniques described have been used by themselves and in combination with others. All the techniques have been successful in some applications and failed in others. Following this decontamination approach, entire structures may still require removal if the remaining contamination is contained in a series of cracks that cannot be removed without compromising the integrity of the structure.

DEMOLITION OF CONCRETE STRUCTURES BY HEAT--A PRELIMINARY STUDY

John M. McFarland

McFarland Wrecking Corporation

Seattle, WA 98108

This paper discusses the possibility of using heat to demolish or dismantle concrete structures. I will approach the subject in this manner:

- Define the problem and tell why demolition by heat should be considered.
- Provide technical background data about concrete and steel and their interactions when heated, together with a calculation of the amount of heat required for a PWR containment structure.
- Tell how this technical data can apply to demolition of a nuclear reactor containment structure.
- Summarize practical applications and recommend further action.

WHY DEMOLITION BY HEAT?

I am projecting out beyond my area of expertise as a wrecking contractor into an area of physical phenomena which is best understood by concrete technologists and heating engineers. I pursue the subject because my experience as a wrecking contractor leads me to believe that heat demolition has possibilities of practical benefit to the nuclear industry.

In the process of demolishing thousands of structures, including fire-damaged concrete and steel buildings, I have used all conventional methods including crane "ball and chain" and explosives. Observing that concrete structures involved in fires are weakened to the extent that the intense heat penetrates the concrete and further observing that the reactor containment structures, which are most resistant to conventional demolition, appear to be natural conservors of heat, this inquiry attempts to find out what happens to the concrete and how that information can be applied as a demolition technique.

The heavily reinforced structures of the nuclear energy industry are by far the strongest buildings I have ever encountered. However, I would like to place the demolition of these structures in perspective. First, the demolition phase does not concern itself with decontamination; that is taken care of by the decontamination teams before the site is turned over to demolition. Second, the installations are so isolated from the ordinary bodily injury and property damage liability risks of downtown city wrecking that taking out a complete plant is, in that sense, a refreshing change. While it is true that none of the structures built or on the drawing boards can resist drilling by a thermite lance or the explosives which can be packed into those drillholes, drilling and blasting is expensive, hazardous and time-consuming work; and it doesn't always work out the way it is planned. We cannot afford to be put in the position of relying on a single solution, for that very reason.

We need alternatives because different circumstances call for different solutions. We search for safer and more economical ways of doing our work. If successful, heat demolition could save up to half a million dollars on a single reactor building demolition and save hundreds of tons of steel for recycling.[1] The benefits for the nuclear industry would be faster, safer, and cheaper ways of getting plants back in operation after shutdowns or getting their sites recycled upon decommissioning.

In 1976 I was asked by Pacific Northwest Laboratory to analyze a pressurized water reactor (PWR) of the Westinghouse design for the purpose of establishing the cost parameters of its removal. I was subsequently asked to do a similar study of a boiling water reactor (BWR) plant.

I found that while the rest of the plant structures can be removed readily enough with established demolition procedures, the massive reactor containment buildings were at the limit of the art for conventional demolition, including explosives.

The PWR containment building (48" thick and 206' high) is at the limit of the art for ball and chain (assuming a 200- to 300-ton crane with a 20,000# battering ram) and would be extremely difficult for explosives except for the fortuitous existence of the post-tensioning ducts. These allow the placement of explosives without extensive drilling.

It is not likely (and certainly not reliable) to expect to continue to find such fortuitous solutions to future demolition poblems. In fact, there may be differences of individual construction within this generic class which would not accommodate this solution.

The point has been reached at which simply leaving demolition to the ingenuity of the demolition contractor is not good enough. Demolition contractors and nuclear plant design engineers need to help each other if they are to properly serve the interests of the owners of nuclear plants and the public. Both need the development of new demolition techniques to cope with the nuclear industry's unique demolition-resistant structures. Heat demolition of concrete is one facet of this overall search for a new generation of demolition techniques.

The path of inquiry has led through the local library, fire department and electrical utility; cement and brick and steel manufacturing firms; manufacturers of large-scale electrical, gas, oil and thermite heating devices; the Portland Cement Association; and the U.S. Patent Office. I have consulted with my peers in the National Association of Demolition Contractors[2] and with some from the European Demolition Association, as well as with personnel of Pacific Northwest Laboratory. I have drawn liberally from the knowledge of others to interpret and apply the information to the demolition of concrete structures.

TECHNICAL BACKGROUND DATA

Concrete is composed of calcium carbonate ($CaCO_3$) cement (20% to 30%) and rock and sand aggregate (70% to 80%). The cement glues the aggregate together. Concrete has some 5% to 6% water by weight, even when dry. Eighty percent of this water is free water and 20% is chemically bound.

When the temperature of concrete is raised to 212°F, the free water is driven off as steam. At about 400°F to 500°F the chemically bound water is driven off. This dehydration causes the cement paste to shrink and loose some of its adhesion. There is a strength loss at this point on the order of 10%.

When the temperature is raised to 1063°F, there is a change in the crystalline structure of quartz (from quartz alpha to quartz beta) which results in swelling and internal cracking. Concrete with quartzitic aggregate loses 50% to 75% of its strength at this approximate temperature. The reactor containment structure should not collapse at this point; but it would be much easier to break the concrete with a breaking ball or to drill it.

When the temperature is raised to 1600°F, portland cement ($CaCO_3$) converts to $CaO + CO_2$, with the CO_2 leaving as a gas. Driving out the CO_2 in this manner is referred to as calcining. Two to three days after exposure to the atmosphere, perhaps longer in the case of a massive containment structure, the CaO absorbs moisture from the atmosphere and converts to $CaOH_2$. $CaOH_2$ is considerably weaker than CaO and will spontaneously disintegrate, with the rocks and sand falling loosely. Such a structure is likely to collapse within two to three days after the heating as a result of this process occurring.

The specific heat of reinforced concrete is a nominal 0.20 Btu per pound. It takes about one-fifth as much heat to raise a pound of concrete 1°F as it takes to raise the same weight of water 1°F. It is noted that the specific heat of concrete decreases with increasing temperature while the specific heat of steel rises. Since specific heat varies with temperature as well as with changes of composition of the concrete, the rounded figure of 0.20 is used as a nominal value.

The combustion process requires two thousand cubic feet of oxygen for each 1,000,000 Btu of heat. Two thousand cubic feet of oxygen weighs 167 pounds. This weight of oxygen necessary to produce 1 million Btu holds true for all hydrocarbon fuels. In addition, it is customary to add 10% excess oxygen to ensure complete combustion.

With the 167 pounds of oxygen necessary for combustion would be 667 pounds of nitrogen if ordinary air were used for combustion. (Air is approximately 20% oxygen and 80% nitrogen.) A large part of the heat required to raise the temperature of nitrogen to the combustion temperature of 2000°F would be going out the stack as exhaust gas. (667 pounds times 2000°F times the specific heat of nitrogen, 0.25 = 333,000 Btu per 1 million Btu produced.) It can be seen that one-third of the heat goes to heating the inert nitrogen. If the exhaust gases are 1000°F, one-sixth of the heat input is lost with the nitrogen gas. Besides the physical heat loss, the nitrogen acts as a dampener on the heat of combustion. Natural gas, for instance, burns at 2100°F with air, while pure oxygen produces a combustion heat of 4200°F.

Pure oxygen is available at an estimated cost of $35 to $50 per ton. It would increase combustion temperatures and reduce stack gases and related heat waste. One million Btu would require 167 pounds. At $50 per ton, that would cost $4.18 per million Btu, plus the cost of the fuel.

Preheaters for incoming combustion gases as well as for fuel are available. By preheating the air used for combustion, for example, to 500°F by using the 1000°F stack exhaust gases for heat, most of the stack heat loss would be eliminated and the combustion efficiency increased greatly.

Radiation heat losses increase in proportion to the fourth power of the absolute temperature and increase rapidly at 200°F above ambient. This tends to limit the outer temperature of the wall to 300°F maximum. Radiation heat losses can be reduced and the steady-state temperature of the outer wall increased by covering the outer surface with insulation. Two inches of insulation would cost in the neighborhood of 50¢ per square foot, and would raise the temperature of the outer wall to some 800°F. Insulation is valuable not only for its conservation of fuel but also for its raising the heat gradient of the wall.

Thermal conductivity is the rate at which heat is passed through a substance. It is expressed in Btu per square foot per degree F temperature difference per foot. The thermal conductivity of concrete is 0.54, and the thermal conductivity of steel is 26.2.[3] Heat goes through steel almost 50 times as fast as it goes through concrete.

Heating Source Options

The heating source options discussed here are electricity, hydrocarbons, and chemical reactions (thermite).

Electricity. Electric space heating elements are limited by the size of the units available. The largest are 40 KVA, which at 3412 Btu per KW equals 136,480 Btu per unit. While it is possible to hook up a large number of units and to design larger units, it is not presently practical to attempt the hourly input required, some 60 million Btu.

Heating by electrical resistance, d.c. or a.c. heating of the reinforcement steel, has the problem in the already-constructed containment vessels of the inaccessibility of the rebar. In addition, the rebar grid would be able to transmit massive amounts of electricity without heating. This method is not presently considered practical for massive heating.

Heating by electrical induction does not have the capacity nor depth of penetration to be a practical method of heating massive structures.

Hydrocarbons. Wood fuel has costs ranging from 20% of the cost of oil for wood chips to $4.00 per million Btu for cord wood. It burns cleanly once it heats up. It could be augmented by other fuels. It is a renewable energy resource and could utilize waste timber. Ecologically as well as economically it seems to have advantages over all other fuels. It costs some $2.50 per million Btu.

No. 2 diesel heating oil is relatively safe and clean burning, with 140,000 Btu per gallon at a wholesale price of $0.666 per gallon on today's market. This calculates out to $4.76 per million Btu.

No. 5 fuel oil is heavier than diesel oil. It requires preheating for combustion and is not volatile enough, even with heating, for high-velocity burners. It has 150,000 Btu per gallon at a wholesale price of $0.5069 per gallon on today's market. This calculates out to $3.38 per million Btu.

Coal has 22 million Btu per ton, more or less, and costs an estimated $3.00 per million Btu. It appears to present insurmountable air pollution control problems for onetime use.

Natural gas is a clean burning fuel; but it has the limitation of requiring a pipeline connection. It has a heat of one million Btu per 1000 cu ft. Its cost is in the area of $3.00 per million Btu.

Propane can be tanked in but it costs more than natural gas. It is unsafe to handle near a heat source. Liquefied natural gas is not deemed safe for the same reason.

Thermite. Thermite is powdered ferric oxide combined with powdered aluminum: $2Al + Fe_2O_3$ transforms to $Al_2O_3 + 2Fe$. The reaction liberates the heat of conversion of iron oxide plus the heat of conversion of aluminum oxide, 198.5 kilocalories plus 181.g kilocalories per mole, respectively.

The $2Al + Fe_2O_3$ reaction produces 1,526 Btu per pound.

The $2Al + Fe_2O_3$ and $2Fe + 3O$ reactions together produce 3,196 Btu per pound.

313 pounds at 3,196 Btu per pound = one million Btu. At per pound, the cost per million Btu is $156.50.

Cost Calculation

To determine the cost of heating a reactor containment structure, it is first necessary to select an example as a reference case. The mass of its reinforced concrete in pounds, times the specific heat of concrete, times the temperature difference in degrees Farenheit, will determine the theoretical heat input in Btu. To the theoretical heat input, add a heat loss factor. This sum times the cost per Btu of the fuel selected will determine the total cost.

The Trojan Pressurized Water Reactor containment structure is used as a reference case in the following calculations:

Volume of reinforced concrete	17,000 cu yd
Weight per cubic yard	4,000 lb
Total weight (mass)	68,000,000 lb

(mass) (specific heat) (temperature change) = Btu

$$68{,}000{,}000 \text{ lb} \times 0.20 \frac{\text{Btu}}{\text{lb}^\circ\text{F}} \times 1540^\circ\text{F} = 21 \text{ billion Btu}$$

Heat loss allowance @ 100% =	21 billion Btu
Total Heat input required	42 billion Btu

Costs of fuel alternatives:

	Cost per million Btu	Total Cost 42 billion Btu
Hydrocarbons:		
Wood	$2.50	$105,000
Diesel heating oil	4.80	201,800
No. 5 fuel oil	3.40	142,800
Natural gas	3.00	126,000
Coal	3.00	126,000
Chemical:		
Thermite	156.00	655,000

APPLICATION OF DATA

There are two separate approaches to destroy a concrete structure by heat:

1. Internal alteration of the concrete (i.e., changing the calcium carbonate to calcium hydroxide by heating to 1600°F). Included in this procedure is the dehydration of the cement paste and differential expansion within the aggregate caused by alteration of the crystalline form of quartzitic components.

2. Differential expansion between the concrete and the steel reinforcement.

Internal Alteration of the Concrete

The principal problem of heating the concrete to 1600°F, aside from the amount of the heat input, is that the temperature of a cross section of wall cannot exceed a linear relationship between the heated inside surface and the unheated outside surface. The unheated outside surface cannot exceed 300°F due to radiation heat loss. With an assumed 2000°F heat in the inner chamber, the midpoint could only reach 1200°F and the three-quarter point could not exceed 800°F.

A possible solution to this problem of thoroughly heating a massive concrete wall is suggested by the existence of vertical post-tensioning ducts in the outer portion of the wall of a PWR reactor containment vessel of the Westinghouse design. This row of vertical ducts, if heated from the lower post-tensioning gallery could theoretically provide what I term a "heat curtain" which could reflect heat back from near the outer wall to the heated inner chamber. The heat gradient between the two sources would tend to become a straight line, with the intervening mass raised to the approximate temperature of the heat sources. The gradient between the heat curtain and the outer unheated surface would be a swift drop, with the outer several inches of concrete left essentially unaffected. This is illustrated by Figure 1.

While it is probable in the Trojan example that the limited heat bearing capacity of those ducts as built would fall short of complete success with an air-combusted fuel, the addition of oxygen to the fuel mix and insulation of the outer wall could increase the heating capability enough to do the job to a substantial degree, if not completely.

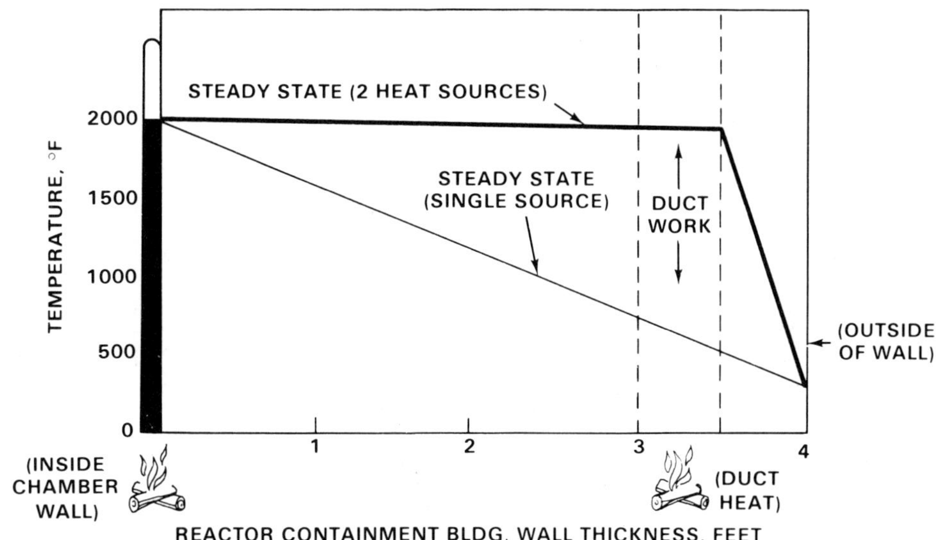

Fig. 1. Heat Curve Through a Wall from Two Heat Sources

In the case of structures which lack ductwork which could provide a heat curtain effect, heat, augmented as justified by insulation and oxygen enrichment of the combustion air, could be effective as an adjunct to conventional demolition. The calcining effect would be augmented by the quartz alteration factor and the dehydration factor. It would be further augmented by the differential expansion factor. This single source penetration is shown by Figure 2.

The thickness of a reinforced wall tends to increase the difficulty of demolition geometrically. Any reduction of the effective thickness of the reactor containment building wall would geometrically decrease the difficulty of demolishing that wall by whatever method used. It could change an "impossible" job into a possible one.

Differential Expansion

The principal obstacle to causing a differential expansion between reinforcing steel and concrete is that their coefficients of expansion are so closely matched, 0.00084 for steel and 0.0008 for concrete. If the concrete and steel is heated together, a significant differential between the two could not be achieved below calcining temperatures.

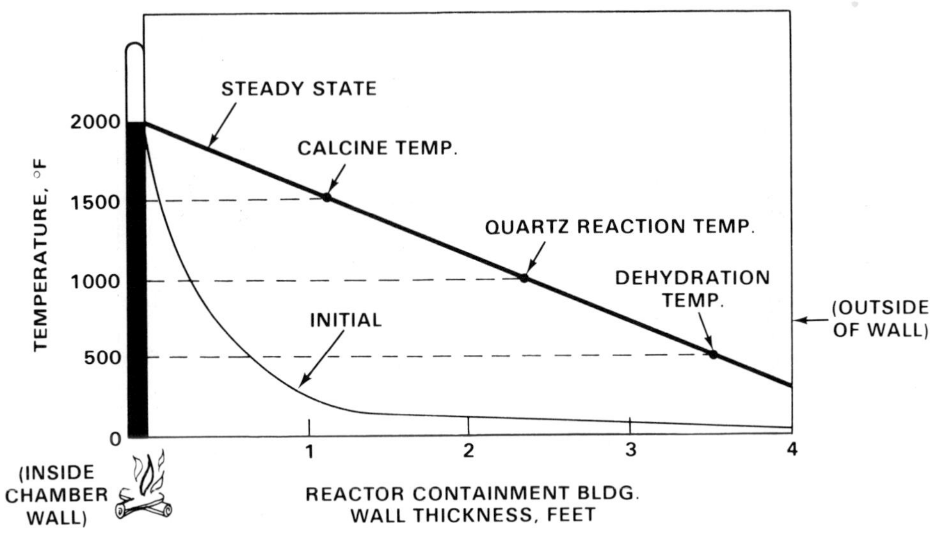

Fig. 2. Heat Curve Through a Wall from a Single Heat Source (decomposition phases shown)

A solution to this problem is suggested by the claim of a Japanese patent, Itoh, et al., U.S. Patent No. 3,727,982,[4] for heating of buried concrete reinforcement by electrical induction heating. It claims that when the reinforcement has been heated 150°C (302°F) above ambient, the reinforcement will break its bond and the concrete can then be readily removed. This appears to be confirmed by J. P. Vidosic, Marks Mechanical Engineering Handbook, 8th edition, p. 5-17,[5] who states:

"When the deformation arising from change of temperature is prevented, temperature stresses arise that are proportional to the amount of deformation that is prevented. . .In the case of steel, a change of temperature of 12°F will cause in general a unit stress of 2,340 $lb/in.^3$."

At 195 $lb/in.^3$ per degree Farenheit, a change of 302°F would generate a stress of some 58,890 $lb/in.^3$. This generally exceeds the full bonding strength of concrete and could reasonably be expected to cause a failure. The practical problem in effecting this solution is that the temperature difference between the steel and the surrounding concrete must approximate 300°F.

Two possible courses are suggested for investigation to attempt to obtain a 300°F temperature difference between the concrete and steel:

1. Due to the extreme difference in thermal conductivity of the steel vs. the concrete, 26.2 to 0.54, a 48:1 ratio, a solution would be to transmit the heat initially through the steel rather than through the concrete. The means of accomplishing this is suggested by the existence of the steel liner covering the inner wall of the reactor containment structure. The liner is used during construction as the inner form for the concrete pour and to support the rebar placement. It has extensive ribbing and direct ties with the rebar network. It would be most difficult and costly to attempt to cut this liner away from the concrete without the heat process. It would be equally or more costly to leave this liner bonded to the concrete and rebar while attempting demolition. This disadvantage can be turned into an advantage by using the liner to transmit heat directly to the rebar network. The heating would free the liner from the concrete and also free the adjoining network of reinforcing steel from the concrete. (With foresight, the liner could be given preliminary cuts before heating so that it could be more conveniently removed when later uncovered.)

2. In the process of heating beyond the inner network of rebar, the heat will probably penetrate the concrete with some irregularity. For example, it will follow the greater concentrations of steel which occur around openings and from there it will reach the outer rebar networks and spread along them. By this action, there will be differential expansion associated with massive heating which is primarily aimed at calcining the concrete.

Heating by electrical resistance on a large scale was not considered practical because of extensive "up front" planning and construction costs.

Heating by electrical induction was not considered practical for massive heating because of the physical impossibility of penetrating concrete to the depths required.

Decontamination by the use of differential expansion was considered from two approaches:

1. Electrical resistance heating of reinforcing bar behind and below contaminated surfaces.

2. Electrical induction heating of shallowly buried reinforcing bar or wire mesh behind and below contaminated surfaces.

Electrical resistance heating of rebar behind and below contaminated concrete would require that the rebar be continuous conductors; that they be placed in a position close to the surface which is to be removed and in a close pattern, with the ends reasonably accessible so that positive and negative leads could be attached onto each bar. It requires preplanning into new construction. By placing the rebar in a pattern some two to six inches behind or below the surfaces of the concrete to be removed, the removal would be relatively safe and easy. An appropriate application of electrical resistance heating would be in the walls and floors of reactor pits.

Decontamination by induction heating of a shallowly buried rebar or wire mesh pattern would avoid the necessity of exposing the ends of the steel for attachment of electrodes and the necessity that the steel be a continuous conductor. Since it has a limited depth of penetration of the induction heat, induction heating should be designed for use in areas not expected to be deeply contaminated. It requires preplanning into new construction to be most effective. It requires a relatively sophisticated device compared to resistance heating. It should be a safe, fast and relatively economical method for decontamination.

SUMMARY

Heat applied to a reinforced concrete structure can weaken it by four actions:

1. Calcining the calcium carbonates to the ultimate state of calcium hydroxide, back to sand and gravel, with no strength remaining.

2. Breaking apart the interal quartz structure of the aggregate, causing the concrete to fracture and lose half or more of its strength.

3. Dehydrating the cement paste, thereby weakening its bond to the aggregate, with a strength loss of about 10%.

4. Causing the steel to expand faster than the concrete, to break its bond with the cement.

These chemical and physical responses of concrete and steel to heat have the following practical applications in the field of nuclear plant demolition and decontamination:

1. Massive amounts of heat applied over a period of time to the interior of reactor containment structures can sufficiently weaken them to be of value as a demolition tool.

2. Electrical resistance heating has the potential to greatly improve the safety, economy, and speed of decontamination of heavily contaminated areas of reactor pits.

3. Electrical induction heating has the potential to greatly improve the safety, economy, and speed of decontamination of shallowly contaminated concrete surfaces.

Heat demolition appears to be a beneficial procedure in that it saves steel for recycling; should cost materially less than drilling and placing explosives or ball and chain; and it appears to be safer in that it minimizes human exposure.

The concept of altering concrete by heat has support in theory, economics, and limited practical experience.

Case studies should be computer modelled to effectively predict the responses and economics of heat demolition of concrete. These computer modellings should be verified by laboratory testing and by field testing. If this is successfully done, the owners of nuclear plants will have established a faster, safer, and cheaper way to recycle plant sites upon decommissioning.

I would appreciate your thoughts, criticisms, and questions, now and later.

REFERENCES

1. R. I. Smith, G. J. Konzek, and W. E. Kennedy, Jr., Technology, Safety and Costs of Decommissioning a Reference Pressurized Water Reactor Power Station, NUREG/CR-0130, U.S. Nuclear Regulatory Commission Report by Pacific Northwest Laboratory, p. G-45, June 1978.
2. J. M. McFarland, "Nuclear Age Demolition," Demolition Age, p. 17, March 1978.
3. American Society of Heating, Refrigeration, and Air Conditioning Engineers Handbook, pp. 37.3 and 37.4, 1977 ed.
4. Itoh, Methods of Electrically Destroying Concrete and/or Mortar and Device therefor, U.S. Patent No. 3,727,982, April 17, 1973.
5. J. P. Vidosic, "Mechanics of Materials," Mark's Mechanical Engineering Handbook, p. 5-17, 8th Edition.

PLANNING STUDY AND ECONOMIC FEASIBILITY FOR EXTENDED LIFE OPERATION OF LIGHT WATER REACTOR PLANTS

Charles A. Negin
Lessly A. Goudarzi
Larry D. Kenworthy

International Energy Associates Limited
Washington, D.C.

Mel E. Lapides

The Electric Power Research Institute
Palo Alto, California

INTRODUCTION

The decision to retire an electrical generating unit from service is usually an economic one based on comparing the refurbishment and operating costs of a unit to the comparable values for new capacity alternatives. Refurbishment (to extend the service life beyond normal anticipated levels) is currently receiving added attention because many of the historic circumstances that favored new capacity additions (e.g., substantial improvement in heat rate, relatively rapid growth of demand, capital availability) are no longer as prominent in future planning as they have been. Given the current emphasis on decommissioning of nuclear power reactors, it is important to point out that the potential exists for operation well beyond the 30 to 40 year period that forms a basis for design and operating licenses. The recognition of this potential is based on experience with fossil fuel plants, some of which have operated 50 to 60 years and which, in general, are not required to be maintained to the same level of physical integrity as their nuclear counterparts.

Nuclear power plant refurbishment warrants attention because it may be in the best interests of the electrical rate payer for plant owners to have the option of extending the life of a reactor plant in lieu of being forced to decommission it if such action is technically unnecessary. This need has, in fact, been recognized by some. State representatives have emphasized this point during NRC solicitation of views for rulemaking. Utility representatives have done likewise in a public meeting for the same purpose. Another example of such recognition is a public news release regarding the decommissioning of Vermont Yankee that states, "Decommissioning would be at the end of the plant's licensing period -- 2007 -- or sometime after that if a license extension is sought."

In order to avoid foreclosing the option for extended life operation, it is necessary to anticipate safety and commercial risks related to equipment aging that may require resolution far in the future. Addressing such issues at this time may at first seem premature; however, we note that the regulatory operating term of the first large LWRs expires in about 25 years. As seen in Figure 1 there is no excess slack when one considers planning lead times of 10-12 years for new capacity decisions, added to a prior reasonable period (10-15 years) for data gathering either to show that degradation, if it exists, is acceptable or to indicate where refurbishment is required.

The purpose of this planning study was to perform an assessment of the engineering and economic feasibility of extended life operation of present nuclear power plant units and to recommend future EPRI programs that may be warranted by the feasibility assessments. This effort concludes, essentially, that there is sufficient economic motivation for refurbishment to warrant more extensive examination for present plants and to identify possible design modifications that would facilitate extended service life in future plants. The costs of replacing the deterioration-prone equipment in a nuclear power plant appear to represent a small portion of the total plant costs, provided downtime is not excessive.

In order to extend service life, it will be important to be able to demonstrate the safety of certain major equipment items (such as the reactor pressure vessel) and to determine the feasibility of replacement or refurbishment. Similarly, verification of the non-deterioration of structures that are close to the vessel, support the vessel, or are an integral part of containment is important. There are also institutional aspects that must be solved, the most important being how to assure license renewal with sufficient lead time to plan on a unit's availability.

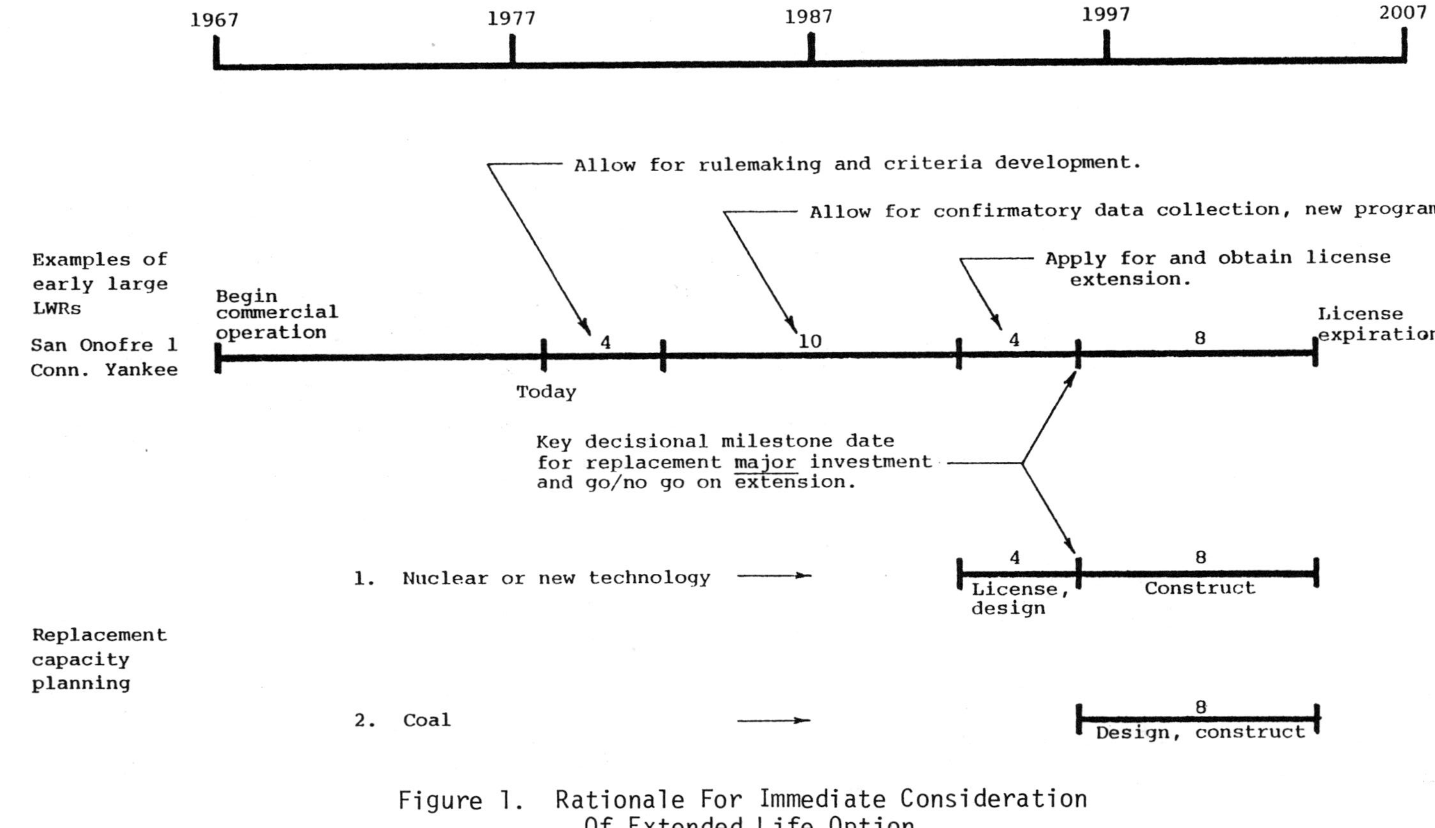

Figure 1. Rationale For Immediate Consideration Of Extended Life Option

RESULTS OF ANALYSIS

Two analyses were conducted: (1) the economics of extended life, and (2) an evaluation of possible equipment refurbishment requirements.

ECONOMIC ANALYSIS

The approach for economic evaluation was to use a relatively simple model to conduct case studies as follows:

(1) A generic study that compares extended life projections with baseline cases of decommissioning after 40 years of operation.

(2) Determination of refurbishment investment that would break even with replacement by a new plant. The effect of refurbishment downtime is included.

(3) Studies of San Onofre Unit 1 and Connecticut Yankee comparing reasonable extended life scenarios with shutdown at the end of the existing license term.

(4) Sensitivity analysis of assumptions regarding replacement power costs, replacement capacity and future fuel costs, decommissioning costs, and extended life capacity factor.

Only the results of (1) and (2) above, the generic analysis and breakeven refurbishment investment, are discussed in this paper.

The relative economic attractiveness of extending the operating life of an LWR beyond its normal operating license period was assessed by (1) calculating the total present value revenue requirements (PVRR) for a range of reasonably expected extended life scenarios, and (2) comparing these requirements with PVRR for baseline cases in which decommissioning occurs at the end of the initial full-term operating licenses.

For the generic study, two baseline cases were formulated that could be used for reference benchmarks to measure the relative attractiveness of the corresponding extended life cases. These cases are used to determine the costs of providing power assuming that the plant being considered is to be shut down and decommissioned. The time duration of the baseline cases are thus chosen to coincide with extended life cases analyzed. Two baseline cases were defined:

- o Baseline Case 1 -- Calculates the cost associated with providing an equivalent amount of energy for 10 years beyond the initial plant life. Additionally, the costs associated with the initial plant decommissioning are included in this case.

- Baseline Case 2 -- This is essentially the same as Case 1 except the duration incorporated was 22 years beyond initial plant shutdown.

The reason for choosing the first case is that extended life duration of 10 years without any extraordinary costs for refurbishment downtime other than would normally be scheduled is the reasonable expectation. That is, a 25% extension of original licensed life is assumed through an in-service maintenance and equipment refurbishment program. As will be shown, the effect of extra costs and downtime is evaluated.

The second case is for comparison with an extended life case of 20 years following an assumed 2-year outage for refurbishment. (Subsequent notation for this case is 2/20.) The effect of a reduction or increase of the 2-year assumption is evaluated. Two years was chosen as a reference value because the replacement of major equipment such as steam generator tubing is anticipated to take one to two years.

The baseline cases account for the costs of (1) decommissioning the plant, and (2) replacing the unit with equivalent new plant capacity.* Several extended life cases were constructed that account for the costs of (1) providing replacement power while the unit was being refurbished/overhauled; (2) generating power over extended life time frame; and (3) decommissioning the plant at the end of extended life.

In each case, the costs were present-valued back to 1978. Comparing the baseline cases with the extended life cases, the initial savings (or losses) of extending the life of the nuclear unit were determined. Sensitivity analyses were performed for assumed excalation rates for fuel costs, downtime associated with refurbishment, assumptions regarding replacement capacity costs, and reduced capacity factor for the extended unit.

The analytical method utilized is a relatively simple one of calculating the PVRR associated with each alternative. Because the object of this study is to ascertain general feasibility and to demonstrate the magnitude of economic benefit/penalties, a simple approach is judged to be acceptable. The PVRR was selected

---

*"Replacement capacity" implies new facilities, while "replacement power" implies utilizing existing reserve margin, whether within the system or purchased.

as a measure-of-merit because it is generally accepted by the electric utility industry and is a normal input to the decisional process of both utilities and rate commissions.

In an actual situation, the decision to extend the operation of any specific plant would have to be based on very specific utility requirements and external conditions that exist at that time -- for example, planning simulations through a generation expansion model that addresses total system costs and reliability.

The results for the generic analysis are presented in two ways. First, the differences between the PVRR of the baseline cases and corresponding extended life cases are presented. This difference was defined as the maximum potential saving or loss associated with the extended life case. Secondly, during the course of the analysis, it was determined that the refurbishment cost and downtime were major parameters affecting the results; thus, they are presented as major results (as opposed to sensitivity results).

Generally, life extension is economically beneficial unless the refurbishment cost and downtime are extensive. Therefore, what is also of interest is the breakeven cost for refurbishment compared with decommissioning and plant replacement. The two major components of refurbishment costs are (1) the direct expenditure associated with replacement hardware, installation labor, maintenance labor, etc., and (2) the indirect downtime cost of providing replacement power.

Thus, in addition to the present value costs, the breakeven costs are presented expressed in equivalent 1978 dollars. From the perspective of the engineer/operator, these breakeven costs can be thought of as the maximum amount that could be spent on refurbishment before it would be cheaper to build replacement capacity. Breakeven is also presented graphically as a function of direct expenditures and downtime.

We emphasize that these results are relative between the baseline case and extended life case; it is the <u>magnitude of the difference</u> between the two that is important. The absolute value for each case is only as valid as the model, assumptions, and inputs, all of which are very general.

Basic Economic Results

The results of the four baseline and extended life cases are shown in Table I. Also shown are the breakeven costs, expressed in equivalent 1978 dollars. The maximum potential 1978 value savings in the 10-year case is \$94 million; in the 2/20-year case it is \$109 million.

Table I

BASIC ECONOMIC RESULTS*
(all values million $)

	Life Extension Period	
	10 years	2/20 years
Present Value Cost		
Baseline Case	166	263
Extended Life Case	72	154
Difference -- Present Value Savings	74	109
Breakdown Cost	575 to 750	645 to 795

Effect of Refurbishment Downtime

One of the major factors influencing the relative attractiveness of the extended life cases is that of the downtime associated with equipment refurbishment and plant overhaul. This factor has a dual impact on the costs. First, it involves replacement power costs while the initial plant is down for refurbishment. This is a first-order effect and at some duration will overrun the economic advantage of the extended life option. Second, the lengthening of the down period will result in increased interest costs during plant downtime. This is a secondary impact for periods of less than two to three years.

Figure 2 shows the reduction in breakeven direct expenditures (equipment, labor, etc.) as downtime increases. Figure 2 reveals that, even with relatively long periods for refurbishment outage, large costs can be incurred before a losing situation would result. That is, the plant owner could afford to spend the indicated value (in 1978 dollars) on refurbishment direct costs and still save money over the replacement alternative.

Replacement Power Costs

In the baseline case, it is assumed that the replacement power costs were equal to the replacement capacity costs (on an annual basis). Assuming the total system is expanding at a reasonably rapid pace, this is not a particularly bad assumption. However, one's intuition would typically lead to a contrary assumption, i.e., that the short-term replacement power cost could be

*A major assumption behind base cases was that a replacement power and capacity cost would be provided by nuclear power generation.

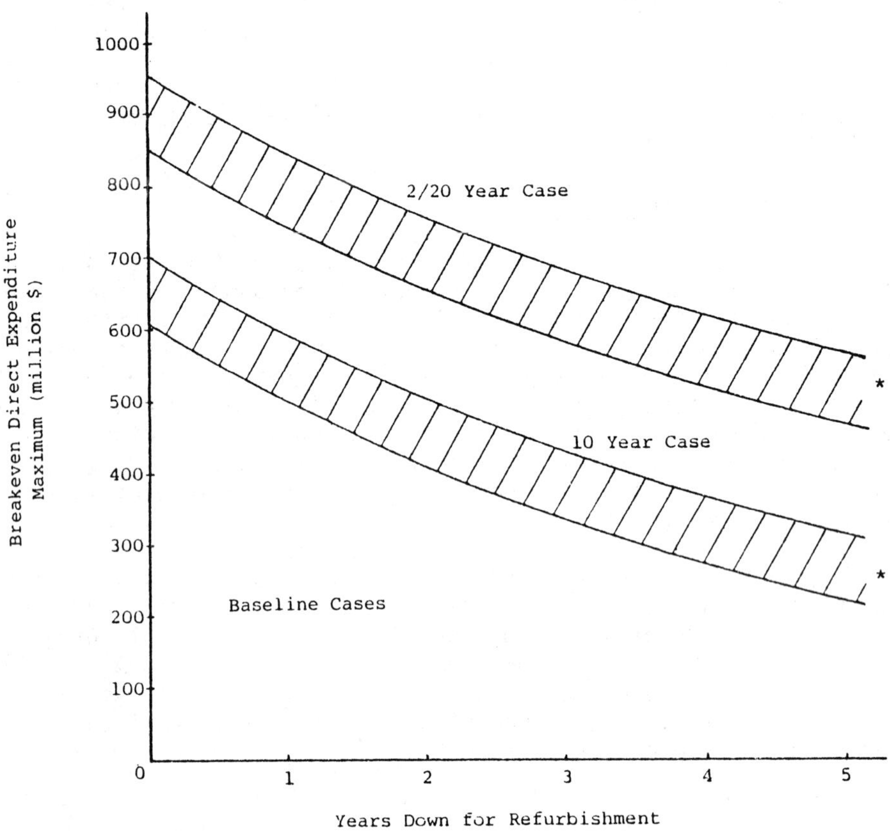

Figure 2. Refurbishment Period Sensitivity Analysis

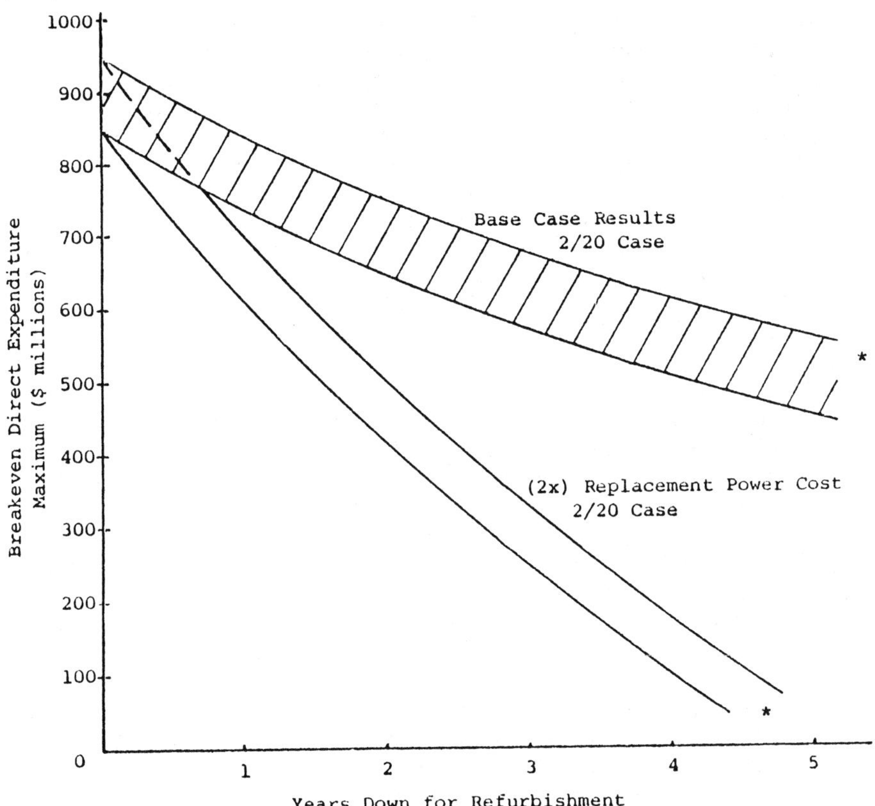

*Band represents authors' qualitative assessment of uncertainty.

Figure 3. Replacement Power Costs and Refurbishment Period Sensitivity Analysis

higher, for example, if a utility must purchase wheeled power. An extreme case was selected to evaluate this situation; replacement power costs were assumed twice as expensive as replacement capacity costs, an assumption that should favor plant retirement.

Figure 2 also presents the results of this analysis. The shaded area is the same as presented in the base case. The solid lines labeled "2x" trace out the impact of the higher replacement power costs on breakeven direct cost for refurbishment under varying periods of downtime. As is apparent in the figure, this change in assumption has a dramatic effect on the results. This substantial impact suggests that a detailed assessment in the context of total system costs (versus our simplified project evaluation) should be undertaken before any final judgments are formed. However, it appears that, with any reasonable outage time (2 years or less), the extended life option would still allow a substantial investment before a net loss would result.

REFURBISHMENT ANALYSIS

The approach for evaluating possible equipment refurbishment requirements was to tabulate plant structure and equipment in five categories, forming a basis for the evaluation. The categories are the following:*

(A) Assured unlimited service life -- items having unlimited service life (more than 80 years), refurbishment or replacement not required, no extraordinary monitoring of performance of condition is required.

(B) Highly probable extended service life -- items having a high probability for extended service life based on presently available methods of refurbishment, conditions of service, and knowledge of physical condition.

(C) Extended life required justification -- items whose service life can possibly be extended assuming reliable information on condition of the item is obtained and feasible refurbishment techniques can be demonstrated.

(D) Infrequent replacement -- items or parts of items likely to require replacement at the end of plant design life or at least once during 40 years, design life replacement items.

*In the report, Category B, C, and D tabulations also comment on repair or replacement difficulty; tabulations are not included due to their length.

(E) Routinely repaired and replaced items -- items requiring recurring replacement or refurbishment during plant design lifetime.

In discussing these tabulations with utility personnel, it became apparent that there were many examples of items in Categories B and C that have, in fact, been refurbished or replaced early in life at one plant or another. In many such cases, design inadequacy rather than age was the cause. Although initial design adequacy is not the subject of this study, a precedent for refurbishment had in fact been established, and the distinctions between Categories B, C, and D lost focus as a result. What became clear from this tabulation exercise was that there are very few specific equipment items that by themselves would swing a decision to terminate operation of a plant. What would be of greater impact is the cost of the total refurbishment requirements.

There are a few equipment items that merit special attention because of combined aspects of replacement difficulty, potential safety concern, and possible commercial risk. These items include the following:

- Primary coolant system within containment and especially the reactor pressure vessel (RPV). Without benefit of a detailed study, replacement of the RPV is roughly estimated to take three years. The rest of the primary coolant system is not as significant. There are precedents set and major work underway on steam generators, RPV nozzles and safe ends, primary system piping, and vessel internals. Thus, the major focus of concern would be of the RPV itself and the ability to move new, large components (such as the RPV or steam generator) into containment.

- Primary shield concrete (neutron shield tank at some plants). These are structures that surround the reactor vessel and for which major refurbishment would be extremely difficult because of limited accessibility. However, there is no reason to believe that significant degradation would occur, even though subjected to neutron irradiation and temperatures somewhat higher than those experienced by other structures in the plant.

- Anchor tendons for pre-stressed, post-tentioned containment. Without further investigation, it is unclear whether this is significant. Accessibility is a concern because anchor tendons at some plants are grouted into bedrock or the base mat.

- o Electrical wire and cable is a concern only if generic aging would result in a requirement to replace a large quantity at one time. Because of the amount and variety of cables and termination points, mass replacement would take a long time. Recovery operations from the Browns Ferry fire have shown such construction to be time-consuming and expensive.

There are many other items that are potential safety or commercial concerns; however, in general they are not as significant as those of the above.

CONCLUSIONS

ECONOMICS

The results of the economic analysis show that extended life of LWRs is economically beneficial if, between now and the first decade of the twenty-first century, nuclear fuel costs remain low relative to other feasible baseload power generation technologies. The results also show that if refurbishment is required, allowable downtime can be of the magnitude of several years and cost on the order of $500 million (in 1978 dollars equivalent) before feasibility would be borderline. These results would be modified for the many specific situations that can be anticipated. However, there clearly is sufficient economic justification to examine operating life extension in more detail.

Impact of Decommissioning Costs

Relative to other costs attendant in the operation of a power plant in the range of 500 to 1500 MWe, the _direct_ cost of decommissioning is a minor one, even based on the most pessimistic estimates to date. The _indirect_ cost of nuclear plant decommissioning, that is having to replace the capacity because it is no longer available, can be very high. This conclusion is reinforced in the case where LWRs are replaced by other capital-intensive plants such as LWRs or high-sulfur coal (or even fusion or solar central stations).

REFURBISHMENT

Everything within a power plant that may degrade with age is refurbishable or replaceable. The cost of such activities could be very significant for large equipment within the containment that would have to be replaced. Precedent already exists, in both fossil and nuclear stations, for replacement or repair of most major components in power plants. Two notable examples within containment are steam generator replacements being planned for Surry and Turkey Point and the replacement of reactor pressure vessel nozzle safe ends at Duane Arnold. The equipment-related decisional factors to extend the life of a plant are likely to be:

- o The cost of the total lot of required equipment refurbishment (as opposed to single items).
- o Not having to replace certain items of equipment (such as the reactor pressure vessel).
- o Licensability, specifically the ability to demonstrate no safety degradation.
- o Assured equipment reliability to satisfy the planned use of the plant (availability).
- o Radiation accessibility.

Based on the economic analysis, several hundred million dollars can feasibly be invested to gain an extra 10 or 20 years operation from a single nuclear power plant. By comparison, published estimates for steam generator replacement at Surry and Turkey Point are in the $50 to $100 million range, and the Duane Arnold safe ends replacement is in the neighborhood of $20 million. Thus, except for major items such as the reactor pressure vessel, the cost of refurbishment would probably not be a deterrent to extended life operation. Furthermore, replacement of the reactor pressure vessel is possible, and this may even be economically feasible.

Licensing and Decommissioning Rulemaking

Nothing has been found in regulations or decommissioning rulemaking intent that specifically prohibits extended life operation. Discussions with the NRC related to decommissioning rulemaking indicate that the NRC is aware of the possibility of extended life. Thus, we presume that forthcoming decommissioning regulations (expected in late 1979) will not preempt the extended life option.

Appendixes G and H of 10 CFR 50 provide a basis for determining the safety of continued operation of the reactor with respect to preservation of ductile properties of the reactor pressure vessel materials. However, it is believed this has never been tested by application. There is no other regulation in Parts 50, 51 or 100 of Title 10 that specifies a basis for extended life.

Figure 1 leads to an interesting observation with regard to licensing lead time and NRC practice that could be classified as a potential "Institutional" problem. That is, in order to extend operations, a plant owner must know approximately eight years in advance of expiration that an operating license will be renewed. Such lead time for this purpose is unprecedented, and as a result it appears that a senior-level dialog between utility managers and regulatory authorities would be appropriate to define this problem in more detail and initiate actions to resolve it.

RECOMMENDATIONS

Because of the potentially substantial economic benefits of extended life operation, several recommendations were made in the report that outline an R&D program to conduct a total requirements assessment to support LWR plant extended life. The assessment would address the requirements in terms of ASME code potential, commercial concerns, data acquisition, improved design, licensing, new methods, and scoping or program definition studies. Some of the detailed recommendations are part of a current EPRI R&D program addressing recommissioning and extended life operations.

ECT REMEDIAL ACTION PROGRAM

Robert W. Ramsey, Jr.

Chief, Nuclear Technologies Branch
Office of Environmental Compliance and Overview
of the Assistant Secretary for Environment
U. S. Department of Energy
Washington, D. C. 20545

The Environmental Control Technology Division (ECT) is a component of the Office of Environmental Compliance and Overview under the Assistant Secretary for Environment of the Department of Energy (DOE). The Division has, since its formation in July 1975, been responsible within the DOE for programs to decontaminate and decommission (D&D) surplus facilities. More recently, the Division was made responsible for determining the need for remedial actions to eliminate potential radiological exposure to people and potential harm to the environment due to residual contaminatin at formerly utilized laboratories and manufacturing or processing sites and at inactive uranium mill tailings storage sites. The purpose of this paper is to describe the scope of various programs conducted by ECT to define the need for remedial action and assist in the implementation of cleanup projects.

Three program activities form the DOE's program of remedial actions. The three programs are:

1. The Grand Junction Remedial Action Program - authorized by Congress under P.L. 92-314;

2. The Uranium Mill Tailings Remedial Action Program - authorized by Congress under P. L. 95-604; and,

3. Formerly Utilized Manhattan Engineer District/Atomic Energy Commission (MED/AEC) Site Remedial Action Program - called FUSRAP.

The paper by Dr. A. F. Kluk of this session describes the related program for D&D of DOE-owned surplus facilities. ECT has been responsible for those facilities that were identified as surplus to further needs prior to October 1, 1976.

Grand Junction Remedial Action Program

The situation that developed at Grand Junction, Colorado, caused a focus of attention on the need for remedial action. Basically, the absence of any control or responsibility for the tailings from uranium milling that was done at Grand Junction led to the inadvertent use of tailings as a construction material and hence the contamination of many residences, schools, and commercial buildings. The sand-like tailings were almost an ideal material for fill or foundation work and, in some cases, found use as fine aggregate for concrete or mortar. By the time that the potential problem that this use of the tailings would cause was recognized, several thousand new structures had been built, many with the use of tailings sand. As a matter of compassionate responsibility, the Congress passed P. L. 92-314 in 1973 to specifically authorize a joint Federal-State program to remove tailings from homes, schools, and commercial buildings which qualify for such action by radon and radon daughter measurements.

The program is a State-operated activity with the DOE now providing 75 percent of the funding and the State 25 percent. To date, it is estimated that about 800 structures will require remedial action and about one-half have been done. About 100 structures are done each year and costs have been about $1.86 million per year. The program has a Federal authorization of $12.5 million for an estimated total of $16.6 million. It is expected to be completed by 1983. In 1977, this program was transferred from the Operational Health and Safety Division of ERDA to ECT where it is now part of the remedial action program.

Uranium Mill Tailings Remedial Action Program

As a follow-up to the concern over the situation that developed with the use of uranium mill tailings in structures at Grand Junction, Colorado, the potential for personnel exposure or environmental contamination due to inactive uranium mill tailings piles has been recognized as a concern. Operating mills which were licensed by the NRC or the States were under their cognizance and had assurances that the tailings control aspects would not be neglected. However, there were a number of uranium mill tailings piles in the West at mills that had closed operations completely and had processed ores only for the AEC and were not under any licensing control or contractual requirements with respect to the tailings. These so-called "inactive uranium mill tailings sites" numbered 22 located in eight Western States. In all cases the tailings were relatively unstabilized, the mills were no longer active, and the companies that still existed had indicated an inability to accept liability for removal of the tailings or their stabilization. What was clearly needed was some authority to allow the situation to be remedied. In April 1978, the DOE proposed legislation to the Congress to authorize remedial action at 21 inactive uranium mill tailings sites in eight Western States and the site of the former Vitro Rare Metals Co. at Canonsburg, Pennsylvania. The program was to be a joint Federal-State undertaking similar to the Grand Junction Remedial Action Program.

The actions of the Congress resulted in the passage of Public Law 95-604, "The Uranium Mill Tailings Radiation Control Act of 1978," on November 8, 1978. Under this Act the Secretary of Energy is to designate sites for remedial action with consultation of the NRC; set priorities for remedial actions with the concurrence of the EPA; arrange agreements with the Governors of States and leaders of impacted Indian tribes, on which some of the tailings exist, with coordination of the Department of the Interior and carry out the needed remedial actions. The bill requires that any permanent disposal site become Federal property, that due consideration be given to the potential processing of tailings for commercial recovery of residual uranium, and that appropriate safeguards against possible windfall profits be instituted in land acquisition and disposal transactions.

The program of remedial action projects has a seven-year life following issuance by the EPA of the criteria for decontamination. A part of the program has been to survey additional known instances of radioactive contamination on public or acquired lands. The survey resulted in a report recently issued to the Congress.[1] The additional sites uncovered in this survey are not expected to result in any addition to the current scope of the program unless the Act is amended in the future to broaden its coverage.

Formerly Utilized MED/AEC Site Remedial Action Program (FUSRAP)

This program has its origin in attempts by the Office of the Assistant General Manager for Environment of the AEC in 1974 to trace the records of radiological conditions at a number of sites and facilities that had been used by the MED and the AEC for nuclear operations. When work at these sites was terminated, most of these facilities or sites were released without restriction.

In some cases the properties were decontaminated by the contractor who owned the facility or by an operating contractor of a Government-owned facility. In almost all cases the record of the results of such decontamination operations were not recorded and no formal certification was made to accompany the release from radiological control. Because of the reductions in allowable exposure and the evolution of more restrictive criteria for release of low-level contaminated property, it was reasonably certain that properties would be identified which exceed current radiological criteria.

Beginning in 1974, a program was undertaken to determine the radiological status at all of the formerly utilized sites. The sites were identified by having the AEC Field Offices screen the past records of contracts or histories of development programs to identify facilities associated with the processing or handling of nuclear materials, including ores and fuel materials. The survey considered 126 such sites as of June 29, 1978, and determined by examination of records that 73 of these sites did indeed require some sort of assessment to determine their radiological condition. As a result of further records searches, conversations with owners, site visits, and in some cases, preliminary radiological surveys, the DOE was able to determine that conditions at 43 of the sites would have to be assessed by full radiological surveys. Special radiological survey teams of the Health and Safety Research Division of Oak Ridge National Laboratory and the Occupational Health and Safety Division of the Argonne National Laboratory have been at work completing these surveys and preparing the radiological survey reports and assessments of health impacts that form the bases of determining need for remedial action.

Based upon radiological surveys performed to date and the assessment of results, it appears that some form of remedial action will have to be performed at about 30 of these sites to render them acceptably stabilized or without radiological restriction of any kind in the future. The most prevalent contamination found in these surveys come from the decay chains of uranium and thorium found in tailings from the milling of ores or residues from the handling or processing fuel materials. The health physics and industrial hygiene measures used in the handling of ores and the criteria for decontamination did not adequately control contamination of facilities and sites when measured by current criteria. In some cases adjoining private property was contaminated; hence, some of the remedial action projects involve operations on impacted private property.

Of high priority among the FUSRAP sites is the former Middlesex Sampling Plant property, Middlesex, N.J., including many adjacent properties which received contamination by airborne dust or surface water runoff. There are also two remote properties where it is evident that contaminated soil was placed as fill or topsoil during construction coincident with decontamination activities at the site. A landfill site in this community was also used to dispose of some contaminated soil under prior arrangement with the community that included burial of the contaminated material and restriction in the deed of the property denoting the presence of radiological contamination.

Following the issuance of the radiological survey report on the Middlesex site,[2] a program of engineering studies and preparation of reports was undertaken for the purpose of more accurately estimating the extent of contamination and volume of soil that would have to be removed within radium 226 concentration limits of 10 and 5 picocuries per gram above background and at background. These limits roughly corresponded with levels of subsurface contamination to infinite depth that would result in radon and radon daughter concentrations in structures at the levels prescribed by the Surgeon General for remedial action or for unrestricted occupancy in the Grand Junction, Colorado, remedial action program. Based upon these additional surveys, estimates of the volume of material involved and the cost of removal and disposal were made. In addition, an analysis was made of the associated environmental effects of the removal project to serve as a source of data for environmental impact documentation required by the application of NEPA procedures to the remedial action projects. Two reports have been prepared on the Engineering Evaluation and the Environmental Analysis.

Various alternative courses of action, mostly related to the disposal of contaminated materials, were considered and costs were estimated for each case.

Table 1 shows two alternatives with the estimated costs and the volume of material that would result from the two levels of criteria for cleanup.

The project at Middlesex is currently planned to begin with the removal of material from two offsite properties to temporary storage on the Sampling Plant site. The Sampling Plant is now in the custody of DOE and would therefore be maintained under surveillance for radiological control of the material. Other private properties surrounding the plant site would also be cleaned up and at the same time suitable onsite measures to prevent recontamination and allow safe interim storage of material would be taken.

As soon as an acceptable permanent disposal site is determined, all material stored on the site plus the contaminated buildings and soil of the Sampling Plant would be removed down to the criteria adopted by the DOE and state and local officials. It is expected that this criteria will be consistent with and based upon guidance being developed for radium in soils by the EPA under the Uranium Mill Tailings Program.

TABLE 1

Estimated Waste Quantities and Costs
For Former Middlesex Sampling Plant
Associated Properties

	Criteria for Decontamination Picocuries/Gmof Radium 226 in Soils	
	5pCi/Gm	Background*
Waste Volume (cubic yds)	54,300	76,850
Waste Weight (tons)	80,600	113,200
Estimated Costs-Millions of Dollars		
On-Site Stabilization	1.5	2.0
Shipment Max Distance of 2,750 miles	14.1	19.5

* 1.0 pCi/Gm

Reference: Engineering Evaluation of the Former Middlesex Sampling Plant and Associated Properties, Middlesex, N.J. - Final Draft, Ford Bacon & Davis Inc., April 1979.

At the conclusion of each remedial action a radiological evaluation is to be conducted by a contractor or laboratory, other than that responsible for the cleanup, to certify that the site is suitably decontaminated for unrestricted use. DOE certification that the site is unrestricted for future use would be noted as part of the property records as a final act of the program.

The Middlesex remedial action project as given above is a specimen scenario that will be repeated in some degree for each of the FUSRAP sites.

Reorganization Within DOE

The program of remedial action has recently been reorganized within DOE to separate the functions of problem definition and environmental overview, as conducted under Environmental purview by ECT, from the functions of implementing remedial action projects and disposing of the wastes, which will now be under the Waste Management Program of the Assistant Secretary for Energy Technology. This arrangement is expected to eliminate potential conflict of interest within the environmental area and provide for truly independent overview assignment, will improve the resources that can be brought to bear on the management of remedial action projects under Energy Technology, and consolidate related activities such as waste disposal.

In keeping with the goals of the DOE to utilize to the fullest the resources of field organizations, both the FUSRAP and the Mill Tailings Programs will be largely delegated to Field Offices at Oak Ridge and Albuquerque, respectively. The Grand Junction Program will continue to be managed as a state-operated program with major coordination with the DOE's Grand Junction Office. Grand Junction will also be a key resource of data for the Uranium Mill Tailings Program.

The total budget for these programs is at present unknown because the disposition component of remedial actions has not been established. However, several of the high priority mill tailings projects are known to be in the tens of millions of dollars as are a few of FUSRAP projects. A rough estimate is about $100-150 million for FUSRAP, $100-200 million for UMTRAK, and $16-20 million for Grand Junction for a total of from $216 to $370 million for the entire program. With a targeted completion of seven years--this averages $30 to $50 million per year.

REFERENCES:

1. Report on Residual Radioactive Materials on Public or Acquired Lands of the United tates, July 1, 1979, DoE/EV-0037

2. Engineering Evaluation of the Former Middlesex Sampling Plant and Associated Properties, Middlesex, N.J., Final Draft, Ford Bacon and Davis Utah, April 1979.

3. Environmental Analysis of the Former Middlesex Sampling Plant and Associated Properties, Middlesex, N.J., Final Draft, Ford Bacon and Davis Utah, April 1979.

MANAGEMENT OF THE DOE INVENTORY OF EXCESS RADIOACTIVELY CONTAMINATED FACILITIES

Anthony F. Kluk

Remedial Actions Program
Office of Nuclear Waste Management
Washington, D. C. 20545

The Department of Energy is the custodian of a large number of facilities and sites which were constructed and used under the Manhattan Engineer District (MED), Atomic Energy Commission (AEC), and the Energy Research and Development Administration (ERDA). These sites and facilities were used in the development and evolution of nuclear energy programs for scientific, defense, and domestic purposes. The increase in nuclear program activities in the 1950's and the subsequent emphasis on domestic electric power generation in the 1960's contributed a large number of contaminated facilities which subsequently became obsolete and were identified as surplus to programmatic needs prior to the formation of the Department of Energy.

Surveys have been conducted under AEC, ERDA, and now DOE to identify the radioactively contaminated facilities that were no longer required to support planned or ongoing programs. The results of the latest survey are documented in the "Preliminary Plan for Decommissioning of Department of Energy Radioactively Contaminated Surplus Facilities,"(1) which was issued in 1978. Issuance of this Plan marked the first time that data on the entire inventory was compiled in a single document. An updated version of this plan will be available shortly. The Plan identifies about 500 contaminated locations which include numerous facilities such as reactor buildings, laboratories, fuel reprocessing plants, and a large number of waste management facilities, such as cribs, trenches and burial grounds, most of which are located at Hanford.

Table 1 shows a breakdown of the inventory by type of facility. The reactors include training, research, production and demonstration reactors; the tanks include both steel and concrete types but not the large million gallon tanks used for high-level waste storage. By far the largest group is waste disposal facilities which consist primarily of contaminated soil locations, most of which are on the Hanford Site. Ultimate disposition must be either removal and consolidation at other disposal areas or stabilization in place.

In addition to the backlog which the Department inherited from its predecessor agencies, currently operational facilities are continuously becoming surplus. Depending on the orignating program, these new facilities can either be decommissioned by the responsible operating program or added to the inventory for future disposition. Our current policy is to manage those identified within the Office of Nuclear Energy Technology (ETN) nuclear programs as part of the surplus facilities inventory and to decommission those identified under other Assistant Secretaries such as Energy Research and Defense Programs through the responsible operating program. At the present time there are 32 facilities which have now been transferred to the inventory from other ETN organizations or will be within the next 2 years.

Figure 1 shows the various locations of the surplus facilities. In addition to those located at the major nuclear sites such as Hanford, Idaho Falls, Argonne National Laboratory and Oak Ridge, there are a number of smaller sites for which DOE is responsible. These include sites such as the Niagara Falls Storage Site (former Lake Ontario Ordinance Works) and the Weldon Spring Site in Missouri, where the entire site is considered surplus. None of the sites shown in Figure 1 are included under the Formerly Utilized Sites or Inactive Mill Tailings Programs.

Historically under the AEC, the management of surplus facilities was a responsibility only of the program division. However, it was found that this arrangement encouraged the accumulation of a large backlog since the priority of decommissioning projects was often subordinated to ongoing programs. Under ERDA and DOE, programs for managing surplus facilities have been centralized within a single organization called the Surplus Facilities Management Program (SFMP). At the present time, the program is under the direction of the Assistant Secretary for Energy Technology (ASET).

The current Headquarters organizational structure is shown in Figure 2. The Surplus Facilities Management Program is part of the Office of Nuclear Waste Management under ASET. Not shown on this chart is the Assistant Secretary for Environment who has responsibility for overview and assessment of the program, including

TABLE 1

CATEGORIES OF SURPLUS FACILITES

FACILITY TYPE	QUANTITY	PERCENT OF INVENTORY
REACTORS	25	5
BUILDINGS/STRUCTURES	124	26
STACKS, TANKS, AND AUXILIARIES	35	7
PONDS, DITCHES, TRENCHES AND OTHER WASTE DISPOSAL	302	62
TOTAL	486	100

TABLE 2

OFFICE OF SURPLUS FACILITIES MANAGEMENT

MAJOR PROGRAM AREAS

- SURVEILLANCE AND MAINTENANCE
- PLANNING AND ENGINEERING
- TECHNOLOGY
- DISPOSITION PROJECTS
- TECHNOLOGY TRANSFER

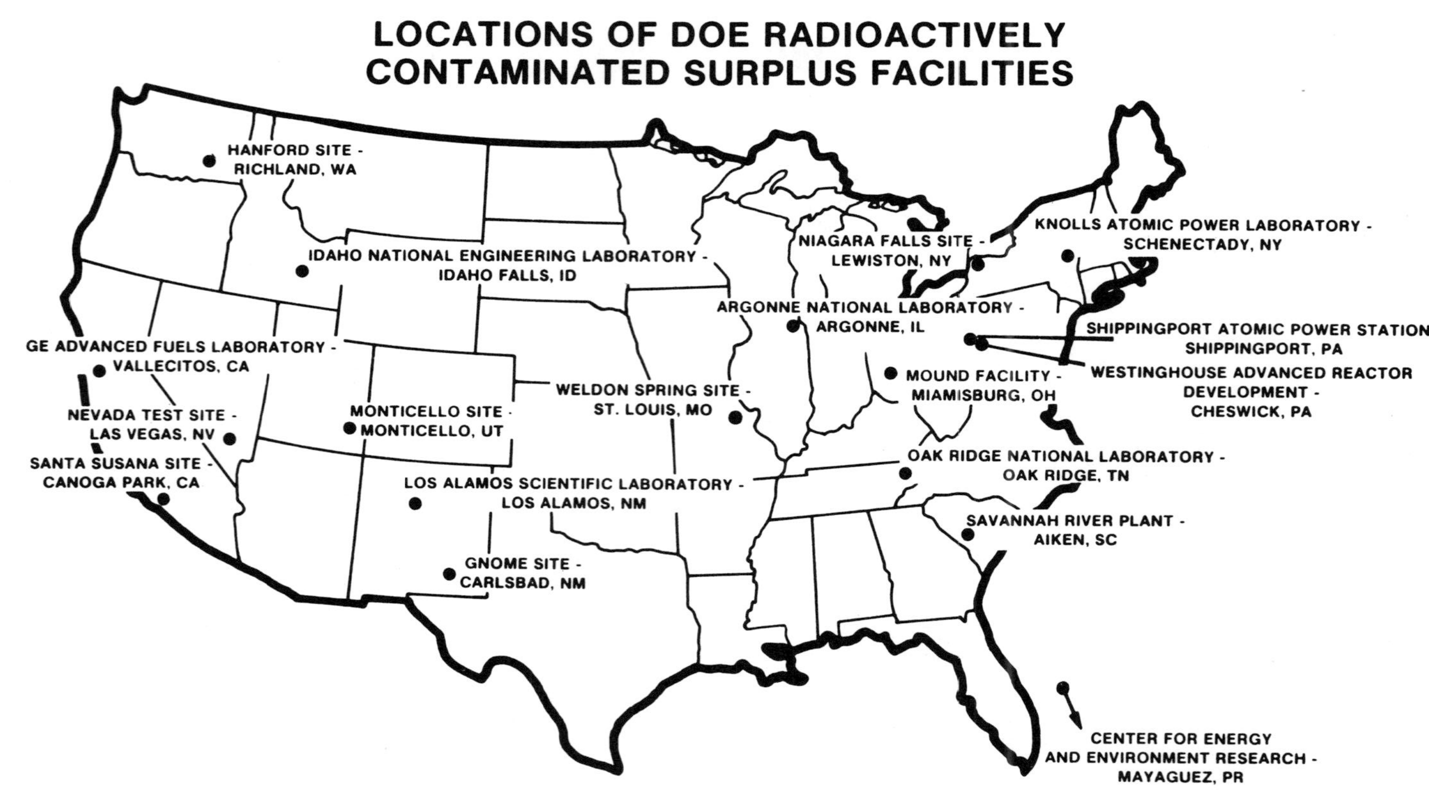

LOCATIONS OF DOE RADIOACTIVELY CONTAMINATED SURPLUS FACILITIES
HANFORD SITE - RICHLAND, WA
IDAHO NATIONAL ENGINEERING LABORATORY - IDAHO FALLS, ID
GE ADVANCED FUELS LABORATORY - VALLECITOS, CA
NEVADA TEST SITE - LAS VEGAS, NV
SANTA SUSANA SITE - CANOGA PARK, CA
MONTICELLO SITE - MONTICELLO, UT
LOS ALAMOS SCIENTIFIC LABORATORY - LOS ALAMOS, NM
GNOME SITE - CARLSBAD, NM
WELDON SPRING SITE - ST. LOUIS, MO
ARGONNE NATIONAL LABORATORY - ARGONNE, IL
NIAGARA FALLS SITE - LEWISTON, NY
KNOLLS ATOMIC POWER LABORATORY - SCHENECTADY, NY
SHIPPINGPORT ATOMIC POWER STATION - SHIPPINGPORT, PA
WESTINGHOUSE ADVANCED REACTOR DEVELOPMENT - CHESWICK, PA
MOUND FACILITY - MIAMISBURG, OH
OAK RIDGE NATIONAL LABORATORY - OAK RIDGE, TN
SAVANNAH RIVER PLANT - AIKEN, SC
CENTER FOR ENERGY AND ENVIRONMENT RESEARCH - MAYAGUEZ, PR

FIGURE 1

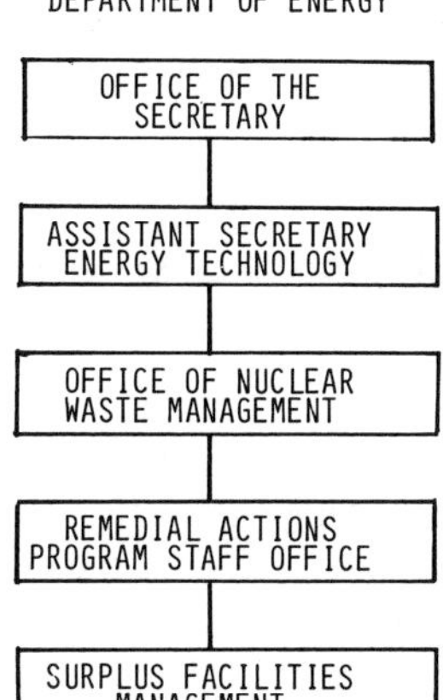

DOE ORGANIZATION
FOR SURPLUS FACILITIES MANAGEMENT PROGRAM

FIGURE 2

DOE-RL ORGANIZATION
FOR SURPLUS FACILITIES MANAGEMENT PROGRAM

FIGURE 3

review of NEPA documents. Under the decentralized management concept, Richland Operations Office has been selected as the lead field office responsible for managing surplus facilities. Their organizational structure is shown in Figure 3. To support the lead field office, UNC Nuclear Industries was selected as lead contractor.

There are 5 major areas of activity under the program for managing surplus facilities including surveillance, development of disposition methods, planning, disposition and technology transfer. These are listed in Table 2.

Surplus facilities require routine surveillance and periodic maintenance to assure that they remain in a safe condition prior to disposition. The SFMP currently provides routine surveillance and maintenance for facilities at Argonne National Laboratory, Idaho National Engineering Laboratory, Hanford Reservation, Oak Ridge National Laboratory, and other sites. These programs will be continued until disposition is completed under the Disposition Projects area of the program.

The planning and engineering category provides for the development of detailed plans and assessments for decommissioning surplus facilities. Programs include acquiring and maintaining a data base and assessment of alternative disposition options leading to selection of disposition mode. Also included is the establishment of overall program priorities for the inventory of surplus facilities based on potential offsite hazard, local site requirements and other factors.

The decommissioning plan(1) which was prepared in FY 1978 and is being updated, proposes a twenty-year program consisting of about 80 projects to eliminate the backlog of surplus facilities at an estimated total cost of over $400 million. Such a program will require an annual funding level of $20-25 million to maintain continuity of contractor staffing for technical planning, supervising, and implementing disposition projects in a cost effective manner.

Another important part of the program is the development of new and improved technology for disposition of nuclear facilities such as decontamination of concrete, volume reduction of process equipment, the decontamination of metals by smelting. These R&D studies are designed to improve: (1) the efficiency of disposition techniques, (2) consistency in planning, (3) the cost estimating process, and (4) design techniques that will assist in eventual disposition of nuclear facilities.

Under the next program category, disposition projects, selected dispositon projects are implemented. Table 3 shows a

TABLE 3

ONGOING DECOMMISSIONING PROJECTS

FACILITY	LOCATION	SCHEDULED COMPLETION
GNOME	CARLSBAD, NEW MEXICO	FY 79
ORGANIC MODERATED REACTOR EXPERIMENT	IDAHO FALLS, IDAHO	FY 79
NEW BRUNSWICK LABORATORY	NEW BRUNSWICK, NEW JERSEY	FY 81
NUCLEAR ROCKET DEVELOPMENT STATION	NEVADA TEST SITE	FY 81
CEER REACTOR	MAYAQUEZ, PUERTO RICO	FY 81
SODIUM REACTOR EXPERIMENT	CANOGA PARK, CALIFORNIA	FY 81
WELDON SPRING SITE	ST. CHARLES COUNTY, MISSOURI	FY 82
PLUTONIUM GLOVEBOXES	ARGONNE NATIONAL LABORATORY	FY 83
PLUTONIUM CONCENTRATION BLDG.	HANFORD, WASHINGTON	FY 83
100-F PRODUCTION REACTOR	HANFORD, WASHINGTON	FY 84
MOUND DASMP AREAS	MIAMISBURG, OHIO	FY 84
NIAGARA FALLS SITE	NIAGARA FALLS, NEW YORK	FY 85

list of 11 ongoing projects, 2 of which will be completed in FY 1979 while the remainder will be completed in FY 1981 or later. Table 4 shows a list of major facilities that are awaiting decommissioning. Projects for these are in the planning stages and will be initiated as program resources permit. One for which plans are well developed is the Shippingport Station west of Pittsburgh which is being handled as a special project. In Table 5, examples of completed decommissioning projects are given. Bonus reactor in Puerto Rico was entombed in 1970 while Elk River was dismantled.

Under the technology transfer category, DOE supports the development and transfer of technology disposition techniques which have application to the commercial nuclear industry. An example of the kind of program being supported is the application of the plasma arc torch technology to cutting thick walled steel pressure vessels. Also an information center is being set up to facilitate information dissemination to the private sector and other government agencies.

As a final note, I would like to briefly review some areas of major concern to the program at this time. These include:

- o Establishment of criteria for residual radioactivity to allow unrestricted use to be defined.
- o Development of a methodology for establishing priorities and for selecting disposition options.
- o Establishment of waste repositories suitable for decommissioning wastes including large volumes of contaminated soils.
- o Establishment of the procedures for certifications to confirm the radiological conditions of a property.

At the present time, criteria are established on a case by case basis for each project that is undertaken by the Department. Factors which are considered include type of contamination present, projected land useage, and State and local requirements. As a result of the recent Report to the President by the Interagency Review Group on Nuclear Waste Management(2), the EPA has agreed to prepare a residual radioactivity standard by the end of calendar year 1981. In addition, NRC will develop D&D standards and regulations for licensed facilities by the end of 1980. These regulations will help to standardize the DOE approach to cleanup activities.

Establishing priorities for decommissioning is a very important part of managing surplus facilities because it determines the order in which the projects will be initiated. At present, the priorities for the backlog are based on such factors as potential off-site hazard, cost of continuing maintenance and surveillance, alternative need for the facility, and resources

TABLE 4

LIST OF MAJOR ETW/OR FACILITIES AWAITING DECOMMISSIONING

ORGANIZATION	FACILITY	APPROXIMATE START OF DECOMMISSIONING
OR	CURIUM FACILITY	1980
ID	MTR REACTOR	1982
OR	MOLTEN SALT REACTOR EXPERIMENT	1982
RL	STRONTIUM SEMI-WORKS	1982
CH	EXPERIMENTAL BWR	MID-1980's
ID	ORIGINAL WASTE CALCINER	MID-1980's
RL	SHIPPINGPORT STATION	MID-1980's
SR	WASTE TANK(S)	MID-1980's

TABLE 5

EXAMPLES OF COMPLETED DECOMMISSIONING PROJECTS

FACILITY	YEAR	FIELD OFFICE/CONTRACTOR	COST
BONUS REACTOR	1970	OR/UNITED NUCLEAR CORP.	$1,600,000
ELK RIVER REACTOR	1974	CH/UNITED POWER ASSOCIATION/ GULF UNITED	6,200,000
EBR-1-COMPLEX	1975	ID/AEROJET NUCLEAR CORP.	775,000
MTR-WORKING RESERVER	1975	ID/AEROJET NUCLEAR CORP.	105,000
KINETICS EXPERIMENT WATER BOILER	1975	SAN/ATOMICS INTERNATIONAL	113,000
SHIELD TEST IRRADIATION REACTOR	1976	SAN/ATOMICS INTERNATIONAL	135,000
HOT CAVE-BLDG. 003	1976	SAN/ATOMICS INTERNATIONAL	149,000
PLUTONIUM EXPERIMENTAL FACILITY (TA-33-21)	1976	AL/LASL	130,000
HANFORD TEST REACTOR	1977	RL/UNC NUCLEAR INDUSTRIES	225,000

available to do the job. For newly declared surplus facilities, many are covered by special transfer agreements which commit resources and establish decommissioning schedules. One of the tasks which we are working on is a revision of the methodology for establishing priorities so that all newly declared surplus facilities can be integrated into the inventory and assigned its proper place in the overall program decommissioning schedule. In this way the resources dedicated to decommissioning could readily be directed to the highest priority projects.

Another concern which must be faced in the SFM program is that decommissioning activities may generate large volumes of contaminated rubble. Reference 1 indicates that a program to eliminate the backlog would generate about 10 million cubic feet of such waste. As the lead field office for low-level waste disposal, the Idaho Operations Office has been tasked by Headquarters with preparing a plan for identifying new sites suitable for handling such waste on a regional basis. The plan is to be available by January 1980.

Finally, we are in the process of establishing internal procedures for certifying that decommissioning activities comply with all applicable requirements. This is especially important for those projects where unrestricted release of property is intended. We anticipate that certification will be part of the overview and assessment function of the ASEV.

References

1. Preliminary Plan for Decommissioning of Department of Energy Radioactively Contaminated Surplus Facilities, HCP/P0701-01, U.S. Department of Energy, Washington, D. C., September 1978.
2. Report to the President by the Interagency Review Group on Nuclear Waste Management, TID-29442, Washington, D. C., March 1979.

SUMMARY OF THE DEVELOPMENT OF A COST/RISK/BENEFIT ANALYSIS FOR THE DECONTAMINATION AND DECOMMISSIONING OF THE HANFORD Z-PLANT

R. A. Sexton and J. P. Melvin

Research and Engineering Function
Rockwell Hanford Operations
Richland, WA 99352

INTRODUCTION

The Hanford Z-Plant complex is a large plutonium facility located on the Hanford Site in Eastern Washington. This study includes eight buildings (see Figure 1) totaling about 240,000 ft^2 of floor space. They contain processing facilities, laboratories, offices, storage space, ventilation equipment, and a 200 foot stack.

The facilities contain contaminated equipment and contaminated structural surfaces. Most of the contamination in Z-Plant is plutonium with some americium.

Z-Plant was designed to convert plutonium nitrate product solutions from the spent fuel reprocessing facilities into plutonium metal for fabrication of weapons parts. The Plutonium Reclamation Facility, 236-Z, was constructed to recover and recycle plutonium from scrap and solutions generated during reduction to metal.

Construction of the facility started in 1948 with production beginning in 1949. Several additions have been made to the plant.

The processing facilities in Z-Plant are currently scheduled for deactivation by 1983, with the laboratory facilities expected to remain operative beyond that.

Consideration is being given to removing the Hanford Z-Plant from service and initiating decontamination and decommissioning (D&D). Various possibilities exist for the level of decontamination to be achieved and the methods of decontamination to be used. These alternatives have been examined in a Cost/Risk/Benefit Analysis.

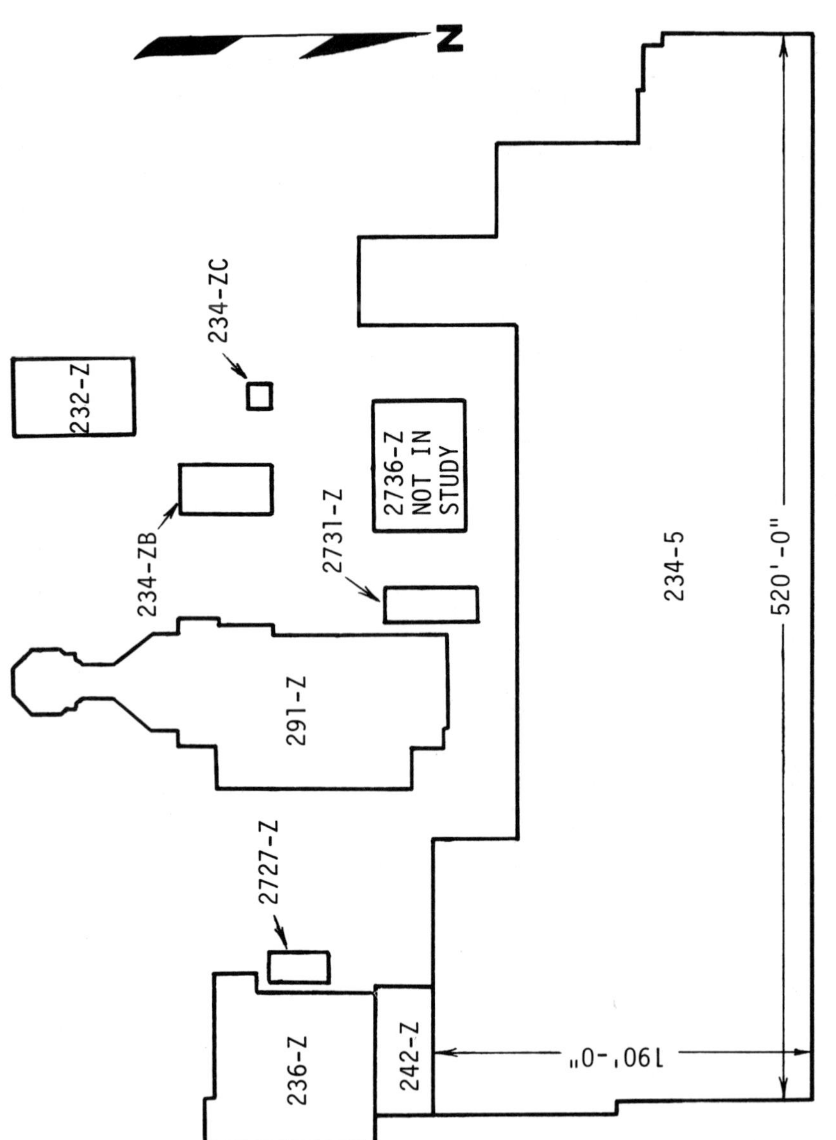

Figure 1 Buildings Included In Cost/Risk/Benefit Analysis

The objective of this study is to quantify costs/risks/and benefits associated with various approaches to the D&D of Z-Plant. It should be understood that subjectivity is involved in comparing radiological hazards to dollars, but this study attempts to provide objective input into the subjective analysis.

The study includes consideration of the alternative of maintaining an inactive facility without undertaking D&D. It also includes analysis of various endpoints, and the impact of using new and developing D&D technology.

In order to maintain flexibility to perform a variety of calculations, and respond to changes in input data or endpoint criteria, a computer program was written to perform the calculations required by this report.

ALTERNATIVES

There are a number of alternatives to be considered in the planning of any D&D project. The major questions addressed in this Cost/Risk/Benefit Analysis are, to what level of residual contamination should the project be carried, and what combination of D&D methods should be used.

Endpoints

The approach taken in this study was to designate residual contamination limits, or endpoints, and then analyze the effects of achieving these endpoints. To facilitate a thorough analysis four different endpoints have been investigated, each of which represents a different level of residual contamination and a different permissible future building use. The study approach permits examination of decontaminating different rooms to different endpoints. The endpoints are defined in Table 1.

All endpoints require both the removal of all contamination above the specified maximum limit, and a means to verify the removal. However, the potential exists for plutonium contamination to migrate into inaccessible areas. For this reason, a distinction has been made between an "unrestricted room" and an "unrestricted site". A room will be considered unrestricted when its surfaces are free of detectable contamination above the criteria. This includes the removal of painted over contamination and of contaminated concrete surfaces. It also includes the removal of contamination contained internally in piping, ductwork, or gloveboxes.

TABLE 1

ALTERNATIVE ENDPOINTS FOR D&D EFFORTS
UPPER LIMITS FOR VARIOUS CASES

	TYPICAL PERMISSIBLE USE	FIXED	REMOVABLE
1. UNRESTRICTED	Office Space	200 dpm/100 $cm^2 \alpha$	20 dpm/100 $cm^2 \alpha$
	Storage Space	.02 mrem/hr*	200 dpm/100 $cm^2 \beta\gamma$
2. CONDITIONAL	Analytical Lab	2000 dpm/100 $cm^2 \alpha$	20 dpm/100 $cm^2 \alpha$
	Storage of Contained Radioactive Material	.1 mrem/hr*	200 dpm/100 $cm^2 \beta\gamma$
3. RESTRICTED (USE)	Low Level Decon-tamination Facility	20,000 dpm/100 $cm^2 \alpha$	20 dpm/100 $cm^2 \alpha$
	Packaging of Low Level Radioactive Waste	.2 mrem/hr*	200 dpm/100 $cm^2 \beta\gamma$
4. RESTRICTED (NON-USE)	No Permissible Use		
	The Facility Is Isolated	Deactivated Facility - No D&D	

*Contact Reading

The site, however, can only qualify as unrestricted when the buildings have been demolished, because only thus can the absence of undetectable contamination be assured.

In defining the endpoints, the position has been taken that only in endpoints #3 and #4 can internally contaminated piping, ductwork, equipment or painted over contamination be left. The reasoning is that for endpoints #1 and #2, the use of the facility will incorporate less frequent radiological surveillance. The rooms, therefore will need to be free of potential recontamination which might go undetected for some period of time.

The Conditional endpoint will provide maximum future utilization of the structure if the Unrestricted endpoint requires dismantlement or is otherwise economically unfeasible. Although restricted to a nuclear oriented use the Conditional endpoint provides the widest choice of options within this constraint. The Conditional endpoint is more economic than the Unrestricted endpoint primarily because scheduled radiation monitoring eliminates the need for absolute assurance that all contamination has been removed.

The Restricted (Use) endpoint provides usable space at a moderate cost. The allowable uses are limited because of the higher residual contamination level. Nevertheless, if housing is needed for activities which are authorized under this endpoint, the decontamination of Z-Plant to this level may be the most economic way to provide it.

The Restricted (Non-Use) endpoint is included to permit a comparison with the other three endpoints. Restricted (Non-Use) is not a gratis option. Although no beneficial use of the facility is permitted, substantial maintenance and surveillance costs will accrue annually.

D&D Methods

The primary trade-off with respect to D&D methods in a plutonium contaminated facility is in the reduction of volumes of waste classified as transuranic. Presently waste containing more than 10 nCi/g of transuranics (TRU) requires packaging and burial in a retrievable configuration. The waste will then be retrieved in 20 years or less, reprocessed and given final disposition. The cost of retrieval and final disposition is expected to be significant enough to justify an extensive effort to reduce volumes of this waste in conjunction with the D&D project.

This study has included consideration of some of the methods being developed for the reduction of volumes of transuranic waste such as vibratory finishing and electropolishing.

The vibratory finishing process consists of placing contaminated material in a bath of abrasive media and vibrating the bath vigorously. The abrasive media are small pieces of ceramic or metal. The abrasive media grates the contamination from the surface of the material, and a fluid flow washes the loosened contamination to the bottom of the tank for collection. Before processing, the contaminated material may have to be cut up either to expose contaminated surfaces to the abrasive media, or to make the pieces small enough to fit in the bath. This process can clean any type of grossly contaminated non-combustible material to below the TRU waste limit of 10 nCi/g, but it cannot clean below a surface contamination level of about 10,000 dpm/100 $cm^2\alpha$.

Electroplishing is actually reverse metal plating. The contaminated object must be an electrical conductor and acts as the anode in an electrolytic cell. An electric current causes anodic dissolution of the surface material. All radioactive contamination on the surface or trapped in surface inperfections is released to the electrolyte.

Vibratory finishing is preferred to electropolishing because it can decontaminate material to below the 10 nCi/g TRU waste limit at less cost. However, two conditions could exist which would favor electropolishing. First, portions of metal materials such as welds may not respond to vibratory finishing because the contamination is trapped within small crevices. Such materials can be cleaned by electropolishing. Second, if regulations ever limit surface contamination levels for non-TRU waste to below 10,000 dpm/100 $cm^2\alpha$, vibratory finishing will be inadequate and electropolishing will be necessary.

DECISION CRITERIA

Three criteria were used to evaluate the available D&D alternatives. These criteria are net cost, onsite personnel exposure, and potential offsite risk. Criteria values calculated consider both the costs and risks incurred during the project as well as the ongoing costs, risks, and benefits after the project. A successful D&D project will create a reduction in annual costs and occupational exposure from ongoing maintenance and surveillance and a reduction in offsite risk.

In quantifying the cost criterion, the study considers the costs of decontamination, managing the waste generated by D&D, and surveillance and maintenance of any remaining contaminated facility, as well as the value of any facility or real estate made available by decontamination.

Occupational exposure includes exposure from the decontamination activity, from management of the generated waste, and from surveillance and maintenance of any remaining contaminated facilities.

Potential offsite risk includes the risk of release from the facility after the project, from the disposed waste, and from any accident which occurs during the decontamination. This is compared to the risk of release from the facility in its present configuration. Risk is expressed as the mathematical product of the consequences of a release in man-rem and the probability of that release.

COMPUTER PROGRAM

A computer program was developed for use in this study. Flexibility was maintained in the design of the program to handle new information on costs, criteria, or efficiency and to adapt to other facilities requiring analysis in the future. Each of the subroutines in the model is briefly discussed below.

Facilities Data

The Facilities Data file is an information storage and retrieval file. More than 100,000 data entries were needed to characterize the almost 400 rooms in Z-Plant. There are two categories of facilities data: radiological and non-radiological. The data stored in this file is used by all of the subroutines.

Offsite Dose

The Offsite Dose Model calculates the radiation hazard to the public associated with each endpoint. The hazard from one postulated event such as an earthquake is the product of three parameters: the probability of the event, the amount of contamination released by the event, and the dose to the public per unit amount of contamination released. The total hazard for each endpoint is the summation of the doses from all considered events.

D&D Tasks

The Generic D&D Tasks were created to organize and standardize all of the diverse physical work activities which are required in the decontamination of Z-Plant. For instance, as many as 19 distinct work activities might be required to remove a contaminated glovebox. These 19 work activities were combined as subtasks to create the Generic Task "Glovebox and Hood Cleanup or Removal."

The D&D Task subroutine includes all six Generic Tasks. Based on facilities data the appropriate tasks and subtasks needed to accomplish the work are chosen. The length of time that each subtask must be performed is calculated and the cost determined. The subroutine also calculates the dose to decontamination personnel, and the volume and disposal cost of the contaminated waste which is generated.

Utility Model

The Utility Model calculates the salvage value of the facility for a given endpoint. Parameters which affect the salvage value are maintenance and surveillance cost, reuse value, and demolition cost if appropriate. Since this subroutine is oriented toward future activities in the facility it is also used to calculated the anticipated occupational exposure of maintenance and surveillance personnel after the D&D project is completed.

CHARACTERIZATION

Characterization of Z-Plant was conducted specifically to provide the data needed to perform the calculations for the study.

Z-Plant's size and heterogeneity make it desirable to divide it into definite areas for the purpose of this study. Eighteen areas were chosen on the basis of structural divisions or use of the area.

The breakdown by area provides preliminary insight into the schedule of decontamination work. The size and heterogeneous nature of Z-Plant may make it advisable to schedule the decontamination work by individual area. A variety of sequences are feasible due to the independence of several of the areas. Furthermore, useful comparisons can be made between areas in Z-Plant and similar areas in other facilities.

There are two possible approaches to characterizing Z-Plant: by system or by room. The four ventilation systems and the eight drainage systems connect many different parts of the complex. The cost of decontaminating such systems can be calculated based on decontaminating the system in its entirety, or by allocating the decontamination cost on a room-by-room basis.

Since it will be desirable to know the cost of decontaminating different areas at different times to possibly different endpoints, the buildings were characterized solely on a room-by-room basis. Some data, such as decontamination levels, is particularly amenable to recording in this manner. Using this characterization method the cost, dose and waste associated with decontamination can be calculated for any room or combination of rooms.

Non-radiological characterization is the accumulation of all necessary data except that which is related to contamination levels. The nature of Z-Plant encourages a breakdown of non-radiological data into these categories: hoods, miscellaneous equipment, piping, conduit, ductwork, and the structure itself.

Radiological characterization is the accumulation of data on the amount and distribution of radioactive contamination in Z-Plant. The objective of the radiological characterization of Z-Plant was to obtain data representative of the conditions that will exist in the facility at the outset of D&D.

The startup and operation of the plutonium recovery facility (236-Z) and the "A" oxide line, and the Terminal Cleanout (TC) of process areas will impact the facility's radiological condition between now and the initiation of D&D. Terminal Cleanout will remove all economically recoverable plutonium and leave the facility in a safe, stable condition. The quantity of plutonium in gloveboxes and other process equipment will be significantly reduced and plutonium bearing liquids will be removed, but no significant dismantling of equipment or decontamination of surfaces outside of process equipment will be included. Therefore, the plutonium inventories which will exist inside process equipment at the start of D&D have been estimated based on the endpoints specified in the Terminal Cleanout Scoping Study. On the other hand, the levels of contamination which now exist outside of process equipment are considered representative of what will be present at the start of D&D.

Existing Radiation Monitoring data were used where applicable and representative interpolations were made to provide additional detail. The log books which record all "incidents" were reviewed to estimate the location of spills which have since been painted over. A limited number of new readings were taken when data was needed which would have a significant impact on the study.

D&D TASKS

The physical activities which must be accomplished to decontaminate the Z-Plant complex were categorized into the following six groups:

1. Glovebox and Hood Cleanup or Removal
2. Miscellaneous Equipment Removal
3. Piping Decontamination or Removal

4. Conduit Decontamination or Removal
5. Ventilation Decontamination or Removal
6. Structural Decontamination

Each of the Generic D&D Tasks was further divided into subtasks. The subtasks for Task 1 include all of the physical activities which might be required to prepare a grossly contaminated hood for shipment to either a burial ground or a decontamination center. All of the subtasks need not be performed on every hood. The decision to perform a subtask is determined from information provided on the data sheets.

The data sheets combined with the subtasks permits accurate assessment of the removal cost of hoods with widely varying characteristics. Some hoods are very large complexes formed by welding together a large number of smaller hoods. The four access gloveboxes to the 236-Z Building canyon each measure 2 ft x 9-1/2 ft x 60 ft. To be removed, such a hood must be cut into small manageable sections, and subtask No. 17, "Cut Partitions", would be operative. On the other hand, a small hood measuring 4 ft x 5 ft x 6 ft would be packaged intact and subtask No. 17 would not be required.

The hood characteristics also dictate the length of time during which a subtask must be performed. For instance, the basic measurement unit for the hood cleaning subtask "Decontaminate Outside" is manhours per square foot of outside surface. The total time required to perform this task is merely the product of the basic measurement unit times the surface area of the hood.

COST ESTIMATING

The subtask as described above is the fundamental cost element. The manhours required to perform a unit amount of each subtask have been estimated. Using the Facilities data the program calculates how many unit amounts of each subtask must be performed. The cost of each subtask is the product of the number of units to be performed, the manhours required per unit, and the labor cost of a manhour.

Each subtask estimate was multiplied by various load factors to account for the conditions listed below:

1. A four man team
2. Workers must wear protective clothing and face masks
3. Additional security requirements in Z-Plant
4. Support by Engineering and other functions
5. Overhead

An appropriate factor is also included to account for the cost of equipment and consumable material.

To calculate the cost to achieve an endpoint the permissible residual contamination limits for that endpoint are put into the program. The computer searches the Facilities Data file to identify any surface, whether on equipment or structure, where these limits are exceeded. The program then activates the subtasks which are needed to bring the contamination levels down to the specified permissible limit.

Excessive contamination levels are treated in two ways. First, certain overriding decision criteria have been specified for each endpoint. For instance, even though they may be acceptably clean on the outside, all gloveboxes and internally contaminated pipe and ductwork will be removed from a room when the Unrestricted endpoint is stipulated.

Second, there is a decontamination factor for each subtask. Absent any overriding decision criteria the computer will search out the subtask which has a sufficiently high decontamination factor to clean the surface to the specified permissible level. For instance, assume that a painted concrete wall is contaminated. If the contamination level on the wall is only twice the permissible level, the computer will choose the subtask "Manual Wash" which has a decontamination factor of 2. If the contamination level on the wall is 50 times the permissible limit, manual washing is not an acceptable subtask. In this case the computer will opt for subtask "Abrasive Blast" with a decontamination factor of 100.

Costs in this study will be calculated in 1979 dollars. This facilitates an objective comparison of the many different alternatives despite potentially different work sequences for each. Once an alternative is selected the work can be scheduled and the appropriate escalation factors applied.

OCCUPATIONAL EXPOSURE CALCULATIONS

The Z-Plant occupational dose is the dose received by personnel employed in the facility. Decontamination influences two components of occupational dose: the dose to maintenance and surveillance personnel and the dose to the decontamination work force. These components tend to offset one another because decontamination will lower the subsequent maintenance and surveillance exposure.

The maintenance and surveillance dose is computed in the Utility Model subroutine as an annual whole body dose rate. First, a residual contamination endpoint level is stipulated from which an average dose rate (man-rem/hour) in the facility can be estimated. Next, the number of hours per year that maintenance and surveillance personnel must spend in the facility for the endpoint is determined. The annual dose rate (man-rem/year) is the product of the average dose rate (man-rem/hour) and the average annual maintenance and surveillance exposure time (hours/year).

The decontamination worker dose is calculated for each subtask by the D&D Generic Task subroutines. Both a whole body dose and an inhalation dose are computed. Each dose is the product of the number of manhours required to perform the subtask times the respective exposure rates (whole body and inhalation) which exist while the work is being done.

The whole body dose is calculated using two dose rates: a room dose rate and a proximity dose rate. The room dose rate, derived from radiation monitoring records, is the average whole body dose rate which exists throughout the room. The proximity dose rate, derived from the radioactive isotope inventory in each glovebox, is the whole body dose rate which exists close to the gloveboxes.

The whole body dose is the sum of the room dose and the proximity dose. The room dose is the product of the room dose rate times the total number of hours that the room is occupied. The proximity dose is the product of the proximity dose rate times the total number of hours spent working near the gloveboxes. Both dose rates are included in the Facilities Data file and are appropriately reduced as decontamination work progresses in order to account for the removal of part of the radiation source.

An inhalation dose is caused by the contamination released to the air in the work area by each subtask. The level of airborne contamination depends upon the concentration of the surface contamination which is disturbed, the fraction of the surface contamination which is released to the air by the disturbance, the volume of the room, and the number of air changes per hour. The inhalation dose depends upon the level of airborne contamination, its isotopic makeup, the breathing rate of the workers, the protection factor offered by their protective masks, and the total time of exposure.

OFFSITE RISK

The various alternatives in the decontamination and decommissioning of Z-Plant can affect the radiological risk of the offsite population in different ways.

Risk is a quantified measure of the radiological hazard to the public resulting from a particular operation or facility. In this case the objective is to determine the effect of the various alternatives considered in the study on offsite risk.

The base case in this study is a contaminated deactivated facility. The presence of radionuclides in the facility and the potential for an accident which would breach the containment and release those radionuclides to the environment represents a hazard to the public.

Decontamination operations will reduce risk by placing the radionuclides in the more stable configuration of buried waste. A complete view of the impact of D&D on offsite risk must consider the reduction of risk represented by the facility, the additional risk created by the D&D operations themselves, and the risk associated with the waste after it is buried. In this study the risk associated with the inactive facility and buried waste in the various final configurations considered is called the innate residual offsite risk. The risk resulting from the D&D operations themselves is called the D&D related offsite risk.

Risk is expressed as the mathematical product of the probability of a given accident and the predicted consequences of that accident. Total risk is the sum of the risks associated with all considered accidents.

Accident probabilities and releases for various accidents were estimated using available literature dealing with Z-Plant, Hanford burial grounds and D&D operations.

Dose conversion factors, which estimate the total population exposure from a given release were obtained using a computer program operated by Pacific Northwest Laboratories at Hanford.

Since Z-Plant is contaminated with plutonium and americium, significant decay of the radionuclides will not occur during the life of the buildings. Trying to predict risk beyond 50 years is difficult since the building will have begun to deteriorate significantly, the local demography will have likely changed, and the meteorology may have changed. This study, therefore, does not attempt to quantify risk in the long term. It is very reasonable however to assume that the building will deteriorate long before significant decay of the radionuclides occurs.

RESULTS

Final results from the analysis are not yet available.

When they are finalized, the values generated by this report will facilitate the following comparisons for the various endpoints.

Annual risk reduction versus one-time risk increase

Annual maintenance and surveillance exposure versus D&D exposure

Net radiological hazard decrease versus D&D cost

Annual maintenance and surveillance cost decrease versus D&D cost.

These comparisons will be useful in making an overall recommendation for the plant. Data of this type should also be particularly useful in prioritizing potential D&D projects where limited funding is available.

OMRE AND HALLAM DECOMMISSIONING PROJECTS AT THE IDAHO NATIONAL ENGINEERING LABORATORY

R. H. Meservey

EG&G Idaho, Inc.

P.O. Box 1625 Idaho Falls, ID 83401

INTRODUCTION

Although we tend to think of decontamination and decommissioning (D&D) as a new program at the Idaho National Engineering Laboratory (INEL), such activities actually date back many years. Several reactors including BORAX-I, SL-I, and EBR-I have been decommissioned in the past. These projects were done individually, however, without the planning of an integrated decommissioning program for all excess contaminated INEL facilities. Such a program has been initiated, and the results of its first two projects, OMRE and Hallam, form the basis of this paper.

Long-range planning for INEL decommissioning projects has been completed and is reported by J. A. Chapin and R. E. Hine in TREE-1250, Decontamination and Decommissioning Long-Range Plan - Idaho National Engineering Laboratory (June 1978).

This document contains the rationale for selecting and assigning priorities to decommissioning projects, and for selecting the proper decommissioning option. It also contains the criteria we use for releasing sites following the decommissioning work.

Hallam Project

The Hallam Nuclear Power Facility at Hallam, Nebraska, was dismantled between 1964 and 1968. At that time, several major components (heat exchangers, primary pumps, evaporators, air eliminators, and superheaters) were shipped to the INEL for storage. Until October 1977, the components were maintained under a nitrogen purge to protect and stabilize the small amounts of contaminated sodium which they contained.

The object of our Hallam D&D project was to design and build a processing system to react sodium in the components and then to remove from them the resulting contaminated caustic solution. The contaminated caustic was to be neutralized and disposed of in a settling pond. If possible, the components were to be decontaminated and returned to the surplus materials list.

The sodium processing system was designed to use the wet nitrogen process. That is, the components were purged with gaseous nitrogen to eliminate any oxygen in the system. Steam was then introduced into the nitrogen purge, thus permitting the sodium-water reaction to take place. Pressures, temperatures, and hydrogen and oxygen content in the exhaust gas stream were monitored throughout the processing operation. Reaction rate was controlled by throttling the steam flow rate. A schematic diagram of the sodium processing system is shown in Figure 1.

Individual components were processed with nitrogen and steam until the hydrogen concentration in the exhaust stream had decreased to a value of less than 4%. At that time, the nitrogen/steam flow was stopped and the component was filled with water. The water filling operation was approached

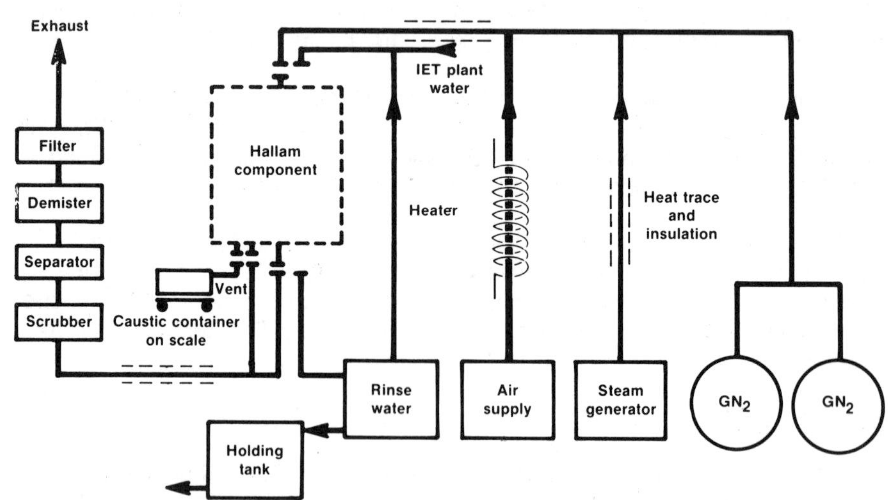

Fig. 1 Hallam sodium processing schematic.

carefully with only small amounts (380 l) added at a time during the initial steps. Monitoring of hydrogen concentration, pressure, and temperature data dictated the rate at which increments of water were added to the vessel. If after several incremental additions of water no further evidence of sodium-water reactions could be detected, the component was slowly filled with water. As a general rule, the components were allowed to stand full of water for several days before an attempt was made to neutralize and drain the caustic solution.

Neutralization of these large amounts of highly caustic (pH 11-13) solutions presented its own unique problems. A system was designed to inject small quantities of concentrated sulphuric acid into a water-filled rinse tank. The rinse tank water was then circulated through the component being neutralized. This process was repeated until samples taken from the circulating water indicated that it had been neutralized. Some components required as much as 380 l of sulphuric acid to neutralize the caustic solution generated durinq processing of this sodium.

Several components contained relatively large amounts (180 kg) of sodium, while others appeared to contain only film residues. Thus, processing parameters varied widely from component to component. This was reflected in the time required to process a component, as well as the pressure, temperature, and hydrogen concentration generated during processing. The average time required to process a component was in the order of 12 hours, although some took as long as 18 hours. This time period included only the sodium/steam reaction portion of the processing, and not the neutralization procedure. Neutralizing required approximately an additional 8 hours per component.

The most troublesome problem that developed during the course of the sodium processing operation resulted from the highly corrosive effects of the caustic. This resulted in repeated leaks throughout the system as the caustic corroded various components. Thus, many parts of the system required repair or replacement, and much time was spent neutralizing and cleaning up caustic. This condition demonstrated the need for welded joints and caustic approved materials when building sodium processing systems.

Following the sodium processing, neutralizing, rinsing, and draining operations, the Hallam Reactor components were opened for inspection. This inspection assessed the effectiveness of the processing operation and determined the final state, with respect to retained sodium and radioactive contamination, of the component internals.

Prior to actually opening the components, they were again filled with water and allowed to stand in this filled condition for several days. The components (vessels) were then drained and cut open. Cutting was accomplished using an air-operated (pneumatic) power saw which automatically traveled along a chain placed around the vessels. Cutting rate for these heavy metal vessels was about 2.5 cm/min.

After the vessels were cut with the pneumatic saw, tube bundles were then cut using a specially built metal cutting band saw. All internal tubes were visually inspected for retained sodium. None of the conventional tubes contained sodium. However, special tube connections in the superheaters and steam generators did contain unreacted sodium. Connections in the form of metal bellows which fit tightly over the end of the tubes also contained sodium trapped between the inside diameter of the bellows and the outside diameter of the tubes. These could not be reached by water during the processing, and thus had not reacted. A photograph of one of these bellows assemblies which has been cut open is shown in Figure 2.

Removal and sectioning of the bellows assemblies revealed small amounts of unreacted sodium such as shown in this figure. With the exception of these small amounts of entrapped sodium, the remainder of the vessel internals were sodium free.

Components such as the heat exchangers and primary pump cases, which were radiologically contaminated, were also inspected to determine the extent of this type of contamination. In general, these components were contaminated with ^{60}Co and ^{137}Cs to levels of less than 1 mR/h. Because of the complex geometry (tube bundles) within these reactor components, decontamination would be either impossible or at least impractical. For this reason, no attempt was made to decontaminate those components which were radiologically contaminated.

As a result of the postprocessing inspection it was concluded that the wet nitrogen process used was adequate for gross removal of sodium. However, sodium trapped in small annuluses may not be reacted. Thus, if equipment having complex geometries or close tolerance components is to be processed with wet nitrogen, a thorough inspection should be performed prior to releasing the equipment for unrestricted use.

Based on the results of the postprocessing inspection, decisions were made relative to the final disposition of the Hallam components. With the exception of some pump components,

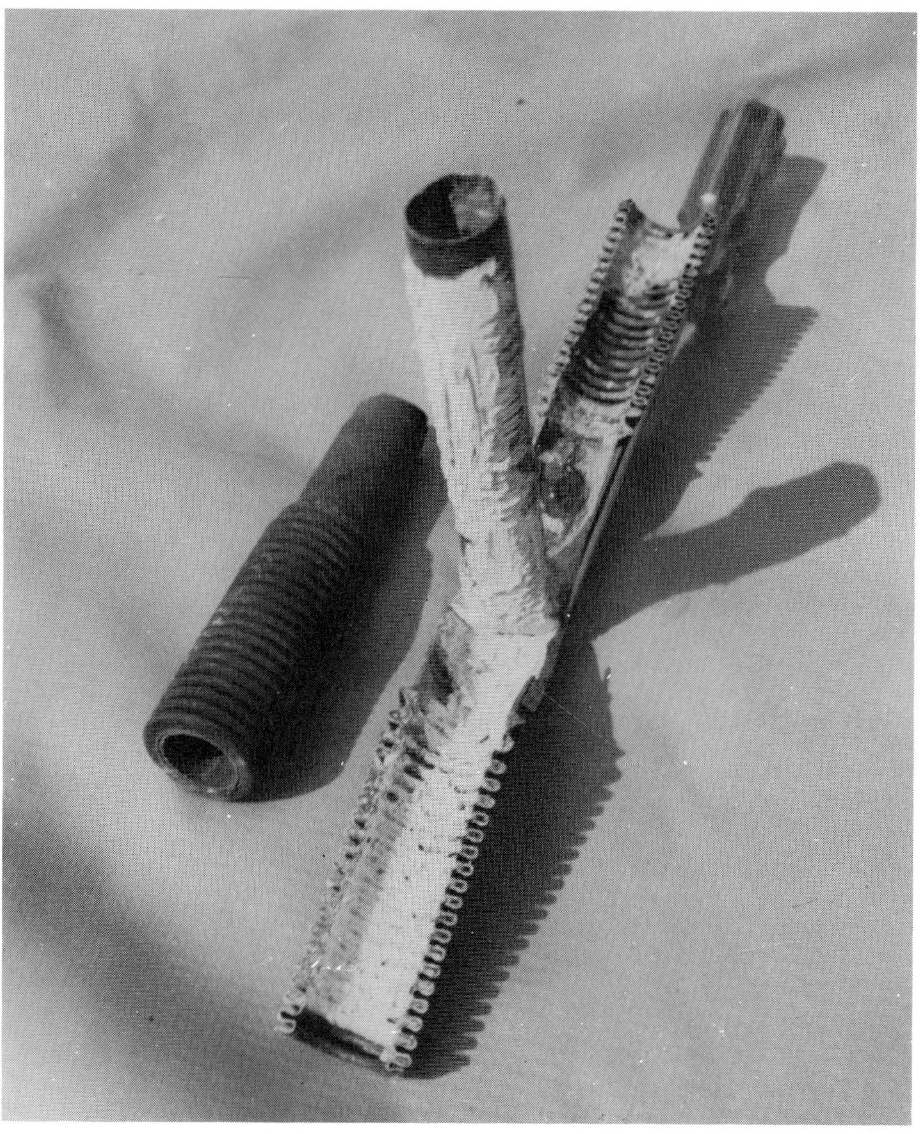

Fig. 2 Bellows assembly sectioned to show unreacted sodium.

all radioactively contaminated components were shipped to the Radioactive Waste Management Complex (RWMC) at the INEL for disposal. This included five intermediate heat exchangers, three primary pump cases, and a pump diffuser and impeller.

Two pump diffusers and two impellers were obtained by Atomics International and shipped to their facility in Canoga Park, California. All of these components contained residual beta-gamma contamination at levels to about 1 mR/h.

The remaining components (three superheaters and three steam generators) did not contain radioactive contamination, but did contain small amounts of residual sodium. Although only small amounts of sodium were present in these components, the location and construction of the components made it impractical to remove. Thus these components were stored at an INEL landfill area and were not released for unrestricted use.

External structures and parts that were not radioactively contaminated were cut from the components and disposed of as scrap metal. This resulted in approximately 45 000 kg of salvaged metal.

Actual costs turned out to be reasonably close (within 7%) of the estimated cost for the project. Our schedule estimates were also very close and the actual processing was completed slightly ahead of schedule. A cost/milestone summary schedule for this project is shown in Figure 3.

<u>Organic Moderated Reactor Experiment (OMRE)</u>

This facility was designed to investigate and develop the organic coolant technology and was operated at the INEL from 1957 until 1963. Following final reactor shutdown, the nuclear fuel and reactor vessel internals were removed and the organic coolant was drained from all systems. The facility remained in this deactivated condition until October 1977. At that time, a decommissioning plan was prepared and disassembly work was started.

Decontamination and decommissioning of the OMRE facility involved removal of the reactor vessel, all associated systems and buildings, all roadways, parking lots, power poles, and fences. The area will be restored by final grading and seeding with crested wheat grass and will be completed during FY 1979. An aerial view of the OMRE facility is shown in Figure 4.

The major portion of the ORME facility contained relatively low contamination levels of only a few mR/h. Only the vessel

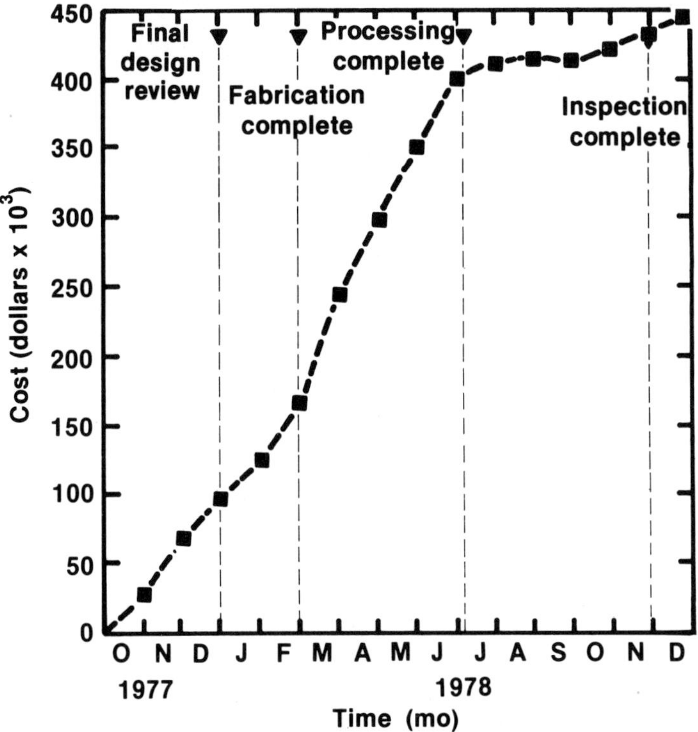

Fig. 3 Hallam Project cost and schedule.

represented a severe radiation hazard (350 R). Other hazards existed, however, in the disassembly of the OMRE facility. In addition to the normal industrial hazards, most of the facility contained the organic coolant Santowax which was flammable. During reactor shutdown, xylene was used in the fuel wash system and could potentially form explosive mixtures. Almost all piping was covered with asbestos which in itself represents a health hazard. Thus, decommissioning of the OMRE facility presented a variety of safety-related problems.

Disassembly work has gone essentially as planned and at the present time, all facilities and components have been removed. Only final grading and seeding remains to complete this project. All noncontaminated material having any salvage value was collected for surplus sale. Nonsalvageable, noncontaminated

Fig. 4 Aerial view of OMRE.

material was sent to a sanitary landfill. All contaminated material was shipped to the RWMC for disposal. Where possible, the contaminated material was cut or broken and placed in 122 x 122 x 244-cm plywood boxes before shipment to the RWMC. A summary of waste volume generated by this project is shown in Table 1.

A full 2 years, including necessary wintertime shutdowns, have been required to decommission the OMRE facility. Although most of the facility was removed during FY 1978, some of the most interesting challenges occurred this year while trying to clean up and release the site. Release criteria from our long-range plan are summarized in Table 2.

Table 1. OMRE Waste Summary

Type	Clean (m^3)	Contaminated (m^3)
Metallic	24.4	605.4
Concrete	3.1	16.0
Soil	--	203.0
TOTAL:	27.5	824.4

Table 2. Acceptance Contamination Levels for Unrestricted Use

Type of Contamination		Acceptable Levels
Surface:		
Removable	α	<20 dpm/dm^2
	β-γ	<200 dpm/dm^2
Fixed	α	<200 dpm/dm^2
	β-γ	<2000 dpm/dm^2
Soil:	α	<100 pCi/g in first 10 cm or 20 cpm above background
	β-γ	<1 nCi/g or <500 cpm

These acceptable levels, as listed in Table 2, will be used as guidelines for unrestricted use and the criteria to be used on specific D&D projects will be evaluated on an individual basis. The criteria, as stated, are still under review and as data are developed these criteria will be revised.

As part of the cleaning effort, high explosives were used to remove concrete and lava rock containing induced activity. Presently, we are processing the release paper and preparing a final project report.

Cost, schedule, and milestone data for the OMRE project are shown in Figure 5. As can be seen, the cost of decommissioning this facility and returning the site to its natural state was about $500,000. A photograph of the site after the major work had been completed, but before final grading and seeding, is shown in Figure 6. A complete report covering this project will be published in the near future.

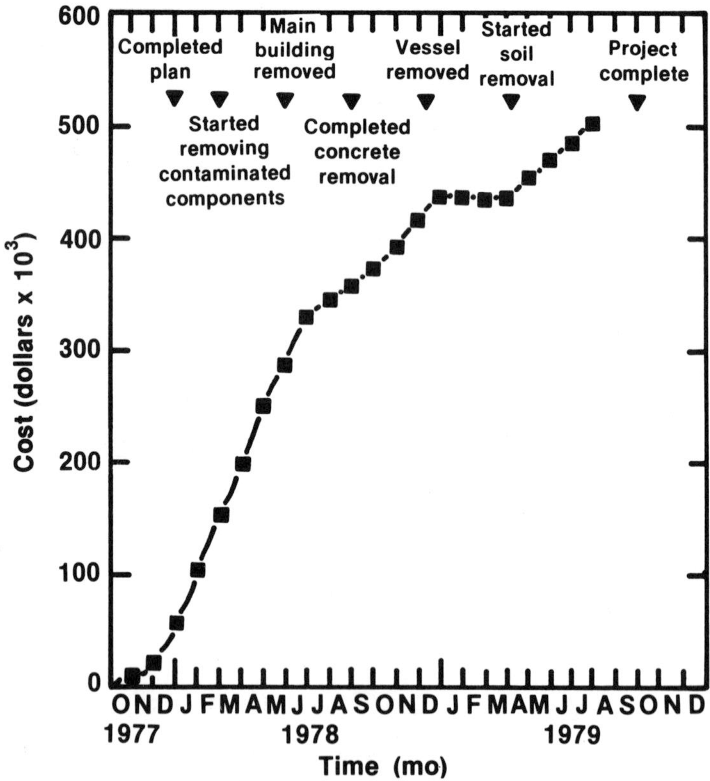

Fig. 5 OMRE Project cost and schedule.

Fig. 6 OMRE area following removal of the buildings and underground equipment.

CONCLUSIONS

We are pleased with the experience gained in performing the OMRE and Hallam decommissioning projects and believe it to be a valuable contribution to the nation's decommissioning effort. The quality of our planning documents and safety evaluations has been improved through the experience gained in these projects. Safe disassembly and handling procedures were developed, thus lending confidence in completing future, more difficult, decommissioning projects. Several new decommissioning plans have been written, and we plan to start new projects as funding becomes available.

THREE VESSEL REPLACEMENTS AT CHALK RIVER

J.W. Logie

Atomic Energy of Canada Limited

ABSTRACT

Since 1947, when Canada's first research reactor, NRX, went critical, CRNL has been faced with the task of partially dismantling nuclear research reactors three times. In each case, twice for NRX and once for NRU, the object was the replacement of the calandria or vessel. Planning and techniques have improved from the hectic days of the first NRX calandria change in 1953, to the relatively effortless NRU vessel change in 1972. Of the many dismantling lessons learned, three stand out:

- careful, detailed planning with time for intensive review, and practice on full-scale mockups

- procedures designed to minimize radiation dose to personnel

- design of special tooling to carry out dismantling operations effectively and safely

We have gained enough confidence in our ability to deal with dismantling operations that power reactor dismantling is viewed without undue pessimism. That is not to say that power reactors hold fewer difficult problems for dismantling teams than do our vessels for us. We are well aware of the much larger radiation fields associated with power reactor components, and the much greater volumes of radioactive material that future teams will have to consider. The application of techniques already at hand, the introduction of methods now being worked out, and the design of future reactors to ease radiation problems, will overcome many of the difficulties we can foresee.

INTRODUCTION

Atomic Energy of Canada Limited's Chalk River Nuclear Laboratories are located on the Ottawa river approximately 200 km northwest of Ottawa. Dismantling operations on research reactors have been carried out three times at this site, each time to replace the calandria or vessel. The calandria of NRX (National Research Experimental) has been replaced twice, and the vessel of NRU (National Research Universal) once. In any nuclear reactor dismantling operation, radiation and contamination must be overcome safely and effectively to get the job done. The accident to NRX in 1952 provided the setting in which the weapons of planning, rehearsal, special tooling for remote handling, and radiation and contamination control, were used to combat these conditions for the first time on a large scale in Canada.

When procedures are carefully planned to reduce personnel radiation dose as much as possible, when extensive mockup work is done to prove out sophisticated tooling, and when training programs leave no important aspect of the job untried, then dismantling of nuclear reactors can be accomplished safely and effectively. Although the scope of dismantling a large power reactor will be wider than that of the research reactors, application of techniques already at hand, introduction of methods being worked out, and the design of future reactors to simplify dismantling, should make this event no more hazardous or difficult.

REPLACEMENT OF NRX CALANDRIA 1952-54

Chalk River's NRX reactor is a vertical heavy-water moderated, light-water cooled, engineering test and research facility that achieved first criticality in 1947, and had operated at power levels up to 30 MW thermal prior to December 12, 1952. On that day, with the reactor shut down, critical height measurements were being made with a number of fuelled sites operating with reduced coolant flow, which was a common but not standard condition.[1] (See Figure 1.) During these measurements, a power surge occurred, causing varying degrees of damage to 22 fuel rods, and leakage of light water coolant into the lower header room. Fuel sheaths, fuel rod flow tubes, and calandria tubes ruptured; uranium metal was exposed to light water coolant, and fission products were leached into the water flowing from the core. Contamination was widespread in the building, and personnel immediately donned full-face army respirators.

The first major obstacle to be overcome was the flooding of the NRX basements with water laden with fission products. A series of operational moves reduced the flood-rate from 1400 L/min to 64 L/min over the two-week period following the accident. Water was pumped to outside temporary storage tanks while a pipe-line was hurriedly constructed to a clay/sand site 2 km inland from the

Ottawa river. Forty-five million litres, containing 10,000 curies of long-lived fission products, were eventually pumped to the disposal area. An intensive study carried out in 1973 indicated no radionuclide movement from this source into a closely-monitored nearby swamp and lake system.[2] Continued testing to this date confirms the findings of 1973.

Repair of the calandria was optimistically anticipated until several inspections of the reactor core, through ruptures in calandria tubes, showed that this would not be possible. Calandria replacement then became the only workable alternative, even though NRX had not been designed with this contingency in mind, and no plan had been devised to carry out such a task.

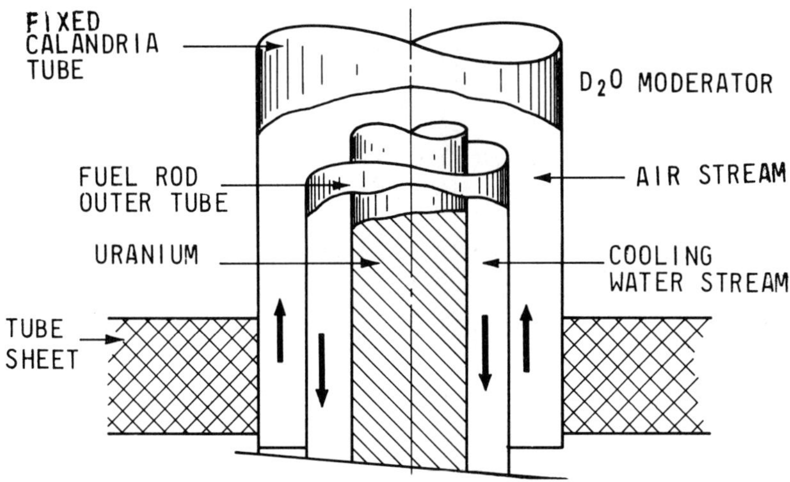

Fig. 1 NRX Fuel 1952

Planning for the removal of the calandria was started soon after calandria inspection, and received support from not only all Chalk River departments, but also from such groups as the U.S. Atomic Energy Commission, and the U.S. Naval Radiological Defence Laboratory. These units shared their expertise in dealing with high radiation fields and contamination, and were integrated with Chalk River and Canadian Armed Forces personnel during both the reactor dismantling and calandria removal stages.

The NRX calandria, an aluminum vessel weighing 3,540 kg, is 2.58 m in diameter, 3.35 m high, and is penetrated by 199 tubes for fuel rods and experimental facilities. These calandria tubes are rolled into top and bottom steel tube sheets, which in turn form part of the calandria assembly. Above the calandria are three 5,000 to 14,500 kg water-cooled thermal shields, four 16,300 kg steel and concrete biological shields, a 10.2 cm thick steel master plate, and the ring-header assembly that supplies cooling water to the fuel rods. A 45.7 cm thick steel deck plate provides the final covering and working platform.

Top shielding components were removed in radiation fields increasing to 200 R/hour on the underside of the last thermal shield; and a rotatable lead shield, constructed with a sliding door in a radial slot, was installed in its place to give access to the damaged fuel. Removal of this fuel was accomplished with great difficulty,

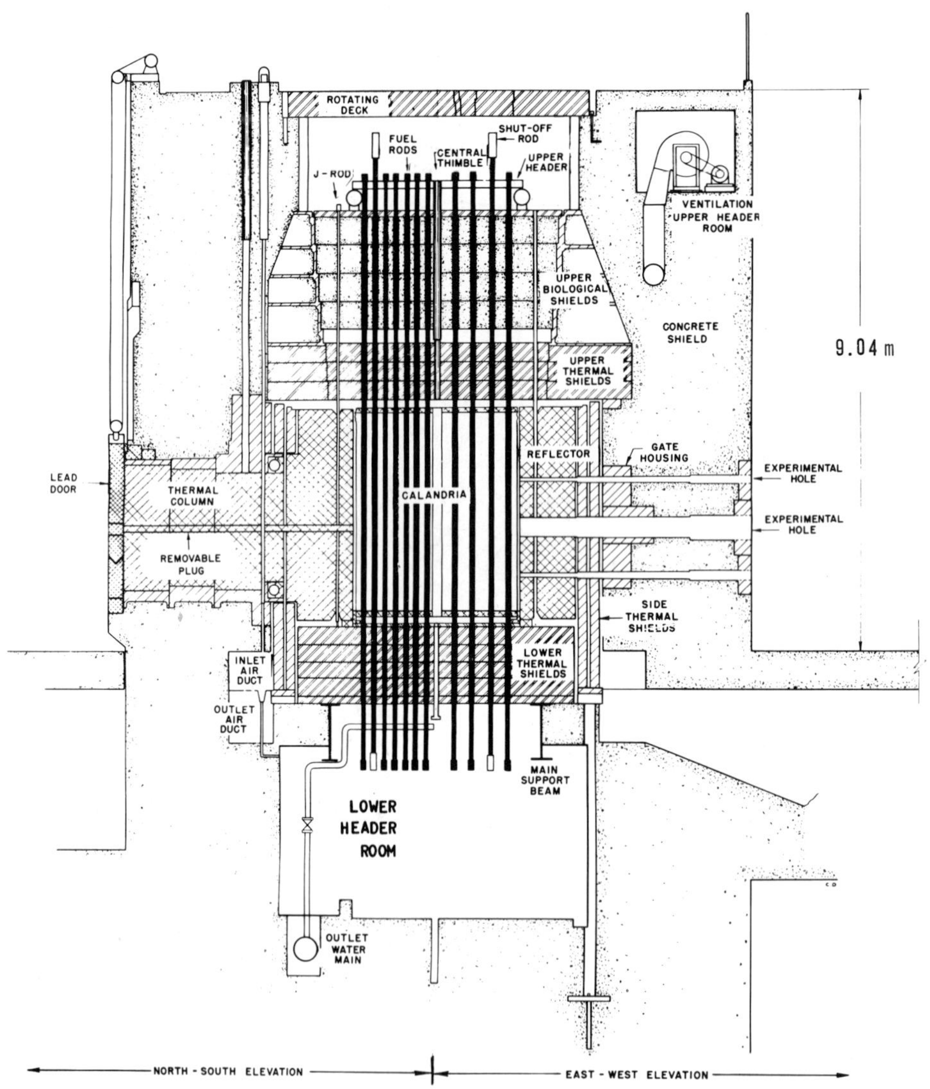

Fig. 2 NRX Elevation

and some unrecoverable debris from the ruptured rods was left to g to disposal with the calandria.[3]

In preparation for calandria removal, plywood mockups of the reactor were constructed and used as training aids. A weighted full-size steel shell, representing the calandria, was then fabricated, and further training and testing were carried out. A final dummy run of the entire removal operation, including transportation to the burial ground, was performed on the night before the real thing. (See Figure 3 for removal schematic.) Calandria removal was completed without difficulty, almost an anti-climax. A lifting adapter, with dogs to grip the underside of the top calandria tube sheet when a lifting strain was applied, was installed in the calandria. The overhead crane, remotely operated, engaged the lifting adapter and removed the calandria into a canvas bag on an upturned skid, which in turn was lowered to the horizontal plane. A grader then towed the whole assembly, including the lifting adapter, to the burial area.

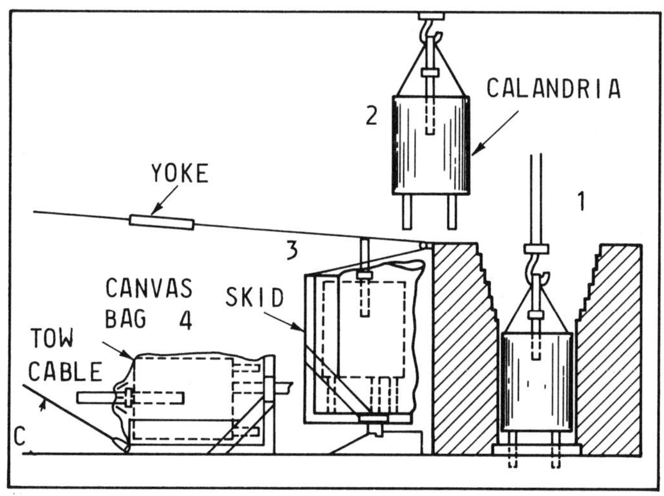

Fig. 3 NRX 1 Calandria Removal Sequence

All material from dismantling operations was bagged in plastic before being transferred to either the site decontamination center for small items such as valves and instruments, or to a large room below the reactor hall main floor for larger components. Storage followed decontamination. The open-pore concrete of the reactor building created a significant problem for decontamination work. Contaminated water was easily absorbed by the concrete, which defied attempts at 100% activity removal. A fresh layer of concrete 15 cm thicker than the depth chipped and ground off was poured to seal the remaining activity, and to act as shielding. Sealant was applied to all building concrete after completion of decontamination work.

In spite of the difficult circumstances surrounding the building cleanup and removal of the calandria, maximum individual radiation dose was held to 17 Rem, and the majority of the work-force received less than 3.9 Rem. All told, approximately 2,600 Rem was accumulated by over 1,100 persons who worked in NRX during this period, although essentially no increase in radiation exposure was incurred until cleanup operations had started.

Table I Selected Radiation Levels NRX 1952

Immediately Following Accident	
Control room door to top of reactor	100 mR/hr
Main floor around reactor	200-1000 mR/hr
Foot of stairs into basement	5-10 R/hr
Directly under reactor at waist level	10 R/hr
During Dismantling	
Bottom of second thermal shield	4 R/hr (c)
Top of third thermal shield	5 R/hr (c)
Bottom of third thermal shield	200 R/hr (c)
During Fuel Removal	
At 10 cm above rotatable lead shield	1-3 R/hr
Over certain holes at top of shield	20 R/hr
Certain holes at top calandria tube sheet	200 R/hr
During Calandria Removal	
Top tube sheet after removal	20 R/hr (c)
Vessel Wall at 3.05 m	65 R/hr

NOTE: (c) indicates contact measurement

Lessons Learned

1. Design: Reactor design should allow for ease of dismantling. Not only will this feature reduce dismantling time and expense at the end of the reactor's life, but perhaps more importantly, return-to-service time can be shortened following major repairs.

2. Planning: Intelligent planning of dismantling operations is mandatory and should come under frequent review. This is especially true in the case of research reactors, where components and superficial structure may be altered almost continually. Full-scale rehearsals, using mockups, must be carried out on difficult operations to uncover unforeseen problems, and to allow accurate procedures to be written. Final full-scale mockup runs should be performed shortly before the actual event, to ensure that disman-

tling crews are confident both in their abilities and the procedures they are expected to follow.

3. Ventilation: Extra ventilation outlets should be installed to cover each area in which dismantling or decontamination operations are to be carried out. Exhaust and filter systems already installed can be used if surplus capacity exists, or special temporary ones will have to be constructed. Logic dictates that both extra outlets and capacity should be designed into the main system, but a survey of available outlets should be a separate item of the pre-dismantling procedures to avoid time- and money-wasting delays.

4. Surfaces: All surfaces in the reactor environment should be sealed so that contamination remains on the surface and therefore easily removable. Particular concern must be paid to open-pore materials such as concrete. Contaminated water easily penetrates this kind of material, but its removal is difficult, costly and time-consuming.

SECOND NRX CALANDRIA REPLACEMENT 1970

The first corrosive failure of an aluminum calandria tube in the new calandria occurred in 1963, and by 1970, the failure rate had become exponential. Failures occur in the helium cover gas space between the top tube sheet and the surface of the moderator. As each tube failed, it was plugged, thus removing it from the fuel loading pattern; and just before reactor shutdown for calandria change, the physics of reactor loading was becoming very difficult, and experiments were suffering. Although individual tubes had been replaced in the past, fully six days of continuous work had been required, and in 1970, when tube failures had risen to 29, and were occurring at an increasing rate, calandria replacement was the only logical recourse.

When the first calandria tube failed, a committee of representatives from involved Chalk River departments was established to write procedures covering the "Calandria Removal Phase". These procedures described step-by-step removal of components and calandria. A critical path diagram was then drawn up, while schedules were continually up-dated to consider changes to components and operating conditions. Training of personnel was carried out during the month preceding the shutdown. Full-scale mockups were used to determine problem areas, and modifications were introduced based on these tests. Television cameras were used during the calandria removal operation itself, and effectively reduced the radiation dose to involved personnel.

The only major difficulty encountered during the calandria removal phase was the initial inability to free one of the upper thermal shields. This was an aluminum component, located just above the calandria, that weighed approximately 5,000 kg when filled with water. Several unsuccessful attempts were made to free the shield, employing such techniques as jackhammer, application of a cutting solvent, hacksawing around the periphery to clear corrosion, and finally, jacking to a limit of 29,000 kg. Success was eventually achieved by circulating a cold water-antifreeze mixture through pipes in the shield. Shrinkage was sufficient to free the component with a pull of about 23,000 kg. Inspection later revealed massive corrosion buildup, and a peripheral weld that stood proud of the shield by two millimeters. This side weld is, of course, non-existent on the new shield.

The second NRX calandria was in all major respects identical to the first one, except that associated radiation and contamination levels were much reduced from those of 1953. A full-scale dummy run was completed 40 days before the actual removal date, and further improvements and modifications were made.

The removal operation was significantly simpler than in 1953. A lifting tool was installed in the central calandria position, and was lowered to engage the bottom tube sheet. The calandria was then removed from the reactor by remote crane operation and deposited on a skid. The lifting tool was disengaged from the tube sheet and removed, and a canvas shroud was lowered over the calandria for contamination containment. Once again the grader towed the assembly to the burial ground. Calandria removal was completed with a minimum of difficulty, and elapsed time from first hooking-on to arrival at the burial site, approximately 2.5 km west of the Ottawa river, was just less than two hours. The calandria was buried under 3 m of sand and earth to reduce radiation to less than 2 mR

Fig. 4 Vessel on Skid

Fig. 5 Canvas in Place

per hour on top of the burial location. Measurements taken immediately after transfer of the calandria indicated no detectable contamination on the roadway leading to the burial site.

Decontamination operations during the second NRX calandria replacement were carried out on two main fronts. All small pieces, such as valves and piping, were bagged, sealed, labeled and sent directly to a decontamination building on the project. All material of this type was processed through a clearing area on the main floor of the reactor building. Meticulous records were kept. Larger components such as the biological shields were cleaned as much as possible in situ, wrapped in plastic on removal, and delivered to the decontamination area for further cleaning.

Radiation Dose Control

The individual's radiation dose was monitored in three ways:

- The normally-carried photo badge was processed every second week as usual.
- A special badge containing a thermoluminescent dosimeter (TLD) was worn by all personnel working on any phase of the replacement. These badges were processed at the end of the program only, to provide an overall picture of radiation dose directly attributable to replacement work.
- A pocket electroscope dosimeter was worn to provide short-term indication of dose received. Information from this source was tabulated daily to reduce the possibility of overexposure. Radiation costs for each phase of the operation were immediately available when the daily dose was recorded with the job number.

Total radiation cost of calandria removal operations was 117 Rem. Only 12 workers, all highly skilled tradesmen, who were involved in the more difficult and demanding operations, received more than the maximum permissable yearly dose of 5 Rem, but none reached 7 Rem.

Lessons Learned or Reinforced

1. Continuous Review of Progress: each week a general meeting was held to discuss the previous week's program, the problems encountered, and the following week's projected work. This method of forced interaction ensured that the various departments involved in the operation were given an opportunity to participate in any discussion touching on their area of responsibility, and brainstorm problems as they arose.

2. Work Schedule: a non-continuous work schedule was evolved, covering two shifts per day, five days per week. This approach allowed time for cleanup and local decontamination,

operations that strongly contributed to safer working conditions, and that reduced personnel radiation dose from low level sources.

3. Shielding Devices: whenever feasible, shielding devices should be over-designed to reduce even further low-level radiation contribution to total dose.

4. Radiation Dose Control: more attention must be paid to the higher-than-normal but still relatively low general fields to be found in a dismantling situation. Worker utilization is more often determined by continuous exposure to background radiation than by a calculated one-time exposure to a relatively high field.

5. Tool Preparation: dismantling tools and procedures for their use must be prepared well in advance of usage to coincide with mockup work and to ensure that time is not wasted during the real operation.

NRU VESSEL REPLACEMENT 1972-1974

NRU is a vertical heavy water moderated and cooled engineering test reactor located at Chalk River Nuclear Laboratories. The reactor went critical in late 1957 and had operated at power levels up to 220 MW thermal prior to June 1972, when it was shut down for replacement of its aluminum vessel. Reactor design includes an annular space between the vessel and a light water reflector shell which in turn forms part of the vessel assembly. This annulus had been penetrated by tubes that had held rods containing various materials for isotope production. A CO_2 atmosphere had been maintained in the annulus until the early 1960's, when annulus integrity was breached by the failure of a thin aluminum window on the reflector side. Leakage of air into the annulus from cooling passages in the reactor structure, coupled with light water leakage from the reflector, water that underwent radiolytic decomposition, created a hostile acid environment that eventually corroded holes in the vessel wall. In early 1971, one such hole was plugged, but in February 1972, more holes were discovered, and although these were plugged to permit reactor operation to continue, it was apparent that extended vessel life was highly doubtful.

Unlike NRX, NRU had been designed with vessel replacement in mind, and preparations were started in 1960, when a contract to write a Vessel Replacement Manual was awarded to an outside engineering firm. Writing and revisions continued until 1968 when the manual was considered essentially complete. Based on the manual, a vessel change arrow diagram was drawn, from which a Program Evaluation and Review Technique diagram was constructed. Continual revision to the PERT diagram produced a final copy in March 1972, just in time for the shutdown.

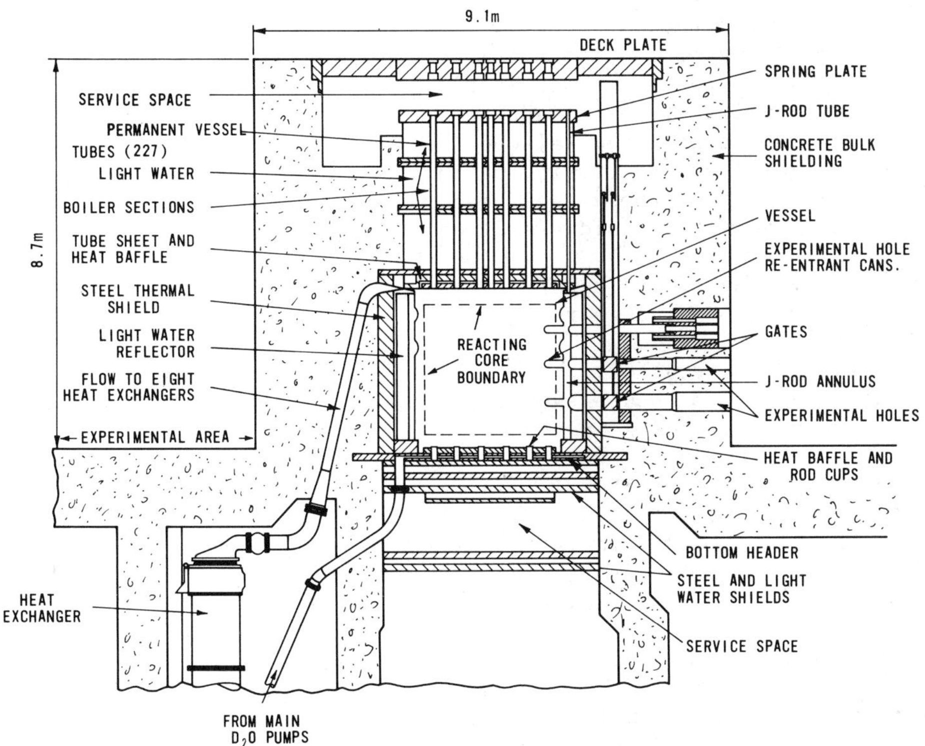

Fig. 6 NRU Elevation

In 1971, the Mechanical Services department, who were responsible for the technical aspects of the operation, gave 16 one-hour lectures to Operations and Radiation and Industrial Safety personnel. Mockups had been built to simulate reactor components, and the first full-scale dummy run was carried out in early 1967, to test transporting the dummy vessel to the burial ground under winter conditions. Review of this and other tests produced major changes to both equipment and procedures, resulting in final acceptance of the overall program in early 1971.

All special tooling had been built and tested by 1967. This large lead-time over end-use allowed extended review and worthwhile modifications. Tools for use in reactors are necessarily unique, and a great deal of effort is expended in the design and proving of them, and the training of operating crews. This was true in the case of NRU and dividends collected repaid the initial effort many times over.

Final preparations for vessel removal went smoothly, following these steps:

- unloading of fuel and removal of other vertical assemblies

from the reactor core
- draining of the coolant-moderator and drying of system piping and the vessel itself
- vacuuming of vessel and annulus to avoid spread of contamination during removal operations
- removal of interfering piping in the lower service space
- breaking of the vessel/header seals
- measurement of old vessel weight and calculation of center of gravity
- lowering of the upper bottom shield/vessel assembly into the vessel change room

Considerable effort had been expended while the vessel remained in the reactor to ensure that the correct lifting point was used. Eight weighings, using instrumented (load cells) lowering rods, were done to produce a final lifting point 44.5 cm from the vessel's geometric center. This compared very favourably with the new vessel's lifting point. Vessel weight was measured at 11,560 kg. The lifting adapter, with its own center of gravity, was factored in, a counterweight was calculated and installed, and a new lifting point was established. Cans had been welded to the top of the lifting adapter in various locations to receive lead shot to correct the calculated lifting point, but their use was not required. During the three weeks immediately preceding vessel removal, final full-scale dummy runs were completed, including a trip to the disposal area, and television and communications systems were commissioned.

Fig. 7 Dummy Run

Fig. 8 Old Vessel Ready

The design of NRU enables the vessel to be withdrawn from the bottom of the reactor, compared to the top in the case of NRX. The vessel, bottom header, and upper bottom shield are lowered together on three large screws, which extend from the top of the reactor to the underside of the upper bottom shield. A carriage in the vessel change room directly under the reactor receives the whole assembly, and moves it to the east end of the room in preparation for vessel removal. The removal operation was carried out without difficulty. In January 1973, on a Saturday to avoid plant disruption, the vessel was lifted from the vessel change room into a nylon shroud that served to contain contamination not previously vacuumed, deposited on a skid and towed to the disposal area, where it was covered with sand. The removal operation took 2 hours and 40 minutes and there was no detectable contamination along the route to the disposal area.

Total radiation cost to personnel for the dismantling and vessel removal operations was 178.5 Rem, and nobody received the maximum permissable yearly limit of 5 Rem. Personnel radiation dose information was tabulated daily, and a computer program was run weekly. This method of control enabled dispersal of the total radiation load evenly throughout the work-force, and prevented approach to individual limits.

Lessons Learned or Reinforced

- in general, techniques evolved during the NRX calandria removals were refined
- emphasis was placed on early preparation of procedures and special equipment, including mockups
- a high degree of control was applied to radiation and contamination
- hoses on high vacuum cleaner systems must be armoured to avoid puncture by high-velocity debris.

SUMMARY OF IMPORTANT CONSIDERATIONS

1. Force Dedication The force leader, his subordinates, and their workers, must all be dedicated to this job. Their normal duties must be turned over, in total, to other members of their organizations.

2. Departure From Safest Approach Only with the greatest reluctance should departure from the safest approach to a situation be considered. Less safe action, where necessary, must contain a safety release mechanism. Planners should always search for the inherently safest procedure, to avoid compromise of personnel or equipment.

3. Deviations From Authorized Procedures Deviations from

previously authorized procedures must have leader approval. The leader must satisfy himself that the change is justified, bearing in mind the criterion of Item 2 particularly.

4. Planning and Training Careful planning of the entire operation should be done before embarking on any phase. Bar charts and critical path diagrams are two possible planning aids. Full-scale rehearsals, using mockups, must be carried out on difficult operations to uncover unforeseen problem areas, and to write accurate procedures.

5. Large Force Where a heavy decontamination schedule is programmed, a large non-nuclear body is required to spread the radiation load, and to ensure that trained station staff do not receive exposures that would prevent them from carrying out tasks requiring their expertise. An intensive training program is mandatory, and instructions to the non-nuclear workers must be simple and explicit. Trained station personnel are invaluable in this role.

6. Special Tooling Make-do tooling and make-shift operations are simply not good enough. Special tooling must be designed, built, and proved out on mockups, where hazardous, time-consuming or difficult tasks are foreseen. Obviously, special tooling requirements should be identified early in the planning phase, to avoid needless holdup later on.

7. Personnel Radiation Dose Control Daily records of personnel dose accumulation must be maintained to avoid over-exposure. This can be achieved by using a pocket electroscope dosimeter, or similar device. Reduction of total radiation dose can be achieved by close crowd control, and by reducing worker participation to the minimum necessary for safe completion of the assigned tasks. Gently elevated (e.g. from 1 mR/h to 5 mR/h) background radiation levels must be appreciated for their contribution to total dose. Worker utilization is more often determined by continuous exposure to background radiation than by a calculated one-time exposure to a relatively high field.

8. Continual Review of Progress Weekly general meetings should be held to discuss the previous week's work, problems encountered, and the following week's schedule. This method of forced interaction ensures that the various groups involved in the operation are given an opportunity to participate in discussion touching on their areas of responsibility. Very often problems can be brain-stormed then and there, thus eliminating the need for further meetings and more time dissipation.

9. Mobile Personnel Decontamination Center This rig, equipped with electrical, water and drain (to containment) connec-

tions, can be located close to the scene of action for initial decontamination work; or it may replace the normal site center if the latter itself is badly contaminated.

Table II
Selected Vessel Removal Data

	NRX 1	NRX 2	NRU 1
First Criticality	1947	1954	1957
Planning time - years	0	7	12
Service time - years	5	16	15
Reactor Flux max. thermal	10^{14}	10^{14}	3×10^{14}
S/D to removal - days	161	28	236
Unload time - days	125	7	17.5
Vessel material - Al	ALCAN 2S	ALCAN 6056	ALCAN 6057
Vessel diameter m	2.58	2.58	3.51
Vessel height m	3.35	3.35	3.60
Vessel weight kg	3,540	3,540	11,560
Max. rad. at 3.05 m	65 R/h	18 R/h	40 R/h
Rem cost (max. individual)	17	7	5
Rem cost (total)	2600	117	176.5

REFERENCES

1. W.B. LEWIS, The Accident To The NRX Reactor On December 12, 1952, Report AECL-232, (1953).
2. W.F. MERRITT and C.A. MAWSON, Retention of Radionuclides Deposited In The Chalk River Nuclear Laboratories Waste Management Areas, Report AECL-4510, (1973).
3. G.W. HATFIELD, Renovation Of The NRX Reactor At Chalk River, Ontario, Report WM-1, (1954).

DECOMMISSIONING THE SODIUM REACTOR EXPERIMENT, A STATUS REPORT

W. D. Kittinger and G. W. Meyers

Rockwell International
Atomics International Division
Canoga Park, California 91304

ABSTRACT

The decommissioning of the Sodium Reactor Experiment near Los Angeles, California is nearing completion. The reactor vessel and all contaminated support systems are being removed in order to return the building and site to unrestricted use. It was found cost effective to preserve the building throughout decommissioning. To date: fuel was removed and declad, tooling and technique development was completed, bulk sodium and sodium films and heels were removed, coolant piping was removed, in-vessel piping was removed by underwater remote explosive and plasma torch cutting, the reactor vessel assembly has been dissected by remote underwater plasma torch cutting, fuel and moderator handling machines were removed intact, and most of the contaminated support systems have been removed. The work has progressed successfully to meet contamination guidelines while maintaining radiation exposures to workers to as low as practicable levels. Noteworthy accomplishments were made in developing decommissioning tooling techniques and in demonstrating the capability to return a nuclear facility to unrestricted use.

INTRODUCTION

The Sodium Reactor Experiment (SRE) was the first nuclear reactor in the U.S. to produce power for supply to a commercial power grid. It was utilized during the period of 1957 through 1964 to develop reactor sodium coolant technology and design. The SRE was located at the Rockwell Santa Susana Field Laboratory near Chatsworth, California, approximately 30 miles from the center of Los Angeles. It served experimental and development programs for liquid metal technology and commercial power generation. The land involved is privately owned and under lease to the Federal Government. Figure 1 shows the exterior of the facility when it was in operation.

Decommissioning of the SRE began with termination of the SRE power expansion program in 1968. The last loading of sodium-potassium bonded, enriched uranium fuel rods was removed about 4 years earlier in 1964. The SRE was placed in the Stage 1* decommissioning mode by: decontamination of the operating areas, removal of unnecessary instrumentation and equipment, establishing maintenance and monitoring programs, removal of the secondary heat transfer sodium, and placement of the primary sodium coolant into controlled storage at the facility. In 1976, as part of the decommissioning program, the fuel rods from core loadings I and II were declad at Atomics International, cleaned of the NaK bonding, and canistered in aluminum containers before being transferred to a Department of Energy (DOE) facility for interim storage.

The SRE was a 20-MWt sodium-cooled, graphite moderated thermal reactor. The core, moderators, reflectors and support structures

*Stages of decommissioning as defines in IAEA 179 Decommissioning of Nuclear Facilities.

Figure 1. SRE Before (Exterior)

were surrounded by a ¼" thick stainless steel thermal vessel liner concentric with a 1½" thick stainless steel reactor vessel. Outside and concentrically surrounding the reactor vessel were: 5½" thick carbon steel thermal rings, a ¼" thick carbon steel outer vessel, and a 1' thick insulation pack, ¼" thick carbon steel cavity liner, and the surrounding 4' thick, high-density concrete biological shielding. The vessel assembly, excluding the biological shield, was approximately 14' in diameter and 29' in depth.

The program objectives were to remove all reactor originated radioactive materials of any significance from the site, as well as other potentially hazardous materials; to preserve the facility structure for new use as long as the approach was cost effective; and to return the site to a safe configuration for future unrestricted use. Performance objectives were detailed in structured documentation for: (1) maintaining exposures as low as practicable, (2) minimizing waste committed for burial to that waste actually contaminated, (3) maintaining good cost control and effectiveness, and (4) maintaining controls for physical safety and environmental protection.

Major planning and engineering efforts for the Stage 3 decommissioning of the SRE were initiated in 1974. Removal of secondary heat transfer and electricity generation equipment began shortly thereafter. Major challenges were posed in complete removal of residual sodium heels and films from all coolant systems and components, and in the remote dissection of components and vessels comprising the reactor. A major tooling and technique development program was initiated in 1975. The development program culminated in techniques for alcohol reaction of sodium under well controlled conditions, design and fabrication of a remotely operated and programmable polar manipulator similar to the Oak Ridge design for Elk River which was equipped with a unique plasma torch cutting tool, techniques for explosively cutting component piping, and techniques for removing all contaminated and activated components while preserving the reactor building and some facility support systems. Where possible, existing technology was adapted to the unique requirements of this facility and program.

Reports of tooling and technique development programs have been made previously by Graves, Streechon and Phillips* which gave detail on remote plasma arc cutting of reactor and support systems components. The intensive planning effort proved beneficial to the technical performance and to effective control of costs.

---

*ANS Transactions Volume 26, TANSA0261-610(1977)ISSN:303-018X; paper presented at the 1977 Annual Meeting held in New York on June 12-16, 1977.

Peripheral and noncontaminated support systems were removed in parallel with the tooling and technique development work. Secondary sodium systems components, nitrogen vault cooling systems, air blast heat exchangers, water systems, and the kerosene cooling systems were the major exterior systems removed.

General guidelines for acceptable surface contamination and concentrations in soil and concrete were established early in the program. The limits reflect an understanding reached jointly by AI and DOE-SAN with interfaces occurring with the California State Department of Health. The guideline concentration for soil and concrete are currently the subject of an environmental analysis which will supplement the initial environmental assessment. The surface contamination limits are shown below in Table 1.

Table 1
Surface Contamination Limits for Decontamination and Disposition

	Total	Removable
Betta Gamma Emitters	0.1 mrad/h at 1 cm thru 7 mg/cm^2 absorber.	100 dpm/100 cm^2
Alpha Emitters	100 dpm/100 cm^2	20 dpm/100 cm^2

The surface contamination limits are somewhat more conservative than those in NRC Regulatory Guide 1.86 and proposed ANSI Standard (N328). The guideline for distributed fission or activation products in soil, concrete, and structural materials is a gross beta value of 100 pCi/g and includes the background activity. It is a guideline value based on several considerations: (1) a local gross beta background in soil found to range from 11 to 48 pCi/g; (2) the predominant isotopes resulting from these operations and now present are ^{90}Sr, ^{137}Cs, and ^{60}Co; and (3) consideration of the expected statistical deviations in sampling results.

A paper by G. W. Meyers and W. D. Kittinger, "Progress Report on Dismantling of the Sodium Reactor Experiment," was presented to the IAEA International Symposium on the Decommissioning of Nuclear Facilities in November 1978. The paper presented SRE decommissioning activities to that date and may be consulted for detail on the summary information which follows.

The safe removal of bulk sodium and sodium residues was accomplished without incident. Primary coolant sodium which was slightly contaminated was removed by heating the system and transferring the sodium into drums under an inert atmosphere of nitrogen. For sodium residues and films, ethyl alcohol was selected to react sodium to passive compounds permitting disposition to burial. Sodium frost and film in the reactor vessel were reacted by installation of an alcohol spray system. Large components such as the heat exchanger were "piped" to an alcohol circulating system. Small piping and other components were cut into sections and subjected to meltout of sodium and drainage of the sodium followed by alcohol flushing or immersion of the components.

Economic and exposure control considerations indicated that the most judicious action for many facility components was the total removal without further dismantling or dissection. The 12' diameter reactor loading face shield plug (about 56 short tons), the 16' diameter reactor ring shield (about 60 tons), 2 fuel handling machines (each about 52 tons), and the moderator handling machine (about 25 tons) were removed totally and shipped by special permit for overload and/or oversize conditions to burial.

Six 5,000 gallon underground gas and liquid waste holding tanks which had provided holdup capability for radioactive decay and sampling were externally cleaned and dispositioned in a like manner. The associated concrete vault was decontaminated by use of a tractor mounted hydraulic hammer which removed the outer contaminated layer of concrete.

The tooling and technique development program indicated explosive cutting would be the most effective technique for removal of reactor vessel internal piping. Main primary and main auxiliary coolant piping and drain overflows were removed underwater by using engineered, shaped charge explosive techniques. Approximately 60 cuts were made. The stainless steel pipes ranged from 0.109 to 0.5 inches in thickness. The use of a remote viewing system for charge placement cutting inspection, coupled with the explosive cutting capability for the curved piping configuration, proved most effective.

The basic approach to dismantling and disposal of the reactor vessel assembly was to remotely remove or dissect components from the water filled vessel, remotely transfer wet components in air to a water storage pit, and selectively package the segments underwater into a disposable cask liner which was then loaded into a shielded cask for transport to the licensed burial site. The bellows connecting the core tank to ring shield was removed by plasma torch cutting with a special fixture. Similarly, the core clamp band and core clamps were removed. The vessel assembly described earlier is shown schematically in Figure 2. Also shown in Figure 2 are inserts of actual operation of the plasma torch system.

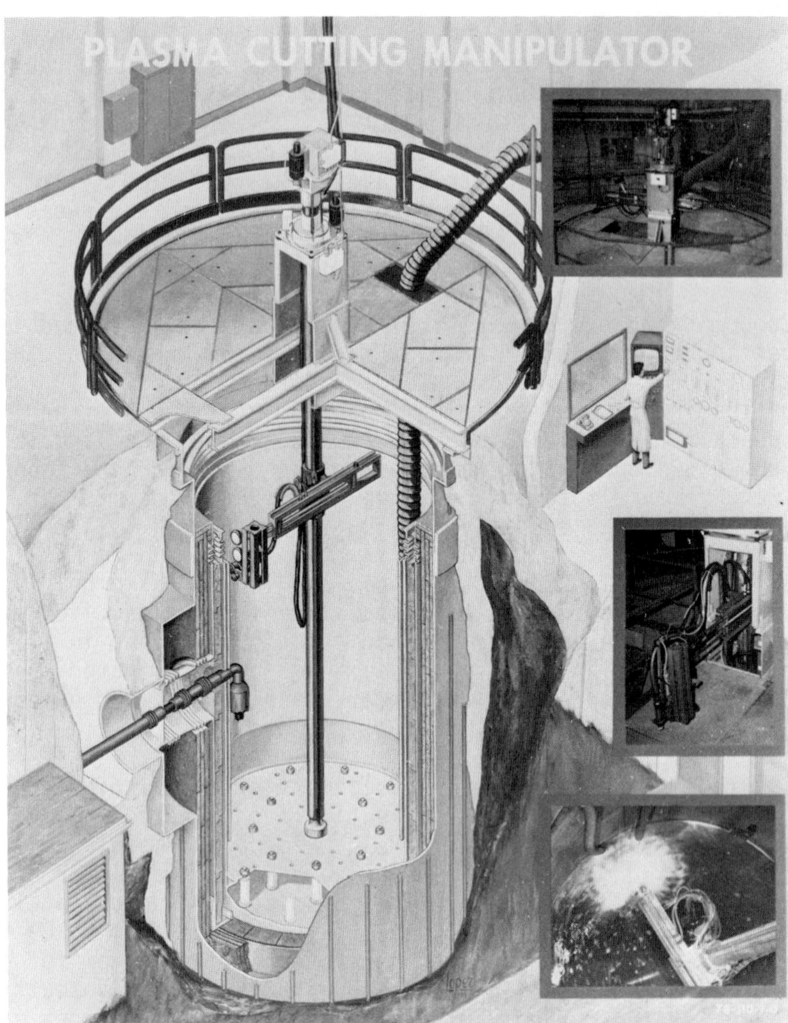

Figure 2. Plasma Torch and Vessel Schematic

An engineering demonstration facility duplicated the SRE dimensions and permitted manipulator checkout and operator training as well as parameter development. It was determined that all vessel components could be segmented by plasma torch with the existing equipment and technology with the exception of the 5½" carbon steel thermal rings and the outer cavity liner which was composed of cooling coils embedded in the concrete biological shield. The thermal liner and core vessel were easily segmented typically into 3 by 3 foot sections, which were held by tensioning devices during the cutting. After preparation, a typical section could be cut in

4 minutes. The 5½" thick thermal rings, weighing about 23,000 lbs. each, were lifted, washed, and painted to control contamination and placed on the operating floor for sectioning. These were cut into 4 arcs by an enclosed, locally exhausted and automated oxy-acetylene torch.

The reactor vessel grid plate was segmented and removed. The core tank bottom was raised and supported off the outer tank bottom with spacers and cut underwater by plasma torch. Water was drained after removal of the core tank bottom. The outer tank and the outer tank bottom were segmented by manipulator and plasma torch, and removed. The successful operation of a water filtration system was necessary for remote viewing which was a tremendous aid in completion of the vessel dismantling.

After cutup of the outer tank, the manipulator was removed. The water was later evaporated and the remaining radioactive sludge was packaged for burial. Insulation was removed manually by direct entry into the reactor cavity. The cavity liner was removed, along with the biological shielding during the excavation and concrete demolition operations.

Major excavation work was necessary in order to gain access to below-grade contaminated systems and to remove some contaminated soil. It was necessary to excavate to approximately 35' in order to remove the reactor biological shielding, the wash cells and piping, and hot cell drain lines. An economic analysis was performed of the costs of accomplishing the necessary excavations: (1) under conditions where the building superstructure was razed, or (2) under conditions where shoring was installed and care taken to preserve the containment superstructure, overhead cranes, associated offices, shops, and basic support services. The second approach, that of preserving the building, was found to be most cost effective and also provides coverage of the exposed areas in the event of inclement weather and aids in the control of any airborne contamination. The shoring design was unique to our experience and consists of slotted, 2' diameter reinforced concrete pilings 6' on center to a depth of 50'. Timber spanners are fitted between pilings as excavation progresses. The shoring which provides personnel safety and structural protection is utilized in the regions where necessary.

The techniques for decontamination and disposition of concrete structures and surfaces are crucial to costs and schedules in most nuclear facility decommissioning projects. The constraints and costs inherent in the disposition of contaminated concrete were significant factors in evaluating methods for handling concrete in the SRE facility. The approaches typically utilized were to first decontaminate concrete surfaces with decontaminating agents in a foam carrier, remove the foam with vacuum or water rinse, monitor the materials and then repeat the process, as necessary, if continued significant contamination was still occurring. This procedure was utilized even for highly contaminated surfaces in order to reduce personnel exposure and airborne contamination potentials even though complete decontamination was not expected.

Where contamination remained, a variety of tools and techniques were utilized to attempt selective removal of only the contaminated portion. Cracks and joints in concrete, as usual, presented difficulties for decontamination. Depending upon the nature of the construction, surface and degree of contaminant penetration; tools found to be effective for selective surface removal were a scabbler, sandblaster, bushing tool, small jackhammer, and the hydraulic ram chisel (Hyram). The tools were all utilized with dust control measures such as HEPA filtered vacuum systems or concurrent water spray and containment structures. The Hyram was also widely utilized for more massive concrete removal.

During the past year, all concrete features at grade level and to about 15' below grade have been decontaminated to meet guideline values by the techniques just cited. Concrete which remained contaminated was packaged per federal requirements and disposed to licensed burial sites. The facility features that were cleaned or demolished by these techniques included: the below-grade sodium coolant piping, pump and heat exchanger galleries (a portion of these galleries was adapted and utilized during decommissioning for reactor vessel segment underwater storage, sorting, and cask loading); the sodium storage, service and pump vaults; the kerosene coolant piping trenches; the fuel and moderator handling machine maintenance bay; instrument and wiring trenches; fuel storage cell monolith; and the reactor bay floor surfaces. Non-contaminated concrete (and soil removed for access purposes), was removed to a location near the facility for later use as excavation backfill.

There were 96 fuel storage cells which accessed to the reactor room floor. By decontamination and removal of the upper 5' of the concrete monolith surrounding these cells, it was possible to pull the cells from the structure and package for disposal. Figure 3 shows utilization of the Hyram in removal of the concrete. The use of this equipment, accompanied by the application of a fine water spray, has proved quite versatile in surface or massive concrete removal.

The ¼" steel reactor vessel cavity liner was scored into sections from the inside under careful control of exposure conditions. The liner sections remained attached to the biological shielding by embedded cooling coils. Removal of the biological shielding from the outside then permitted freeing the liner sections as the

Figure 3. Removal of Fuel Storage Cells with Hyram

work progressed. By sampling, it was found that only the inner 10" annulus of the biological shielding at the maximum flux location remains activated to levels above the guideline. The Hyram is being utilized to remove the concrete shielding from the outside in a controlled and monitored manner. This approach has minimized generation of superfluous waste for packaging and shipment to the burial site.

A recent exterior photograph of the SRE is shown in Figure 4. The degree of excavation necessary to gain access may be noted in the photograph.

The experience for external radiation exposures to personnel is detailed in a paper to the ANS Winter Meeting of 1978 by R. J. Tuttle.* Maximum exposure periods occurred during residual sodium removal and reactor vessel system removal. There were no individual annual exposures above 3 rem. The exposures during the past year were very low since most of the high radiation sources had been removed.

Activities that will follow are: removal of the wash cells and hot cell, removal of the bottom portion of the biological shield and cavity liner, final structure decontamination, removal

*ANS Transactions, Volume 30, TANSAO30 1-814 (1978) ISSN: 0003-018X, Page 608.

Figure 4. SRE (Exterior)

of air cleaning and other support facilities, site repairs, final radiation surveys, and completion of the final report.

In summary, it is believed that the most noteworthy accomplishments in the SRE dismantling project are:

1. Proving and refining the technology for remote cutting of highly radioactive structures by plasma arc torch,

2. Proving and developing the technology for remote explosive cutting of radioactive components,

3. Increasing the experience base and technology for removing and reacting large amounts of contaminated sodium,

4. Showing and applying concepts for economic removal of large items of equipment,

5. Developing and applying concepts which dismantle a nuclear facility and yet preserve the site and superstructure for immediate alternate use,

6. Dismantling a major nuclear facility with minimal radiation exposures to the workers and without adverse impact on the environment,

7. Providing a data base of information for worldwide use on other decommissioning projects, and

8. Possibly the most important is the actual demonstration that there are no insurmountable technical problems to decommissioning to any Stage.

HANFORD PRODUCTION REACTOR DECOMMISSIONING

R. K. Wahlen

UNC Nuclear Industries - Operations Division

Richland, Washington 99352

INTRODUCTION

There are nine plutonium production reactors located at the Hanford Site in Richland, Washington. Eight of these reactors have been shutdown, defueled, deactivated and placed in retirement. The nineth reactor is N Reactor which remains in operation. The combined thermal energy developed during operation of all nine reactors was in excess of 20,000 MWt. This required large and complex systems to dissipate the heat generated and to monitor the many parameters incidental to safe and efficient operation of the nuclear reactors.

During the last several years, the Department of Energy has been involved in a study to determine the final disposition of the retired government-owned facilities on the Hanford Reservation. Two programs are in progress and are directed towards final disposition of these facilities. First, in 1974, a site cleanup program was started which is involved in the elimination of potential industrial and radiological hazards and the final disposition of "clean" (free of contamination) facilities. Second, in 1977, UNC Nuclear Industries was authorized to begin the planning necessary to implement full-scale decommissioning of the contaminated portion of one reactor complex. The facility selected for this demonstration program is identified as 100-F.

The progress made in these programs, along with some of the techniques being considered in the Decontamination and Decommissioning (D&D) work is included in this report.

DESCRIPTION OF A TYPICAL HANFORD REACTOR FACILITY

A typical retired Hanford Production reactor facility[1] includes the reactor and several support systems and activities which were housed in 32 buildings located in an approximately six square mile area. There was a pump station located on the bank of the Columbia River through which river water was pumped to a storage reservoir, then pumped from the reservoir through a chemical treatment facility, flocculators, sedimentation basin, anthracite filters and into a 10,000,000 gallon underground storage tank. From storage, the water was split into three different systems to meet three distinct needs; fire and sanitary, service, and process. The process water was used as reactor coolant and required about ninety percent of the total process capability. The process water was pumped from the underground storage tanks to four 1,200,000 gallon tanks located above ground at the head of the main process pumps which supplied the coolant flow to the reactor. The coolant passed through the fuel process tubes in the reactor, removing the heat from the fission process. The effluent was held up in retention basins to allow for heat decay and then flowed through an outfall structure and was discharged into the center of the river.

A large coal-fired steam plant with full-load capacity of 400,000 lbs/hr was used to provide backup power for the process pumps and other emergency equipment.

The reactor building contains the reactor block; the reactor control room; a spent fuel discharge pool; fuel storage basin and associated fuel handling equipment; fans and ducts for ventilation and confinement systems; and support offices, shops and laboratories. Figure 1 shows a cutaway of a typical reactor building.

Dismantling, packaging and disposal of the reactor block is the greatest challenge in the D&D program at Hanford. The reactor is a graphite-moderated, water-cooled reactor used for producing weapons-grade plutonium. The core of the block is 28' thick, 36' high and 36' wide. The stack is made up of high-purity graphite blocks 4-3/16" square by 48" long. The criss-cross stacking arrangement gives the stack stability. The stack has 2004 penetrations running horizontally for process tubes which contained fuel elements during operation. Also, there are nine horizontal side-to-side penetrations for the control rods, nine horizontal penetrations for the experimental test holes, and 29 vertical penetrations for the safety rods. See Reactor Block Construction Details in Figure 2.

---

[1]*N Reactor, which continues to operate, has many different design features when compared with the retired reactors. One of the main differences is that the reactor coolant system at N is a closed recirculating loop as compared with the once-through cooling system at the retired reactors.*

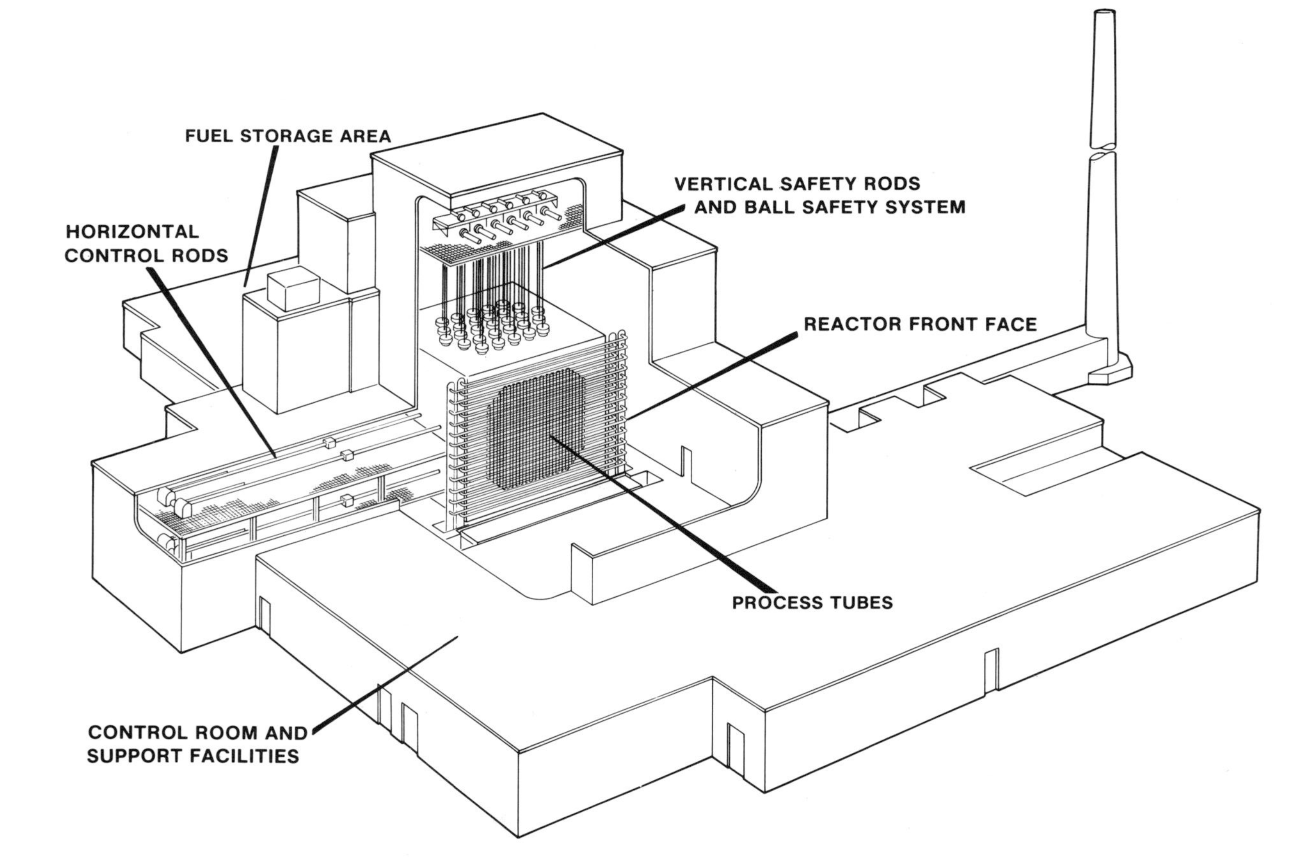

Figure 1 - Cutaway of a typical reactor building showing the reactor block, reactor control room, storage basin, etc.

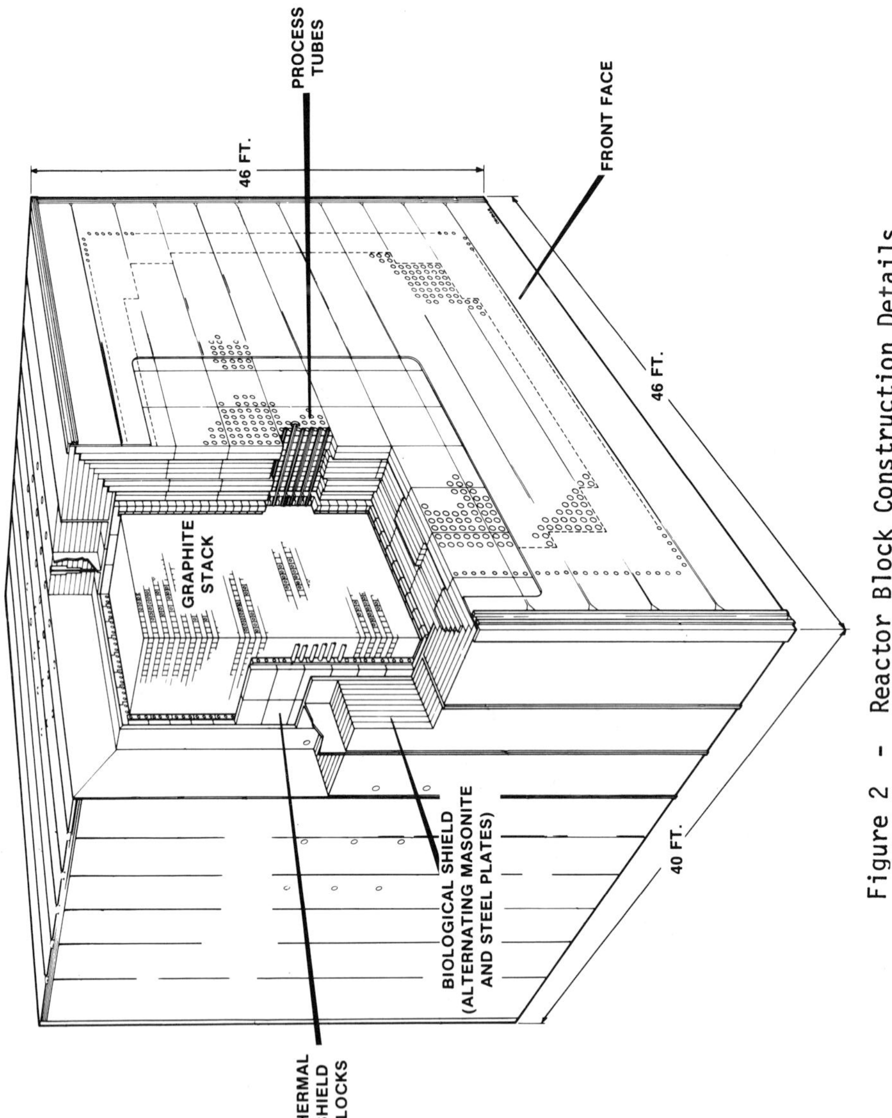

Figure 2 - Reactor Block Construction Details

A thermal shield surrounds the graphite moderator stack on all six sides. The shield is fabricated of cast iron blocks of varying sizes and thickness depending upon location in the reactor assembly.

A biological shield surrounds the thermal shield on all sides, except the bottom. The shield is made up of alternate layers of steel and masonite and is 52" thick. This shield absorbed heat and neutrons that escaped the moderator and thermal shield.

The concrete base of the reactor is overlayed with steel plates which are welded together and to the bottom of the biological shield. While this steel is not a part of the biological shield, it does complete what amounts to a leak-tight steel box which completely encloses the reactor. Neoprene expansion joints are provided at the corners of the shield to allow for thermal expansion of the reactor. The 2004 fuel process channels are lined with aluminum process tubes, to which nozzles and connectors are attached to supply cooling to the reactor. The cooling system consisted of a supply manifold made up of 36" supply piping and 39 four-inch cross headers.

The cooling water effluent system is essentially a duplication of the supply piping.

The nine horizontal control rods (HCR) are 75' long. The rod tip or neutron absorbing section is 29.5' in length. The tip is that portion of the HCR which enters the reactor and is therefore irradiated. The rods were water-cooled and hydraulically driven.

The 29 vertical safety rods (VSRs) were provided to maintain the reactor subcritical after complete Xenon decay and to effect a rapid shutdown under emergency conditions. The VSRs entered the reactor through the top biological shield. A step-plug arrangement limited the escape of radiation around the rod.

The backup emergency shutdown facility is provided to support the VSR system in case of VSR system failure. The system is identified as a Ball 3X system. Boron steel balls were used and dropped vertically into the VSR channels upon a trip of the Ball 3X system.

A fuel storage basin and transfer area served as a collection, storage and transfer facility for the irradiated fuel elements after they were discharged from the reactor. The basin, which is 81' long by 74' wide and 20' deep, is basically a large water tank which was filled with 18' of water during fuel storage. The basin is currently filled with dirt to control the spread of activiation products which remain in the storage basin.

HANFORD SITE CLEANUP PROGRAM

The objective of the site cleanup program is to eliminate potential industrial and radiological safety hazards and to dispose of surplus facilities which are free from radioactive contamination. This program was started by the Department of Energy in 1974 and has continued since that time on a year to year limited funding basis. A total of 69 structures, varying in size from 97,000 square feet to 200 square feet, have been removed. Figure 3 shows 100-F Area before site cleanup was started and Figure 4 shows 100-F Area as it now stands. Similar progress has been made in the 100-H Area and about 50% of the buildings in 100-D Area have been removed.

The techniques used in site cleanup include dismantling the structures piece-by-piece where salvage is of importance, and the use of explosives and a wrecking ball where salvage is not economical.

DECOMMISSIONING PROGRAM

The objective of this program is to develop and demonstrate specific methods and techniques for dismantling, packaging and disposing of a representative production reactor facility. It will establish realistic disposition cost, manpower, and exposure commitments based on sound practical techniques.

There are five buildings (the main reactor building, gas recirculation building, filter building with ventilation exhaust stack, liquid waste lift station and a lab facility which was used in biological research programs) included in this initial D&D Program. Also included are the underground waste burial trenches and cribs in the 100-F location.

The planning required to carry out the decommissioning of 100-F Area has included site characterization, preparation of a facility/site description, an environmental assessment, quality assurance plan, management plan and a disposition plan.

The site characterization study was carried out over the last three years. Samples were taken and analyzed to establish the radionuclide inventories and concentrations in the facilities and ground disposal sites. The maximum contamination levels for the five buildings ranged from 5000 c/m (counts-per-minute) to 100,000 c/m. The maximum exposure rate in Rem-per-hour range from .02 to 300. The total inventory of radioisotopes is approximately 37,250 curies. The radioisotopes are primarily Cobalt 60 and Carbon 14. See Figure 5 for additional details.

Figure 3. 100 F Area Before Site Cleanup.

Figure 4. 100 F Area After Removal of Noncontaminated Facilities.

FACILITY/ COMPONENT	MAX. CONTAMINATION LEVEL (C/M)	MAX. EXPOSURE RATES (R/hr)	APPROXIMATE INVENTORY (Ci)
REACTOR BLOCK	5,000 (EXTERIOR)	300	35,000
CONTROL SYSTEMS	100,000	5	50
FUEL HANDLING/BASIN	100,000	1	100
RETENTION BASIN SYSTEM	50,000	.020	90
GROUND DISPOSAL FACILITIES	80,000	20	2,000
OTHER MISC. SYSTEMS	100,000	50	10
			37,250

Figure 5 - Radiological Conditions 100-F Area

A detailed facility and site description was deveoped to be used by the planners in support of their planning and design efforts.

An environmental assessment which identifies the long-range environmental impact of decommissioning activities, as well as the impact of any residual contamination left on the site after decommissioning, has been completed in draft form, circulated for comments and is currently being revised in a final draft.

The quality assurance plan was developed and describes the quality assurance activities to be performed by UNC to assure the development and compliance with procedures for health, safety, environmental protection, safe disposition of hazardous materials and restoration of facilities and grounds to conditions which will allow restricted occupation of the area. The plan is based on the general requirements of DOE Manual Chapter 0820, Quality Assurance. Its primary objective is health, safety and environmental protection, rather than quality of products.

The project plan has been developed in draft form and identifies responsibility of participants in the project, and establishes the work breakdown structure, performance, reporting and control requirements.

The disposition plan identifies the major tasks required for accomplishment of the 100-F Decommissioning Program; recommends or proposes methods for accomplishing each one of these tasks; and discusses the general health and safety requirements which will be imposed on the project. The disposition plan is divided in six major activities.

Site Preparation

This activity involves restoring the necessary services and revising existing facilities in 100-F Area to prepare the site for Decontamination and Decommissioning.

Reactor Building Decontamination

This activity removes all the reactor support systems such as reactor inlet and outlet piping, vertical safety rods, horizontal control rods, process tubes, etc. When the activity is completed, the only equipment remaining in the building will be the reactor block. This activity also includes decontaminating the surfaces inside the building where radioactivity is found. Following the removal of the reactor block (see next activity) the reactor building will be demolished and removed from the site.

Reactor Block Removal

Two alternate methods for removal of the reactor block are being planned. The preferred method is to seal all the openings in the reactor, install a small water recirculating system and fill the reactor with water to be used as a radiation shield for the workers during dismantling and to facilitate contamination control. A bridge crane will be installed over the top of the reactor as illustrated in Figure 6, to handle the parts of the reactor as it is dismantled. A temporary ventilation control enclosure will be installed as illustrated in Figure 6, for a loadout station. The station will accommodate a railroad car or a truck to transport the contaminated material to the burial site. The enclosure will also have an airlock compartment with a decontamination facility for cleaning the railroad cars and/or trucks before they leave the facility.

The top biological shield will be cut in small sections using an arc-saw and acetylene torch. The sections of shield will be packaged and transferred to the railroad car.

Following removal of the top biological shield, a moveable working platform, as shown in the sketch in Figure 7, will be mounted on top of the reactor for the crane operator to work from. Removal of the thermal shield, graphite stack blocks and the biological shield will be in a selected sequence to insure that there is water over the graphite and thermal shield at all times to provide radiation shielding to the workers.

The alternate plan for dismantling the reactor block follows the same general procedure, except the work is done without flooding the core of the reactor block. This plan would require a shielded control room for the crane operator and an enclosure over the top of the reactor for contamination control.

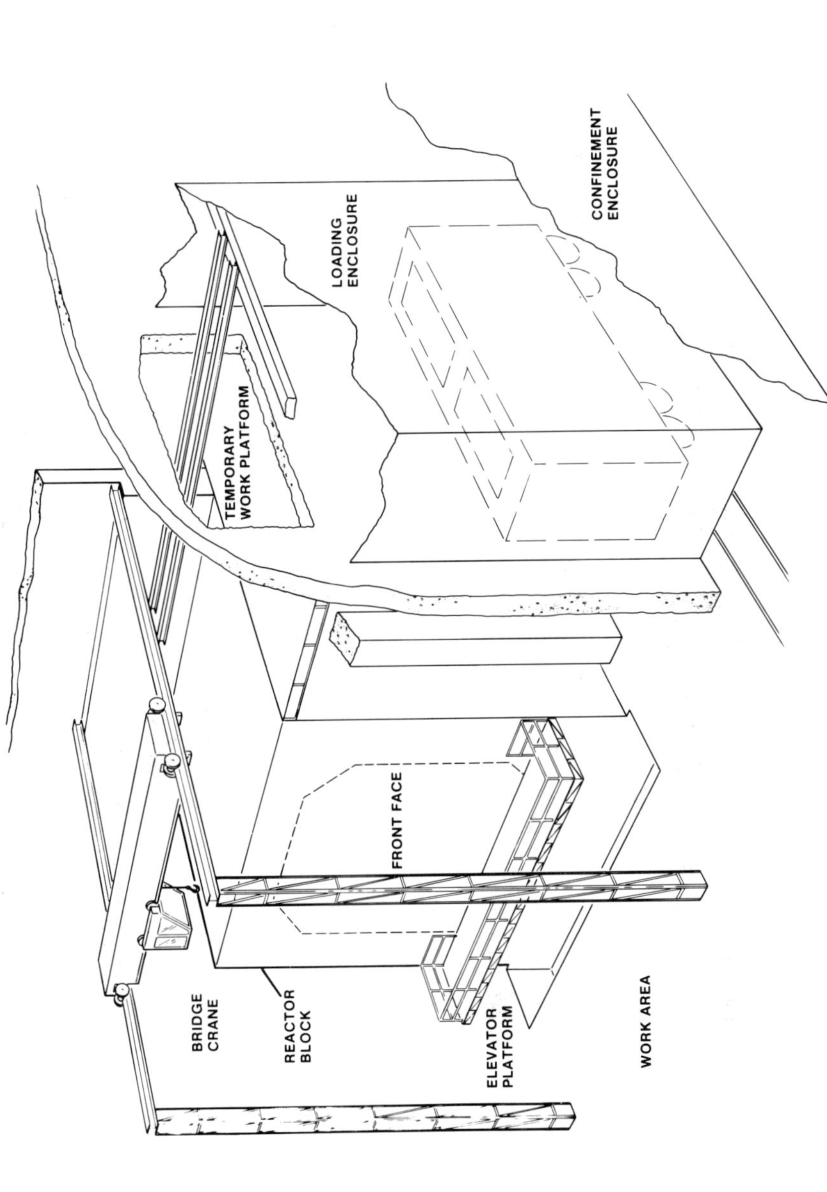

Figure 6 - Shows the Bridge Crane and temporary ventilation control enclosure for a loadout station.

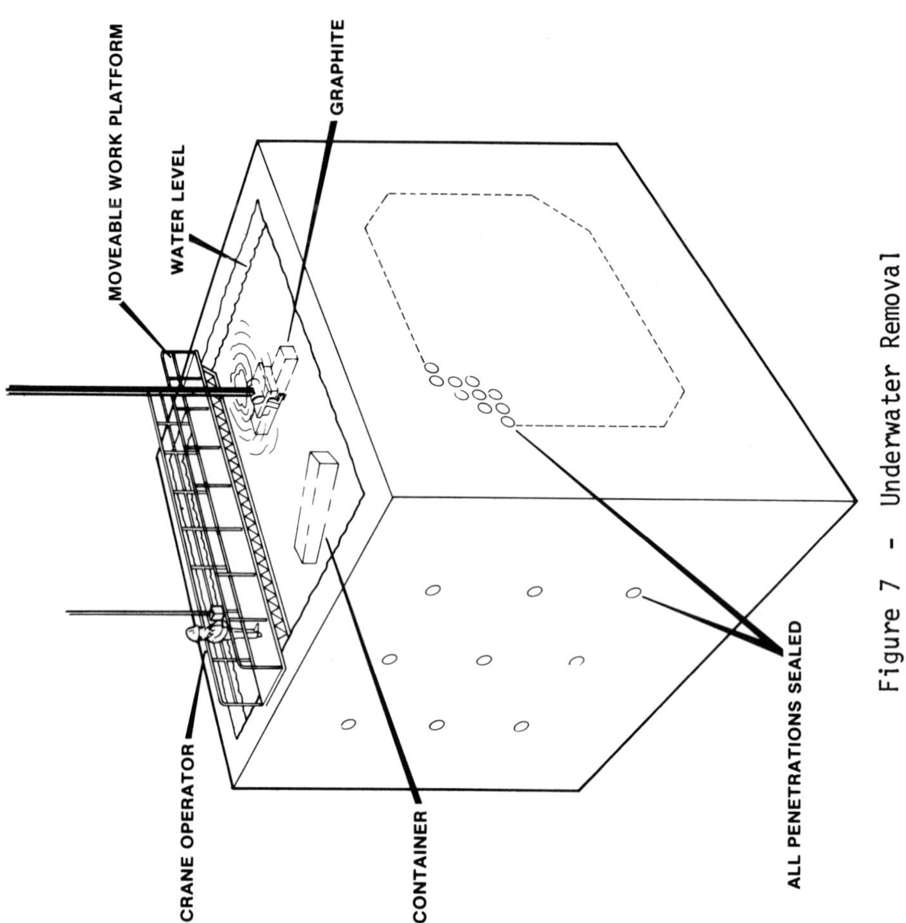

Figure 7 - Underwater Removal

Removal of the Support and Research Facilities

There are four facilities which supported the reactor operation included in this activity: gas recirculation facility; ventilation exhaust and confinement system; contaminated liquid waste lift station; and a large biological laboratory building. The equipment and buildings will be decontaminated and disposed of by conventional techniques.

Stabilization of Burial Grounds, Cribs and Trenches

The waste facilities at 100-F Area were utilized during the operating period for disposal of radioactive wastes. There were three different types of disposal sites. The cribs were open-bottomed trenches filled with coarse rock and covered with dirt. They were used to disperse liquid effluent. The leach trenches were open-topped, long, narrow, unlined excavations used to dispose of large volumes of low-level liquid wastes. The burial grounds were excavated trenches in which solid waste was placed and covered with a minimum of four feet of clean earth.

The treatment given the waste facilities in the disposal program will be in two categories:

> All long-lived radioactive material and transuranics in excess of unrestricted release limits will be packaged and relocated to approved off-site disposal areas.
>
> All material with half-lives less than approximately 30 years will be left in place. A suitable bio-barrier will be installed over the waste to prevent the intrusion of the contaminated area by plant and animal life. Figure 8 shows a cross section of the bio-barrier concept. Alternatives to the above, such as removal of all contaminated material will be developed as necessary.

Project Closeout

The project closeout will provide guidance to ensure the disposition has been satisfactorily completed. A final comprehensive and detailed radiation survey of the area will be conducted to ensure that it meets the specified release criteria. A closeout document will be prepared, providing a summary of the disposition project activities and describing the final status of the site.

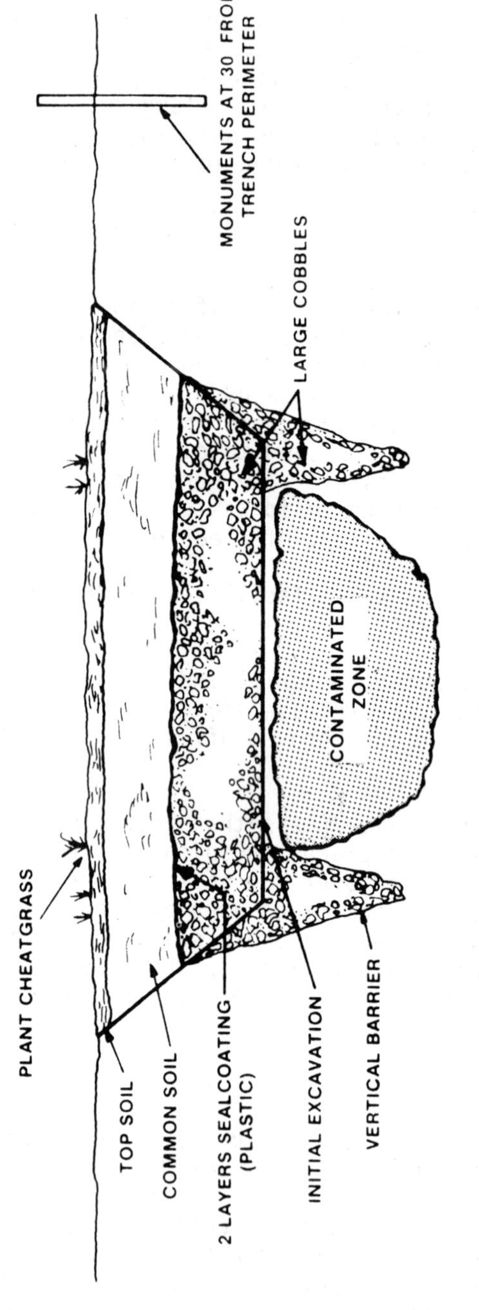

Figure 8 - Bio-Barrier

Engineering evaluations in support of each one of the activities has already been done on all major activities, except Project Closeout. Each engineering evaluation provides recommended methods and alternatives, estimates of personnel required, equipment requirements, identification of tool development needs and occupational exposure, and waste volume estimates.

The six major activities are being divided in small work packages which will be used by the work force to proceed with the actual D&D work. The current schedule calls for Site Preparation work to start early in FY 1981 and the Project Closeout to be completed by late in FY 1984.

THE IMPACT OF DECONTAMINATION ON LWR RADIOACTIVE WASTE TREATMENT SYSTEMS

G. R. Hoenes, L. D. Perrigo, J. R. Divine
and L. G. Faust

Pacific Northwest Laboratory
Battelle Memorial Institute
Richland, Washington, U.S.A. 99352

INTRODUCTION

Increased radiation levels around certain reactors in the United States and accompanying increases in personnel exposure are causing a reexamination of available options for continued operation and performance of required maintenance and inspections. One of the options is decontamination, which could have profound effects on the radwaste treatment system. An assessment of the impact of decontamination on radwaste treatment systems involves a careful study of the various effects that would result from the use of such a process.

THE IMPACT OF DECONTAMINATION

The following discussion will consider the impact on the radwaste treatment system of each type of decontamination process, i.e., mechanical, dilute and concentrated.

THE IMPACT OF MECHANICAL DECONTAMINATION

Mechanical decontamination processes are characterized by use of fluids, fluids with abrasives or abrasive techniques. For any sizable operation, one or both of the first two techniques would be employed.

The first unit operation in the liquid system is a settling tank to remove the large particles. These larger particles are highly abrasive and could damage piping and components. After passing the pump, the waste would be filtered. The filtrate would enter the standard liquid radwaste system. It is not expected to have any special properties that would affect the system.

At the present time, there is not enough information to do a quantitative analysis of the radiation exposures associated with mechanical decontamination. Because this method of decontamination may be used on an isolated area of the reactor primary system, the waste stream will probably be less in volume than that for a concentrated decontamination. The mechanical decontamination process, however, will likely be used on areas which have high levels of radioactive crud buildup. This will likely result in a very high concentration of radioactivity in the waste stream. The radiation exposure rates and the associated problems for the mechanical process will be similar to those associated with either of these processes.

THE IMPACT OF DILUTE PROCESS DECONTAMINATION

The majority of dilute process decontamination agents contain weak acids or salts of weak acids, and organic chelating agents. Other dilute processes entail modification of the oxidation characteristics of the coolant to effect a change in the deposit structure which causes deposited crud to release from the surface.

Materials

Corrosion in decontamination solution is kept low by the use of mild chemicals or inhibitors. Ayres[1] reports that oxalic acid corrodes carbon steel at a rate of about 0.6 mg/cm^2-hr. He also noted that 400 series stainless steels corrode at about the same rate as carbon steel in common, inhibited decontamination solutions.[1]

In the Can-Decon process,[2] which uses very dilute (0.1 wt% total chemicals) solutions, the Canadians observed quite low corrosion rates, even on carbon steel (see Table I).

TABLE I. Corrosion During Can-Decon[2]

Material	Average Penetration Rate, μm/h
Carbon Steel	0.28
410 Stainless Steel	0.08
316 Stainless Steel	0.02
Monel-400	0.03
Nickel	0.02

The addition of oxygen to the primary system for decontamination will increase corrosion, though exposure time periods are so short that little total corrosion would be expected. In Canadian tests at Douglas Point Reactor[2] little corrosion was noted even though redox cycling, low pH, thermal shock, and mechanical shock were used.

Waste Solidification

The principal processes used for waste solidification are: cement, urea-formaldehyde (UF) asphalt (bitumen), and vinylester such as the Dow process.[3] The primary advantage reported for the vinylester is its ability to hold water. The urea-formaldehyde process has been plagued by free-standing water problems.[3]

It appears that bitumen and cement are suitable for resin solidification. In addition, discussion with some of the suppliers of various plastic processes suggest that the plastics may be excellent means for solidifying ion exchange resins. From discussions with reactor operating personnel, it appears that all solidification systems are difficult to maintain.

Radiation Exposure

During a reactor decontamination using a dilute process, such as the Can-Decon, radiation exposures in the radwaste treatment system occur as a result of working with or near process resin columns and filters.

Three 100 ft^3 resin beds are needed, each being preceded by a filter for removing particulates. Since the fuel will be in the reactor during this decontamination, we will assume that approximately 10^4 Ci of ^{60}Co will be deposited on the resin and the prefilters. For exposure rate calculations, it is assumed that 500 Ci of ^{60}Co is deposited on each filter and 3000 Ci on each resin bed.

Waste from the dilute process will consist mostly of spent resin and filter cartridges. The resin and filters may be shipped in a cask or placed in 55-gallon drums. The amount of ^{60}Co in a shipping cask (150 ft^3) may vary from about 2700 Ci to almost 5000 Ci. The amount of ^{60}Co in a 55-gallon drum filled with resin may vary from about 130 to 220 Ci.

Table II lists the exposure rate for radwaste from dilute process decontamination and from typical spent resin from operating plants. These dose rates indicate that a remote facility is probably necessary to work with unshielded spent resin. During the site visits it was evident that in most cases adequate shielding was not in place. This indicates that design should include more shielding for decontamination.

TABLE II. Exposure Rates for Spent Resin in 55-Gallon Drums or 150 Cubic Foot Casks

Waste Type	Container	Typical Surface (R/h) Exposure Rate	Average
Decontamination	Drum	210-900	
Decontamination	Cask	400-1300	
PWR	Drum	13-56	
PWR	Cask	0.5-1000[4]	615[9]
BWR	Drum	6.2-27	
BWR	Cask	0.33-200[4]	90[4]

Radiation exposure will also be encountered from handling the spent filter cartridges. At operating PWRs, radiation exposure rates near unshielded filter cartridge packages range from 0.1 to 40 R/h with an average reading of 6.5 R/h.[4] Since filters used during decontamination are likely to have exposure rates of 200 R/h, remote handling will be required.

THE IMPACT OF CONCENTRATED PROCESS DECONTAMINATION

Concentrated decontamination consists of the application of one or more solutions, generally of 5-10% concentration, to remove activated corrosion products and/or rupture debris from reactor primary systems. The three chemical agents that have been most frequently used, or are scheduled for use are phosphoric acid, AP-Citrox*, and NS-1 (Dow Chemical Company proprietary solution). These processes require extensive rinsing to remove the chemicals from the primary system between steps and before restoring normal coolant chemistry conditions.

Materials

Phosphoric acid. Data shows that hot inhibited phosphoric acid should not be allowed to remain in the reactor for extended periods. At ambient temperature, however, the corrosion rate of various stainless steels in uninhibited phosphoric is less than 25 μm/yr.[5]

The effect of inhibited phosphoric acid on carbon steel is shown in Table III. Because of its corrosiveness, storage of uninhibited phosphoric acid in contact with carbon steel cannot be recommended, even at normal room temperature

* A two step process using alkaline permanganate as the first step and citric acid and oxalic acid and their salts as the second step.

TABLE III. Carbon Steel Corrosion Rates[6,7] in Phosphoric Acid (μm/d)

Material	Average Corrosion
ASTM A-216 WKA	41.6
ASTM A-105 GR II	41.2
SAE 4140	58.3
ASTM A-515 GR 70	39.0
ASTM A-615	51.7
ASTM A-234 CRWP II	67.4
ASTM A-1603	65.2
Time of exposure	15-30 min.
Temperature	67-82°
H_3PO_4 concentration	7-10%

AP-Citrox. The alkaline permanganate (AP) step is required for decontamination in certain systems, such as those containing stainless steel and having a reducing environment.

Ayres[1] notes that AP is not very corrosive to stainless steel (300 or 400 series) or to carbon steel. The corrosion rate of carbon steel at 105-110°C is about 6 μm/day.

There are several areas of concern with AP.[1] First, chromium plate and hard-facing alloys are attacked and require water flushing for protection. The second problem with AP is that any residual may cause caustic cracking when the system returns to temperature.[1] Also, alkaline permanganate is an excellent oxidizing medium and as a consequence, must not contact organic components.

The Citrox step of the AP-Citrox process can be used alone or as the individual citrate or oxalate components. The data for ammonium citrate shows that carbon steel corrodes at rates as high as 20 mg/cm^2-hr (25 μm/hr). The use of inhibitors can greatly reduce this (see Table IV).

TABLE IV. Effects of Inhibitors in Dibasic Ammonium Citrate Solutions[(a)] on Corrosion Rates of Carbon Steel[1]

Inhibitor	Concentration (g/liter)	Average Corrosion (mg/cm^2-hr) 60°C	85°C
None	---	0.14	5.1
Phenylthiourea	5	0.2	1.0
Acridine	1	0.2	0.7

(a) 8 wt % $(NH_4)_2HC_6H_5O_7$

TABLE V. Corrosion Data from Component Decontamination Using Inhibited Citrate-Oxalate Mixture[8]

Material	Corrosion Rate (μm/d)
304 Stainless Steel	0.6
410 Stainless Steel	2.4
Aluminum Bronze	6
17-4 PH	10.8
Tungsten Carbide	6
Nickel	8.4
Chrome Plating	0.6

During a four-hour decontamination of a fueling machine, Ontario-Hydro observed the corrosion rates given in Table V.[8] This decontamination was performed using a 6% solution of Decon Turco 4512A (an inhibited proprietary citrate-oxalate mixture) at 85°C.

NS-1. Results of the tests showed that for 1020 carbon steel, some low alloy steels (2-1/4 Cr-1 Mo) and the 400 series stainless steels, the corrosion rates were about 25 to 125 μm/yr.[9] The 300 series stainless steels yielded general corrosion rates of less than 125 μm/yr.[9] Copper, nickel, chromium, and specialty alloys (such as Hastelloy and Stellite) gave, generally, similar or lower corrosion rates.[9] The effect of temperature in some cases was significant. Incoloy 800 at 205°F had corrosion rates of 175 μm/yr whereas at 275°F the reported rates were ten times as large.

Significant effort was also placed on studying stress corrosion cracking. The study showed that Fe^{+3} or Cr^{+6} in concentrations greater than 0.12 wt% were required to cause intergranular stress corrosion cracking in descaled sensitized 304 stainless steel.[9] Cracking also occurred with sensitized, scaled, 304 stainless steel in the same solutions without the added iron or chromium.

Waste Solidification

Phosphoric acid. Phosphoric acid, in 8-10% concentration is strongly acidic. It can be solidified by mixing with an absorbant (such as Zonolite) and cement. There are also other proprietary processes used.

AP-Citrox. The most probable action with waste AP-Citrox is to mix both chemicals together which neutralizes them. The order of mixing is important to prevent possible evolution of ammonia. In addition to neutralizing the chemicals, the mixing reduces permanganate to manganese dioxide which precipitates and removes much of the radioactivity from the liquid.

No difficulties are anticipated in mixing this product material with cement. The use of bitumen, however, can be hazardous due to the presence of manganese dioxide, a strong oxidant.[3]

Chemical similarity of urea formaldehyde and vinylester suggests that the plastic processes will be acceptable for AP-Citrox waste.

NS-1. No information has been obtained on the compatibility of NS-1 with bitumen. In various discussions it was found that NS-1 is not compatible with urea formaldehyde or cement and is one reason that the Dow vinylester system was developed.

Liquid Storage Capacity

For concentrated processes decontamination, the volume requirement is large for liquids. At all times during a decontamination there needs to be at least one system volume of rinse water available to remove the corrosive chemicals from the equipment.

For a single step decontamination, a minimum of two system volumes would be required with larger amounts of storage depending on the system and desired procedures. For two step decontamination processes, a minimum of six system volumes storage capacity is needed.

Radiation Exposure

During a reactor decontamination using a concentrated process, such as that proposed for Dresden-1, radiation exposures in the radwaste treatment system occur as a result of work associated with maintenance and with the concentration, solidification, and disposal of the used decontamination solution and the contaminated rinse water.

If, as indicated by our survey of reactor operators, there are solidification facilities that are not operating, a means of solidifying waste must either be brought in or built onsite.

The amount of liquid storage needed for a one-step decontamination process and first rinse is approximately 200,000 gallons, assuming that the amount of decontamination solution and rinse water for the first rinse is about equal to one primary system volume. At least one additional rinse may be necessary which raises the total liquid used to 300,000 gallons. Decontamination solution and rinse water may be directed to separate tanks.

Some holdup of corrosion products can be expected in the radwaste treatment system. If the system is new, a rapid buildup of radioactivity could occur. A study done by Babcock and Wilcox found that radiation fields increased steadily during initial stages of operation in which only normal solutions were processed.[10] Decontamination solutions are expected to behave similarly.

Contamination in the form of deposits, scale and sludge forms on the evaporator shell walls, heat exchanger tube bundles and bottom well. The major portion of the exposure results from these deposits.[10,11]

Radiation exposure rates near piping in the radwaste treatment system may be five or six orders of magnitude greater for the decontamination waste than for normal waste. Shielding of piping is, therefore, an important consideration in the design of a radwaste system which is to handle decontamination wastes. Additional exposure may occur as a result of plateout or deposition of corrosion products in low-velocity flow areas such as instrument lines, near bends in the pipe, or low points in the system or deadlegs.

Plateout or deposition of corrosion products is of concern during any maintenance which may be required during or following treatment of decontamination waste. For normal radwaste system operations, the deposited corrosion products in evaporators, tanks and piping contribute almost all the radiation field near these components. In some cases, the deposited corrosion products may even result in a higher exposure rate when the system is drained. If this also holds true for decontamination wastes, it may be impossible to do any maintenance in some cases and necessitate removal of equipment by remote operations before maintenance can begin. For example, pumps are not expected to accumulate much radioactivity and could be maintained with hands-on operations. However, if pumps are located in close proximity to waste evaporators, bends in piping or deadlegs, they may need to be removed from the system before any maintenance can be conducted. Such factors must be considered in design of radwaste treatment systems for decontamination wastes.

The evaporator concentrate is placed in 55-gallon drums and solidified. Radiation exposure calculations for the concentrated process are based on removal of 6000 Ci of ^{60}Co. It is likely that the waste will not be thoroughly mixed.

Exposure rates for solidified wastes are listed in Table VI. Radiation exposure rates that are this much greater than ordinarily encountered may lead to problems not only for handling the waste, but for shipment and/or disposal of the waste.

One other waste problem is associated largely with the concentrated process. During the decontamination operation, there will be some leaks in the primary and the radwaste system due to action by the solution. The liquid from these leaks are best caught on some sort of absorbing media so that concentrate in the containment vessel does not absorb the solvent. This absorbed solvent must somehow be removed from the containment vessel and entered in the waste process stream. Significant radiation exposure of workers could result during this operation.

TABLE VI. Exposure Rates from Solidified Waste in 55-Gallon Drums

Description of Waste	Exposure Rate (R/h) Surface
Decontamination Waste	10-25
Normal Solidified Radwaste	0.00005-1[4] (Average 0.185)[4]

COST CONSIDERATIONS

The cost of preparing for decontamination is dependent on many factors that have no firm value and on a few factors that are relatively easily defined. Some of these factors are discussed in this section.

When considering a decontamination effort, the reason for the decontamination must be defined and the number of decontaminations which may be done should be considered. This will help define the economics of decontamination and the radwaste treatment system.

Licensing

The areas of licensing and environmental impact must also be considered. The degree to which these topics must be addressed will significantly affect the cost of decontamination. Discussions with some of the contacts suggested licensing costs can range from 25 to 100% of the facility cost. No licensing costs are included in the estimates given below.

Design

The estimates provided here are preconceptual estimates, sometimes referred to as scoping estimates. These estimates are typically prepared on the basis of dollars/ft^2 and were obtained from recently completed, similar nuclear construction projects. These data are available from the Department of Energy.

Dollar/ft^2 data are typically presented in the form of ranges. Considering that the facilities and equipment described in this report would have to meet the appropriate seismic and tornado criteria in the Nuclear Regulatory Guides, the high end of the range was selected. The figure is about \$650/ft^2.

Mechanical Decontamination. Mechanical processes are used on a localized basis and, therefore, will generate a relatively small quantity of waste. It is assumed that existing tanks will be used; these may require additional shielding.

Because the size of equipment decontaminated and the amount of contamination is unclear, no effort is made to estimate costs of facilities or shielding.

Dilute Process Decontamination. It has been estimated that three ion exchange columns are needed for system cleanup plus one for chemical regeneration during decontamination. With only limited capacity available in the existing facility, and to keep it from becoming badly contaminated, it is assumed that a small shielded facility will be constructed adjacent to the existing radwaste facility. The new facility will contain room for four columns plus associated pumps and filters. The new facility will also contain storage space for hot radioactive resin.

Two concepts are considered -- a conventional, hands-on maintenance facility, as shown in Figure 1, and a facility designed for remote* operation, as shown in Figure 2.

For the hands-on operation, such as in Figure 1, the estimated facility cost, including process equipment, is about \$2.6M. In addition, \$0.8M is required for containment penetrations and \$0.5M for in-reactor piping.

The remote facility, as shown in Figure 2, which would be expected to have lower radiation exposure costs, has a higher capital cost of about \$4.4M. Additional costs are for containment penetrations and piping.

Concentrated Process Decontamination. Present radwaste systems will not meet the needs for concentrated decontamination for reasons discussed above concerning materials, throughput, storage constraints and shielding. It is assumed, therefore, that a new facility will be constructed. This new facility will contain six tanks (each 100,000 gallons), an evaporator, and associated pumps. Space is allocated for solidification and drumming.

Both a hands-on maintenance facility and a remote facility are estimated (see Figures 3 and 4). The hands-on facility is a schematic made from drawings of the new Dresden-1 radwaste system. Using \$650/ft^2 as a facility cost, the Dresden facility is estimated at about \$29M, which is close to the \$28M reported in our informal industry survey.

The remote facility, as shown in Figure 4, is based on the Dresden facility but with a revised layout. This facility is estimated to cost about \$38M.

---

* As used here, remote means that there are no hands-on steps in either normal operation or maintenance operations.

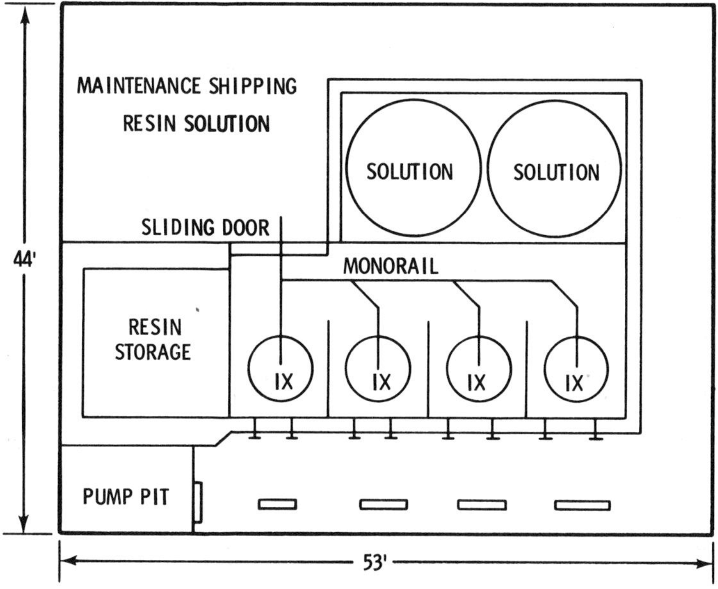

FIGURE 1. Four Column Hands-On Radwaste Facility
Dilute Process

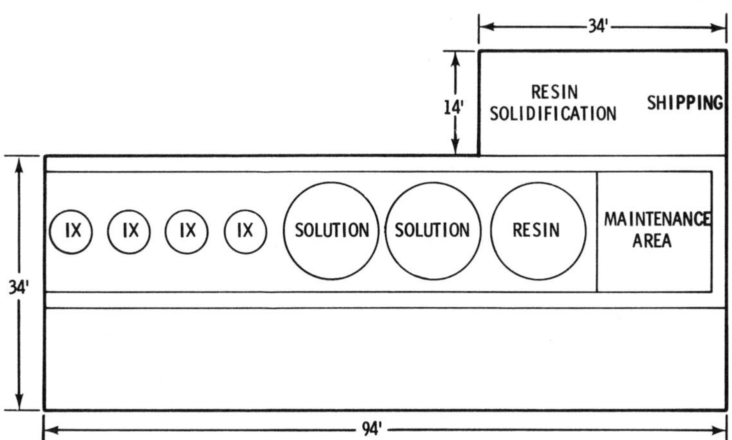

FIGURE 2. Four Column Remote Radwaste Facility
Dilute Process

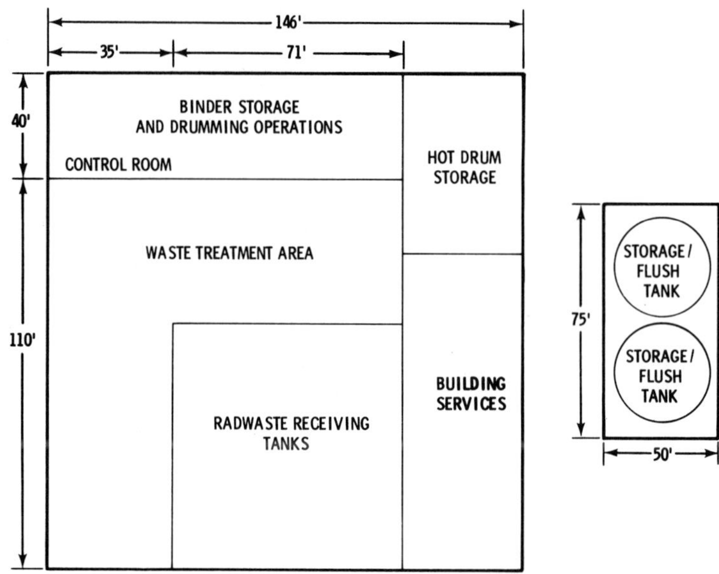

FIGURE 3. Radwaste Facility at Dresden-1[15]

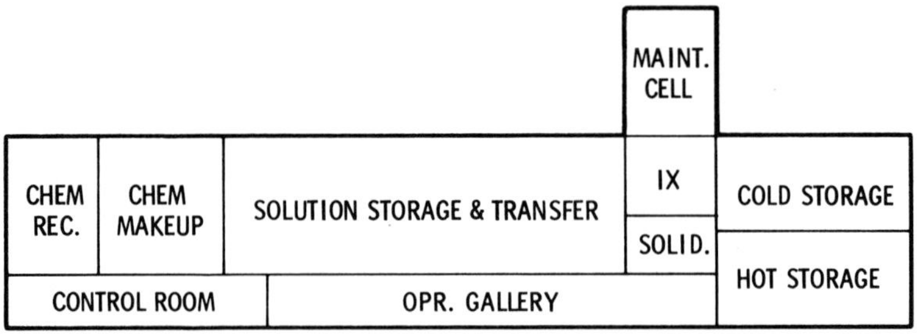

FIGURE 4. Remote Radwaste Facility
Concentrated Process

The cost of an add-on facility is the only cost considered. The cost of including the needed facility in the original reactor construction is not estimated but it would be less.

SAFETY CONSIDERATIONS

In this subsection, factors concerned with safety that would arise from the use of decontamination are considered.

General Considerations

The information presented on general considerations will emphasize concepts and provide a "roadmap" to consider the impact of decontamination on safety. OSHA and state requirements will control and direct these types of analyses and the conditions that should be imposed upon such operations.

Chemicals and Chemical Toxicity

When chemical decontamination of a reactor primary system is undertaken, sizable quantities of chemicals are involved leading to a change from power generation to chemical processing. The operators of nuclear reactor systems are trained to handle the mechanical, electrical, health physics and nuclear tasks necessary to generate electric power. When appreciable amounts of chemicals are introduced, chemical processing and other backgrounds are required. This shift in background needs and experience has often led conventional and nuclear power utilities to choose chemical service organizations to undertake these operations. When that is done, the service personnel have to be familiarized with site conditions, procedures and operations.

Whenever a safety assessment is made there is a need to involve knowledgeable safety professionals. Industrial hygienists will be able to assess what the major impact will be on personnel from the chemicals that may be used for decontamination, including the need for special toxicity measurements.[12]

Fire Prevention

The primary emphasis on fire prevention techniques is expected to be focused on chemicals and chemical operations that are a part of a decontamination operation. The decontamination materials or their reaction products that are discharged to the radwaste system are potential explosion and fire hazards.

Fire prevention has another facet. The most significant impact of temporary or modified facilities and equipment is on the existing fire prevention procedures at the plant undergoing decontamination. Because there are unusual or off-standard operations, access to buildings and facilities may be blocked or impeded.

Working Conditions

Decontamination, because it is not a routine operation, will have a significant safety effect on working conditions. Since decontamination is an off-standard operation, all of the routines and approaches to reactor and radwaste treatment system operation will have to be rethought. Detailed operating procedures must be prepared and approved for those new conditions. A very detailed training program is needed to teach the operational crews to follow approved procedures. Otherwise, the tendency for the operators is to fall back on standard procedures.

Probably the most important effect on working conditions is stress on personnel. Any kind of sustained campaign will place stress on personnel if they are required to function in a new and unusual manner or if they are retained at work for prolonged periods of time. The work of Morgan, et al.,[12] showing the difference between performance and capacity as a function of successive periods of continuous work is illustrated in Figure 5. We recommend that any decontamination operation have at least two different crews to undertake the operation.

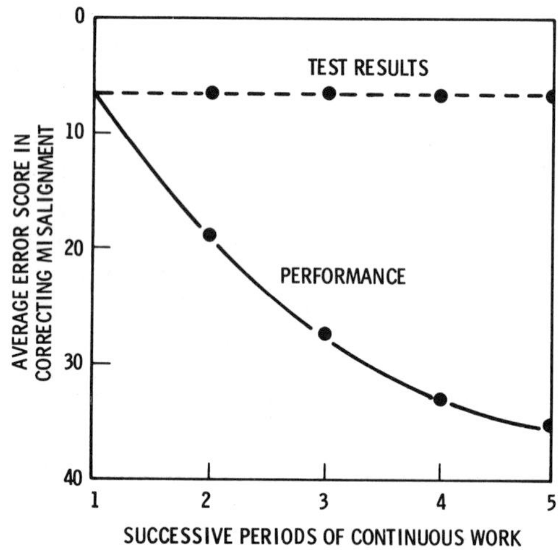

FIGURE 5. Impact of Successive Periods of Work on Work Errors[14]

Organization

It is especially important that off-standard operations have organization structures that clearly show lines of authority and areas of responsibility. Because different types of people, observations, risks and procedures are encountered, revised organizational structures are needed to ensure that safe conditions are maintained. Regular reactor and radwaste operating maintenance crews should be included in such an organization. Others, such as the chemical operators, who are involved in these special operations must also be included.

Specific Considerations

A discussion of generic problems is presented with conditions that might be encountered during the decontamination of a nuclear reactor and the handling of the waste from that operation. This treatment is not inclusive but is intended to focus on the implications of generic problems.

Mechanical Decontamination

One of the most important problems arising from mechanical decontamination is the possible reaction of metal particles that are generated during such processes with acidic solutions in the radwaste treatment system. Such reactions result in the generation of hydrogen. Design of these facilities must be done in such a way that it can accommodate hydrogen evolution.

Another safety impact is the activation of metal particles that are left in the primary system. These will become extremely radioactive as they pass through the core, and may find their way into the radwaste system. These highly radioactive particles will likely enter the system after the decontamination has been completed and will cause much higher radiation levels than would normally be expected.

Dilute Process Decontamination

The primary safety concerns for dilute processes will be those having to do with chemicals and chemical toxicity, but specific care should be addressed to chelating compounds and oxalic acid that are frequently used in such processes. Chelating agents such as EDTA, if ingested, are known to strip calcium from the body causing specific physiological problems. The ingestion of oxalic acid is very dangerous leading to the precipitation of calcium in the kidneys, causing irreparable damage to the body.

Concentrated Process Decontamination

Probably the most important safety problems for concentrated chemical decontamination comes from the large amounts of chemicals that are used in such operations. The 40 tons or more that are expected for such operations must be properly stored, mixed, introduced into the reactor system, removed from that system, introduced into the radwaste treatment system and processed for disposal in a safe and acceptable manner. Some processes may generate sizable amounts of ammonia gas. This gas must be accommodated in the design of the system or arrangements made to control the production of ammonia gas.

WASTE DISPOSAL ALTERNATIVES

Handling and disposal of wastes generated by decontamination is similar for any of the processes which have been discussed. Perhaps the only significant differences are the waste volumes generated and the radioactive content of the resultant concentrates.

There are currently only three operating disposal sites. Any of these sites would be capable of disposing of wastes from decontamination, providing it has been solidified.

For waste received at a disposal site, restrictions are based on being physically able to handle the waste. The disposal companies want to keep occupational exposures as low as reasonably achievable and assure that offsite doses remain well below existing standards. If a shipment exceeds the normal capacity of the disposal operation in terms of size, weight or exposure rate, increased operational costs will be passed on to the customer for any special equipment or procedure which must be used.

It does not seem likely that there is any great obstacle to disposing of decontamination wastes given the current conditions. There is an indication, however, that the states in which the disposal sites are located could establish more stringent guidelines regarding radioactive waste disposal. How this situation would impact disposal of decontamination wastes must await decisions by the state governments.

One alternative to immediate disposal of decontamination wastes is to store the waste onsite. Temporary storage onsite is not entirely a new concept. All waste must be kept onsite for some time before it is shipped to a disposal site and, at Surry Power Station, steam generators containing many curies of ^{60}Co are being stored onsite until the plant is decommissioned.[13] Licensing problems may exist, but these can probably be overcome.

If waste is stored onsite the amount of shielding used should be sufficient to ensure that doses are within regulatory limits for the site. Location of the storage building onsite would have to be such that normal traffic flow (pedestrian and motorized) would not pass near the structure with great frequency.

The particular option for waste disposal that a specific site chooses depends on the site location, regulations affecting the site, the total inventory to be stored, availability of burial space, costs for the various options, land available onsite for a storage building, and many other factors. The important issue is that there be a safe and reasonably economic way to eventually dispose of the waste permanently.

CONCLUSIONS

The Pacific Northwest Laboratory study of the influence of decontamination on LWR radwaste treatment systems produced several key conclusions and recommendations. These are listed below:

1. Only at N-Reactor is there a means to accommodate radwaste produced during decontamination. The Dresden system is expected to be ready to accommodate such solutions by the summer of 1979. Other radwaste treatment systems at power reactors would require extensive modifications to accommodate radwaste from decontamination.

2. Solidification of the processed decontamination waste may be a significant problem. Solidification agents may not be compatible with the chemicals in decontamination wastes.

3. The disposal of the large amounts of concentrated wastes resulting from the operation of the radwaste treatment system may be a problem. Such waste has to be handled and stored onsite, transported and ultimately disposed of in an acceptable site.

4. Based on discussions with utilities, there is doubt that the materials of construction in current radwaste treatment systems are capable of handling the chemicals from a concentrated process.

5. The total storage volume, for concentrated decontamination, is not sufficient in existing radwaste treatment systems. It is also clear that the ion exchange systems are neither large enough nor properly shielded to handle highly radioactive dilute process waste. Current radwaste systems do not have remote operating capability. Such a capability may be required for decontamination wastes.

6. Greater attention should be placed on designing reactors and radwaste treatment systems for decontamination. Such design should consider the proper equipment layout and spacing to accommodate access, maintenance and modifications. In addition, adequate specification of materials and means for introducing and removing chemicals from reactors and radwaste treatment systems must be made.

7. A means of handling waste material resulting from leaks in the primary system during the decontamination must be developed. The radwaste system must be capable of accepting and handling these wastes.

8. Onsite storage of solidified decontamination wastes may be a viable options, but license amendments will be necessary.

9. To determine the cost incentives for decontamination long-term benefits associated with reduced occupational exposure and increased operating efficiency should be balanced against the immediate costs for the decontamination, including building modification, reactor downtime and personnel exposure.

10. Decontamination safety must be addressed. Analysis should consider chemical toxicity, changes in operating procedures, fire prevention and work conditions. Probably the most important effect on working conditions is stress on personnel and human fatigue.

REFERENCES

1. J. A. Ayres, "Decontamination of Nuclear Reactors and Equipment," Ronald Press, New York, 1970.
2. P. J. Pettit, J. E. LeSurf, W. B. Steward, R. J. Strickert, and S. B. Vaughan, Decontamination of the Douglas Point Reactor by the Can-Decon Process, paper delivered at the National Association of Corrosion Engineers' National Conference, Houston, Texas, March 1978.
3. A. H. Kibby, H. W. Godbee and E. L. Compere, "A Review of Solid Radioactive Waste Practices in Light-Water-Cooled Nuclear Reactor Power Plants," NUREG/CR-0144 (ORNL/NUREG-43), prepared by ORNL for the NRC, October 1978.
4. T. B. Mullarkey, T. L. Jentz, J. M. Connelly, and J. P. Kane, "A Survey and Evaluation of Handling and Disposing of Solid Low-Level Nuclear Fuel Cycle Wastes," NUS Corporation, Report No. AIF/NESP-008, October 1978.

5. Corrosion Engineering Bulletin CEB-4, "Corrosion Resistance of Nickel-Containing Alloys in Phosphoric Acid," International Nickel Col, New York.

6. W. K. Kratzer, "Final Report, Production Test N-383, 1975 N Reactor Internal Decontamination," United Nuclear Report UNI-355-A, Richland, Washington, December 31, 1975.

7. W. K. Kratzer, "Final Report: Production Test N-442, 1978 N Reactor Internal Decontamination," United Nuclear Report UNI-1018C, Richland, Washington, December 14, 1978.

8. C. S. Lacy, W. B. Stewart, A. B. Mitchell, "Decontamination Experience at CANDU-PHW Reactor," Paper No. 43, Water Chemistry of Nuclear Reactor Systems, BNES, London 1978.

9. Dow Chemical Company, "Technical Study for the Chemical Cleaning of Dresden-1," Dow Chemical Company Report No. DNS-D1-016, June 15, 1977 (Vol. 1-8).

10. P. J. Grant, D. F. Hallman and A. J. Kennedy, "Oconee Radiochemistry Survey Program, Semi-Annual Report," [D. L. Uhl, ed.], Babcock & Wilcox, Report LRC-9042, March 1975.

11. P. J. Grant, A. J. Kennedy, D. F. Hallman, E. T. Chulik and D. L. Uhl, "Oconee Radiochemistry Survey Program," Babcock & Wilcox, Report No. RDTP1 75-4, May 1975.

12. Clifford T. Morgan, Alphonse Chaparies, Jesse S. Cook, III and Max W. Lund, eds., "Human Engineering Guide to Equipment Design," McGraw-Hill, New York, 1963.

13. Virginia Electric Power Company, "Steam Generator Project, Surry Power Station Units Numbers 1 and 2," USNRC Docket Nos. 50-280 and 40-281, June 1978.

OCCUPATIONAL EXPOSURE AND ALARA

Charles S. Hinson and Thomas D. Murphy

Office of Nuclear Reactor Regulation
U.S. Nuclear Regulatory Commission
Washington, D. C. 20555

INTRODUCTION

In order to maintain radiation doses to maintenance, contractor and operating personnel as low as are reasonably achievable, it is necessary to provide design features and operating procedures to reduce radiation fields around reactor coolant system components, to reduce the time needed for maintenance operations, to reduce the frequency of such operations, and to reduce the number of people involved in such operations. Plant design features intended to reduce occupational doses during plant operation also serve to lessen personnel doses during the eventual plant decommissioning. Recent recommendations by the NRC Regulatory staff,[1] under consideration by the Commission, may require that industry provide additional means to reduce occupational radiation exposure. This paper outlines the regulatory considerations for further controlling doses to personnel at nuclear power plants.

OCCUPATIONAL EXPOSURE DATA AT COMMERCIAL LWRS

The NRC has requirements for reporting occupational radiation exposure data for licensee power reactors in both 10 CFR Part 20 and in Regulatory Guide 1.16. The analyses in this section is drawn from data received in response to these NRC requirements and reported annually by the NRC.[2] Table 1 shows that the average man-rem per reactor per year has been increasing for reactors since 1969 at a rate roughly equivalent to the increase in the average megawatt (electric) years generated per reactor per year and the average rated plant capacity. This data is shown graphically in figure 1. Average exposure per individual has remained relatively even in the same time period while the average number of personnel per reactor with measurable exposure has increased, as shown in Table 2. Table 3 provides a breakdown of the 1977 data reported to the NRC by work and job function. This table clearly shows that the majority of the dose is received by maintenance workers performing routine and special maintenance activities. Table 4 indicates that these observations have been reasonably consistent over the four years that the NRC has received data in a format which allows a breakdown of exposures by work and job function.

CAUSES OF INCREASING COMMERCIAL REACTOR PLANT EXPOSURES

The staff has observed that the primary causes of the increasing radiation exposures at power reactors are: (1) increasing radiation fields around reactor components, (2) a relatively constant need to perform routine and special maintenance activities at reactor plants, and (3) several recent, widespread unexpected repair efforts on major reactor components.

(1) Increasing radiation fields

Several recent studies have documented the increasing radiation levels associated with reactor coolant components at U.S. nuclear power plants. Based on extensive surveys of operating plant shutdown radiation levels, Sawochka, et. al. have plotted radiation level increases on PWR (Figure 2) and BWR (Figure 3) reactor coolant piping.[3] The important corrosion product radionuclides contributing to these increasing radiation levels are shown in Table 5. The nuclide contributing most to the dose is Co^{60}.

(2) Maintenance activities

The reasonably constant forced outage rate over the past several years for such reasons as steam generator tube leakage and pump and valve failure and maintenance lead us to believe that increasing radiation exposures to plant personnel may be more influenced by increasing radiation levels than by increased size or increased failure rate of equipment. Typical activities for which routine exposures are required by reactor plant personnel are listed in Table 6.

(3) Unexpected major reactor plant repair efforts

Although U.S. power reactors have had a reasonably constant outage rate in the last several years, there have been several major repair operations in recent years that have caused unexpected high collective radiation dose accumulations at several reactors. These are feedwater nozzle, pipe crack and control rod drive return line nozzle crack repairs in boiling water reactors and repair of steam generator tube degradation in pressurized water reactors.

TABLE 1

MAN-REM SUMMARY

	Number of Reactors	Average Rated Capacity* (MWe)	Average MW-Yr Generated	Average Man-Rem/ Reactor	Yearly Average Man-Rem/ MW-Year
a. SUMMARY - LIGHT WATER REACTORS					
1969	7	247	184	178	1.0
1970	10	300	189	350	1.8
1971	13	367	248	280	1.1
1972	18	408	311	364	1.2
1973	24	496	299	582	1.9
1974	34	575	319	404	1.3
1975	44	630	404	475	1.2
1976	53	668	413	499	1.2
1977	57	682	464	570	1.2
1978	64	702	494	497	1.0
b. PRESSURIZED WATER REACTORS					
1969	4	349	274	165	0.6
1970	4	349	245	684	2.8
1971	6	399	320	307	1.0
1972	8	446	318	463	1.5
1973	12	533	314	783	2.5
1974	20	619	341	331	1.0
1975	26	643	461	318	0.7
1976	30	684	444	460	1.0
1977	34	707	510	396	0.8
1978	39	723	509	428	0.8
c. BOILING WATER REACTORS					
1969	3	112	64	195	3.0
1970	6	267	152	127	0.8
1971	7	339	187	255	1.4
1972	10	434	306	286	0.9
1973	12	459	283	380	1.3
1974	14	513	290	507	1.7
1975	18	611	321	701	2.2
1976	23	647	373	549	1.5
1977	23	645	396	828	2.1
1978	25	668	471	604	1.3

*Maximum Dependable Capacity (Net) - the dependable gross electrical output as measured at the output terminals of the turbine generator during the most restrictive seasonal conditions (usually summer) <u>less</u> the normal station service loads.

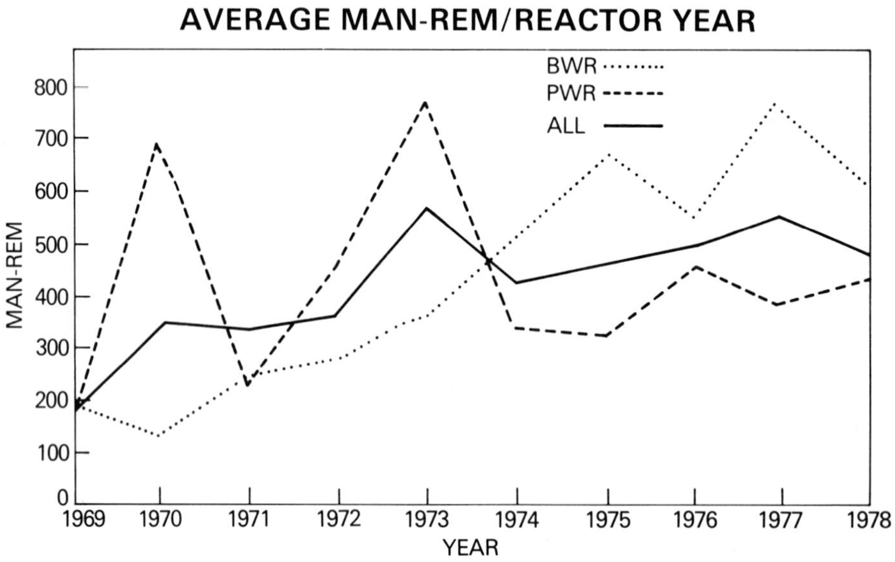

AVERAGE MAN-REM/REACTOR YEAR

BWR ·········
PWR - - - - -
ALL ———

MAN-REM

0
100
200
300
400
500
600
700
800

1969 1970 1971 1972 1973 1974 1975 1976 1977 1978

YEAR

FIGURE 1

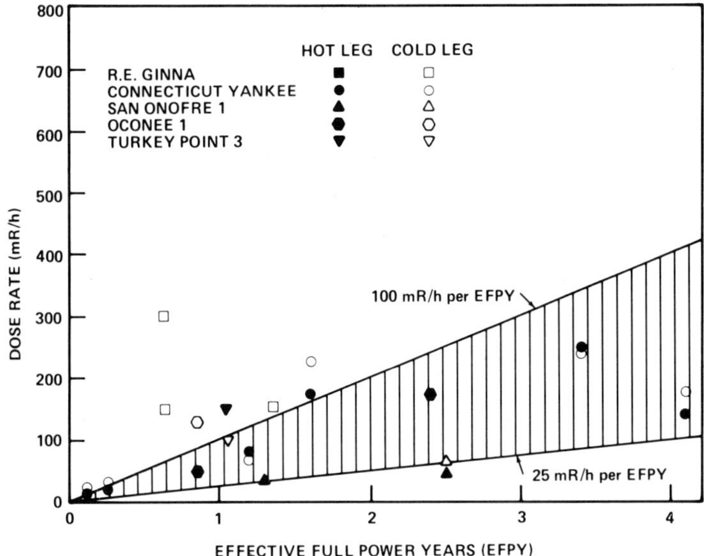

SHUTDOWN RADIATION LEVELS ON CURRENT GENERATION PWR PRIMARY SYSTEM PIPING

FIGURE 2

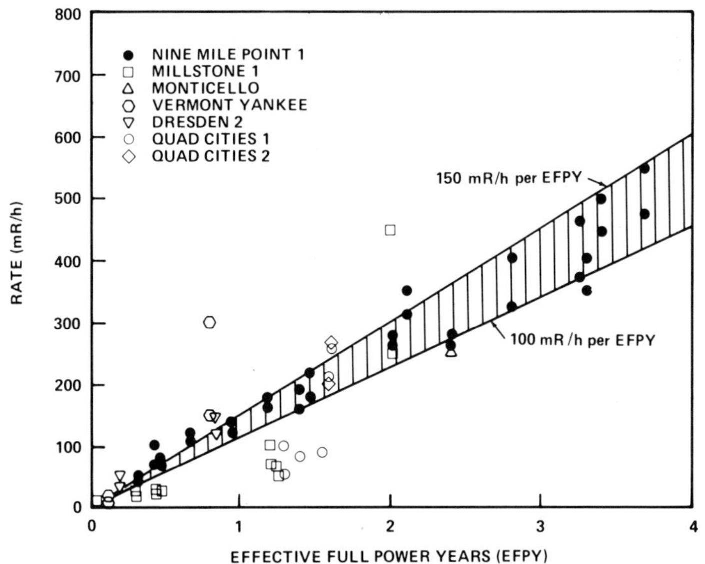

SHUTDOWN RADIATION LEVELS ON BWR RECIRCULATION PUMP SUCTION PIPING

FIGURE 3

TABLE 2

AVERAGE OCCUPATIONAL RADIATION EXPOSURE PER INDIVIDUAL

Year	Average Exposure Per Individual (Rems)	Average Number of Personnel With Measurable Exposure Per Reactor
1969	0.9	149
1970	0.6	380
1971	0.7	309
1972	1.0	345
1973	0.9	607
1974	0.7	543
1975	0.7	625
1976	0.7	669
1977	0.8	742

TABLE 3

PERSONNEL DOSE BY WORK FUNCTION (1977) PER DATA SUPPLIED BY LICENSEE IN ACCORDANCE WITH R.G. 1.16

TABULATION OF MAN-REM BY WORK FUNCTION

Work Function	Station Employees	Utility Employees	Contract Workers and Others	Total	
Reactor Operations	2,626	224	427	3,277	(10.5%)
Routine Maintenance	3,470	1,544	3,708	8,722	(28.1%)
In-Service Inspection	205	245	1,536	1,986	(6.4%)
Special Maintenance	1,700	2,123	6,398	13,221	(42.5%)
Waste Processing	1,207	36	554	1,797	(5.8%)
Refueling	852	390	829	2,071	(6.7%)
Totals	10,060	4,562	16,452	31,074	

TABLE 4

PERCENTAGES OF PERSONNEL DOSE BY WORK FUNCTION

Work Function	Percent of Dose 1974	1975	1976	1977
Reactor Operations and Surveillance	14.0%	10.8%	10.4%	10.5%
Routine Maintenance	45.4%	52.5%	31.7%	28.1%
In-Service Inspection	2.7%	2.9%	5.7%	6.4%
Special Maintenance	20.4%	19.0%	39.5%	42.5%
Waste Processing	3.5%	6.9%	4.8%	5.8%
Refueling	14.0%	7.7%	7.9%	6.7%

TABLE 5

IMPORTANT CORROSION PRODUCT ISOTOPES

Nuclides	Half-Life	Production Reaction
Co 60	5.3 Years	Co 59 (n,γ) Co 60
Co 58	72 Days	Ni 58 (n,p) Co 58
Cr 51	28 Days	Cr 50 (n,γ) Cr 51
Fe 59	45 Days	Fe 58 (n,γ) Fe 59
Mn 54	310 Days	Fe 54 (n,p) Mn 54

TABLE 6

TYPICAL ACTIVITIES CAUSING EXPOSURE

- Valve Repair, Maintenance and Repacking
- Eddy Current Testing of Steam Generator Tubes
- Repairs to Spent Fuel Pool, Liner and Upender
- Steam Generator Tube Plugging
- Solid Waste Drumming
- Reactor Internals Inspection
- Fuel Sipping and Inspection
- Pump Maintenance, Repair and Seal Replacement
- Feedwater Sparger Replacement
- Steam Leak Inspections
- Reactor By-Pass Line Replacement
- Control Rod Drive Mechanism Repairs and Overhaul
- Inservice Inspection
- Routine Instrument Maintenance
- Demineralizer Element Replacement and Resin Removal
- Piping Modifications
- Incore Instrumentation Replacement
- Refueling, Including Head Removal and Reinstallation
- Condensate Demineralizer Work
- Cleanup of Spills

NRC ACTIONS TO CONTROL RISKS ASSOCIATED WITH OCCUPATIONAL RADIATION EXPOSURES

Confronted with present uncertainties regarding radiation dose-health effect relationships, some of which are not likely to be resolved for many years, the staff has considered further regulatory strategy to assure that workers are adequately protected. In brief, the NRC staff has concluded that (1) a reasonable additional effort should be made to control further the overall risks associated with occupational radiation doses, and (2) appropriate control of risk can be achieved through regulatory action which places additional emphasis on maintaining occupational doses ALARA by making radiation protection programs inspectable and enforceable. It should be noted that such regulatory action has been recommended by the NRC staff irrespective of the question of whether dose limits should be lowered.

The NRC staff has considered many alternatives, and combinations thereof, for such additional regulatory action that would reduce radiation risk. The principal alternatives are summarized below:

Alternative 1: Amend 10 CFR Parts 20, 30, 40, 50 and 70 to require licensees to develop and implement occupational ALARA programs, with guidance on the program content to be given in regulatory guides tailored for the various types of licenses activities.

Alternative 2: Require these licensees to perform cost-benefit analyses for major tasks and to provide all safety procedures, equipment, and facilities that are shown to be cost-effective, using a dollars-per-man-rem criterion.

Alternative 3: Establish annual design and operational collective dose (man-rem) objectives (goals) for various types of licensees.

Alternative 4: Continue to implement the occupational ALARA concept through the license review process and the issuance of regulatory guides that address ALARA in different types of facilities, training of workers, surveys, etc., but without amendment of the regulations requiring licensees to develop and submit ALARA program for NRC evaluation and incorporation into the license.

Alternative 5: Impose additional specific design criteria in 10 CFR Part 50, Appendix A, to reduce occupational radiation doses at power reactors.

It is the NRC staff's proposal (SECY-78-415 of July 29, 1978) that implementation of Alternative 1 will result in control of the current trend of increasing collective dose in a reasonable and effective manner. As mentioned above, this alternative would require licensees to develop and implement individual occupational ALARA programs, with guidance on the program content to be given in regulatory guides tailored for the various types of licensed activities. In developing these occupational ALARA programs, licensee-management would have to consider the ALARA aspects of each radiation work effort, treating specific factors involved in the effort within the guidance to be provided in regulatory guide format. The NRC staff would evaluate the ALARA programs using the regulatory guides and commonly accepted approaches to worker protection to judge their acceptability. These regulatory guides will be developed after evaluation of public comments on the proposed rules and will be available in final form when the amendments are published to be effective. Six months after the effective date of the new regulations, IE inspectors would begin to determine that the licensee has developed an adequate ALARA program and is implementing the program. The NRC staff proposes that these amendments be applied to all licensees who are required by the regulations, technical specifications, or license conditions to perform personnel monitoring, bioassays, or air sampling.

In addition to whatever alternative or combination of alternatives is selected, the NRC staff has proposed certain related amendments. These amendments would eliminate the Form NRC-4 exposure history and impose annual dose-limiting standards while retaining quarterly standards, including 5 rems per year, 3 rems per quarter to the whole body. Related amendments would conform other sections of the regulations to express, in terms of the new annual standards, the standard for dose to minors, the requirements for the provisions of personnel monitoring equipment, and the requirements for control of total dose to transient and moonlighting workers.

REGULATORY GUIDE 8.8 AND REACTOR DECONTAMINATION/DECOMMISSIONING

The document presenting ALARA design guidance is Regulatory Guide 8.8, Information Relevant to Ensuring that Occupational Radiation Exposures at Nuclear Power Stations Will be As Low As Is Reasonably Achievable, Revision 3, 1978. For reactors, the NRC will use this guide as the primary basis of acceptability of licensee's ALARA programs. For reactors undergoing construction permit and operating license review, the NRC staff concentrates on the design considerations for reducing radiation exposures and radiation fields. The guidance provided in Regulatory Guide 8.8 for the control of activated corrosion products (crud) is of particular interest:

> Design features of the primary coolant system, the selection of construction materials that will be in contact with the primary coolant, and features of equipment that treat primary coolant should reflect

considerations that will reduce the production and accumulation of activated corrosion products (crud) in areas where it can cause high exposure levels The following items should be considered in the crud control effort:

(1) Production of Co-58 and Co-60, which constitute substantial radiation sources in crud, can be reduced by specifying, to the extent practicable, low-nickel and low-cobalt bearing materials for primary coolant pipe, tubing, vessel internal surfaces, heat exchangers, wear materials, and other components that are in contact with primary coolant. Alternative materials for hard facings of wear materials of high-cobalt content should be considered where it is shown that these high-cobalt materials contribute to the overall exposure levels. Such consideration should also take into account potential increased service/repair requirements and over-all reliability of the new material in relation to the old.

(2) Loss of material by erosion of load-bearing hard facings can be reduced by using favorable geometrics and lubricants, where practicable, and by using controlled leakage purge across journal sleeves to avoid entry of particles into the primary coolant.

(3) Loss of material by corrosion can be reduced by continuously monitoring and adjusting oxygen concentration and pH in primary coolant above 250°F and by using bright hydrogen-annealed tubing and piping in the primary coolant and feedwater systems.

(4) Consideration should be given to cleanup systems (e.g., using graphite or magnetic filters) for removal of crud from the primary coolant during operation.

(5) Deposition of crud within the primary coolant system can be reduced by providing laminar flow and smooth surfaces for coolant and by minimizing crud traps in the system to the extent practicable.

In addition, Regulatory Guide 8.8 addresses the removal of activated corrosion products (decontamination).

Potential doses to station personnel who must service equipment containing radioactive sources can be reduced by removing such sources from the equipment (decontamination), to the extent practicable, prior to servicing. Serviceable systems and components that constitute a substantial radiation source should be designed, to the extent practicable, with features that permit isolation and decontamination. Station design features should consider, to the extent practicable, the ultimate decommissioning of the facility and the following concerns:

(1) The necessity for decontamination can be reduced by limiting, to the extent practicable, the deposition of radioactive material within the processing equipment--particularly in the "dead spaces" or "traps" in components where substantial accumulations can occur. The deposition of radioactive material in piping can be reduced and decontamination efforts enhanced by avoiding stagnant legs, by locating connections above the pipe centerline, by using sloping rather than horizontal runs, and by providing drains at low points in the system.

(2) The need to decontaminate equipment and station areas can be reduced by taking measures that will reduce the probability of release, reduce the amount released, and reduce the spread of the contaminant from the source (i.e., from systems or components that must be opened for service or replacement). Such measures can include auxiliary ventilation systems, treatment of the exhaust from vents and overflows, drainage control such as curbing and floors sloping to local drains, or sumps to limit the spread of contamination from leakage of liquid systems.

(3) Accumulations of crud or other radioactive material that cannot be avoided within components or systems can be reduced by providing features that will permit the recirculation or flushing of fluids with the capacity to remove the radioactive material through chemical or physical action. The fluids containing the contaminants will require treatment, and this source should be considered in sizing station radwaste treatment systems.

As power reactors increase in age, the buildup of crud within the primary system leads to increasing radiation fields. The subsequent increase in occupational exposures is a problem which can impede operational maintenance and inspection functions. High plant radiation levels can also have a negative impact on end-of-plant-life decommissioning operations.

The Commission is continuing its studies on plant design features to control exposure and facilitate plant decommissioning. The decontamination operations at Dresden 1, the Surry and Turkey Point steam generators replacement programs, and the Indian Point 1 secondary side steam generator chemical decontamination will all yield valuable information that can be used to complement existing exposure reduction and decommissioning techniques. In FY 1975 and FY 1976, the Commission initiated studies of the decommissioning safety and costs of nuclear fuel cycle facilities and of light-water-cooled reactors.[4] Battelle Pacific Northwest Laboratories was selected to perform the greatest part of these studies. Initial findings from these studies suggest that the following general criteria be considered in the selection of plant design features:[5]

- Decreasing decommissioning cost
- Improving occupational or public safety
- Reducing total decommissioning time
- Creating less volume of radioactive wastes
- Improved ease of performing the decommissioning

SUMMARY

As discussed above, the NRC has under consideration certain additional actions to be taken to control the collective dose to NRC licensee's radiation workers to levels as low as is reasonably achievable. The actions that are being considered will be subjected to significant public scrutiny after publication in the FEDERAL REGISTER, including a hearing. In the meantime, we are concerned that the trend of increasing radiation fields and increasing collective doses at reactor plants may lead to unacceptable risks to radiation workers at nuclear power plants. For this reason, it is necessary that the industry and licensees continue to implement the ALARA concept with the utmost vigor.

One approach that may be effective in reducing risks to radiation workers is to reduce radiation fields in areas where significant reactor plant maintenance work is required. In addition, continued review of the technology associated with design and layout of nuclear plants is required to assure that the latest, cost effective and exposure reducing approaches are used. Finally, in the operation of nuclear power plants, consideration of dose reducing activities must constantly be a priority of plant personnel.

REFERENCES

[1]U.S.N.R.C. Commission Paper, SECY 78-415, Further Actions to Control Risks Associated with Occupational Radiation Exposures in NRC-Licensed Activities, R. B. MINOGUE, July 31, 1978.
[2]NUREG-0482, LINDA JOHNSON PECK, Occupational Radiation Exposure at Light Water Cooled Power Reactors, 1977, U.S.N.R.C., May 1979.
[3]SAWOCHKA, S. G., N. P. JACOB, W. L. PEARL, Primary System Shutdown Radiation Levels at Nuclear Power Generating Station, EPRI-404-2, December 1975.
[4]NUREG-0436, Plan for Reevaluation of NRC Policy on Decommissioning of Nuclear Facilities, U.S.N.R.C., December 1978.

[5]NUREG/CR-0130, R. I. SMITH, G. J. KONZEK, W. E. KENNEDY, JR., Technology, Safety and Cost of Decommissioning a Reference Pressurized Water Reactor Power Station, Report for U.S.N.R.C. by Pacific Northwest Laboratory, June 1978.

RADIATION CONTROL IN PWR PRIMARY SYSTEMS

Y. Solomon and J. D. Cohen

Westinghouse Electric Corporation
P. O. Box 355
Pittsburgh, PA 15230

INTRODUCTION

In early 1977, the Electric Power Research Institute (EPRI) and Westinghouse (W) entered into a contract to examine the long-term radiation trends in PWR reactor plants. The motivation for EPRI support of this type of research, stems from an interest in improving plant availability and reliability. In recent years, utilities have recognized that radiation buildup in primary coolant loops create difficulties in plant maintenance and repair.[1,2,3,4] The problem of radiation exposure to maintenance workers is becoming more severe every year.

Many approaches can be taken to alleviate this continuing problem. For example, remote tooling, temporary shielding, close supervision, additional mock-up training, strict control of access, local decontamination, etc. However, the most lasting way to eliminate this difficulty is to reduce the source. Therefore, the industry encouraged EPRI to fund a program to evaluate the long-term buildup or at least to develop techniques which might temporarily modify radiation sources during a shutdown period. Westinghouse was chosen as one of the technical investigators in a total industry wide research program.[5] This paper will describe the EPRI-W program and briefly report some of the significant results which have been obtained in the past 2 years.

PROGRAM OVERVIEW

One of the major considerations in developing a radiation control program is that the techniques to be used must not affect plant availability in any significant way. The objective is to identify the effect on radiation fields of various chemistry and operation factors.

A review of radiation field data revealed that the available data were not consistent. Therefore, the first objective was to develop standard radiation monitoring procedures which would then be used as a tool in the studies. Other objectives were to identify the effect on radiation fields of various types of plant operation, different primary coolant chemistry treatments and materials selection.

The total program is planned to cover a multi-year period and will investigate the source of corrosion products and the effect of operation on their transport in the Reactor Coolant System (RCS). Specific items to be studied include pH control, hydrogen peroxide additions, boric acid control, in-plant operational radiation levels, laboratory work regarding corrosion product solubilities and mathematical modeling of transport and deposition.

STANDARD MONITORING PROGRAM

A Standard Radiation Monitoring Program was defined to enable consistent sets of data to be obtained from various reactor plants. This program (in operation at eight operating plants, planned for twelve additional plants - domestic and foreign) defines the location and techniques of radiation level measurements after reactor shutdown. Data are obtained at the location of permanent markers on the outside surface of reactor coolant system components. Radiation surveys are also specified for the steam generator channel head internals when maintenance or inspection activities are planned.

In addition to the radiation surveys, certain chemistry data which may be important to the deposition, release, and transport of corrosion products are recorded. Table 1 defines the chemistry analysis and radiation surveys required for this monitoring program. Figure 1 shows the key locations of the standard radiation survey.

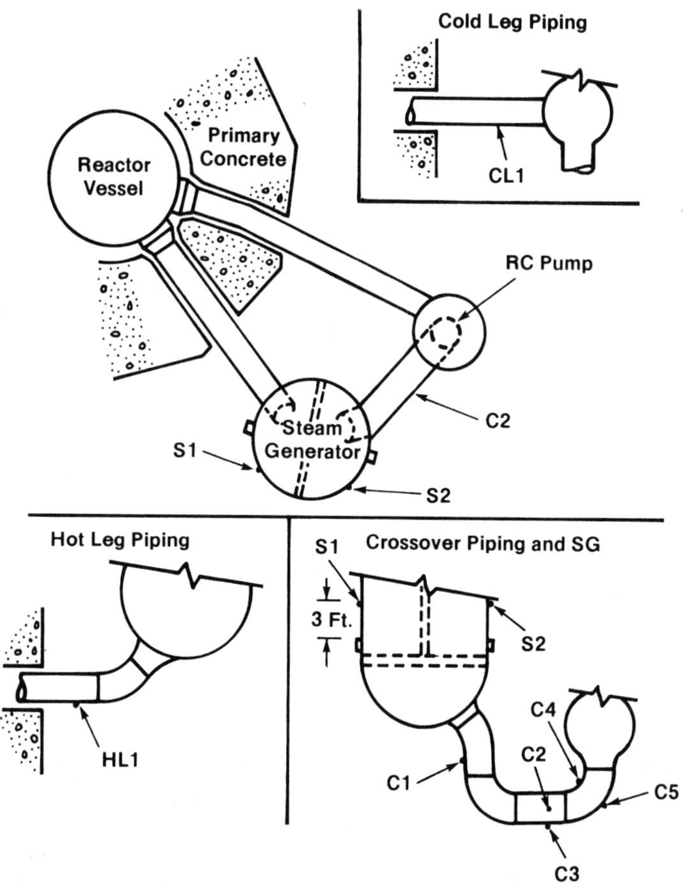

FIGURE 1. REACTOR COOLANT LOOP SURVEY MARKER LOCATIONS

TABLE I EPRI-WESTINGHOUSE STANDARD RADIATION MONITORING PROGRAM DATA REQUIREMENTS

COOLANT ANALYSES DURING NORMAL OPERATION	
Parameters	**Recommended Frequency**
Boron concentration	Daily
Lithium concentration	Three times per week
SHUTDOWN RADIATION SURVEYS	
Location	**Recommended Surveys**
Reactor coolant loop	Once after reaching hot shutdown conditions and/or at least 24 hours after reducing to zero power
	Once after reaching cold shutdown conditions. Plant should be at cold shutdown for at least 1 week prior to survey.
Steam generator channel head	Prior to primary side inspection or maintenance of steam generator
	Additional surveys as required by plant staff or plant procedures

Although this standard technique improves the consistency of after shutdown radiation level data, there are some difficulties with inter-plant comparisons of loop data. For example, local arrangements and shielding are not identical, component sizes (therefore sources) are different in plants with different numbers of loops, etc. However, we believe that the data obtained represents the best and most consistent post shutdown radiation level data available on large, similar power plants. We expect that this data will be a tool to assist us in our overall objectives.

Some samples of our analysis of this data were reported by M. Crotzer and R. Shaw[6] and F. Frank and J. Sejvar.[7] Dr. Crotzer reported the preliminary results of studies concerning the radiation buildup in two nearly identical plants with differing operational chemistry control schemes. Figure 2 shows the results of this particular study. F. Frank reported on the apparent buildup and levelling off of radiation fields as a function of the time of plant operation. Figure 3 illustrates this behavior for Steam Generator Channel Head data.

FIGURE 2. REACTOR COOLANT LOOP RELATIVE CONTACT DOSE RATE VERSUS PLANT OPERATING TIME

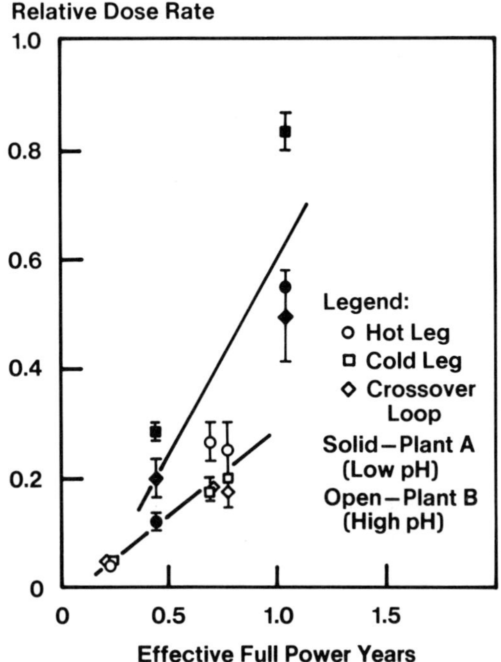

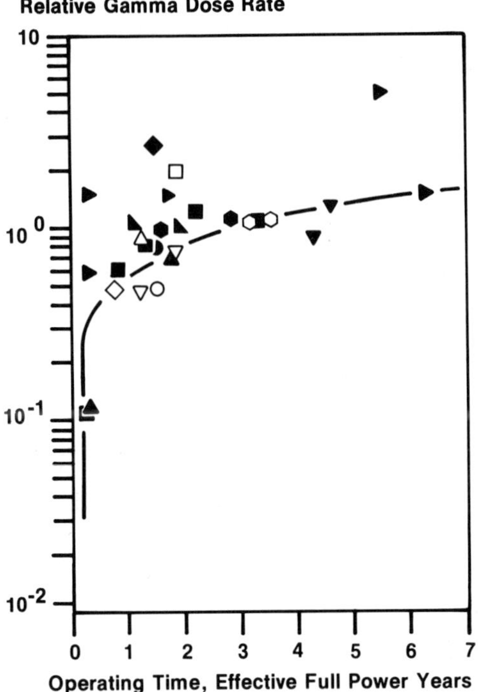

FIGURE 3. AVERAGE S.G. CHANNEL HEAD DOSE RATE VS. PLANT OPERATING TIME

In other related measurements internal to a steam generator, some detailed work has been done on the radiation level distribution inside the primary channel head. Thermoluminescent dosimeters (TLDs) imbedded in lead bricks were used for this purpose. The data has been reported and compared to calculations in S. Anderson's work.[8] The results indicate that the tube sheet surface is the largest contributor to the radiation field. Calculations and measurements compared well.

SPECTRAL MEASUREMENTS

Gamma spectra measurements of the reactor coolant suspended and deposited activities can assist in assessing the importance of operational changes to the radiation source. A number of spectral measurements are included in the program. These include (1) the Reactor Coolant Activity Monitoring System (RCAMS), (2) an in-plant measurement on loop piping and (3) measurement of removed components.

The RCAMS consists of a GeLi detector, a minicomputer and gamma-ray spectormeter. The GeLi detector[9] is focused on a bypass stream of the reactor coolant sampling system. The spectrometer output is recorded and continuously monitors the coolant-borne activity. Its most effective use has been to monitor the coolant during shutdown and other relatively fast transients. In this way, some assessment has been made of differing modes of shutdown as they affect the transport of source material. The software is programmed to evaluate 30 or more radionuclides and to print isotopic activities at predetermined intervals.

Another spectral measurement is made using a GeLi detector focused on a portion of the reactor coolant system piping (After Shutdown). This calibrated system, together with analysis of water-borne activity, can determine the crud layer activity on the inside surface of the piping. M. Crotzer[10] reports some data and the techniques used in this measurement. He also reports spectral measurements made on a component removed from contact with the reactor coolant. This component is a diaphragm covering the man-way of a steam generator channel head. These spectral measurements have been used to identify the nuclides of interest and to demonstrate differences in deposits from plants operating with different pH control of the reactor coolant. Our results show that post shutdown radiation levels in Westinghouse PWR plants are largely (60% to 90%) due to Co-58 and Co-60.

SHUTDOWN CHEMISTRY OPERATIONS

High releases of Co-58 and Co-60 to the reactor coolant have been observed during shutdown operations of PWR plants.[11] Our objective is to obtain these releases quickly, while facilities for coolant purification are operable. The oxygenation of the primary system by aeration and by hydrogen peroxide addition have been studied.

Data from ten refueling shutdowns at seven different Westinghouse plants have been reviewed. In addition the effects of hydrogen (delayed oxygenation), boration, and reactor coolant pump operations on crud releases have been studied. In several recent tests, the effect on steam generator radiation fields has been monitored.[12]

The general results are as follows:

A. Some release of the cobalt isotopes occurs as a result of borating the coolant and reducing its temperature. This is due to increasing the coolant acidity.

B. A faster release occurs upon oxygenation, air and peroxide being equally effective. The peroxide, however, is easier to control and is recommended.

C. The total release can be cleaned up via system demineralizers.

D. The materials released are primarily from core surfaces and partly from plant surfaces.

E. The total process appears to neither contaminate nor decontaminate plant surfaces.

EFFECTS OF OPERATING CHEMISTRY

A simplistic model of core crud buildup was formulated for the purpose of looking at past data to determine if a correlation exists between reactor coolant chemistry and steam generator internal radiation fields.[13] This, in effect, is an attempt to reconcile the scatter in the observed results (See Figure 3). For this purpose, it was necessary to define some pH, at temperature, against which the specific reactor coolant chemistry could be compared. We defined the dividing line as that pH for which the temperature coefficient of solubility of magnetite (Fe_3O_4) changed from negative to positive.

The Sweeton-Baes values for Fe_3O_4 solubility were used.[14] "High pH" and "Low pH" are, therefore, values (at temperature) above and below this value, respectively. For commercial PWR coolant, the pH at temperature is a function of both the boric acid (added as a burnable soluble poison) and the lithium hydroxide (added to control pH) concentrations. Thus, the lithium concentration needed to achieve the "High pH" condition decreases during a core cycle as the boron concentration decreases.

TABLE II HOW WELL DOES THE THEORY FIT?

	RADIATION LEVEL TREND		
pH Trend	Low Increase or Decrease	Average Increase	High Increase
Low	1		1
Low to High	2	4	9
High to Low		1	1
High	5	2	1

Evaluation was made of nine plants operating for 27 cycles. A summary of the results is shown in Table II. The ten results showing large increases in radiation field due to low pH or low to high pH operation and the five results showing low radiation field increases due to high pH operation are consistent with our expectations. The five results at the opposite corners (low radiation field increases at low pH and high radiation field increases at high pH are contrary indications of our presumed pH effect). The other seven results are inconclusive. Additionally, the program included a two-plant test, one operating an entire core cycle in a "High pH" chemistry and one in a "Low pH" chemistry. Early results indicate that the direction of radiation field buildup is as expected, as shown previously in Figure 2.

CORROSION PRODUCT DEPOSITION MODEL

A computer program[15] for corrosion product transport and deposition has been developed by Westinghouse. This program, CORA, has been improved[16] under the EPRI contract, to account for certain parameters which are described above. For example, the change of solubility for non-isothermal systems has been included. Also the reactor coolant system pH as a function of lithium and boron concentrations and system temperature has been developed. At this point, little analysis of the actual data collection in the EPRI-W program has been accomplished. Future calculations using this improved CORA will be compared to plant data.

LABORATORY WORK

The basic properties of crud composition are being examined in the laboratory.[17] The existing solubility data base for materials, coolant composition, and temperatures will be extended The solubility of cobalt from nickel cobalt ferrite materials under differing chemistry, hydrogen, oxygen and other water parameters are being investigated. The low temperature (20° to 60°C) oxidation is being investigated. To this date, the results have indicated that:

A. The Fe solubility from nickel ferrite is less than that in equilibrium with magnetite in a simulated reactor coolant.

B. The Ni solubility is less than that of Fe, and has little, if any, temperature coefficient at reactor temperatures. (See Figure 4)

C. The solubility of both species increases with the increasing acidity that occurs when boric acid is cooled. (See Figure 5)

D. On oxygenation of the cooled solution, the solubility of Ni (and presumably Co) increases dramatically.

E. The expected variation of Fe solubility with dissolved H_2 concentration ($[Fe] \propto [H_2]^{1/3}$) is exhibited to relatively low H_2 concentrations, but Fe concentrations higher than expected are observed at $[H_2] \leq 5$ cm^3(STP)/kg H_2O.

SUMMARY

W and EPRI have cooperated in a multi-year research program directed towards controlling post shutdown radiation levels in PWR plants. The program includes monitoring radiation level trends and defining the responsible nuclides. Various shutdown techniques are evaluated for their effects on radiation level buildup. Finally, some detailed corrosion product transport mathematical models and laboratory studies are being supported.

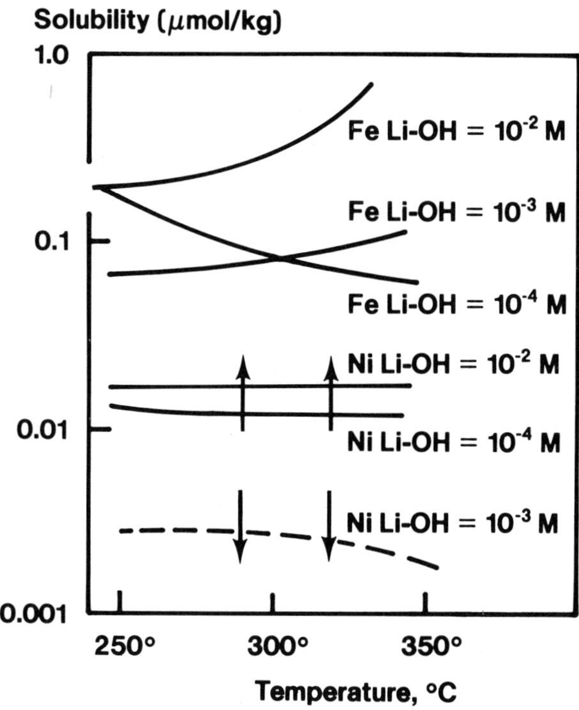

FIGURE 4. SOLUBILITY OF IRON AND NICKEL FROM $Ni_{0.6}Fe_{2.4}O_4$ IN THREE DIFFERENT SOLUTIONS OF 0.06 M $B(OH)_3$

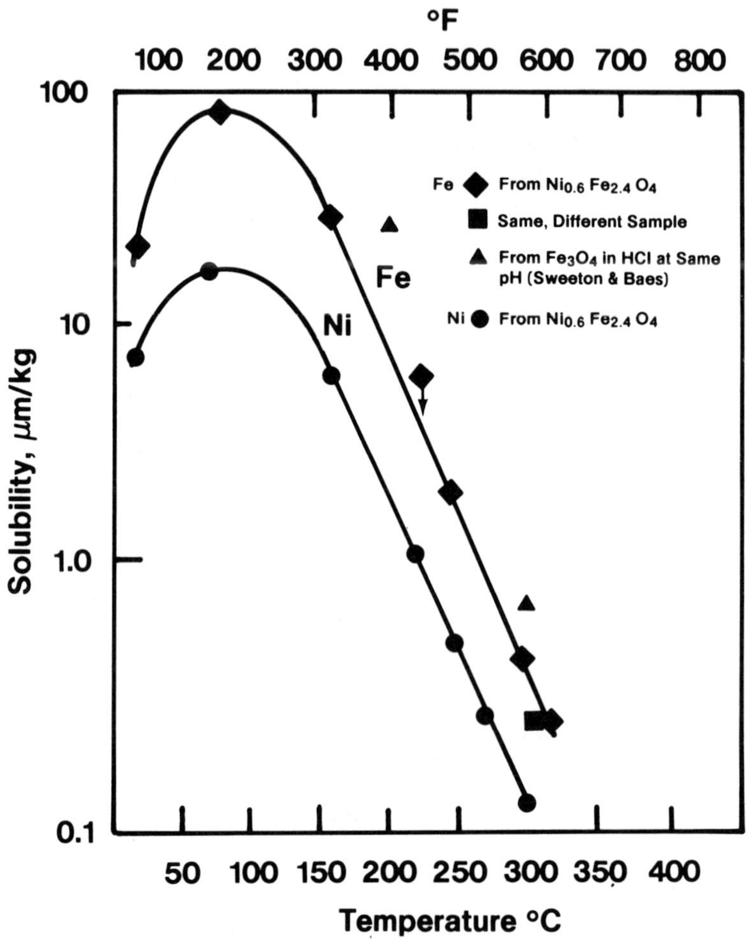

FIGURE 5. SOLUBILITY OF IRON AND NICKEL FROM $Ni_{0.6}Fe_{2.4}O_4$ IN 0.2 M $B(OH)_3$

LEGAL NOTICE

This report was prepared by Westinghouse as an account of work sponsored by the Electric Power Research Institute, Inc. (EPRI). Neither EPRI, members of EPRI, nor Westinghouse, nor any person acting on behalf of either:

a. Makes any warranty or representation, express or implied, with respect to the accuracy, completeness, or usefulness of the information contained in this report, or that the use of any information, apparatus, method, or process disclosed in this report may not infringe privately owned rights; or

b. Assumes any liabilities with respect to the use of, or for damages resulting from the use of, any information, apparatus, or process disclosed in this report.

REFERENCES

1. P. M. Garrett, "Current Trends in Occupational Radiation Exposures at U.S. Commercial Power Plants," Nuclear Engineering Int., 23, 51 (April 1978).

2. H. W. Dickson, et al, "Controlling Occupational Radiation Exposures at Operating Nuclear Power Stations," Nuclear Safety, 18, 492, (August 1977).

3. Verna, F. J., "On-Line With Verna - Occupational Exposure," Part 1, Nuclear News, 20, 9 (July 1977); Part 2, Nuclear News, 20, 11 (Sept. 1979).

4. Warman, E. A., "Occupational Exposure at Nuclear Power Plants," Vol. 44, Nuclear Technology, (June 1979).

5. Robert A. Shaw, "Getting at the Source: Reducing Radiation Fields," Nuclear Technology, Vol. 44, (June 1979) 97.

6. M. E. Crotzer, R. A. Shaw, "The Apparent Influence of PWR Coolant pH on Radiation Field Buildup," ANS Meeting (Atlanta) June 1979.

7. F. J. Frank, J. Sejvar, "Radiation Level Trends in W PWR's," British Nuclear Engineering Society, London, December 1978.

8. S. Anderson, et al, "Measurement of Gamma Ray Dose Rate Distribution Internal to a PWR Steam Generator Using LiF Thermoluminescent Dosimeters," BNES, Dec. 1978, London.

9. M. E. Crotzer, W. J. Bestoso, "An Automatic On-Line Reactor Coolant Activity Monitoring System," ANS Transactions, Winter 1978, Washington, DC.

10. M. E. Crotzer, "Measurement of Radionuclide Surface Activity in a Pressurized Water Reactor," ANS Transactions, Winter 1978, Washington, DC.

11. J. W. Kormuth, "Shutdown Operational Techniques for Radiation Control," EPRI Report NP-859.

12. J. W. Kormuth, "Refueling Shutdown Studies at Point Beach Unit 2," EPRI Report NP-860, to be issued.

13. Y. Solomon, "An Overview of Water Technology for PWR's," Proceedings of the BNES International Conference on Water Chemistry for Nuclear Reactors, Bournemouth, England, October 1977.

14. F. H. Sweeton, et al, "The Solubility of Fe_3O_4 in Dilute Acid and Base Solutions to 300°C," ORNL-TM-2667, July 1969.

15. J. Sejvar, "Revision of CORA - A Computer Code for Calculating the Activation of Corrosion Products in Reactor Systems," WCAP-7708, May 1971.

16. S. Kang, "Report to EPRI on CORA Improvements," to be published as a W-EPRI report, 1979.

17. Y. L. Sandler, "Structure and Solubility of PWR Primary Corrosion Products," Presented at NACE Meeting, Houston, TX, March, 1978.

COOLANT CHEMISTRY CONTROL DURING PWR SHUTDOWN-COOLDOWN

J. C. Cunnane and
W. R. Stagg
Babcock & Wilcox Company
P. O. Box 1260
Lynchburg, VA 24505

R. A. Shaw
Electric Power Research
Institute
P. O. Box 10412
Palo Alto, CA 94303

INTRODUCTION

Because of the current uncertainty and controversy surrounding the biological consequences of exposure to low levels of ionizing radiation, significant efforts are underway to control occupational radiation exposure within the nuclear industry. There is a growing recognition that an important component of these efforts is the work which is aimed at reducing the source of the radiation in those areas of the plant which require inspection, maintenance and repair operations. Decontamination is clearly one approach to source control. Another approach is to control the buildup of radioactive deposits on out-of-core surfaces. For modern pressurized water reactors (PWRs), which operate with very low failed fuel levels, this reduces to controlling the buildup of activated corrosion products. Fortunately, data collected from operating plants indicate that even with present designs and materials better control is possible. For example, some sister plants with nominally similar design show wide variations in their radiation fields, indicating that some plants have, fortuitously, succeeded in controlling the activated corrosion product buildup better than others. This point is illustrated in Figure 1 which shows the trends in the upper sheet contact radiation fields at the three Oconee units. However, despite these clear differences, the identification and development of control techniques are difficult since many of the pertinent phenomena involve a complex interaction of system and operational variables and are, as yet, poorly understood.

One of the operational variables which is generally believed to play an important role in corrosion product generation and transport and in the consequent buildup of the shutdown radiation fields

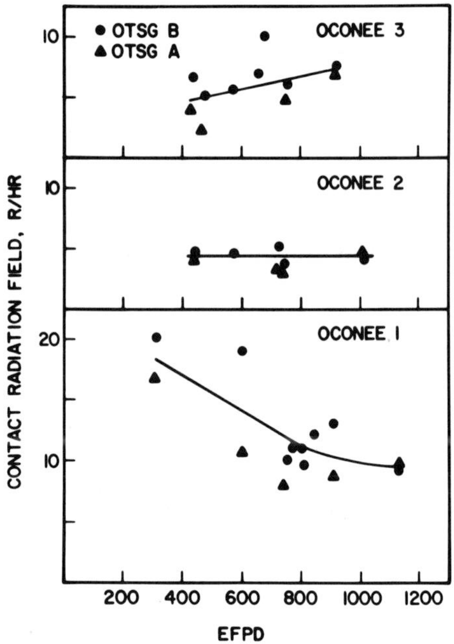

Figure 1. Behavior of the Tube Sheet Contact Radiation Fields at the Oconee Units.

is the primary coolant chemistry. Studies[1,2] which have followed the radionuclide concentration in PWR primary coolant through various phases of plant operation have shown that this activity increases by orders of magnitude during the shutdown-cooldown sequence. This implies that the total amount of activity which passes through the steam generators during the shutdown-cooldown sequence is comparable to the amount which passes through during a normal cycle. Hence, it was felt that there was a significant potential for activity deposition on the out-of-core surfaces as a result of the crud transport and redistribution which occurs during this phase of plant operation. These considerations, together with the realization that there were opportunities to control the primary chemistry during shutdown and cooldown, led us to devote a significant effort to the study of the activated corrosion product transport during PWR plant shutdown and cooldown. Most of the discussion in this paper deals with the transport of soluble or nonfilterable species.

THERMODYNAMIC STABILITY AND SOLUBILITY BEHAVIOR OF PWR CORROSION PRODUCTS

Some insight into the observed behavior of PWR nonfilterable corrosion products can be gained by:

1. Examining the thermodynamic stability of corrosion product oxides in contact with the primary coolant as the temperature is reduced.

2. Examining the solubility behavior of the thermodynamically stable oxides as the temperature is reduced.

Macdonald, et al.[3-8] have used thermodynamic data to calculate which oxides are stable in an equilibrium system containing known concentrations of hydrogen and oxygen at elevated temperatures. The stability of the oxides can be expressed in terms of the equilibrium dissolved hydrogen concentrations (or the Henry's Law equivalent hydrogen partial pressure) for the reduction reactions. In Figure 2, the equilibrium dissolved hydrogen concentrations are plotted for the reduction reactions of $NiFe_2O_4$, CoO and NiO. The cross-hatched area represents the range of the normal PWR dissolved hydrogen specifications during operation. At hydrogen pressures which are greater than the calculated equilibrium values, the reduction reactions should proceed spontaneously, i.e., the reduced forms of the oxides which are shown on the right of the reaction equations in Figure 2 are stable in contact with the primary coolant. Since $NiFe_2O_4$ is believed to be a reasonable model oxide for the PWR primary crud, it is instructive to examine its stability behavior. Figure 2 shows that at PWR operating temperatures $NiFe_2O_4$ is stable with respect to Ni and Fe_3O_4 for all dissolved hydrogen concentrations within the range of the normal specifications. However, it is also clear from Figure 2 that it is possible, depending on the relative rates of cooling and degassing, for the $NiFe_2O_4$ which is in contact with the primary coolant to be reduced to Ni and Fe_3O_4 during cooldown. The extent to which this reaction will occur depends on the cooldown sequence and the kinetics of the reduction reaction. The implication of this stability behavior is that while the crud solubility behavior during the initial part of the cooldown is similar to that of $NiFe_2O_4$, its subsequent solubility behavior could be determined by a relatively small inventory of Ni and Fe_3O_4 at the crud-coolant interface.

Since $NiFe_2O_4$ is believed to be a reasonable model substance for PWR crud and since it is expected to determine the solubility behavior of PWR crud during operation and at least the initial part of the cooldown, we have attempted to determine its solubility behavior. A calculational technique based on available thermodynamic data was employed. The method of Macdonald, et al.[9] for calculating the solubility of binary metal oxides provided the basis for the calculational technique used here. Because of the incongruent nature of $NiFe_2O_4$ dissolution and the thermodynamic stability of solid nickel phases under operating conditions, it was assumed

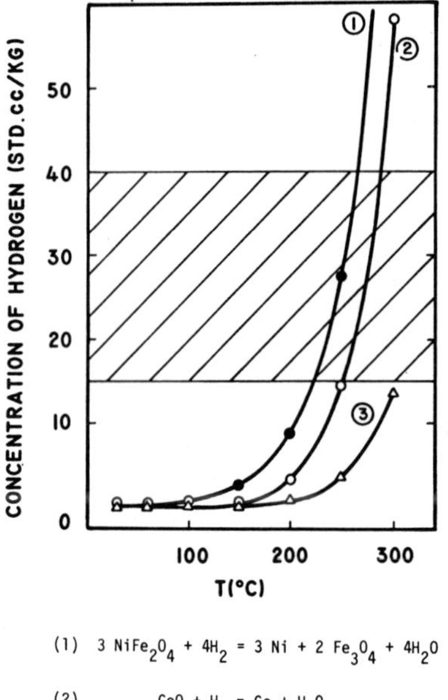

(1) $3\ NiFe_2O_4 + 4H_2 = 3\ Ni + 2\ Fe_3O_4 + 4H_2O$

(2) $CoO + H_2 = Co + H_2O$

(3) $NiO + H_2 = Ni + H_2O$

Figure 2. Thermodynamic Stability of Corrosion Product Oxides in Contact with Primary Coolant.

for calculational purposes that the dissolution of nickel ferrite could be described by:

$$\text{Nickel Ferrite} \longrightarrow \underline{Ni} + \text{soluble iron species}$$
$$\underline{Ni} \rightleftarrows \text{soluble nickel species}$$

The reactions which were included in this model system for the dissolution process are shown in Table 1. Comparison of the model based calculations with available experimental solubility data showed reasonable agreement at least in the qualitative behavior of the solubility as a function of the pertinent coolant chemistry variables. Hence, such calculations proved to be a useful tool in the interpretation of plant data and in the development of chemistry control techniques.

TABLE 1

$NiFe_2O_4$	$+\ 4H^+$	$+\ 2H_2$	=	$2Fe^{2+}$	$+\ Ni + 4H_2O$
$NiFe_2O_4$	$+\ 6H^+$	$+\ H_2$	=	$2Fe^{3+}$	$+\ Ni + 4H_2O$
$NiFe_2O_4$	$+\ 2H^+$	$+\ 2H_2$	=	$2Fe(OH)^+$	$+\ Ni + 2H_2O$
$NiFe_2O_4$	$-\ 2H^+$	$+\ 2H_2$	=	$2HFeO_2^-$	$+\ Ni$
$NiFe_2O_4$	$+\ 4H^+$	$+\ H_2$	=	$2Fe(OH)^{2+}$	$+\ Ni + 2H_2O$
$NiFe_2O_4$	$+\ 2H^+$	$+\ H_2$	=	$2Fe(OH)_2^+$	$+\ Ni$
$NiFe_2O_4$	$+\ 4H^+$	$+\ H_2$	=	$Fe_2(OH)_2^{+4}$	$+\ Ni + 2H_2O$
$3NiFe_2O_4$	$+\ 10H^+$	$+\ 3H_2$	=	$2Fe_3(OH)_4^{5+}$	$+\ 3Ni + 4H_2O$
$NiFe_2O_4$	$-\ 4H^+$	$-\ 2H_2$	=	$2FeO_4^{2-}$	$+\ Ni - 4H_2O$
$NiFe_2O_4$	$-\ 4H^+$	$+\ 2H_2$	=	$2FeO_2^{2-}$	$+\ Ni$
$NiFe_2O_4$	$-\ 2H^+$	$+\ 2H_2$	=	$2Fe(OH)_3^-$	$+\ Ni - 2H_2O$
$NiFe_2O_4$		$+\ 2H_2$	=	$2Fe(OH)_2(aq)$	$+\ Ni$
Ni	$+\ 2H^+$	$-\ H_2$	=		Ni^{2+}
Ni	$+\ H^+$	$-\ H_2$	=		$Ni(OH)^+ - H_2O$
Ni	$-\ H^+$	$-\ H_2$	=		$HNiO_2^- - 2H_2O$

PWR PLANT SHUTDOWN-COOLDOWN DATA AND INTERPRETATION

As mentioned earlier, the shutdown-cooldown sequence in PWR plants is accompanied by a dramatic increase in the nonfilterable activity in the primary coolant. This presumably reflects a corresponding increase in the solubility of the corrosion product oxides. The data for the Oconee 3 end-of-cycle (EOC) 2 shutdown which are shown in Figure 3 are fairly typical. Based on the thermodynamic calculations outlined above, the data in Figure 3 can be interpreted as consisting of two components:

1. An initial increase in crud solubility associated with cooldown and boration,

2. A subsequent component which reflects the solubility behavior associated with the development of oxidizing conditions in the primary coolant.

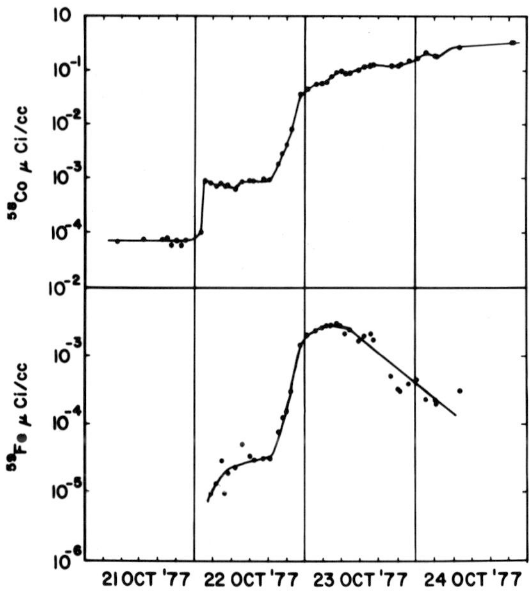

Figure 3. Behavior of the Soluble Cobalt-58 and Iron-59 During the End-of-Cycle Shutdown at Oconee, Unit 3.

The contrasting behavior of the ^{58}Co and ^{59}Fe, which is associated with the development of more oxidizing conditions in the primary coolant beginning on October 23, is instructive. The thermodynamic calculations predict a much lower solubility for iron in the absence of a reducing environment. This expectation is consistent with the observed ^{59}Fe behavior. On the other hand, the ^{58}Co activity continues to increase slowly on October 23 and 24. Again this behavior is consistent with a slight net production of hydrogen peroxide by radiolysis and a resulting increase in nickel solubility.

From an examination of such plant data and the expected solubility behavior based on thermodynamic calculations, it appeared as if the solubilization during shutdown and cooldown could be minimized by a combination of proper pH control and maintaining reducing conditions during the shutdown sequence.

SOLUBILITY CONTROL DURING SHUTDOWN AND COOLDOWN

Programs which were designed to demonstrate the value of pH and redox control for controlling the solubility behavior during shutdown and cooldown were implemented during the Rancho Seco EOC 2 and Oconee 2, EOC 3 shutdowns, respectively. The results of the redox program will be discussed at some length to illustrate that crud solubility can be controlled by suitable coolant chemistry control. The results of the pH control program will be discussed more briefly.

The thermodynamic calculations which were discussed earlier showed that the crud solubility behavior during the initial stages of cooldown is sensitive to the lithium concentration in the primary coolant. At relatively high lithium concentrations (2.0 ppm), calculations indicated that the crud dissolution may be minimized during the initial stages of the cooldown. A test of this prediction was made possible by the planned shutdown for Rancho Seco, EOC 2. Relatively high lithium concentration was maintained in the primary coolant up to the end of cycle 2. Thereafter it was allowed to dilute down as makeup water was added during cooldown. The overall results which were collected during the cooldown sequence indicated that some control of the solubility behavior is possible by pH control during the initial stages of cooldown.

The redox control program which was implemented during the Oconee 2, EOC 3 shutdown was designed to see if, as predicted by calculations, the ^{58}Co solubilization which is associated with the development of oxidizing conditions could be delayed by maintaining reducing conditions in the primary coolant. Reducing conditions were maintained by a combination of dissolved hydrogen retention and a hydrazine addition program. During the shutdown all key plant performance, primary coolant chemistry, and primary coolant radiochemistry data were monitored. Figure 4 shows the observed behavior of the filterable and nonfilterable ^{58}Co. The behavior of the filterable ^{58}Co will not be discussed here except to point out that most of the particulate crud bursts are correlated in time with plant operating events such as reactor shutdown, switchover to low pressure injection (LPI) cooling, reactor coolant pump cycling, etc. The behavior of the nonfilterable ^{58}Co is just what was expected based on prior shutdown data and the calculated effects of maintaining reducing conditions in the primary coolant. The significant features of the observed behavior are:

1. A rapid increase in the nonfilterable ^{58}Co with the increased crud solubility which accompanies cooldown and boration.

2. A subsequent hold in the nonfilterable ^{58}Co associated with the small solubility changes during the latter stages of the cooldown while reducing conditions were maintained in the primary coolant.

3. A dramatic increase in the ^{58}Co associated with the solubility increase which accompanies oxygenation of the primary coolant.

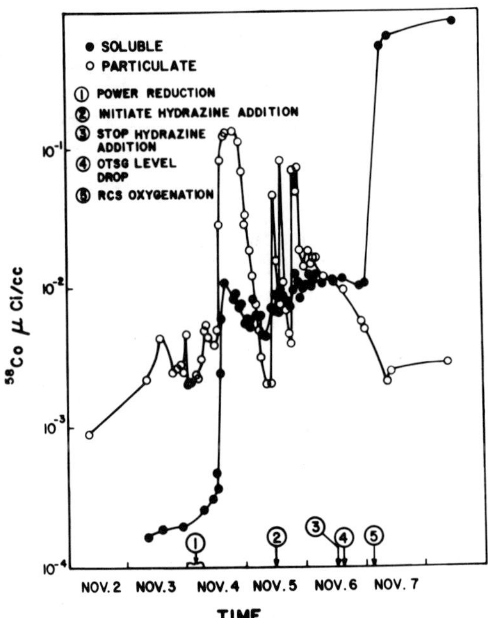

Figure 4. Behavior of the Soluble and Particulate Cobalt-58 During the Oconee, Unit 3, End-of-Cycle 2 Shutdown.

These results indicate that the dissolution which is associated with oxygenation can be delayed by maintaining reducing conditions in the primary coolant. A combination of hydrogen retention and a hydrazine addition program is suitable for maintaining the necessary reducing conditions.

SUMMARY AND CONCLUSIONS

Although many questions still remain, it is clear that progress is being made in understanding the behavior of corrosion products in the primary system of modern PWRs. The Oconee 2, EOC 3 shutdown study showed that the oxygenation component of the solubilization can be delayed during the cooldown process by maintaining reducing conditions in the primary coolant. The combination of hydrogen retention and a hydrazine addition program is an effective way of maintaining bulk reducing conditions. Also, it appears that there is some opportunity to control the cooldown-boration component of the shutdown solubilization by appropriate pH control. Both of these techniques could be optimized to minimize the crud dissolution during this phase of plant operations. However, further work is needed to show that solubility control during shutdown is important in controlling the shutdown radiation fields and to develop the optimum solubility control strategy.

ACKNOWLEDGEMENT

The authors would like to acknowledge the support of the Electric Power Research Institute (EPRI) in performing this work as part of the EPRI program RP 825-1.

REFERENCES

1. E. T. Chulick, P. J. Grant, D. F. Hallman, A. J. Kennedy, Edited by D. L. Uhl, "Oconee Radiochemistry Survey Program Semiannual Report," LRC 9047. Babcock & Wilcox (1975).
2. S. G. Sawochka, P. S. Wall, J. Leibovitz and W. L. Pearl, "Effects of Hydrogen Peroxide Additions on Shutdown Chemistry Transients at Pressurized Water Reactions," EPRI NP-692 (1978).
3. D. D. Macdonald and T. E. Rummery, "The Thermodynamics of Metal Oxides in Water Cooled Nuclear Reactors," AECL-4140 (1972).
4. T. E. Rummery and D. D. Macdonald, "Prediction of Corrosion Product Stability in High Temperature Aqueous Systems," Journal of Nuclear Materials, 55: 23-32 (1975).
5. T. E. Rummery and D. D. Macdonald, "The Thermodynamics of Selected Metal Ferrites in High Temperature Aqueous Systems," AECL-4577 (1973).
6. D. D. Macdonald, G. R. Shierman and P. Butler, "The Thermodynamics of Metal-Water Systems at Elevated Temperature, Part 2: The Iron Water System," AECL-4137 (1972).
7. D. D. Macdonald, G. R. Shierman and P. Butler, "The Thermodynamics of Metal-Water Systems at Elevated Temperatures, Part 3: The Cobalt Water System," AECL-4138 (1972).
8. D. D. Macdonald, "The Thermodynamics of Metal-Water Systems at Elevated Temperature, Part 4: The Nickel-Water System," AECL-4193 (1972).
9. D. D. Macdonald, G. R. Shierman and P. Butler, "The Thermodynamics of Metal-Water Systems at Elevated Temperatures, Part 1: The Water and Copper-Water Systems," AECL-4136 (1972).
10. Y. L. Sandler and R. H. Kunig, Nucl. Sci. Eng., 64: 866 (1977).

EFFECT OF HIGH-TEMPERATURE FILTRATION ON PWR PLANT RADIATION FIELDS

M. Troy and S. Kang

Nuclear Technology Division
Westinghouse Electric Corporation
Pittsburgh, Pennsylvania 15230

and

G. T. Zirps and D. W. Koch

Nuclear Equipment Division
Babcock and Wilcox Company
Barberton, Ohio 44230

INTRODUCTION

The most important source of radiation exposure to pressurized water reactor (PWR) plant maintenance personnel is associated with the activated corrosion product deposits (crud) on excore plant surfaces. Therefore to reduce radiation exposures to a level consistent with ALARA (As-Low-As-Reasonably-Achievable) control of crud transport and deposition is essential.

Both analysis and experiment indicate that hot filtration of the reactor coolant can effectively reduce radiation fields in the plant and help significantly in control of radiation exposures. This paper reviews that analysis and the experimental work, and reviews the status of a current hot-filtration research program sponsored by the Electric Power Research Institute (EPRI).

ANALYSIS OF HOT FILTRATION EFFECTS

Model Predictions of Performance

On a best estimate basis high temperature filtration of reactor coolant, at the rate of 0.5% of the core full-power flow rate, has the potential for reducing PWR radiation exposure by a factor of 2. The results of the supporting analyses are plotted in Figure 1 which shows the effect of filtration flowrate on the expected annual average personnel radiation exposure.

In predicting overall exposures, a basic assumption made here is that the deposited radioactive corrosion products contribute 70% of the total annual exposure to plant operating and maintenance personnel. Beslu et al (1979) estimate the contribution closer to 80%. Data from operating PWR experience indicates that the corrosion product radionuclides contribute at least 50% to the dose while it is obvious that this contribution must be less than 100%. Therefore, the assumed 70% is a conservative estimate of the average of these two bounds. The remaining 30% is attributed to all other sources including fission products and dissolved corrosion products in the coolant.

It is also assumed that the filtration directly affects only the particulate corrosion products in the coolant. Thus the upper curve in Figure 1 would become asymptotic to a value of 70% even if full flow filtration were assumed.

The above discussion is further illustrated by Figure 2 which depicts the partition of total exposure between deposited crud and other sources. The 46% average reduction due to filtration at the design flowrate of 0.5% is indicated in the left-hand bar graph. The top 30% of the left-hand bar graph represents that part of the exposures unaffected by filtration, and therefore comes over directly to the right-hand graph.

Selection of the 0.5% filtration flow fraction as the design point for a proposed filtration prototype is based on engineering judgment as to the relative cost-to-benefit ratios for different sized systems. The curves in Figure 1 indicate that by doubling the filtered flow fraction at perhaps 1.5 - 2.0 times higher cost would net an increase in benefit of only approximately 20%.

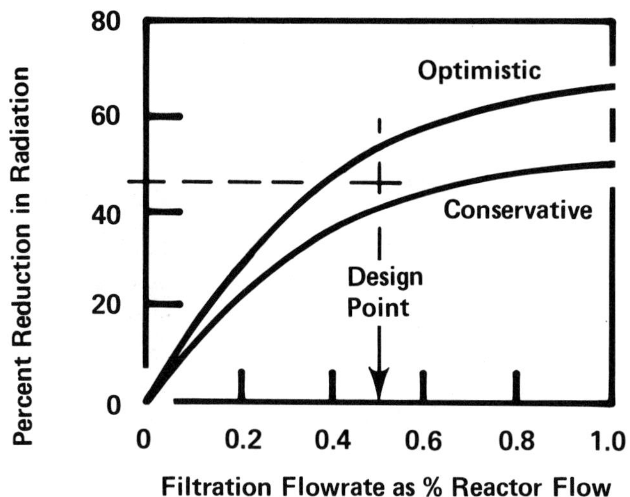

Fig. 1 Best Estimate Radiation Dose Reduction vs. Filtration Flowrate

The CORA Model

The analysis described above was based on a calculation of radiation sources using CORA, a multi-mechanism model of system activitation which reflects the transport, neutron activation and deposition of corrosion products from plant structural components. The latest version of CORA was published by Sejvar (1971).

In CORA the reactor coolant system (RCS) is described by a number of nodes which represent the corrosion products in different stages and locations in the overall crud transport phenomenon. Incore and excore regions are represented and treated separately.

A two-layer structure for the incore and excore crud deposits is assumed. These are identified by the nodes labled

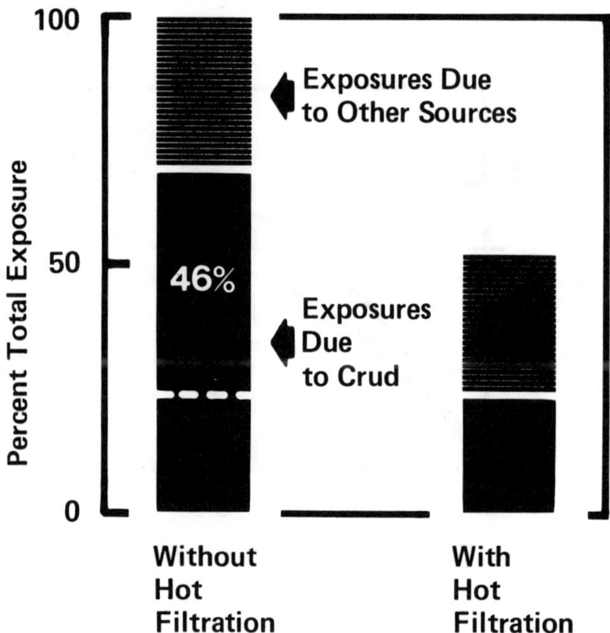

Fig. 2. Effect of Hot Filtration on Total Personnel Exposures

"transient" and "permanent" respectively. It is assumed that particulates in the coolant exchange only with the transient layers while the dissolved species exchange directly with both transient and permanent layers. A transition from transient to permanent morphology is allowed to occur. Two nodes represent respectively the dissolved and particulate corrosion product species in the coolant, and serve as the transport paths between the incore and excore regions.

It is important to realize that the current CORA model is an "isothermal" model while the RCS in an operating PWR involves non-isothermal surfaces both in-core and ex-core. In actual practice solubility-controlled processes can not occur simultaneously at the same location in the real plant.

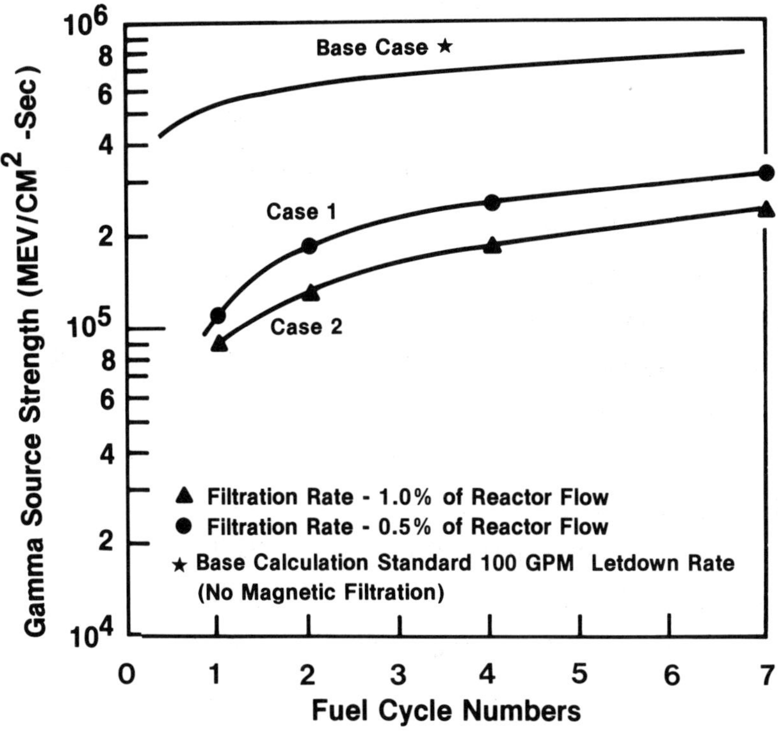

Fig. 3. Radiation Source Strength of Corrosion Product Deposits vs Core Cycle

Predictions of Crud Radiation Source Strength

CORA was used in the present study to calculate the radiation source strength for the deposited crud as a function of time. Two sets of calculations were made from which the "conservative" and "optimistic" bounds of Figure 1 were generated. The "conservative" calculation is shown in Figure 3 for the base case with standard purification only and for two rates of additional high temperature filtration.

A "clean plant" or zero initial source level was assumed for the calculations. If hot filtration were inserted as a backfit to a plant with a net accumulation, then the initial value would be represented by the top line in Figure 3 at the

Table 1. Potential Effect of Filtration on Manpower and Personnel Exposures

	Without Filtration	With Filtration
Productive time required (Hrs)	4	4
Radiation field strength (R/Hr)	4	2
Dose allowed per worker (Rem)	1	1
Allowed time in-field per worker (Min)	15	30
Prep. time in-field per worker (Min)	10	10
Productive time in-field per worker (Min)	5	20
Number of workers required for job	48	12
Prep. time out-of-field per worker (Min)	120	120
Total man-hours expended for job	108	30
Total radiation exposure accumulated (Rem)	48	12

time of insertion. Expectedly the source level would then decrease with time approaching the lower curve representing the appropriate filtration rate. Whether or not the permanent crud layer thickness can be reduced because of the filtration is not known. However, a decrease of about 50% due to Co(58) decay would obtain if filtration did reduce crud buildup significantly.

The CORA analysis was made under two sets of assumptions to yield a "conservative" and an "optimistic" bound on the best estimate exposure. The assumption in the optimistic case is that the excore crud deposit build-up is dominated by particulate deposition processes, while the conservative assumption is that the solubility processes are dominant.

Benefits of Filtration to Plant Maintenance and Operations

In the previous arguments it was assumed that the relationship between radiation dose reduction and total radiation exposures reduction is linear. In general, this is not so and the indicated factor-of-two savings due to hot filtration could be much more as shown by the following hypothetical but reasonable example.

Table 1 illustrates a maintenance operation requiring 4 hours of productive time by a single worker in a 4 R/hr radiation field. It is assumed that there will be preparation time both in-field and out-of-field for each participating worker. A limit of 1 Rem exposure per worker has been imposed. Note that the real effects of the predicted factor-of-two reduction in radiation source by filtration is a factor-of-four reduction in total exposure and almost a factor-of-four reduction in total man-hours expended on the job. The key to what is happening is the ratio of the allowable time-in-field to the non-productive time-in-field. The latter is some minimum constant value depending on the nature of the technology. The allowable in-field time decreases as the field strength increases and as the allowable exposure limit decreases. Both trends are real in the current and foreseeable environment.

It is easily seen how the above trends affect the direct cost of maintenance itself and possibly plant availability. As the radiation field strength increases, the number of workers needed for a job increases, the proportion of non-productive time increases, and the total direct cost of the maintenance esculates in a non-linear fashion. When workers of appropriate qualification became a premium, plant availability will suffer directly as the outage is extended because of the more elaborate maintenance procedures adopted to minimize exposures.

EXPERIENCE TO DATE

Hot filtration of PWR reactor coolant has received evaluation in loop tests in Canada (AECL) and in France (CEA). These programs included testing of etched-metal-leaf filters, deep-bed graphite filters and magnetic filters. In all, no commercialscale PWR experience with reactor coolant hot filtration has been reported.

Canadian Test Program

In the early 1970's, as reported by LeSurf (1977), the Atomic Energy of Canada Limited (AECL) realized that particles of undisolved corrosion products play an important role in the transport of activity, although their precise role was not fully appreciated until much later. Particles may be removed by filtering devices in conventional cooled purification systems, but the relatively high purification flowrate (with letdown in pressure and temperature) needed to reduce radiation levels would result in a significant loss of heat energy and increased plant costs. Therefore, a filtering device was sought that could operate effectively at full system pressures and temperatures.

AECL has reported primary coolant hot filtration tests run on in-pile and out-of-pile loops and in tests run in connection with the 25 MWe Nuclear Power Demonstration reactor (NPD) at Chalk River, Ontario.

Thexton (1975) reported that in extensive earlier AECL loop tests an etched-metal disk mechanical filter was evaluated. Efficiency for crud removal was acceptably high but the filter tended to plug prematurely apparently due to the growth of corrosion products in the narrow passages of the filter. These crystals of corrosion product were not easily removed by back-flushing. This filter type was not included in the later NPD tests.

In the NPD tests of chief interest here, two types of electromagnetic filter (EMF) (one with a sphere-matrix and one with a steel-wool matrix) and a deep-bed graphite filter were evaluated. The initial phase of the NPD testing was with the graphite filter only. The latter tests were conducted with the graphite filter downstream of the EMF. Each of the three types of filter proved efficient alone in removing circulating crud particulates from the NPD coolant.

Heathcock et al (1977) reported that tests with the deep-bed graphite filter alone began early in 1973 and were successful in demonstrating its capacity for removing circulating crud from the NPD coolant. There was later evidence that the graphite filter was less efficient than the EMF's in removing submicron-size crud particles, however. One relative operational disadvantage of the deep-bed graphite filter was identified in the need to valve out the filter, cool and to dispose of the contaminated graphite after loading.

Moskal et al (1977) reported that the NPD tests of the stainless-steel-wool EMF began late in 1973 and continued intermittantly through early 1975. During the first 6-month period NPD ^{60}Co activity (associated with crud) decreased by 30%. At this time the NPD reactor loop was chemically cleaned and the NPD coolant ^{60}Co concentration rose to a 5-fold maximum during the next 6-month period. The EMF was in operation only about onethird of this period. Continuous EMF operation was resumed at the beginning of 1975 at which time the coolant ^{60}Co activity decreased sharply to about 10% of its peak value.

The sphere-matrix EMF was introduced in the fall of 1975 and was on stream continuously thereafter until late 1976. During this period the coolant ^{60}Co concentration (associated with crud) continued to decline to ∿5% of the peak 1974 value and to remain at this level until the end of the test period.

After start of high-temperature filtration following a chemical decontamination in 1973, the radiation fields in the NPD boiler room declined from an average of ∿200 mR/hr to ∿80 mR/hr and appeared to stabilize at this level. Prior to hot filtration the boiler room fields had steadily risen. It is evident that high-temperature filtration was quite effective in removing the crud-related activity released after the decontamination and that generated during the period after.

Although all these filters tested in the NPD loop were effective in reducing and maintaining at a low level the crud-related activity, the sphere-matrix EMF was judged to be superior to the others in its operational characteristics and in its suitability in meeting CANDU system design requirements. This filter, was operated at 1070 psi and 482°F for extended periods, the longest of which was 135 days, without cleaning/flushing. An almost inperceptible increase in the pressure differential across the filter was observed although the radiation levels increased significantly. The flushing of the filter was completely effective in discharging the collected crud.

As a result of the Canadian program, Ontario Hydro has purchased a sphere matrix electromagnetic filter from the Babcock & Wilcox Company. This 1,000 gpm filter has been delivered and will be installed initially in their reactor coolant pump test loop to train the operating personnel. The Babcock & Wilcox Company has been advised that Ontario Hydro intends to purchase 8 more filters to be installed in the Reactor Coolant Systems in the Darlington 854 MWe Nuclear Power Plant which will begin operation in the late 1980's.

French Test Program

The French studies, reported by Darras, et al (1977) and Dolle, et al (1976), were conducted with a light-water PWR coolant. Tests were conducted with only the ball-matrix type of EMF and a granular graphite deep-bed filter. Both filter types were deemed to have general merit for the application, but there were operational differences to be weighed in making a choice between them. For example, the graphite bed has to be cooled and sluiced out of the system for cleaning or replacement while the EMF was flushed in situ and hot. The French and Canadian experiences with both filters were similar.

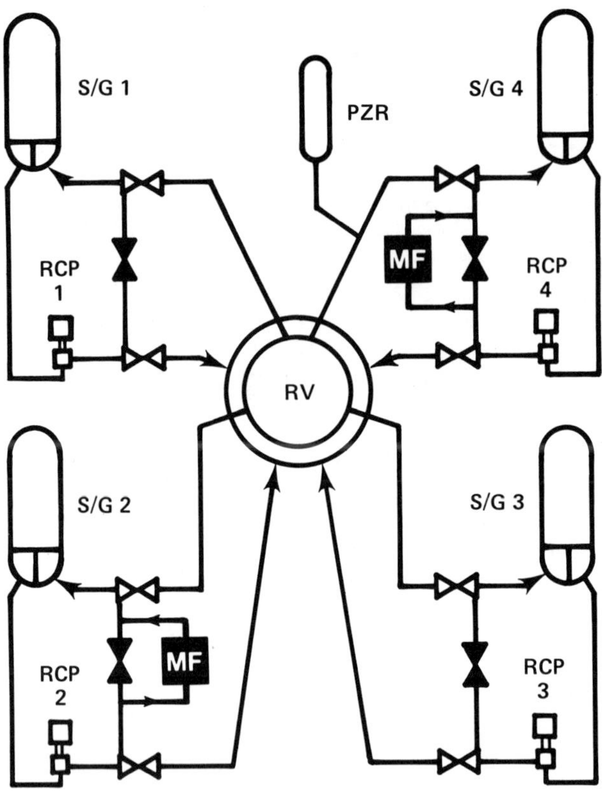

Fig. 4. Schematic of a PWR Plant with Reactor Coolant Hot Magnetic Filtration

Experience Confirms Analytical Prediction

The reported experience with reactor coolant tends to support the model predictions described above, that significant reduction in circulating corrosion product load can be achieved. The ultimate effect on plant radiation exposure levels can only be inferred. One needs to demonstrate that a hot filtration system can be installed in a commercial PWR nuclear plant, a system that will be both effective in reducing total plant personnel exposures and practical to operate within the constraints of normal PWR nuclear power plant practice.

EPRI PROTOTYPE PROGRAM

A 1977 feasibility study was sponsored at Westinghouse by the Electric Power Research Institute (EPRI) to evaluate the potential of magnetic filtration as a technique for the

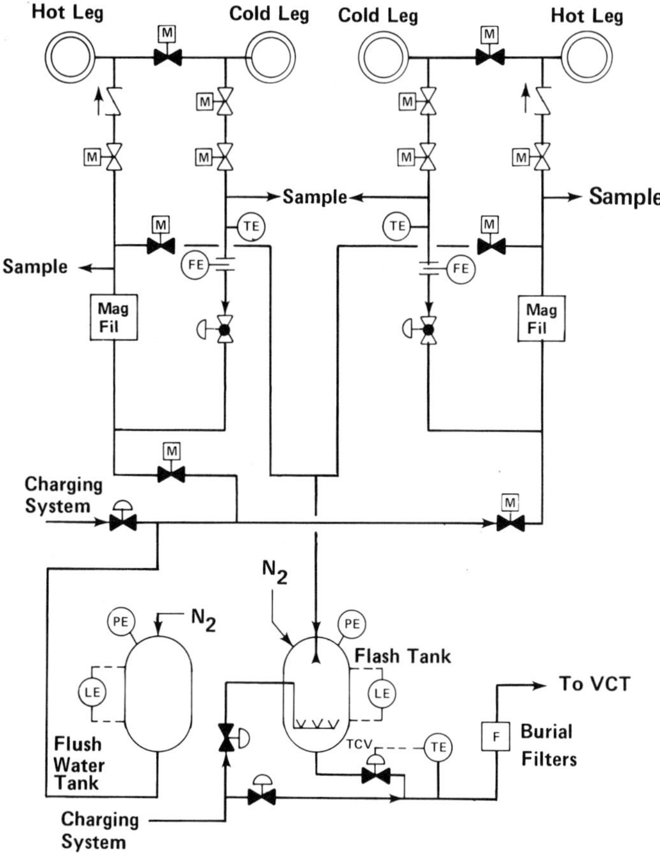

Fig. 5 Reactor Coolant Magnetic Filtration System

removal of corrosion products from PWR power plant coolant streams. This study, Troy, et al (1978), led to a recommendation for a prototype RCS high temperature filtration (HTF) program to demonstrate under actual plant conditions the predicted 50% reduction in personnel radiation exposures discussed above.

Late in 1978 EPRI accepted a joint proposal by Westinghouse and a PWR utility for such a prototype HTF program, and a preliminary system design was generated. The original utility partner has since found it impractical to continue so efforts are now underway to find another utility participant.

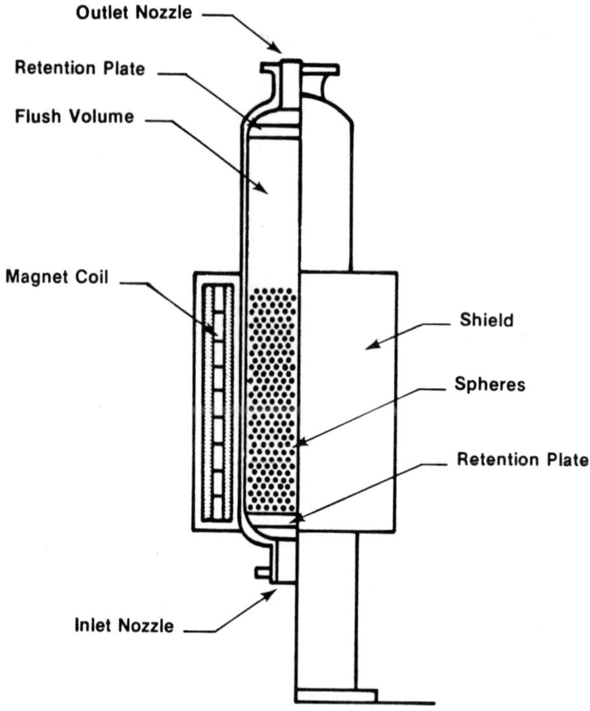

Fig. 6 Schematic of Sphere-Matrix Electromagnetic Filter

In scope, the EPRI program will provide for the design, installation and operation of a full scale high temperature filtration system (HTFS) in an operating PWR plant. Evaluation of the HTFS performance will be based on a comparison of the filtration-equipped plant with other PWR's not equipped with RCS hot filtration. Criteria of performance will be (1) levels of radiation fields at key locations, (2) accumulated personnel exposures (3) reduction in core crud buildup, and (4) plant availability and operational impact. The program will lead to a cost-benefit analysis of the HTFS and recommendations as to its usefulness in meeting ALARA objectives. In addition, data from the program will lead to a better understanding of crud transport and its role in radiation exposures.

The installation of a prototype HTFS in an operating PWR will require an estimated investment of \$3-4 million, of which the EMF's represent $\sim$ 15-20%.

Prototype Design Concept

Early negotiations in search of an operating PWR plant, whose availability schedule was consistent with the objectives of the program, produced a candidate plant equipped with RCS main loop stop valves. The 8-inch bypass lines associated with the loop stop valve configuration afford a convenient access to the RCS without the necessity of direct penetration into the largediameter loop piping. Although not essential to the success of the HTFS, for a first-effort this convenience is

an asset. The initial system design has therefore been based on using the bypass line approach.

Figure 4 illustrates the basic concept of installing a filtration loop in each of two RCS bypass lines. The flow bypassed through each filter will be 0.25% of the total reactor flow.

Figure 5 shows in more detail the HTFS, consisting of the two filtration loops served by a single filter flushing and waste disposal system. Filter flushing would be sequential.

The two-filter concept minimizes the volume of flush water and the volume of waste slurry per flush, hence minimizes the volume of the shielded flash tanks. The size of piping and valves needed to handle the design total flow is also minimized. Other advantages include more layout flexibility, redundancy of filtration capability, dispersion of the total volume of radioactive solids collected by the filter matrix, reduced impact of bypass flow on a single loop, and flexibility to compare the operational characteristics of filters of different designs.

A typical EMF consists of a non-magnetic pressure vessel containing a matrix of magnetic stainless steel. Surrounding the vessel is a water-cooled solenoid electromagnet which provides the magnetizing field. The matrix elements distort this field producing gradients and the attractive force on the crud particles.

After evaluation of the relative filter design features and a review of the previous experimental work, it was the concensus that the sphere-matrix EMF offered the best opportunity currently available for achieving the goals of reactor coolant hot filtration with a minimum impact on the existing plant operation. Therefore, the Babcock and Wilcox sphere-matrix EMF, shown schematically in Figure 6, was selected for the EPRI prototype HTFS.

The B&W electromagnetic filter requires a flush flow of water in the forward direction. With the matrix degaussed, the flush water fluidizes the ball matrix exposing the complete surface of the balls to the water resulting in an effective removal of particles.

Water for the flush is taken from a flush tank located inside the containment. Prior to flushing, the tank is filled with hot water (∼543°F) from the RCS which compresses the nitrogen gas in the tank to RCS pressure(∼2300 psig). After the EMF is isolated from the RCS and the matrix degaussed, the flush is initiated. Crud from the matrix is transported to a flash tank, located close to and below the EMF's. An orifice at the flash tank inlet limits the flush flow to ∼ 2/3 of the filter design flow. The flush flow is terminated automatically on low level in the flush tank and the EMF is returned to service after its matrix is re-magnetized.

When the flash tank is in thermal equilibrium with the flush water the temperature will be less than 250°F. The temperature reduces further to <200°F by heat losses to the ambient. When it is desired to transfer the crud to the disposal system, water is added to the flash tank from the Chemical and Volume Control System (CVCS). This stirs up the crud and allows for easy transfer. A small overpressure of nitrogen provides the driving force.

Meeting the overall goal of radiation exposure reduction with the HTFS depends critically on the design of the waste disposal system, since personnel exposures from this source could offset gains in reduction of sources on the RCS surfaces. This part of the total design, however, is highly plant specific. Since selection of the candidate PWR plant has not yet been resolved, the details of the waste disposal subsystem have not been established.

CONCLUSIONS AND RECOMMENDATIONS

The conclusion to be drawn from the experience and analyses to date is that hot electromagnetic filtration of PWR reactor coolant is a very promising technique for radiation exposure reduction and should be subjected to a realistic prototype demonstration. The EPRI demonstration program would directly benefit the particular plant involved but the results and experience gained would also benefit the entire electric utility industry.

REFERENCES

1. Beslu, P., Frejaville, G., "Occupational Radiation Exposure at French Power Plants: Measurement and Prediction", Nucl. Technology, Vol. 44, June, 1979.
2. Darras, R., Dolle, L., Chenouard, J. and Laylavoix, F., "Recent Improvements in the Filtration of Corrosion Products in High Temperature Water and Application to Reactor Circuits," Paper No. 43, Int. Conf. on Water Chem. of Nuclear Reactor Systems, Boumemouth, England, October 1977.
3. Dolle, L., Rosenberg, J. and Darras, R., "Etude de Procedes de Filtration de l'Eau a Haute Temperature dans les Circuits Primaires", Bull. D'Information Scientifiques et Techniques du CEA, No. 212, March 1976.
4. Heathcock, R. E. and Lacy, C. S., "Graphite Beds for Coolant Filtration at High Temperature, Paper No. 39, Ibid.
5. LeSurf, J. E., "Control of Radiation Exposures at CANDU Nuclear Power Stations," J. Br. Nucl. Energy Soc., Vol. 16, January, 1977.
6. Moskal, E. J. and Bourns, W. T., "High Flow, High Temperature Magnetic Filtration of Primary Heat Transport Coolant of the CANDU Power Reactors," Paper No. 40, Ibid.
7. Sejvar, J., "Revision of CORA - A Computer Code for Calculating the Activities of Corrosion Products in Reactor Systems," WCAP-7708, Westinghouse PWRSD, Pittsburgh, PA, May, 1979.
8. Thexton, H. E., in "Materials Research in AECL," AECL 5227, Summer, 1975.
9. Troy, M., Dallas, D. E., Arvidson, B., Dawson, A. M., Tower, S. N., Waskiewicz, R. W., Cuscino, G. P., Calderwood, A. S., Sejvar, J. and Kang S., "Study of Magnetic Filtration Applications to the Primary and Secondary Systems of PWR Plants," EPRI Report, NP-514, May, 1978.

THE FRENCH PROGRAM ON ELECTROMAGNETIC FILTRATION

P. Verdoni, M. Dubourg

FRAMATOME
Tour Fiat - 92084 Paris La Defense - FRANCE

R. Darras, L. Dolle

CEA Saclay
BP N°2 92190 Gif Sur Yvette - FRANCE

M. Arod

CEA - Cadarache
13115 Saint Paul Lez Durance - FRANCE

INTRODUCTION

The elctromagnetic high temperature filtration has a powerful capability for preventing the harmful contamination of primary circuit components by the deposit of activated corrosion products, and can be applied at relatively high flow rates in light water reactor technology.

The activated corrosion products are based for PWR on magnetic (Fe 304) and nickel ferrite formed at elevated temperatures and contain cobalt 58, cobalt 60, maganese 54, and other species which contribute to the activation of internal surfaces. Owing to the very slight solubilities of these various species in the primary coolant at elevated temperatures, the major part of activity is carried as particles in suspension.

By using high temperature electromagnetic filters located on a by-pass line of a primary loop, it is possible to compete with the deposit formation of corrosion products and to prevent their activation on the fuel assembly surfaces by fixing the corrosion products on a magnetized ball matrix at high temperature without any thermal degradation of the derivated primary coolant. In addition, electromagnetic filtration has a potential capability to selectively fix activated corrosion products on the ball matrix, which are released from the reactor during thermal and power transients of reactor operation.

For demonstrating these anticipated benefits of an electromagnetic filtration to the water reactor purification system, the Commissariat at l' Energie Atomique (CEA) and Framatome have jointly undertaken a large R & D program, the objectives of which are the following:

1. Reduce the activity of the primary coolant during plant operation.

2. Minimize the activated corrosion product built up on primary piping.

3. Demonstrate the economical interest of an electromagnetic filtration system integrated into a PWR NSSS.

For reaching these objectives the French R & D program an electromagnetic filtration is backed by the following work:

1. Experimentation of an electromagnetic filter prototype on out-of-pile and in-pile test loops. Limit performances of the prototype are measured on the NADINE test loop (out-of-pile loop) and efficiency of the filter with regard to non magnetic products has been measured on the IRENE loop (in-pile loop).

2. Long range experimentation with an electromagnetic filter built by Framatome and located on a by-pass circuit of an integrated PWR (Chaudiere Avancee Prototype CAP reactor).

3. Development of analytical methods (code PACTOLE) and models for predicting the surface contamination and the benefits brought by industrial filters to a PWR plant.

4. Design and construction effort of industrial filters for integration into primary coolant purification loops.

EXPERIMENTAL STUDIES OF PROTOTYPE FILTERS

Studies which have been undertaken to develop an industrial electromagnetic filter, have involved experimentation of prototype filters both on out-of-pile and in-pile loops and on the CAP PWR reactor located at the Cadarache Nuclear Center.

Out-Of-Pile Loop Studies

Most of the out-of-pile studies have been carried out in the DOLMEN loop which is designed to circulate pressurized water at high temperatures.

Filtration Efficiency Measurements. Dolmen loop has been used to measure filtration efficiency of a prototype electromagnetic filter on a reduced scale. This filter, which comprises three water cooled magnetic coils, consists of a non ferromagnetic steel tube packed with a matrix of 13% chromium steel balls.

Main characteristics of this filter are:
- filtration tube diameters: ID = 74 mm; OD = 89 mm
- ball matrix: 530 mm height; weight = 11 kg
- ball diameter = 6 mm
- operating magnetic field = 1800 gauss

Filtration efficiency measurements have been done at 250°C at a flow rate of 2.5m^3/h (11 gpm) in the main circuit corresponding to a linear velocity of 30 cm/s in the ball matrix (16cm/s in the unpacked filtration tube).

Filtration efficiency E is defined as the ratio:

$$E = \frac{C_i - C_o}{C_i}$$

where C_i and C_o represent the suspended solids concentrations at the inlet and outlet of the filter respectively. For this measurement, a suspension of magnetite was injected at a constant flow rate upstream from the filter. The magnetic suspended solids concentration upstream from the filter was maintained between 0.7 and 0.04 ppm. Filtration efficiencies measured during these experiments are plotted on Figure 1 versus magnetic concentration of the filtered water. As can be seen on Figure 1 the amount of dissolved oxygen has a great influence on the filtration efficiency. For oxygen concentration in the range of 0.01 ppm, the efficiency is higher than 99%. The detrimental effect of dissolved oxygen, specially evident for low magnetic concentrations, is attributable to the formation of Fe_2O_3 which is less magnetic than Fe_3O_4. During the experiment, the chromium release rate of the balls, measured by neutron activation, was found to be very low.

Matrix Pressure Drop. Among the parameters which direct the optimization of an industrial electromagnetic filter intended for the primary system of a PWR reactor, the matrix pressure drop is of great importance. Measurements that have been done at 255°C in the DOLMEN loop have given the following values:

- 50 mbar/m for a specific flowrate of 1.7 m^3/h (apparent velocity 30 cm/s in the filter).
- 120 mbar/m for a specific flowrate of 5 m^3/h-dm^2.

The pressure drop is only slightly affected by the saturation degree of the filter.

Capacity of the Filter. Another important consideration for the design of an industrial filter is its capacity; i.e., the mass of impurities which can be retained by 1 kg of ball matrix before saturation. Capacity measurements taken at 250°C on the Dolmen loop have indicated the average value of 2g/kg matrix for the purification of nickel ferrite in the concentration range between 0.03 and 0.7 ppm; capacities were determined by calculation of mass of impurities which had been retained once the filter had been saturated to 95% efficiency. Similar experiments have been carried out on the Nadine loop which is another pressurized water out-of-pile loop.

In the concentration range of 0.05 to 0.15 ppm impurities, a somewhat higher value of 2.3 g/kg matrix has been determined at 280°C.

Critical Velocity. Theoretical developments on the antagonistic strengths exerted on a particle crossing through the filter - the magnetic attraction force and the dragging force in the direction of the liquid flow - indicate a critical value for the flow velocity. Particles entering into the filter at higher velocities than the critical value are not picked up by the electromagnetic field. A given electromagnetic filter critical velocity depends on various different parameters, among which are the impurities concentration of the effluent, the electromagnetic field amplitude and the saturation degree of the filter. Critical velocities measurements taken in the Nadine loop at 280°C with a 0.5 meter ball matrix height are summarized in Figure 2.

Critical velocity varies in the range between 20 and 30 cm/s. Increasing the electromagnetic field enables the operation of the filter with higher fluid flow velocities. Figure 2 outlines the dependency of critical velocity on the saturation degree of the filter; however, in the conditions of the present testing there is virtually no influence of the saturation degree for magnetic fields higher than 250 000 At/m.

Other tests aimed at studying the influence of filtration temperature, impurities concentration and ball matrix height on the critical velocity are presently in progress on Nadine loop.

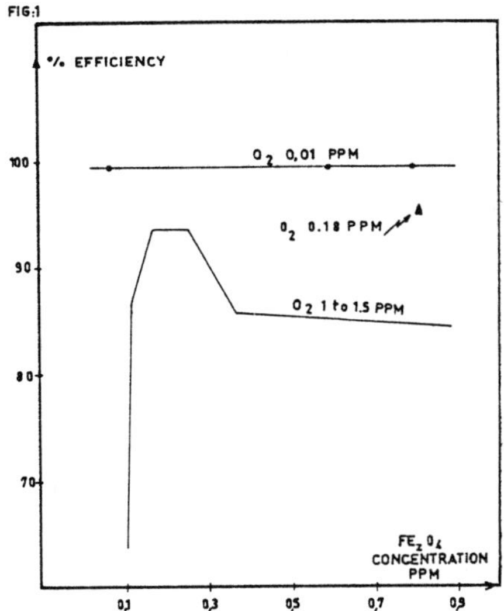

Filtration Efficiency Versus Fe_3O_4 Concentration And Water Oxygen Content

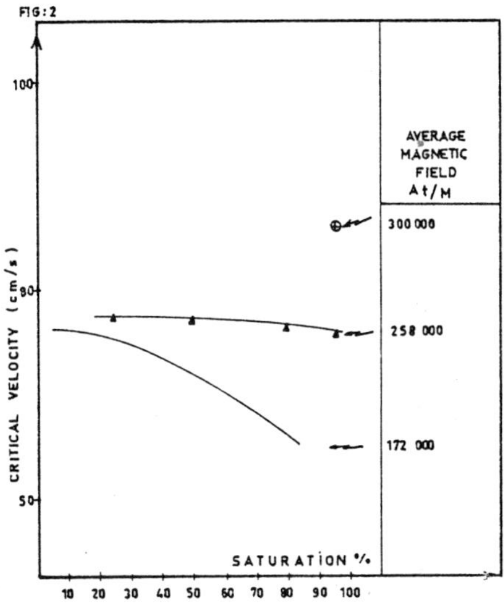

Critical Velocity Versus Saturation Degree At Various Magnetic Fields

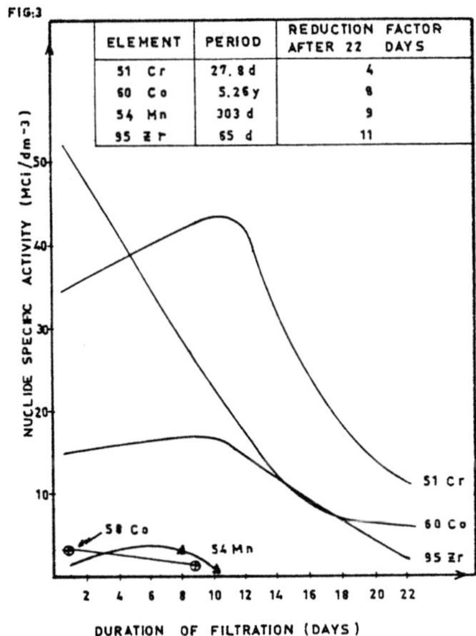

Irene Loop - Filtration Efficiencies

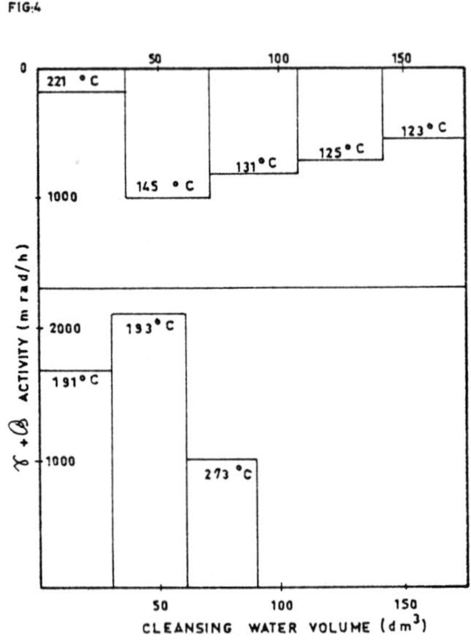

Specific Activities Of Individual Cleansing Fractions Determined During Two Typical Declogging Operations On the In Pile Irene Loop

FIG 5

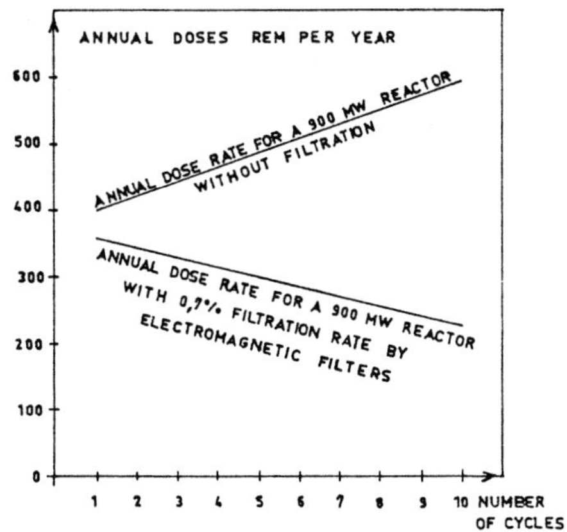

EFFECT OF ELECTROMAGNETIC FILTRATION
0,7 % OF RATED FLOW RATE

REDUCTION OF - 10 % AT THE END OF 1st CYCLE : 40 rem
REDUCTION OF - 20% AT THE END OF 3rd CYCLE : 80 rem
REDUCTION OF - 40% AT THE END OF 5th CYCLE : 200 rem
REDUCTION OF - 52% AT THE END OF 10th CYCLE : 320 rem

FIG 6

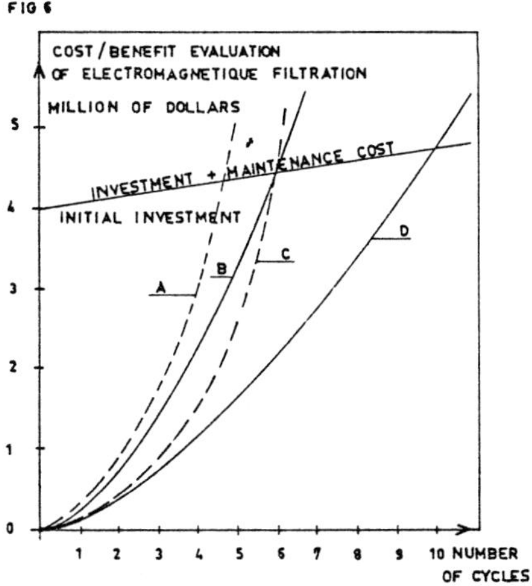

A - MAN REM 10,000 $ REEVALUATION COST 10 % PER YEAR
B - MAN REM EVALUATED AT 10 000 $ AT CONSTANT COST
C - MAN REM 5000 $ REEVALUATION COST 10 % PER YEAR
D - MAN REM EVALUATED AT 5,000 S AT CONSTANT COST

In-Pile Loop Studies

Purification of Corrosion Products. In addition to the filtration efficiency measurements which have been carried out by means of magnetite injection in the Dolmen loop, electromagnetic filtration of fine radioactive particles typical of a reactor coolant has been studied. For this purpose, the in-pile Irene loop is roughly representative of those released in a PWR primary coolant. Flow rate in the filter was about 1 m^3/h (apparent velocity 7cm/s) which represents 10% of the total loop flowrate.

Specific activities of different nuclides present in the 300°C lithiated-borated water circulated in the loop were measured at the inlet of the electromagnetic filter after 1, 8, 10, 15 and 22 days of experimentation. Measured values are listed in Figure 3 which shows the reduction factor of specific activities observed for ^{51}Cr, ^{60}Co, ^{54}Mn, and ^{95}Zr.

Evolution of specific activities is controlled both by the filtration efficiency and by the nuclear decreasing of each nuclide. Filtration efficiencies related to these measurements vary with concentrations of studies species; typical values are 91% for ^{51}Cr, 92% for ^{60}Co and 95% for ^{95}Zr. The excellent retention of zirconium, although this element is not ferromagnetic, demonstrates the ability of the electromagnetic filter to retain non-magnetic elements like manganese, zinc and cesium which have been found in the sludge cleansed out from the filter.

At the end of the experiment, the total specific activity of the loop was as low as 20 $\mu Ci/m^3$.

High Temperature Back Flushing. One of the particularities of the electromagnetic filter developed in France is its ability to be back flushed at its operating temperature. In the high temperature back flushing process, the filter is primarily insulated from the system and flushed with water at operating temperature. The ball matrix is then demagnetized through several inversions of the electromagnetic coils excitation circuit, where the intensity is decreased. The steam water mixture is then expanded in a water cooled condenser which is subsequently insulated and allowed to cool to a sufficiently low temperature so the liquid can be blown-down with the load of suspended solids that it contains.

The thorough back flushing of the filter requires several cleansing operations as the one described above. The back flushing efficiency has been determined at the end of a series of purification tests carried out on the inpile Irene loop. In fact, it is possible to define a partial efficiency for each individual cleansing operation.

Figure 4 represents the specific activities of individual cleansing fractions as determined during two typical back flushing operations. As can be seen, the efficiency increases during the first operations and generally reaches its maximum at the second.

High temperature back flushing methods appear to be preferable to medium temperature ones as the same efficiency can be achieved through the use of less water. Minimizing the amount of back flushing water is of great concern to reduce the volume of radioactive effluents that are to process.

CAP Reactor Experimentation

Another filter, (described hereafter in this paper), designed to purify the primary coolant in the nominal operating conditions of a commercial reactor, has been installed on a bypass circuit of the CAP reactor which is an integrated PWR. Preliminary results which have been obtained until now bring out:

- a satisfactory behavior of the filter on the mechanical electrical and thermal points of views,
- very appropriate thermohydraulics, particularly during the high temperature back-flushing process,
- a good efficiency for the retention of suspended corrosion products of the loop, more particularly for ^{58}Co, ^{60}Co and ^{54}Mn, despite their very low concentration in the coolant (some ppb).

This experimentation is still running.

INDUSTRIAL FILTERS

In view of possible integration of electromagnetic filters into operating reactors, FRAMATOME has undertaken calculations and design efforts for outlying the benefits brought by electromagnetic filtration on the overall dose exposure received by operating and maintenance personnel.

Evaluation of the Benefits Provided by Electromagnetic Filtration of the Primary Water of a PWR

The effect of an electromagnetic filtration located on the side arm loops of a primary circuit has been computed with the CEA computer code Pactole for a 1300 MW type of reactor. The calculation of dose reduction brought by electromagnetic filtration has been performed, taking into account the following assumptions:

- a large reactor has been selected (1300 MW). This reactor is equipped with two electromagnetic filters located on the side arm loops and having a total filtration flow rate of 2000 gpm (100 gpm per filter) which corresponds to a filtration flow rate of about 0.5% of the rated flow of a 4 loop or 0.7% of the same flow of a 3 loop reactor (900 MW).

- filtration efficiency of insoluble magnetite and nickel ferrite is taken close to 100%. This assumption is consistent with the results obtained on the CAP reactor where the efficiency of the electromagnetic filter has been evaluated on the total amount on iron and nickel (soluble and insoluble species) to be 80% for ferrite concentration in the range of 5 to 10 ppb.

- comparative calculations between a reactor not equipped with electromagnetic filter and a reactor equipped have been carried out for covering an operating period of the reactor of 10 cycles. These calculations have taken into account the primary chemistry variation through the fuel cycle. This variation plays an important role on the solubility of species.

- dose reduction and primary coolant activities have been calculated on four radioactive species: ^{58}Co, ^{59}Co, ^{59}Fe, ^{54}Mn. The main conclusions of these calculations are the following:

Activity Build Up on Primary Circuit. Electromagnetic filtration reduces the activity build-up on the primary circuit. This reduction is non uniform, and in some cases, a slight activity build-up increase can be noticed on the hot leg where erosion phenomena occur. The reduction of dose rates is small for the first cycles, but this reduction increases greatly with the number of cycles.

The reduction is estimated to be:
15% at the end of the first cycle
25% at the end of the third cycle
35% at the end of the fifth cycle
60% at the end of the tenth cycle

Primary Water Activity Effects on Auxiliary System. The activity of primary water is reduced by electromagnetic filtration by a factor of 3 at the end of fifth cycle and 8 at the end of the tenth cycle.

From the measurements performed on a reactor, it can be noticed that a large portion (∿ 50%) of the man rem number is due to various interventions on auxiliary circuits. The activity of these circuits is mainly caused by the activity of primary water and therefore, the reduction of the primary water activity will have a large impact on the activity of auxiliary circuits.

Table I gives the dose breakdown on an operating 900 MWe reactor according to the buildings (reactor building and auxiliary building) and the nature of operation: Reactor shutdown work, refueling operation, maintenance, inservice inspection. The beneficial effect of electromagnetic filtration is illustrated by reductions of:

20% at the end of the third cycle
40% at the end of the fifth cycle
52% at the end of the tenth cycle

If a 900 MW plant is considered and an initial dose rate of 400 man rems per year is assumed and this dose slowly increases on a continuous basis to reach 600 man rems at the end of the tenth cycle of service, (see Figure 5), a cost/benefit can be established according to the cost of man rems. The cost/benefit curve (Figure 6) shows that the initial investment and maintenance cost of a filtration system is redeemed in 5 or 6 cycles, if a cost of man rems is assumed to be:

- $ 5000 with an escalation rate of 10% per year
- $ 10 000 at constant cost

It must be said that the cost/benefit evaluation has been performed on an average reactor which does not present any specific maintenance problems nor requires major repair work. The benefit of electromagnetic filtration increases if such a case occurrs because the dose reduction is proportional to the exposure of personnel; i.e., if a major repair work has to be done on a reactor, this repair work will be carried out with less personnel or in less time on a reactor equipped with an electromagnetic filtration system.

Industrial Filters for PWR

FRAMATOME is studying the possibility of using electromagnetic filters on purification loops in operating PWR plants. Size of the filter and its capacity is mainly governed by the availability of space in the containment, and by the necessity of having connection branches on the primary circuit.

Cost/benefit studies show that the optimum size for hot filtration of the primary water corresponds to a filtration flow rate between 0.5 and 1% of the total flow rate. For higher filtration flow rate (4 to 5%), the benefit obtained by the greater reduction in dose is largely compensated by the cost of energy lost for cooling the core due to the bypass flow rate and by the extra investment cost for implanting filters in the containment.

FRAMATOME is presently designing a high temperature purification loop of the primary circuit equipped with 1000 gpm filters which could be installed in the containment of PWR plants with a minimum of disturbance. Filters of this size could be installed on:

- New plants in which the purification loop will be built on a side arm of primary pumps.
- Operating plants (as backfitting) in which the purification loop will be connected on existing auxiliary piping connections.

A schematic diagram of a purification system on a 900 MW PWR is shown (see Figure 7 and 8). The filtration system consists of:

- Two electromagnetic filters.
- High pressure piping for connecting the filters to the primary circuit and double isolation by means of motor operated valves.
- Safety filters which constitute a secondary barrier for preventing balls from going to the primary circuit.
- Flashing tank and the associated equipment.
- Filters cartridges for collecting the radioactive slurry and oxides.
- Storage tank for the liquid effluent.

The FRAMATOME electromagnetic filter is characterized by the fact that it operates continuously under PWR conditions and this equipment is capable of being regenerated remotely during plant operation from the control room of the reactor. The electromagnetic filter itself consists of a vertical pressurized vessel made of austenitic stainless steel which contains the ball packing and coolant to be purified. The ball packing is magnetized by means of 4 external C shape electromagnets, vertically disposed outside the pressurized tank. In normal operation, the filter is connected to the primary circuit by means of class 1 piping equipped with a double isolation valves.

When the filter is saturated in iron oxides, it can be isolated from the primary circuit and regenerated by two-phase-flow back-flushing in a high-pressure flashing tank. During this operation, the water contained in the filter is partly transformed to steam which washes and wipes out the deposits on the demagnetized steel balls.

TABLE I
TYPICAL DOSE BREAKDOWN
FOR A 900 MW REACTOR

REACTOR	REACTOR BUILDING AERA	OPERATION	DOSE WITH NO FILTRATION	DOSE WITH FILTRATION OF FLOWRATE OF 0,7 % END 3th CYCLE	5th CYCLE	10th CYCLE
REACTOR SHUT DOWN	REACTOR BUILDING	REFUELING	48	48	48	48
		REPAIRS AND MAINTENANCE	122	100	80	73
		INSERVICE INSPECTION	80	60	52	32
	AUXILLIARY NUCLEAR BULDING	REPAIRS AND MAINTENANCE	90	70	40	30
REACTOR IN OPERATION	REACTOR BUILING	INSERVICE INSPECTION	18	18	18	18
	AUXILLIARY NUCLEAR BUILDING	REPAIRS AND MAINTENANCE	190	140	90	60
		TOTAL	548	436	328	261
		DOSE REDUCTION REM	0	112	220	287
		DOSE REDUCTION		20 %	40 %	52 %

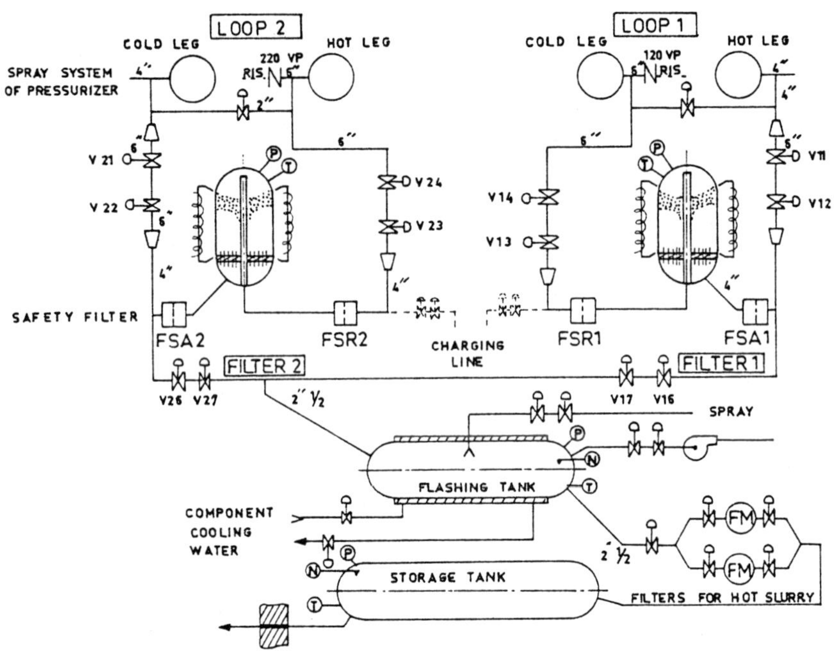

FIG 7 FLOW DIAGRAM OF AN ELECTROMAGNETIC FILTRATION SYSTEM

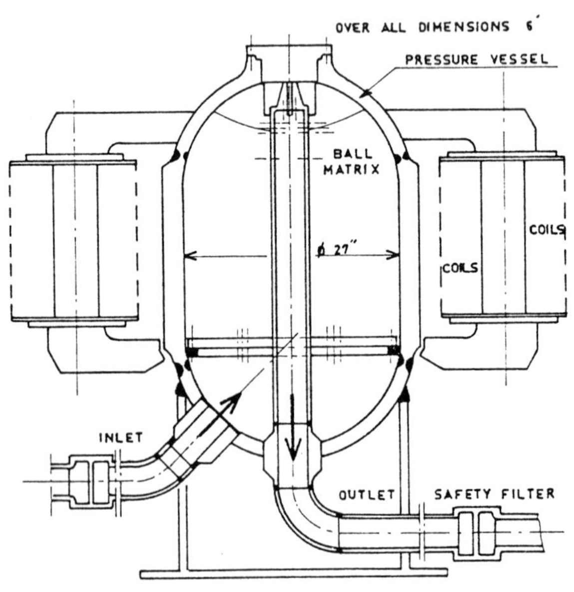

FIG 8 ELECTRO MAGNETIC FILTER

Design efforts toward the application of the system to 900 MWe, 1300 MWe PWR plants have started, and the construction of an industrial filter designed to PWR conditions will start before the end of this year.

CONCLUSIONS

The program carried out jointly by CEA and FRAMATOME for several years follows the main objectives which are:

- To obtain endurance testing of the performance in PWR conditions of a small size electromagnetic filtration system in order to properly qualify components fixed to the primary and regeneration systems,

- to maximize operational characteristics of filters on high-pressure loops, in order to improve the efficiency and flow rate of a given size of filters, keeping in mind that the size of the equipment which must be located in the containment building of a reactor is of major importance,

- to evaluate, through design and calculation efforts, the impact of such a filtration system on the plant performance.

The calculated and experimental results of this program show the benefit of electromagnetic filtration on the reduction of the overall dose exposure. Model development work has to be pursued for a better appreciation of electromagnetic filtration during crud release generated by power and thermal transients.

Another benefit, which has not at this time been properly evaluated, is the reduction of the amount of crud deposit on fuel rods which could lead to an increase of thermal transfer between fuel rods and primary cooling and also minimize the impact of crud deposit on corrosion behavior of fuel cladding.

As result of this joint effort, FRAMATOME is now in a position to propose electromagnetic filtration equipment specially conceived for nuclear application in light water reactors.

REFERENCES

1. P. Besly, G. Frejaville, and A. Lalet, "A computer code PACTOLE to predict motivation and transport of corrosion products in a PWR," CEN - FRANCE, International Conference on Water Chemistry, Bournemouth, October 26-28, 1977.

2. R. Darras, L. Dolle, and J. Chenouard, "Recent improvements in the filtration of corrosion products in high temperature water and application to reactor circuits," CEA - CEN Saclay, Internation Conference on Water Chemistry, Bournemouth, October 26-28, 1977.
3. R. Darras, L. Dolle, J. Chenouard, and M. Dubourg, "Filtration electromagnetique des particules vehiculees dans les circuits primaires des centrales electronucleaires a eau sous pression," Colloque AIEA - Gestion des dechets en provenance des reacteurs de puissance sur le site Zurich, March 26-30, 1979.
4. L. Dolle, P. Grandcollot, J. Chenouard, P. Darras, and P. Basler, "Extraction of insoluble corrosion products from water circuits of Nuclear Power stations by Electromagnetic Filtration," 2eme Congres Mondial de la Filtration, Londres, Sept. 18-20, 1979.
5. L. Dolle, J. Rozenberg, and R. Darras, "Etude des procedes de filtration de l'eau a haute temperature dans les circuits premaires," BIST CEA - #212, March 1976, p. 65-78.
6. "Study of Magnetic Filtration Applications to the Primary and Secondary System of PWR Plant," EPRI Report NP 514, May 1978.

BWR RADIATION CONTROL THROUGH OPERATIONAL PRACTICES

Gerard F. Palino

General Electric Company
175 Curtner Avenue
San Jose, California 95125

ABSTRACT

The rate of radiation buildup on out-of-core Boiling Water Reactor (BWR) primary system surfaces can be reduced by reasonably straight forward changes in plant operating practices. The operational practices considered most relevent include: optimized operation of condensate and reactor water demineralizer systems; control of dissolved oxygen in the feedwater system to minimize corrosion product input into the primary system; and changes in shutdown/ lay-up/ and startup practices. This presentation includes a discussion of the current radiation levels and isotopics observed on out-of-core surfaces of operating BWRs. Evidence for successful implementation of these radiation control practices is presented along with the current status of our modeling development. Finally, the General Electric (GE) program for radiation control through operational practices is introduced.

INTRODUCTION

Data generated in-part as a result of the Electric Power Research Institute (EPRI) funded Boiling Water Reactor Radiation Assessment and Control (BRAC)[1] program indicate that reductions in the rate of activity buildup on out-of-core surfaces, and consequently control of general radiation fields, may be achieved through a combination of plant operational procedures which closely control the water quality in the primary system. The type of procedures which appear to lead to reduced rate of activity buildup include:

1) optimized operation of the condensate and reactor water cleanup demineralization systems;

2) control of dissolved oxygen in the feedwater to minimize the corrosion product input into the primary system; and
3) changes in shutdown/ layup/ and startup procedures.

The introduction of plant operational control appears to offer a relatively inexpensive means of controlling or reducing personnel exposure in accordance to ALARA.

RADIATION ASSESSMENT

The buildup of radiation levels and specific isotopic activities on primary system piping at several BWRs has been determined as part of a long term program to assess the buildup of shutdown radiation levels in the BWR. The program of data collection began in 1970 and was part of a cooperative program between GE and five utility companies to increase the understanding of the water chemistry in the BWR, and is continuing today under the sponsorship of EPRI[1]. Figure 1 shows the average shutdown radiation level seen on the recirculation lines for many of the operating BWRs in this program. It is apparent that early in the life of the BWR the general trend is toward an increase in the radiation level on the recirculation piping in the range of 50-150 mR/h per effective full power year (EFPY).

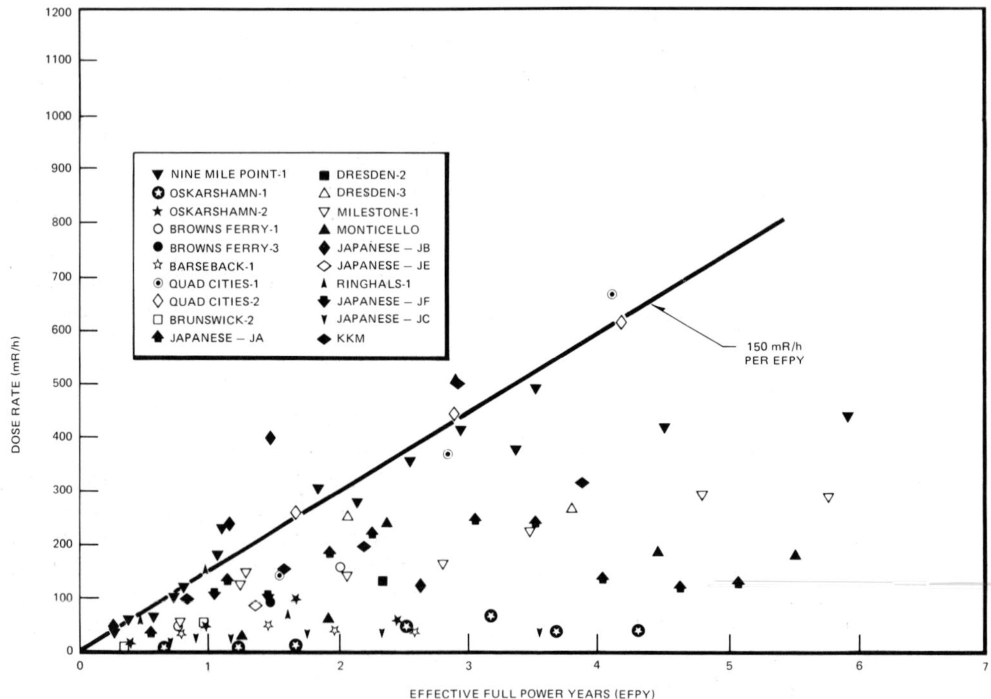

Fig. 1. Radiation Field Buildup on Recirculation Piping

The radioisotopic composition of the deposits on the inside of the primary system piping has been determined using a shielded and collimated Ge(Li) or IGe detection system, Figure 2. Calibration of the GE detector/shielding system for specific pipe geometries and procedures for conversion of counting data to $\mu Ci/cm^2$ of a given radioisotope on the inner wall of the piping has been reported earlier[2,3]. Measurements are usually taken on the recirculation piping inside the drywell and on the piping of the reactor water

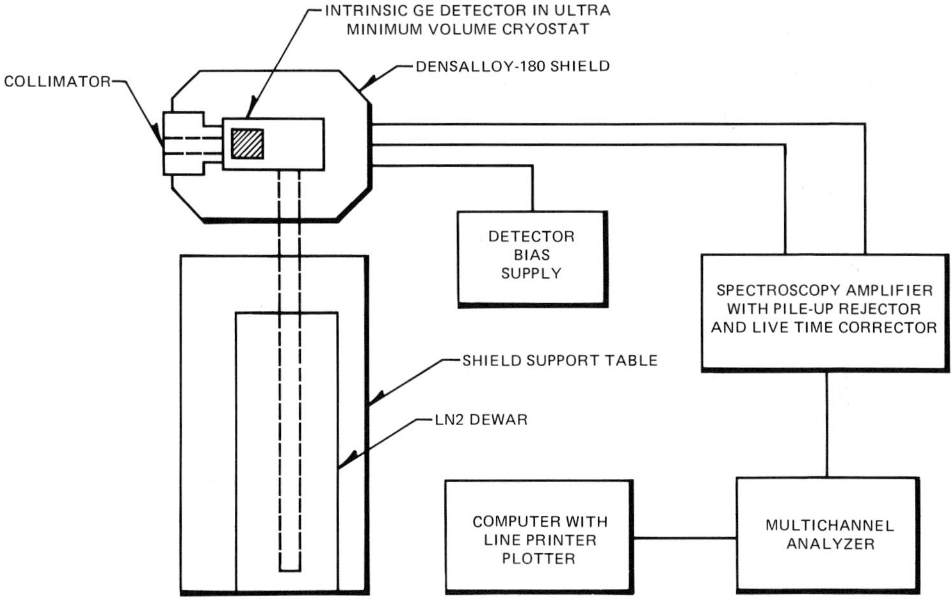

Fig. 2. Block Diagram of the GE Pipe Gamma Scanning System

cleanup (RWCU) system. These gamma scan data have been summarized in a recent GE report[4]. A summary of the average $\mu Ci/cm^2$ data for selected radioisotopes deposited on the primary system recirculation line is presented in Table 1. These radioisotope data are supplemented in most instances by gross dose rate readings of the same targets. In the early phases of this program, the dose rate data were taken with available site instrumentation. Since 1977 all measurements have been taken with GE provided instrumentation for which there is consistent and traceable calibration. These most recent measurements have been taken with a Cutie-Pie (CP) type instrument and with a directional probe gamma survey meter (DPGSM)[5]. The latter instrument has been demonstrated to provide a more reliable reading in confined quarters, such as the cleanup heat exchanger room, where neighboring sources would influence any reading taken with the standard type CP instrument.

Table 1. Average Co-60, Co-58, Mn-54, Zn-65, Zr-95, and Ru-103 Deposited in the Primary System Recirculation Line

EFPH	Co-60	Co-58	Mn-54	Zn-65 (μCi/cm^2)	Zr-95	Ru-103	AVG (mR/h)/ SURVEY METER
BRUNSWICK-2							
2839	0.3	0.9	0.2	0.2	0.05	N.O.	13/CP
8408	2.2	1.7	0.9	0.8	N.O.	N.O.	56/DPGSM
MILLSTONE-1							
10409	3.4	1.8	1.5	0.1	0.2	0.2	130/PIC-6
11185	3.3	0.7	1.0	0.1	0.04	0.1	150/PIC-6
17830	7.1	1.7	3.1	0.6	0.4	1.2	140/PIC-6
23810	5.8	1.1	1.6	0.1	0.9	2.2	160/PIC-6
30144	8.9	1.5	2.3	N.O.	0.8	3.2	220/CP
MONTICELLO							
10560	0.5	0.4	0.1	1.0	0.2	0.2	32/CP
16450	1.2	0.3	0.3	1.2	1.0	1.2	65/CP
20560	4.0	1.0	1.0	3.5	12.4	12.5	240/CP
24810	7.9	1.3	1.1	4.5	7.8	9.8	510/CP
38598	8.5	0.6	0.5	1.3	0.7	2.0	190/DPGSM
NINE MILE POINT-1							
15755	12.6	1.5	1.5	0.3	0.6	0.7	300/PIC-6
21854	13.6	1.4	1.3	N.O.	0.8	0.9	450/CP
30358	14.5	1.3	1.5	0.7	1.5	2.7	490/CP
38887	19.5	1.5	1.2	N.O.	1.7	2.7	420/DPGSM
50996	21.0	1.5	1.4	N.O.	N.D.	N.O.	440/DPGSM
QUAD CITIES-1							
24447	25.3	1.7	1.3	N.O.	N.D.	0.4	400/DPGSM
35571	34.0	1.5	2.1	N.O.	3.2	4.3	670/DPGSM
QUAD CITIES-2							
25000	27.7	0.9	1.0	N.O.	N.O.	0.2	440/DPGSM
36185	34.2	1.5	1.4	N.O.	2.1	2.6	620/DPGSM
JAPANESE – JA							
30210	5.7	0.6	3.4	N.O.	0.8	1.5	145/CP

DPGSM – DIRECTIONAL PROBE GAMMA SURVEY METER
CP – CUTIE PIE TYPE INSTRUMENT
PIC-6 – EBERLINE INSTRUMENTS
N.O. – NOT OBSERVED
N.D. – NOT DETERMINED

The analysis of the data indicate that from the viewpoint of radiation control that Co-60, with its two energetic gamma rays and long half-life, will ultimately dominate the radiation field for all of the BWRs surveyed to date. Early in plant life the shorter lived nuclides, Co-58, Mn-54, etc.., will dominate and there is at least one incidence (Monticello between 20000-25000 EFPH) where fission products contributed significantly to the dose rate. The data on the average Co-60 $\mu Ci/cm^2$ on the recirculation piping of BWRs are presented in Figure 3. The data show that there is a wide range of levels of Co-60. Gamma scan measurements were recently performed at both the JA (53344 EFPH) and JC (31073 EFPH) reactors. These data are not yet available for release, but preliminary analyses indicate that for both plants the radiation levels and the Co-60 levels in the primary system remain at these low levels.

A. Favorable Plant Experiences

Recent favorable BWR experiences in maintaining low levels of activity on out-of-core surfaces or trends in the reduction of the rate of radiation buildup include Nine Mile Point-1, Monticello, Millstone-1, Oskarshamn-1/-2, and the JA and JC reactors. These data are shown in Figure 4. Each plant has its own individual characteristics that relate to the ultimate buildup of radiation levels and the total system interactions are quite complicated and do not yield to simplistic solutions. Nevertheless, some generalities appear to be common and suggest a route to radiation control that may yield success. Selected plants are discussed below in terms of the design or operational practices that are believed to impact on the rates of radiation buildup.

Monticello

In recent years the radiation level on the recirculation piping has virtually leveled off. Periods of high radiation levels at this reactor in 1975 were traceable to faulty fuel that has since been removed. This plant has undergone extended periods of steady operation and utilizes the capacity of the condensate treatment and RWCU system (2% of feedwater flow) to the fullest extent. Present practice is to operate all seven of the condensate demineralizer units in parallel, whereas before one or two of the units were held in reserve. This change in operation has resulted in a reduction in the net flow rate through each unit which allows more efficient removal of soluble species. This effect is most readily verified from the reduced activity of Zn-65 detected in the most recent gamma scan measurement program, and in an improvement of the quality of the feedwater.

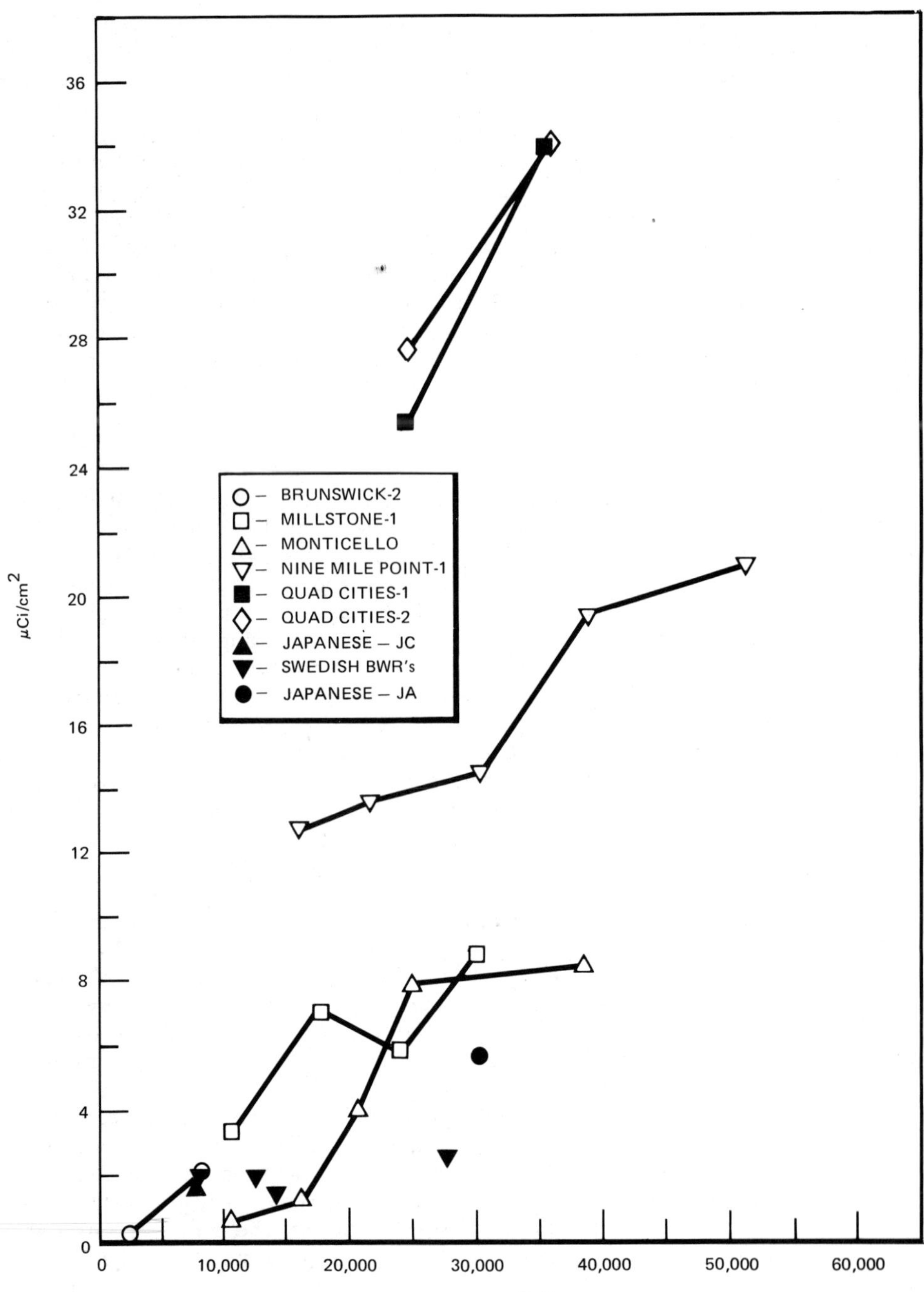

Fig. 3. Average Co-60 Concentration on BWR Recirculation Lines

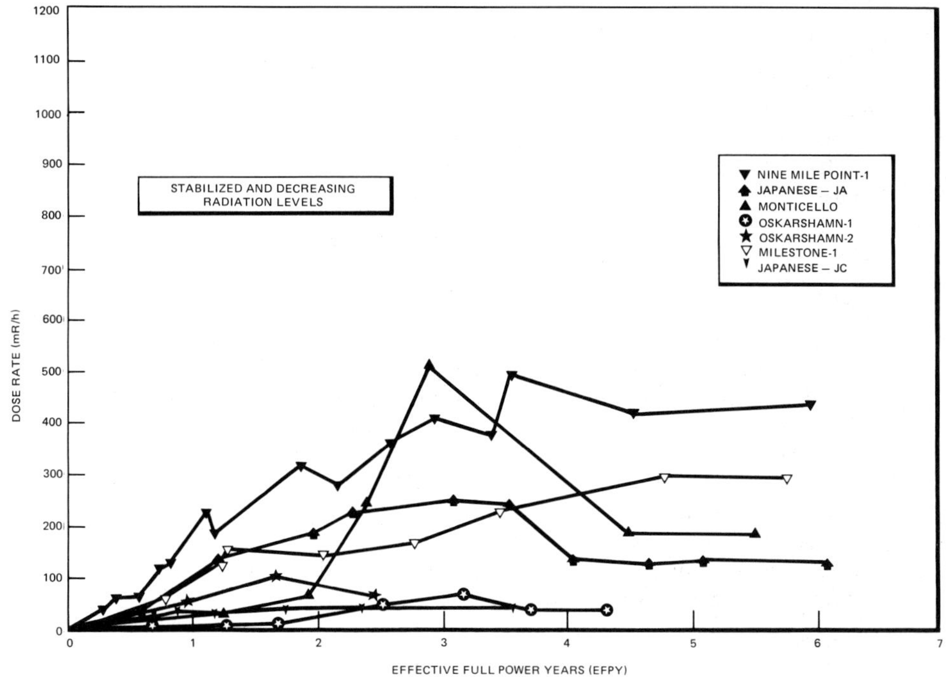

Fig. 4. Stabilized and/or Decreasing Radiation Levels on BWR Recirculation Piping

Nine Mile Point-1

The $\mu Ci/cm^2$ of Co-60 on the recirculation piping at NMP-1 has in general been higher than most of the studied BWRs and has undergone a relatively rapid increase after ∿30,000 EFPH. The most recent measurements (April 1979 @ 50996 EFPH) indicate that the rate of radiation buildup has decreased markedly. During this most recent cycle a number of changes have been initiated at NMP-1 that impact favorably on the water quality in the feedwater and in the primary system. The following significant changes in past operation were noted:

- the spare condensate demineralizer was placed in regular service (six instead of five units on stream);
- the time period between regeneration of the deep-bed condensate demineralizer was decreased to 60 days;
- the control rod drive water suction was switched to the demineralizer outlet rather than the condensate storage tank, thus

lowering the conductivity in the primary system;

- the RWCU system flow was roughly doubled from 2% to 5% (system design capacity) for the last month before shutdown;
- the feedwater dissolved oxygen level was increased from 15-25 ppb to 40-50 ppb (unintentionally) which resulted in a decrease of the feedwater iron concentration from ∿10 ppb to the current level of ∿2 ppb; and
- the periods of continuous operation have increased.

JA Site

The JA reactor site has had a remarkable turnaround in radiation levels as shown by the data presented in Figure 4. In the early periods of plant operation the radiation level increased at a rate of ∿100 mR/h-EFPY until it reached a maximum of ∿300 mR/h in early 1975. During this period of early operation the feedwater iron input was quite high and the first evidence of crud induced fuel failures were noted. Since 1975, there have been a number of changes in the operational practices that are believed to contribute to the observed reduction in the radiation field. The following operational changes were noted:

- higher quality condensate demineralizer resins were utilized which resulted in an improvement of the decontamination efficiency and the feedwater quality;
- oxygen was injected at the demineralizer outlet to maintain the dissolved oxygen level at 60-80 ppb in the feedwater, which resulted in a reduction of the feedwater system corrosion and, therefore, a reduction in the crud input (to ∿1 ppb total iron in the final feedwater) to the primary system;
- during periods of extended shutdown the feedwater system is hot drained and put in dry layup to reduce the system corrosion;
- the feedwater system is flushed to the hot well prior to startup, which reduces the incidence and severity of crud bursts during startup; and
- the condensate and RWCU system (7%) are operated at their design capacity.

JC Site

The radiation buildup rate in the primary system of the JC reactor is less than 15 mR/h-EFPY. This plant is an example of a BWR where basic system design and careful operational practices have been married to yield enviable results. The program at the JC site involved the following items of interest relevant to radiation control through design or operational practices:

- the site has dual treatment systems (resin powdered filters plus deep-bed demineralizers in series) for both the condensate and RWCU (7%) systems. All systems operated at design capacity.

- the dissolved oxygen in the feedwater is closely controlled at a level between 20-25 ppb;
- the feedwater is recirculated at one-third feed flow during periods of shutdown;
- the feedwater system is flushed prior to startup; and
- the site has had an exceptionally long period of continuous operation.

The soluble iron in the final feedwater is reported to be ∿0.1 ppb and the insoluble iron is also quite low at ∿1 ppb.

The following conclusions relative to radiation control through operational procedures can be made from the observations discussed above and from other data that have not been included in this presentation.

1. The operation of the condensate treatment and RWCU system should be optimized to provide the highest water quality attainable under system capacity and operational limitations.
2. The dissolved oxygen in the feedwater system should be controlled to minimize the corrosion product input into the primary system.
3. During periods of extended shutdown, consideration should be given to hot drain and dry layup of the feedwater system. At a minimum, constant recirculation of the feedwater should be considered. Prior to startup the feedwater system should be flushed to reduce the incidence of startup crud bursts.
4. Extended periods of plant continuous operation appear to be beneficial.

Additional system operational changes which impact on radiation control through operational practices contain aspects of the above program. These include rapid repair of condenser leaks to reduce the load on the condensate system and possible piping changes that might lead to improved water quality in the primary coolant.

B. Modeling of Radiation Buildup in the BWR Primary System

Implementation of any radiation control through operational practices program requires a reasonably detailed understanding of the basic phenomenon responsible for the radiation buildup. The development of a cohesive model of radioisotope activation, transport, and deposition has proved to be a formidable task. To be useful, a model should have predictive capabilities so that future shutdown radiation levels can be predicted for a BWR whose operational practices and radiation levels are monitored early in life. In addition, the model should be able to predict the effect of any proposed changes in design or operational practices so that the various techniques for radiation control can be studied prior to implementation in the actual program.

The General Electric Company is in the final stages of development of such a model[6]. The basic features of the interactions considered in the GE model are shown in Figure 5 and are outlined for discussion below.

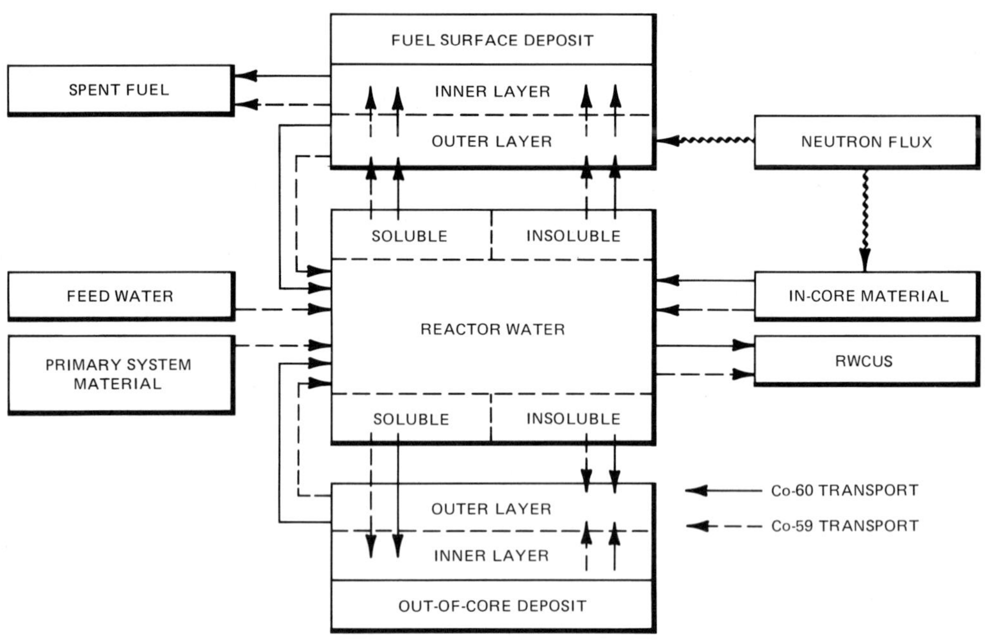

Fig. 5. Block Diagram of GE Co/Co-60 Transport Model

Coolant

- Both "soluble" and insoluble Co/Co-60 species are considered in the primary system coolant.

Fuel Surface

- The deposit on the fuel surface is considered to be a double layer whose magnitude and composition is related to the iron concentration.

- The loosly attached layer is considered to be transformed into the tenaceous inner layer with a majority of the Co/Co-60 being released from the outer layer.

In-Core Cobalt Sources

- Activation and release of activated species from in-core materials (primarily stellite) are considered.

Out-of-Core Surfaces

- On the out-of-core surfaces double layer formation is considered where the inner layer is formed by base metal corrosion and the outer layer is formed mainly by water-borne crud deposits.
- Transformation of the inner layer to the outer layer is considered to occur.
- Stable Co and Co-60 species are incorporated into the inner layer during corrosion.
- Isotope exchange is considered.

The input parameters to the model include total cobalt and iron in the final feedwater, reactor water conductivity, nuclear activation parameters, and variable reactor parameters such as the type of condensate treatment system, reactor design parameters, RWCU flow and

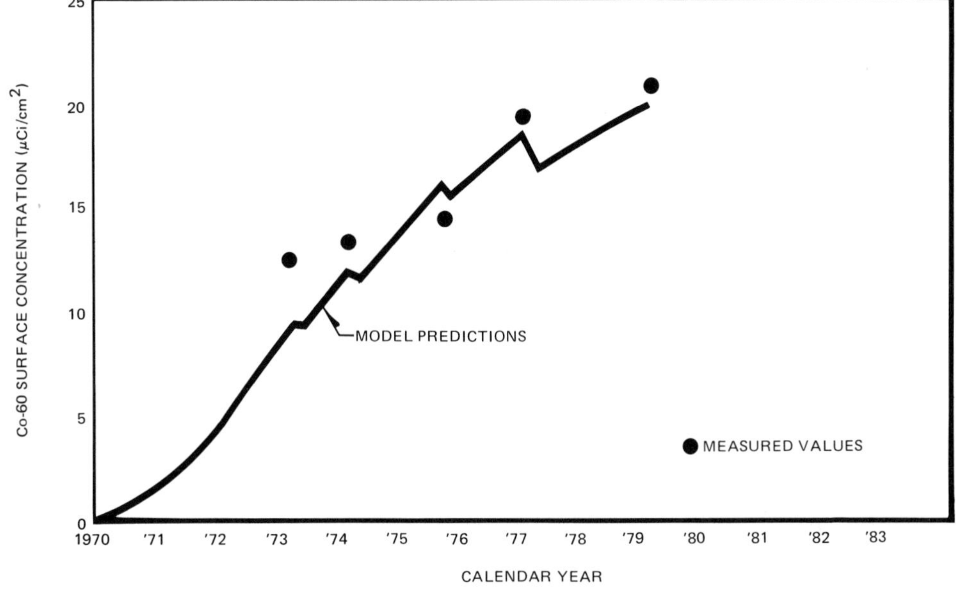

Fig. 6. Model Predictions of Co-60 Activity Concentration on Recirculation Piping Wall (NMP-1)

efficiency, fuel cycle data, etc... The model contains on a phenomenological basis, a considerable amount of information gained from studies of BWR water chemistry, from studies of BWR fuel and out-of-core deposit composition, radioactivity, morphology, and crystal structure, from studies of corrosion and release rates of materials such as stellite, carbon, and stainless steels, and from studies of deposition of corrosion products in in-reactor test loops.

This model has been successfully used to reproduce the observed radiation buildup on NMP-1, Monticello, Brunswick-2, and the JA and JC reactors. Figure 6 illustrates the model application to NMP-1. In addition, parametric studies are underway to evaluate the impact of the adjustable parameters that are believed to be reasonable considerations in any practical radiation control program. To date, the results of these studies are encouraging.

C. GE Program on Radiation Control Through Operational Practices

Since each site is unique in the application of any radiation control program, the General Electric Company has established a program to provide a more detailed evaluation of each plant situation[7]. The radiation control program consists of eight different tasks as outlined below:

1. Evaluation of Present Radiation Levels and Isotopic Makeup;
2. Assessment of Current Plant Operating Practices;
3. Definition of Recommended Radiation Control Program;
4. Consultation on Utility-Generated Action Plan;
5. Fuel Deposit Analysis;
6. Oxygen Injection System Development;
7. Plant Water Chemistry Analysis Upgrade; and
8. Plant Water Chemistry Data Base Management.

While each task may be conducted essentially independently of the others, the maximum benificial effects will be obtained if all tasks are performed in an integrated manner by a single organization.

CONCLUSION

In-plant radiation control through operational practices has been demonstrated on a number of currently operating BWRs. Consciencious implementation of the program results in positive control and can result in a reduction of the radiation fields. The proposed program should not be considered the only radiation control/reduction methodology; decontamination may be necessary. It is suggested that both options need consideration and should be considered together.

ACKNOWLEDGMENTS

The author would like to acknowledge the partial support of EPRI and the efforts of the Chemical Engineering Subsection at the General Electric Company for providing the extended data base from which material for this presentation is derived. He is especially indebted to Dr. C.C. Lin for his contributions in the area of modeling and Dr. W.R. DeHollander and W.L. Walker for their formulation of the integrated GE program for radiation control through operational practices.

REFERENCES

1. The program "BWR Radiation Assessment and Control" (BRAC), RP-819 -1, was established in November 1976 by EPRI and the General Electric Company. Following the inception of this program, several organizations have joined the program through information-exchange, Technology Development Agreements. These organizations include: ASEA-ATOM, Hitachi, Toshiba, Kraftwerk Union, and the Japan Atomic Power Company (JAPC). Prior to the BRAC program, two other programs existed: Nine Mile Point Water Chemistry Program (started in September 1970) and the Water Chemistry Program Extension (WCPE) (March 1973 to September 1976). Participants in the WCPE included Northern States Power Company, Detroit Edison Company, Northeast Utilities, Tennessee Valley Authority, Empire State Electric Energy Research Company, ASEA-ATOM, and JAPC.

2. G. F. Palino, "Radioisotope Activities on BWR Primary System Piping, Part I. Calibration of the G.E. Ge(Li) Pipe Gamma Scanning System", General Electric Report NEDC-12646-1, Class I, October 1976.

3. G. F. Palino and E. G. Brush, "Radioisotope Activities on BWR Primary System Piping, Part II. Calibration Curves for Interpreting In-Plant Gamma Scanning Measurements", General Electric Report NEDC-12646-2, Class I, October 1976.

4. G. F. Palino and J. Blok, "Radioisotope Activities on BWR Primary System Piping, Part IV. In-Plant Gamma Scanning Data through April 1979", General Electric Report NEDC-12646-4, Class I, July 1979.

5. The DPGSM is a combination of the Eberline PRS-1/2 "RASCAL" and a modified HP 220A probe. The HP 220A probe provides a directionality to the measurements by exhibiting a 9:1 front to back ratio for cobalt-60 radiation. This combination instrument has been recommended by GE for use in the GE-EPRI initiated "BWR Radiation Level Surveillance" program; General Electric Report NEDC-12688, Class 1, December 1977.

6. The efforts in modeling "activity transport" had its beginnings with the Water Chemistry Program Extension (Ref. 1) and continued into the EPRI/GE funded BRAC program. This effort culmanated with the publication of GE Report NEDC-13461, Class 1; J. Blok and J. Younger, "Contribution of Fuel Rod Deposits to the Buildup of Radiation on Out-of-Core Surfaces of the Nine Mile Point-1 BWR", in March 1977. Since that time, GE has undertaken, on its own, an extensive program of "activity transport" modeling. The model discussed in this presentation is but a small part of this effort.

7. Additional detailed information on this program can be obtained from BWR Technical Services, General Electric Company, 175 Curtner Avenue, San Jose, California 95125.

RADIATION EXPOSURE, RADIATION CONTROL AND DECONTAMINATION

Robert A. Shaw, Michael D. Naughton and Alan D. Miller

Electric Power Research Institute
P.O. Box 10412
Palo Alto, CA 94303

INTRODUCTION

The radiation exposure received by nuclear power plant personnel is determined by the radiation fields to which the personnel are exposed and the length of time of exposure. This paper will discuss only the former, the radiation fields. In particular the design and operational features of nuclear power plants which most influence the buildup of these radiation fields will be explored, together with the difficulties of ascertaining the effects of changes in such design and operational features on radiation fields and exposure.

The plant features most influential in controlling radiation buildup include materials' selection, coolant chemistry control, filtration and purification techniques and decontamination. Plant testing and monitoring are underway to further elucidate the radiation effects of controlling or changing these various features. The eventual goal is to reduce radiation exposure received by plant personnel in carrying out their duties of operations and maintenance through a reduction of the fields that are present.

RADIATION EXPOSURE

The exposures experienced at nuclear power plants in the U.S. are documented annually by the NRC. The trends of these exposures are useful in evaluating radiation control techniques. The average annual exposure per plant for personnel employed at nuclear power plants in the U.S. for boiling water reactors (BWRs), pressurized water reactors (PWRs) and all light water reactors (LWRs) are

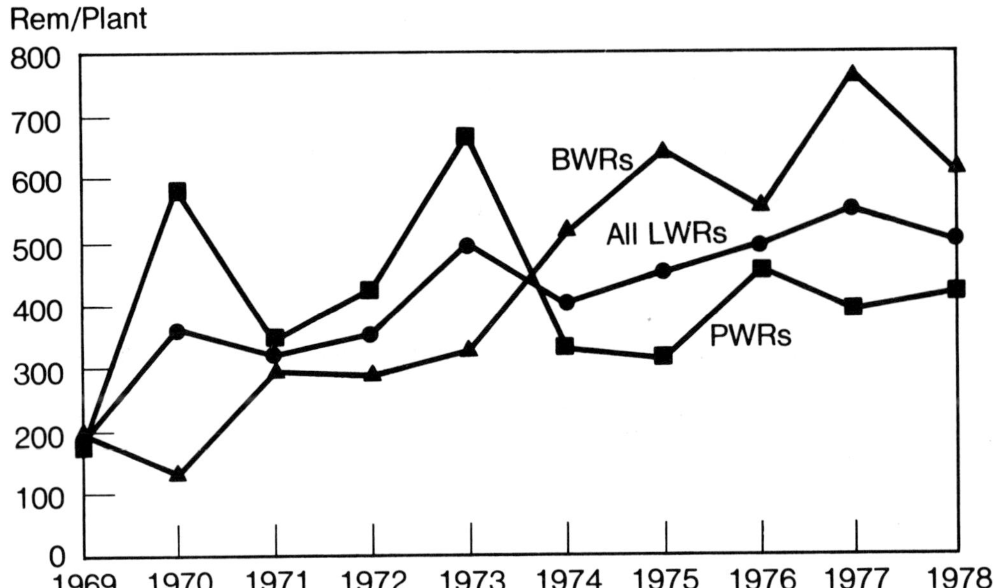

Fig. 1. Average annual U.S. nuclear power plant occupational radiation exposures.

shown in Figure 1. Each of these curves is erratic, making extrapolations and trends uncertain. In searching for the causes of these irregularities a number of instances of plant exposure much higher than the great majority were found. Using these data it was clearly demonstrated that these sets of exposures do not constitute a normal distribution. From this it follows that the average of these exposures is not an appropriate descriptor of plant exposures and is not suitable for the description of trends and extrapolations. Analyses of these sets of data show that the logarithms of these exposures do follow a normal distribution, or in other words, these exposures follow a log-normal distribution. In the log-normal distribution the geometric mean is analogous to the average for the normal distribution, and the geometric dispersion is analogous to the standard deviation. Hence the geometric mean is the appropriate parameter to describe trends in plant exposures, which are shown in Figure 2. The geometric means are less erratic than the averages. The trends over the last few years for the geometric means are increases of roughly 60 and 35 rem/plant/yr for BWR and PWR plants respectively.

There are a number of factors which influence plant exposures. There are four of particular interest: regulatory requirements and equipment failure, both of which are difficult to quantify,

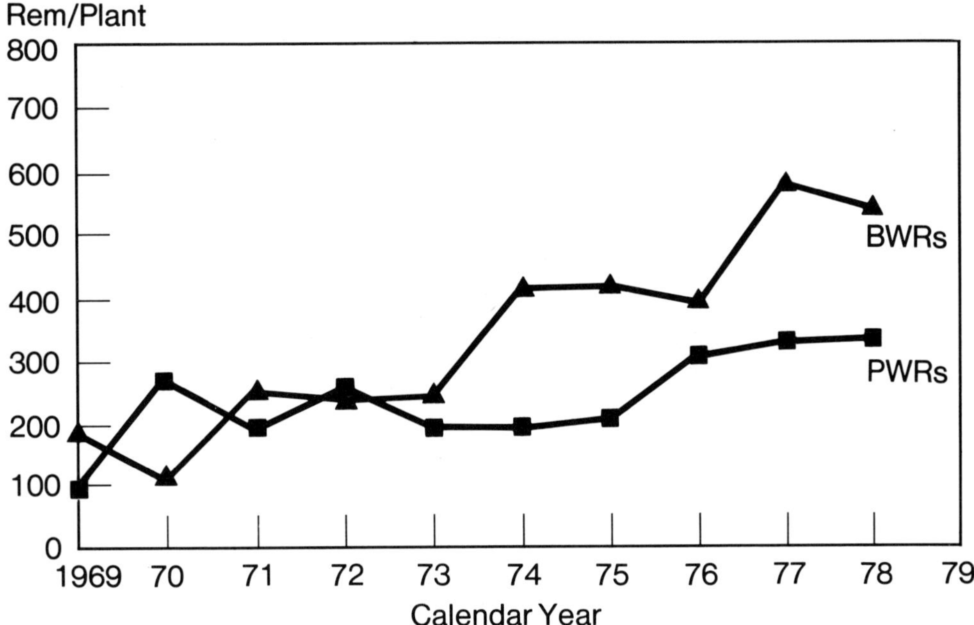

Fig. 2. Geometric means of U.S. nuclear power plant annual occupational radiation exposures.

and plant capacity and plant age. These four factors are not independent and any correlations with these factors must bear this in mind. For example, the trends shown in Figure 1 include plants with much higher capacities in recent years than in earlier years.

The last factor, plant age, and its influence on exposure patterns can be analyzed, although it is also influenced by variations in average plant capacities. The geometric means of annual plant exposures increase in early life to a peak at about 3-5 years, followed by, for the most part, a decrease to a lower values in subsequent years as shown in Figure 3. This pattern is predicted when equipment failure rates and radiation field growth are considered. The combination of these two factors, the "bathtub" shaped variation of equipment failure rate and the asymptotic growth of Co-60 activity shown in Figure 4 does result in a prediction of exposure somewhat similar to that shown in Figure 3. This latter feature of nuclear power plants, radioactivity, and its control, are the emphasis of the remainder of this paper.

BWR RADIATION CONTROL

Radiation fields are of course the source of nuclear power

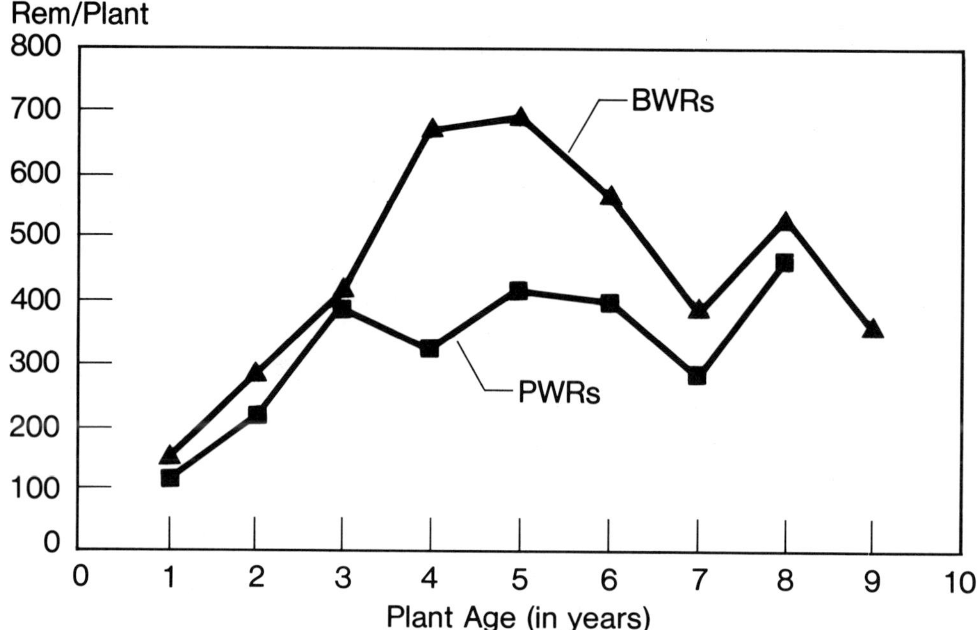

Fig. 3. Geometric means of U.S. nuclear power plant annual exposures by plant age.

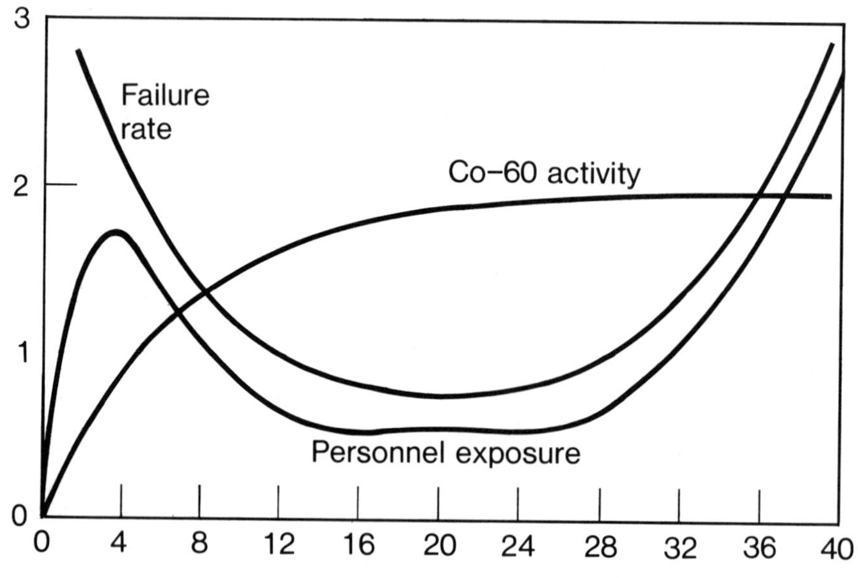

Fig. 4. Personnel exposure pattern predicted from equipment failure rate and Co-60 radioactivity buildup.

plant personnel exposure. Radiation fields from BWR recirculation pipes have generally increased with plant operations, as shown in Figure 5. The broad scatter of data shows a rate of increase of up to 150 mr/h for each full power year of operation (FPY). The trend continues to be an approximately linear increase in these fields with plant operation. The dominant source of these radiation fields is Co-60, as shown in Figure 6. Here the contributions of the various radioisotopes to the radiation fields emanting from recirculation pipes at three different BWRs are shown. Gamma spectroscopy of these pipe fields with GeLi detectors was used to determine pipe surface activities. These activities were then adjusted for the gamma ray emission energy for each radioisotope. The emission energies compare well with the independently measured fields indicated at the bottom of the figure. Each measurement was taken at about 3.5 FPY. Results of this type have consistently shown the dominance of Co-60 independent of BWR plant age. This conclusion can be further tested by observing the correlation of the deposited Co-60 on recirculation pipe internal surfaces to the

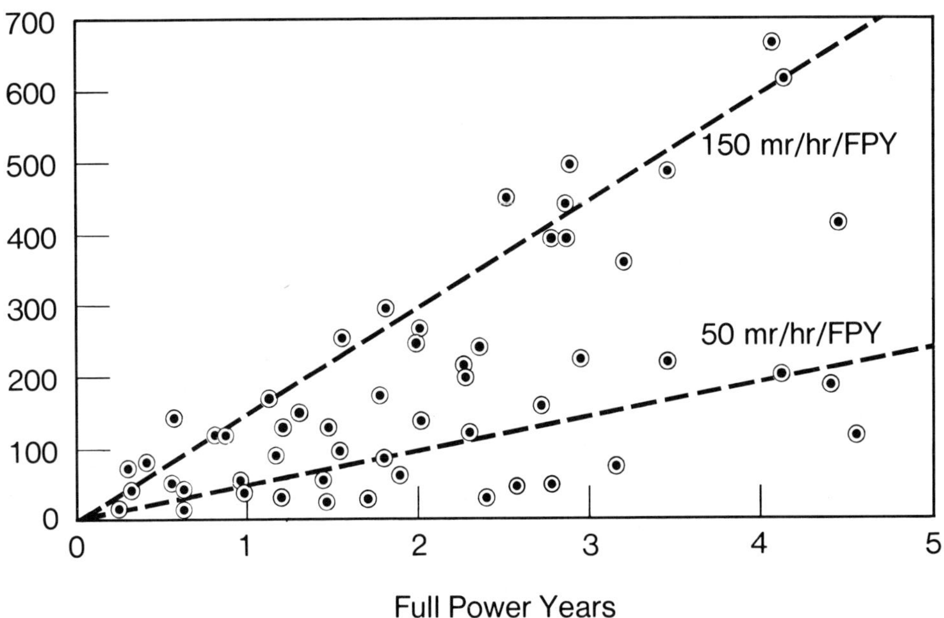

Fig. 5. BWR recirculation pipe radiation field variation with plant energy generation.

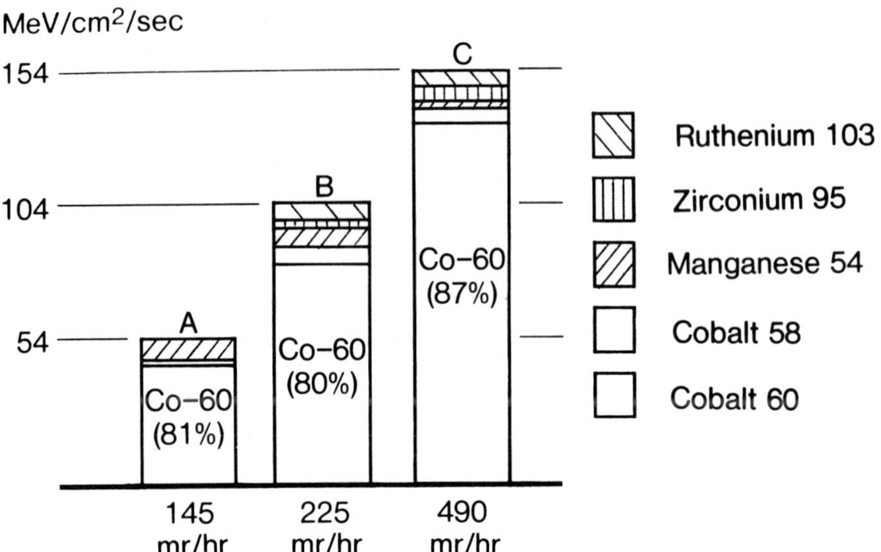

Fig. 6. Radioisotope contributions to BWR recirculation pipe radiation field gamma-ray energy.

radiation fields emanating from them, as is done in Figure 7. These data corresponded to plant operations of up to 4.5 FPY. The line is, in a sense, a calibration curve of fields to deposited Co-60 or vice versa. Points significantly above this line indicate substantial contributions from other radioisotopes. The one point which is far above the curve was a case of substantial fission products from fuel failures. There are three BWR plants with especially interesting radiation field histories, Shimane, Tsuruga and Oskarshamn-1. These are highlighted in Figure 8. Shimane and Oskarshamn-1 have fields much lower than those in U.S. BWRs. Tsuruga has produced a dramatic decrease in their fields. From these and other BWR experiences we conclude that the techniques listed in Table I are the best candidates for effective radiation field control in BWRs. Feedwater contaminant control through feedwater oxygen control, effective operation of condensate purification systems and/or the use of high temperature filters is expected to reduce fuel deposits and the associated production of Co-60. Similarly, Co-60 production can be reduced by replacement of high cobalt alloys in the system with low cobalt materials. Tests are presently underway to qualify alternate low cobalt alloys for use as pins and rollers in BWR control blade assemblies. Decontamination solvents have been identified which can effectively reduce BWR reactor coolant system fields. However, materials testing and chemical processing design for the use of these solvents have not

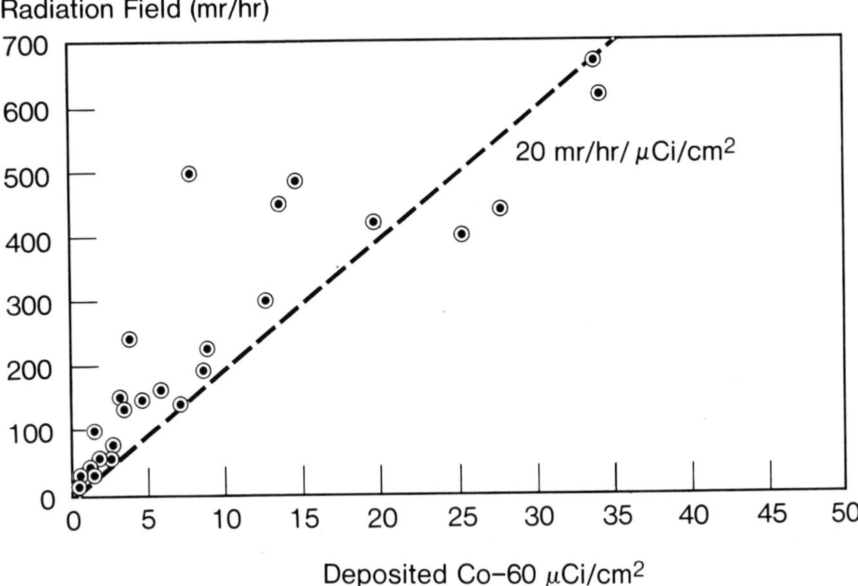

Fig. 7. BWR recirculation pipe radiation field variation with Co-60 surface deposition.

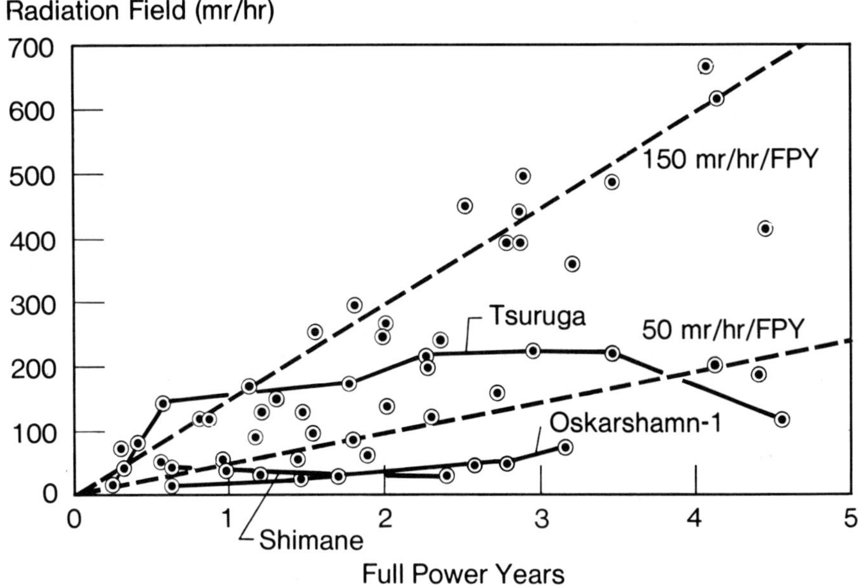

Fig. 8. BWR recirculation pipe radiation field variation with plant energy generation.

Table I. BWR Radiation Control Techniques

- Feedwater Contaminant Control
  - Oxygen control
  - Condensate purification
  - High temperature filtration
- Cobalt Reduction
- Decontamination

been demonstrated. A corrosion product deposition loop has been installed at Hatch-2 BWR, shown in Figure 9, which should further clarify the effectiveness of some of these BWR radiation control techniques. Hot reactor coolant is drawn from the cleanup pump discharge and is routed through the three legs of the loop. The bottom leg conditions tubular coupons with reactor coolant. In the top section, previously conditioned coupons can be exposed to various chemical additives, such as decontaminating solvents, oxygen, ammonia, hydrazine or acids or bases, to determine their effects on deposited radioisotopes and surface morphology. In the center leg, coupons will be exposed to reactor coolant which has been filtered in a high temperature particulate filter (HTF). Direct comparison of the deposits on these coupons with those from the conditioning section will aid in assessing the effectiveness of controlling radioisotope deposition with HTFs. To further test the use of HTFs an electromagnetic high temperature filter (EMF) has been installed to treat feedwater in Isar (or Ohu), a German BWR. A testing program to evaluate the performance of this EMF is being conducted by KWU under contract to EPRI. This 6500 gpm filter processes 1/3 of the total feedwater flow reaching the feedwater tank as illustrated in Figure 10. The drains of the high pressure heaters are forward pumped into the right hand side of the baffled feedwater tank, which is the side from which the inlet to the EMF is drawn. The test will assess the ability of EMFs to reduce BWR feedwater contaminants in forward pumped drain systems. It should also be possible to deduce the effect of such EMFs on BWR radiation fields.

In the area of decontamination a subject of present concern is recontamination. The word "recontamination" is used to describe

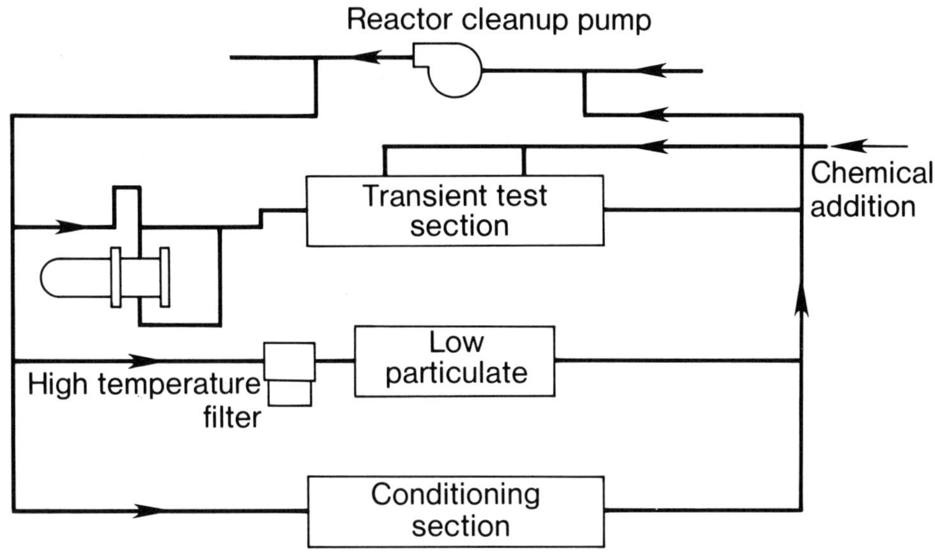

Fig. 9. BWR corrosion product deposition loop at Hatch-2.

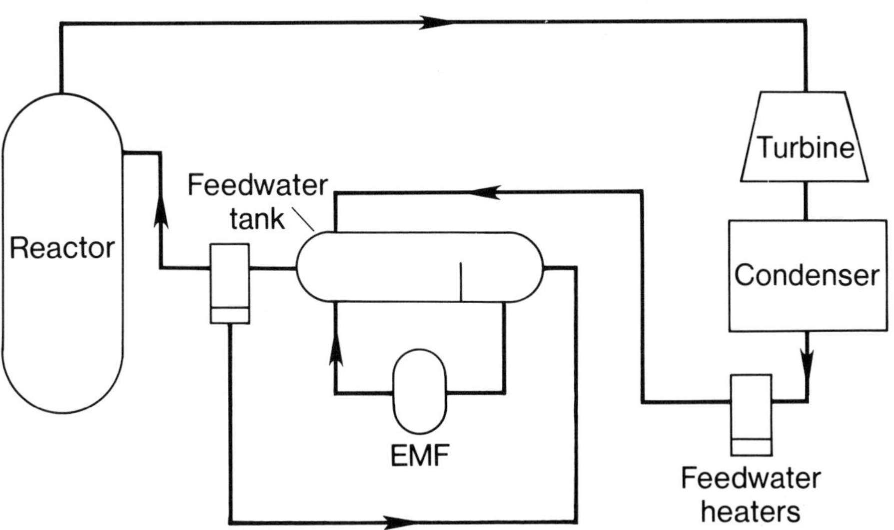

Fig. 10. BWR high temperature filter installation at Isar.

the buildup of radioisotopes on a surface following the decontamination of that surface. The decontamination of the reactor water cleanup heat exchangers at the Peach Bottom-2 BWR using the Dow NS-1 solvent gave us an opportunity to study their recontamination. There are three adjacent sections of piping associated with these heat exchangers: original pipe, pipe newly installed following the decontamination and decontaminated pipe, as shown in Figure 11. Deposited Co-60 activities on these surfaces shown on the left of the figure for October, 1978 were taken less than a day after the system was returned to service. On the right side of the figure are the measured deposited activities following 4 months of service. This shows that in those 4 months, the activity on the new pipe has already reached that on the original pipe, and that the activity on the decontaminated section has increased phenomenally. It is now more than 2 times greater than both the original and the new pipe sections. This suggests that the decontaminated surface is left in a state of high thermodynamic activity, a situation of great concern for operation subsequent to decontamination.

BWR RADIATION CONTROL

In PWR plants, radiation fields within steam generator channel heads are the field measurements which are most available and are

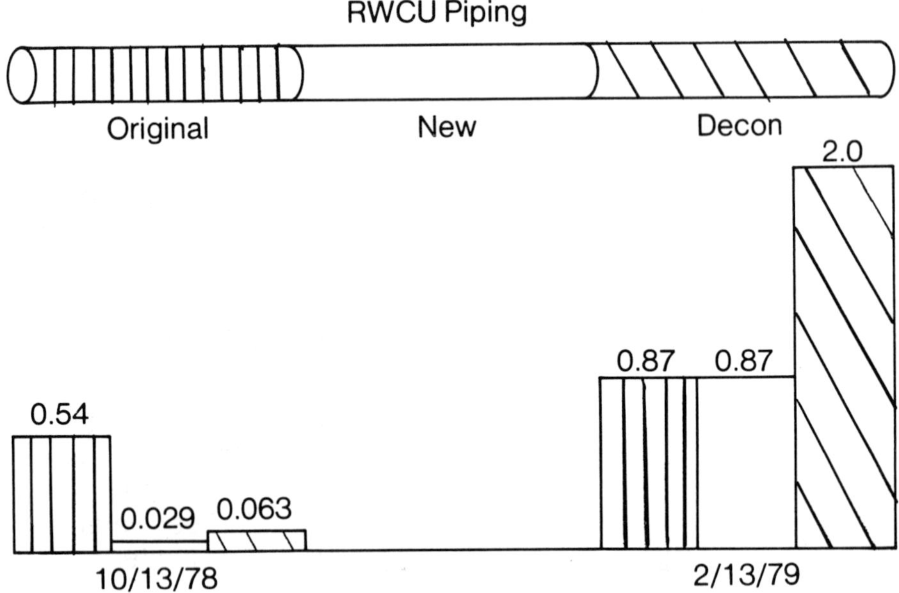

Fig. 11. Co-60 decontamination in μ Ci/cm^2 after decontamination of BWR reactor water cleanup heat exchanger.

best for inter-plant comparison. These fields at the center of the steam generator channel heads, about 3' from the tube sheet, show a broad scatter when correlated with plant operation for U.S. PWR plants, as in Figure 12. In assessing the radioisotopes responsible for coolant pipe and steam generator radiation fields Co-58 and Co-60 are far and away the isotopes most responsible for the fields in PWR plants. When their activities are adjusted for their gamma-ray emission energy, as is done in Figure 13, it becomes evident that after a few full power years Co-60 becomes increasingly the dominant isotope.

A number of PWR radiation control techniques presently under serious consideration are listed in Table II. The first, reactor coolant pH control during operation, makes use of the postulated influence of reactor coolant pH on fuel deposits and the transport of radioisotopes throughout the coolant circuit. It is based upon the coordinated adjustment of LiOH with the reactivity-controlled changes in boric acid concentration.[1,2] Tests of the influence of operational pH control on radiation fields are presently underway at Beaver Valley-1 and Trojan.

Short-term pH changes have been tested at Rancho Seco and do not appear to have significant influence on radiation fields.

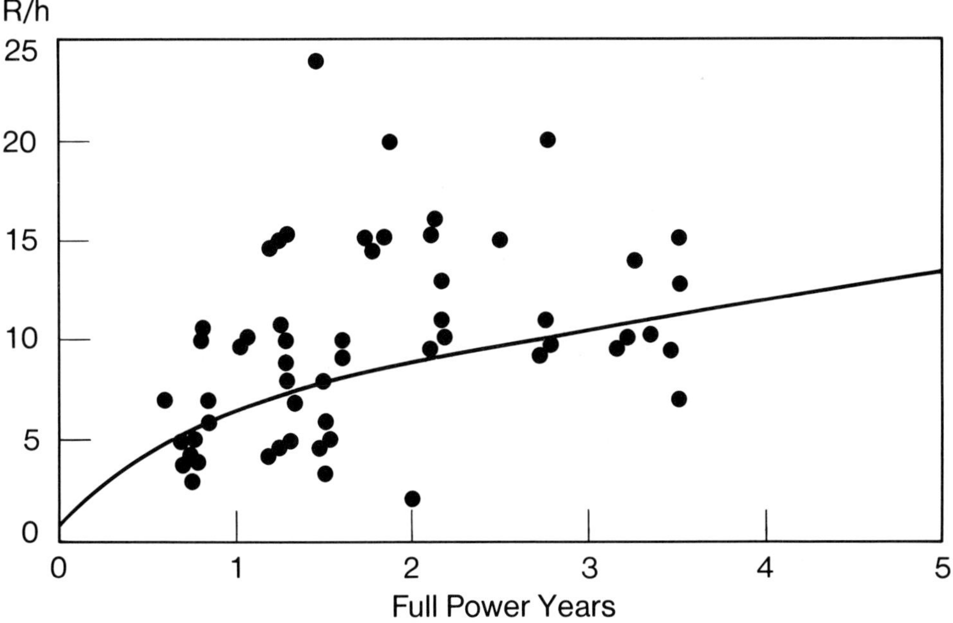

Fig. 12. PWR radiation fields at steam generator channel head center with plant energy generation.

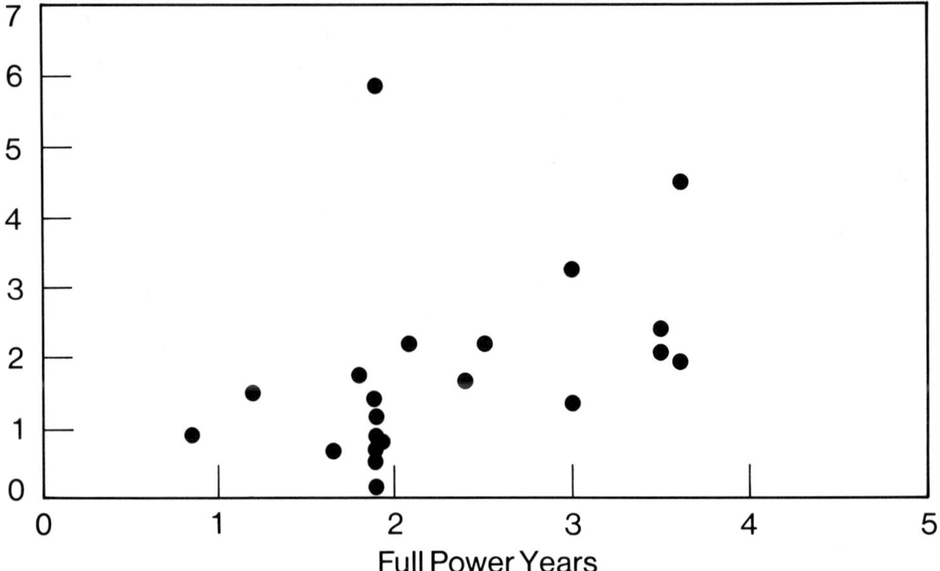

Fig. 13. Co-60 to Co-58 gamma ray energy ratio vs. PWR plant energy generation.

Table II. Radiation Control Techniques

- Operational pH
  - Full cycle
  - Short-term
- Cooldown chemistry
  - H_2O_2
  - Redox potential
  - pH
- Decontamination
- High temperature filtration

Likewise, H_2O_2 addition during cooldown has little effect on pipe and steam generator fields, although the removal of the Co-58 released by the H_2O_2 addition can significantly reduce the reactor cavity fields and associated exposures during PWR refueling outages.[3] The effects of redox potential control, using H_2 and hydrazine, and of pH changes during cooldown are presently under investigation, particularly at the Oconee plants of Duke Power.

The search for decontaminating solvents has been much more elusive for the tenacious films in PWR systems than for BWR systems. Nonetheless the search for such solvents continues, primarily through testing on PWR artifacts.

In the area of high temperature filtration, the design for a high temperature filtration test is illustrated in Figure 14. This filter would process reactor coolant by tapping off the by-pass line from the cold leg to the hot leg. Two such filters, each on a different steam generator loop, would process a total of about 0.5% of reactor flow. Discussions for such an installation are presently (September 1979) underway with a U.S. utility.

This type of filter, an EMF, primarily removes contaminants in the particulate form, as do most types of HTFs. In corrosion product transport particulates are only one of three forms in the

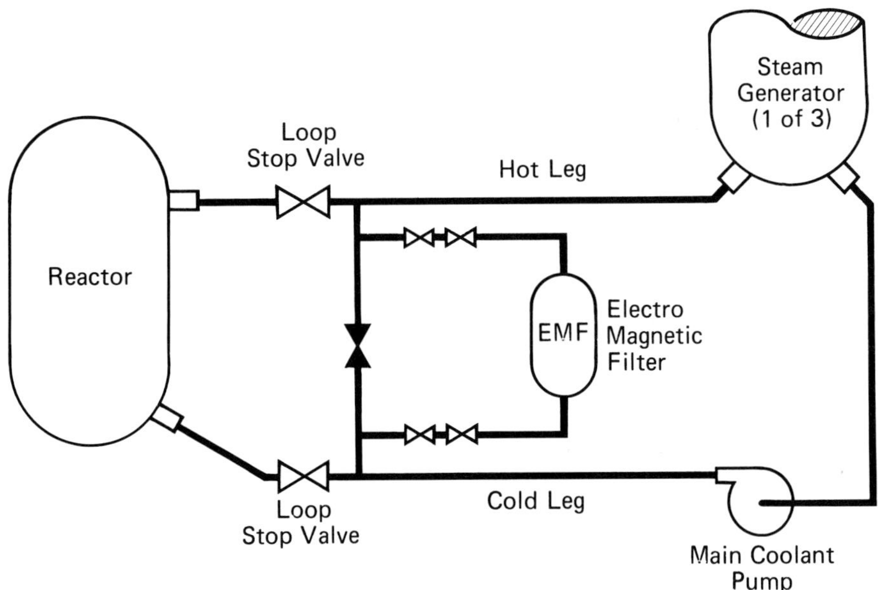

Fig. 14. Design for PWR high temperature filter test installation.

the water, with solubles and colloids also present, as illustrated in Figure 15. It is not presently possible to accurately assess whether the removal of particulates from reactor coolant water will have significant influence on the radioactive contaminants retained on surfaces. Such assessments are a difficult part of radiation control, and complicate the evaluation of techniques for radiation exposure reduction. Since radiation exposure reduction is a primary goal, monitoring such exposures would assess our success, if plant exposures varied consistently from year to year, which they don't. Secondarily our goal is to reduce radiation fields. But monitoring of steam generator fields from a single plant shows wide variations in fields between shutdowns even in the absence of any radiation control testing as is illustrated in Figure 16. The monitoring of fields and isotopics on coolant piping, coolant contaminants and fuel deposits all aid in completing the picture and provide us with the most consistent and dependable, but least available, information. Nevertheless it is important to emphasize the number of processes involved (see Figure 15) and the difficulty in ascertaining which of these is controlling and under what operating or cooldown conditions. There are presently underway a number of projects aimed at increasing our understanding of these complexities.

In summary, one of the techniques of reducing occupational radiation exposure is to control the radiation fields which cause such exposures. Radiation field control techniques which are presently being tested focus on the control of corrosion product transport by chemical control, filtration and decontamination. The evaluation of these techniques require a sound basic knowledge of radiation fields and sources and of radiation exposure prior to the application of such techniques. The fulfillment of this requirements is presently the most difficult aspect of the control of radiation fields and radiation exposures in light water nuclear power plants.

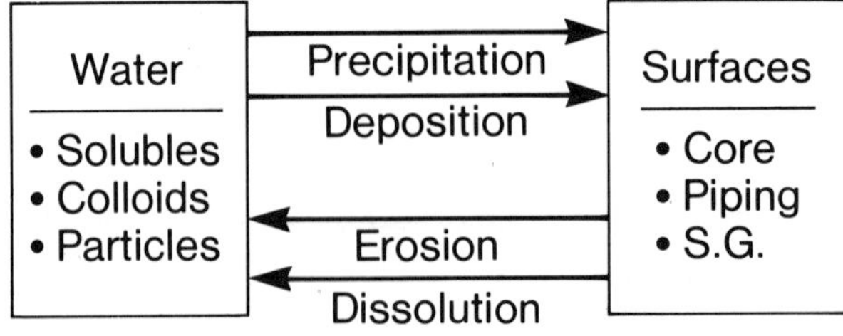

Fig. 15. Corrosion product forms and transport mechanisms in LWRs.

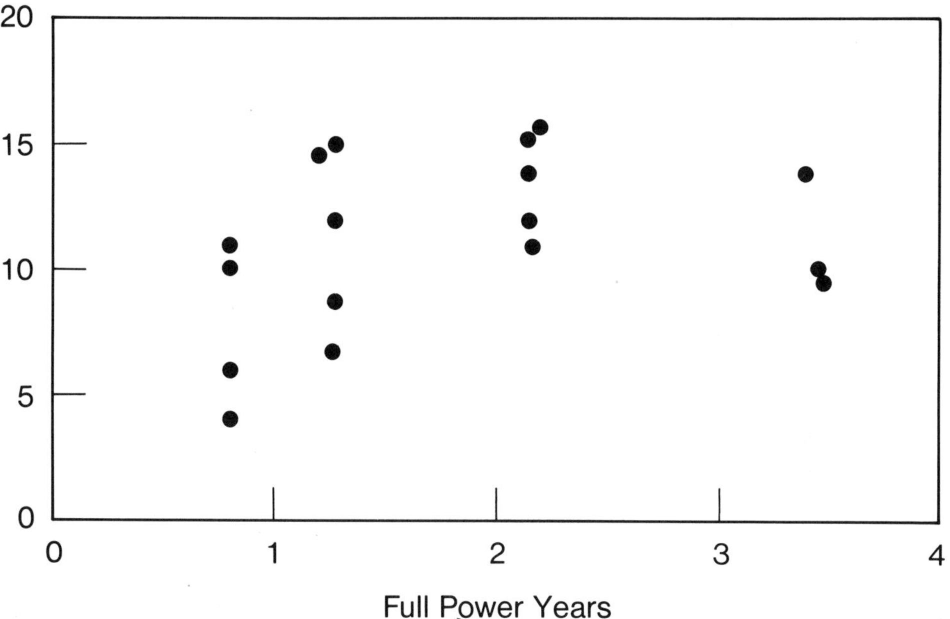

Fig. 16. Variation of steam generator radiation fields at a single PWR.

REFERENCES

1. Y. Solomon "An Overview of Water Chemistry for Pressurized Water Nuclear Reactors", Proceedings of the British Nuclear Energy Society Conference on Water Chemistry of Nuclear Reactor Systems, Bournemouth, England, October 24-27, 1977.
2. M. E. Crotzer and R. A. Shaw, "The Apparent Influence of PWR Coolant pH on Radiation Field Buildup" Transactions of the American Nuclear Society, Vol. 32 (757-759), June 1979.
3. "Effects of Hydrogen Peroxide Additives on Shutdown Chemistry Transients at Pressurized Water Reactors", EPRI Report NP-962, April 1978.

EVALUATING DECOMMISSIONING COSTS FOR NUCLEAR POWER PLANTS

Richard R. MacDonald

San Francisco Power Division
Bechtel Power Corporation
San Francisco, California 94105

INTRODUCTION

The ultimate decommissioning of today's large nuclear power plants has become a subject of increasing interest and importance throughout the industry. Frequent debates have centered on the numerous studies that have been conducted to determine the costs associated with various decommissioning alternatives. These cost estimates are widely quoted and compared, and cost figures are reported ranging from $25 million to more than $130 million. Sometimes these comparisons fail to give adequate regard for the basic differences in scope or in the bases of the estimates. For example, variations exist in the extent of structural demolition considered and in the base years of the estimates. Occasionally estimates are even quoted that have been escalated to the year of final decommissioning (more than 130 years in one case for a delayed dismantlement option as reported in Reference 1). Comparison of the dollar values alone is of course meaningless without recognition of the impact of differences in the bases of the estimates.

This paper presents an overview of the economic aspects of decommissioning of large nuclear power plants in an attempt to put the subject in proper perspective. This is accomplished by first surveying the work that has been done to date in evaluating the requirements for decommissioning. It is intended to review the current concepts of decommissioning and discuss a few of the uncertainties involved. The paper identifies the key factors to be considered in the economic evaluation of decommissioning alternatives and highlights areas in which further study appears to be desirable.

It is important to recognize at the outset that in reviewing the economic aspects of decommissioning, one must fully understand the factors involved in selecting a decommissioning approach. Establishing an estimate of the cost and a financing approach requires a balancing of somewhat conflicting objectives, such as minimizing personnel exposure and minimizing cost. This paper provides a brief discussion of the key considerations in such an evaluation to tie together the technical and economic aspects of decommissioning.

The comments provided in this paper result partially from studies performed for the Department of Energy as reported in Reference 2, and from subsequent work in this area.

BACKGROUND

As indicated by the bibliography prepared by Battelle Pacific Northwest Laboratory for the NRC (Reference 3) numerous reports and papers have been published relating to the subject of nuclear power plant decommissioning. A substantial part of this literature reports on prior decommissioning experience that should be considered in the evaluation of decommissioning requirements. More than 65 nuclear facilities have been decommissioned. The approaches used have included mothballing, entombing and dismantling, with the selection based heavily on current economics. This experience, however, is somewhat limited in scope. The reactors decommissioned to date have been relatively small, ranging from less than 1 MWth to 260 MWth. Most were small research reactors with comparatively short operating histories. The largest reactor to be dismantled was at Elk River, Minnesota, a 58 MWth boiling water reactor. The Elk River dismantling, as reported in Reference 4, included total removal of the reactor and containment and the remainder of the plant was converted for non-nuclear power generation. This experience forms the most significant element of the experience base for the subsequent U.S. decommissioning studies. This dismantling effort and the current dismantling work at the Sodium Reactor Experiment in California (a 30MWth reactor that is being removed by Atomics International while the structures are to be converted for further use as a manufacturing facility, Reference 5), have provided valuable experience with application of the techniques involved in decommissioning. This includes experience with remote plasma arc cutting of reactor vessel components; controlled explosive demolition and explosive pipe cutting; and component, structure, and site decontamination. It is readily apparent that the designs and construction of these plants are significantly different from today's commercial nuclear power plants. Therefore, direct application of the techniques developed for the Elk River and SRE decommissioning is not possible. This is particularly true in regard to cutting of the RPV and demolition of the structures.

It is clear that the techniques will have to be further developed or new equipment adapted to the specific tasks. However, with a reasonable extension of prior experience in decommissioning and application of experience from other industries, there are no apparent technological barriers to successful decommissioning.

As mentioned above, the literature also includes several studies conducted throughout the world to review the requirements for decommissioning of large commercial nuclear power plants. References 6 through 9 are perhaps the most significant of these studies and are frequently referenced in other studies or papers. The primary objective of these studies was to assess the feasibility of decommissioning with current technology and to arrive at an order of magnitude estimate of the costs involved. The level of detail and scope of the studies, and the approaches taken, vary widely. There is considerable variation in the assumptions on which the various studies are based. Therefore, detailed comparisons are difficult. It is important, however, to gain an appreciation for the effects of variations in key parameters, or base assumptions, such as:

1. Definition of alternatives. A wide variety of alternatives are considered, as discussed below.

2. Base year of cost estimate. The base years range from 1975 to 1978 with some studies even projecting the costs to the expected date of decommissioning.

3. Extent of demolition and removal of structures. The studies range from removal down to 3 feet below grade to total removal including foundations. The characteristics of the reference plants further complicate comparisons. For example, the Battelle study (Reference 6) assumes removal to one meter below grade. However, the reference plant, Trojan, is constructed with the top of the containment basemat at grade level.

4. Reference plant characteristics. Some studies are based on a specific plant and site while others assume a "generic" set of characteristics. This results in wide variations in the nature of the plant being considered.

5. Radwaste volume reduction, packaging, and disposal requirements. Future requirements are somewhat uncertain and a variety of assumptions have been made, ranging from requiring all waste to be packaged in 200 liter or 400 liter drums (Reference 7) to use of large wooden containers for up to 2500 cu. ft. of low specific activity (LSA) material (Reference 9).

6. Labor rates. The decommissioning of nuclear plants is extremely labor intensive and the overall cost is heavily dependent upon the labor rates used. Comparison of the rates used in the various studies is difficult.

Operating experience is also extremely useful in evaluating decommissioning requirements. For example, numerous efforts have been conducted with respect to decontaminating operating facilities. This decontamination has been undertaken either to minimize exposure to operating personnel or to reduce contamination of equipment to be maintained or replaced. The most extensive program for decontamination of an operating facility to date is the program now being developed for Dresden 1. A similar program is being investigated for Fukushima Unit 1 in Japan. A recent report by the Electric Power Research Institute (Reference 10) reviews the literature on dilute chemical decontamination of light water reactors.

DECOMMISSIONING ALTERNATIVES

The current concept of decommissioning offers a wide range of alternative approaches. The Code of Federal Regulations (10CFR 50.82) requires that termination of an operating license be accomplished by dismantlement and disposal of the facility in a manner which "will not be inimical to the common defense and security or to the health and safety of the public." NRC guidance for such disposal is contained in Regulatory Guide 1.86 "Termination of Operating Licenses for Nuclear Reactors," which suggests four options:

1. Mothballing
2. In-place entombment
3. Removal of radioactive components and dismantling
4. Conversion to a new nuclear system or a fossil fuel system.

Each of these approaches has been used in a past decommissioning effort. However, within this framework several alternatives for interim or final disposition of the retired plant are conceivable, involving a wide range of levels of isolation, decontamination, or removal of components or structures. Approaches that require continuing control of radioactive or contaminated materials should be considered as interim dispositions. Complete decontamination must be achieved to a level permitting release of the site for unrestricted use before the radiological concerns can be considered as fully resolved. Nevertheless, these "interim" solutions should be recognized and evaluated as potentially viable alternatives at the time of decommissioning. The duration of the interim solution might range from less than one year to several decades. It does not appear reasonable to consider protective storage for longer periods, such as the thousands of years estimated to be required for total decay of radioactivity to permit unrestricted access to the site, due to the difficulty of assessing the durability of the plant structures. Also, only a few components within the reactor would require isolation for such long periods. Such isolation could be more efficiently achieved by other storage means.

Protective storage solutions, as discussed above, would necessarily include continuing surveillance and periodic reassessment to ensure that the health and safety of the public is protected, and that any modification of the status of the plant is achieved at the optimum time. This decision would be based on the most current information regarding the plant contamination levels, state of technology, regulatory requirements, and economic considerations. Figure 1 depicts the wide range of possible solutions to decommissioning of a nuclear power plant, including both interim and final measures. Each of these options should be investigated at the time of final planning for decommissioning.

In relation to economic evaluations and financial plans, one of the most significant of the interim solutions to be considered is the possibility of extending the operating life of the plant beyond the currently accepted 30 to 40 years, or "recommissioning" the plant for an extended period. This might involve refurbishing or replacing components that cannot be qualified for continued operation, or implementing design modifications to extend the operating life. It appears that there is a significant probability that this alternative will be applied frequently in the future. A few of the factors that may encourage the recommissioning of plants at the end of 40 years are listed below:

a. A large portion of the capital cost of a nuclear plant is associated with elements that are likely to have economic lives considerably longer than 40 years (e.g. structures, cooling towers, electrical cables, etc.).

b. There is a strong indication that the size of future units will begin to stabilize, offering less effect of economies of scale than in the past periods of rapid growth.

c. The design of nuclear units is also expected to stabilize as the technology matures. This may result in less change in the regulatory requirements and therefore more economical extension of the operating license, since less upgrading would be required. Unless dramatic technological breakthroughs occur, operating costs for the older plants may not be significantly higher than for newer plants.

d. Nuclear fuel cycle costs are expected to remain a small proportion of the total power generating cost for a nuclear plant. Therefore, in comparison to other alternatives, continued operation is likely to be attractive.

e. As more sophisticated analytical techniques, and additional operating experience and test data become available the large margins of conservatism will be recognized in the design of components and structures for seismic and hydrodynamic loads, radiation protection, and the wide range of postulated events. This conservatism may permit qualification of the plant for longer service.

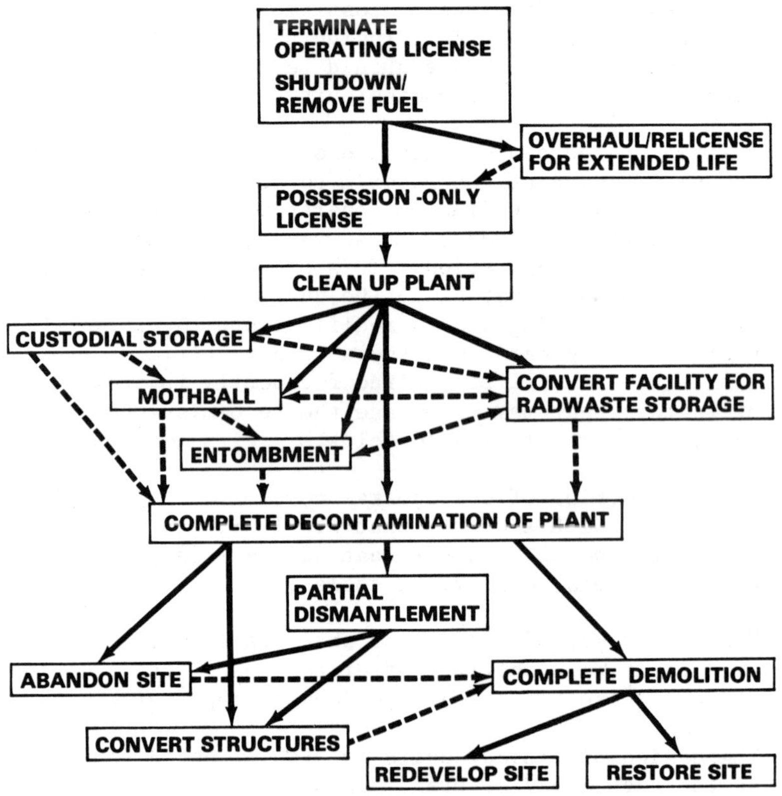

Fig. 1 Decommissioning Alternatives

SELECTING A DECOMMISSIONING APPROACH

The selection of the appropriate decommissioning approach for a particular plant must be based on a balanced evaluation of several factors, including the following:

1. Personnel exposure (ALARA)
2. Safety (during decommissioning and long term)
3. Environmental impacts
4. Regulatory requirements
5. Land value, potential reuse of the site
6. Unique utility and site requirements
7. Technological feasibility
8. Economics

A complete economic analysis can only be performed after the other factors are understood. This type of assessment cannot be adequately performed 30 to 40 years prior to the termination of reactor operation. Therefore, the final selection of a specific decommissioning mode should not be made until shortly before shutdown. However, the utilities and regulatory agencies must, of course, consider the decommissioning questions much earlier in the cycle. It is important to explore the scope of potential approaches, particularly on an industry-wide basis.

DECOMMISSIONING PLANNING

Decommissioning plans should be developed in sufficient detail to identify the unique characteristics of each plant and site and the unique utility requirements. An order of magnitude estimate of the potential cost of decommissioning should then be developed as a basis for long term financial planning. Research and development programs should continue to be pursued on a national and industry-wide basis to assure a coordinated effort. This should include consideration of the critical areas of the decommissioning effort that might be simplified by design modifications in future plants, or by revisions to the standard operating practices.

Due to the inherent uncertainties in any advance decommissioning planning, it appears reasonable that such plans should be based primarily on a "maximum approach" to decommissioning, i.e. total decontamination of the site, removal of the facility, and restoration of the site. At the same time, maximum flexibility must be maintained to accommodate the changing technology, regulations, and priorities. The less severe alternatives discussed above must, of course, be given careful consideration before electing to totally dismantle the plant and demolish the structures.

The plant specific studies and cost estimates should build from the work that has been done to date and should include consideration of all of the following elements of the decommissioning program:

1. Detailed planning and engineering
2. Licensing
3. Quality assurance
4. Project management and administration
5. Decontamination of equipment and surfaces
6. Health Physics support
7. Dismantling of equipment and demolition of structures
8. Waste management and disposal
9. Environmental controls
10. Personnel exposure and industrial safety
11. Security
12. Labor productivity effects
13. Site restoration or redevelopment alternatives.

Future studies of decommissioning should carefully distinguish the efforts required for release of the site for unrestricted use, restoration of the site, and redevelopment.

Financial plans should be developed with appreciation for the wide range of decommissioning possibilities and recognition of potential changes in the requirements as discussed below. Financial plans should be flexible and frequently reviewed to ensure continuing applicability.

DESIGN PROVISIONS FOR DECOMMISSIONING

One of the key incentives for early investigation of decommissioning requirements and costs is to provide a basis for identification and evaluation of potential modifications to the design of new plants to facilitate their ultimate retirement. The work accomplished to date indicates that decommissioning of today's large nuclear power plants is clearly feasible. However, the continuing study of the alternatives and techniques for decommissioning, in conjunction with additional decontamination and decommissioning experience, will help to ensure reasonable consideration of the requirements during the design phase. The important task is to identify the most critical aspects in the decommissioning process and consider areas where minor modifications will yield substantial benefits.

Caution must be exercised, however, in implementing major design changes solely for the sake of decommissioning. Any advance planning conducted during the design phase must be considered as preliminary, and maximum flexibility must be maintained. Therefore, it is unlikely that significant or costly changes could be justified on a cost/benefit basis. The discounted present value of a benefit to be realized 30 to 40 years in the future would alone probably not justify any substantial expenditures for modifications. This is particularly true since the future benefit cannot be precisely predicted. Therefore, a detailed cost/benefit analysis of most modifications is not currently feasible during the design phase.

It appears more reasonable to approach the consideration of design changes by first investigating those provisions that will likely yield other significant parallel benefits. There are three primary areas to be considered:

1. <u>Layout</u> - Accessibility, equipment removal paths and transport distances, separation of contaminated systems, maintenance space.

2. <u>System and Equipment Design</u> - Material and equipment selection, decontamination provisions, replaceable radioactive components, separation of systems required during decommissioning.

3. <u>Structural Design</u> - Decontaminable surfaces, prefabricated structural units, simplified designs for ease of assembly and disassembly.

The objective of any design modification should be to minimize personnel exposure or simplify equipment removal or structural demolition. In this regard, one important conclusion might reasonably be drawn from the work that has been performed to date: most measures taken to reduce operating personnel exposure, simplify construction or facilitate operation of the plant are likely to have a favorable effect on decommissioning costs. Potential decommissioning requirements should be considered during the design phase to at least ensure that features are not incorporated that might inhibit the decommissioning process.

Decommissioning should also be included as another general criterion in the evaluation of alternative design concepts. During the design process many decisions must be made between reasonably equivalent or equally acceptable alternatives. Often no substantial economic benefit can be attributed to any of the alternatives, and the selection must be made on a technical judgement basis. The consideration of decommissioning requirements might on many occasions swing the decision where cost is not a significant factor. For example, a separate radwaste facility may not be clearly justifiable on an economic basis (particularly for a single unit plant). However, when the benefits of a separate radwaste facility during decommissioning are considered, the decision may become more obvious. Another example might arise in the initial selection of the site for a new plant. In some cases the differences between two sites may have no distinguishable effect on the design or construction costs of the plant. However, one of the sites may have apparent advantages when decommissioning is considered.

Each of these decisions must be made on a case by case basis for each plant. Evaluation of such modifications to facilitate decommissioning requires a keen appreciation of the design, construction, and operating requirements, as well as an understanding of the potential decommissioning options.

TRENDS AFFECTING DECOMMISSIONING COSTS

Before establishing a financial plan for accumulation of funds for decommissioning, it is important to recognize and consider the technological, regulatory, economic and social trends that might have substantial impact on the ultimate cost of disposition of a retired facility. A few such trends are discussed below to provide a framework for evaluation of alternative funding methods.

Personnel Exposure Limits

At the present time there is a definite trend toward increased emphasis on reduction of personnel exposure. This objective could have a bilateral effect on the cost of nuclear power plant decommissioning. First, a reduction in the acceptable level of personnel exposure during the decommissioning process will tend to increase the overall cost by requiring increased use of remote techniques for removal of contaminated equipment and shorter exposure times for personnel. It might also affect the decision regarding the appropriate decommissioning approach, since delayed dismantling might become more attractive.

On the other hand, increased emphasis on control of personnel exposure might also have a reverse effect on decommissioning costs. Encouraging the continued improvement in equipment design and plant layout will facilitate the operation and maintenance of the plant. These more stringent requirements will also encourage the use of periodic decontamination programs, such as those now under development, and the adoption of more extensive cleanliness control practices. These changes would result in a simplification of the decommissioning process through a reduction in the total radioactive inventory, as well as through improved access and equipment removal provisions.

Technology

Numerous programs are now being conducted that will refine the technology to be applied to decontamination, dismantling, and demolition of nuclear power plants. These programs include investigation of decontamination techniques to reduce operator exposure during the life of the plant; decontamination of equipment to permit disposal as nonradioactive waste; surface decontamination of structures; advanced metal cutting techniques (arc saw, plasma torch, etc.); and demolition techniques for heavily reinforced concrete structures. In addition, as the need for decommissioning of larger plants arises, it is probable that tremendous advances will be made. Past decommissioning experience at Elk River, SRE and other plants has shown that technology can be rapidly adapted from other industries and improved when the need arises. For example, the adaptation of the plasma arc torch and the refinement of the techniques for reactor vessel cutting were very successful at both Elk River and SRE. Other examples at SRE include the adaptation of offshore oil industry experience with explosive pipe cutting inside the reactor vessel and the use of shoring adapted from experience of firms engaged in the addition of basements to existing high rise buildings (Reference 5).

Waste Disposal Requirements

Most of the decommissioning studies performed to date have assumed that the radioactive wastes resulting from a decommissioned nuclear plant would be disposed of by shallow land burial. (The study performed by Battelle Pacific Northwest Laboratory for the NRC (Reference 2) also gave consideration to the increased cost of deep geologic disposal of certain highly activated reactor internals.)

There are currently six licensed shallow land burial sites for disposal of commercial radioactive wastes (Beatty, Nevada; Barnwell, South Carolina; Richland, Washington; Sheffield, Illinois; Morehead, Kentucky; and West Valley, New York). Of these, only the first three are presently open and accepting additional radioactive wastes. Waste burial requirements and transportation requirements are likely to become more restrictive. Burial fees and transportation costs are increasing. However, the national requirements for radwaste disposal from all sources is currently receiving considerable attention. The Department of Energy Task Force for Review of Nuclear Waste Management has provided several recommendations regarding the development of a coordinated national program for management of nuclear wastes (Reference 11). The task force recommendations included the following:

1. Consolidation of responsibility for the six commercial and fourteen government low level waste disposal sites. This would permit the establishment of uniform criteria and encourage the development of improved capabilities.

2. Reduction of waste volumes by minimizing waste creation at the source and by processing for volume reduction prior to disposal.

3. Further investigation of alternatives to shallow land burial.

The volume of radioactive wastes resulting from decommissioning of commercial power reactors is not expected to reach a significant level until after the year 2000. There are approximately twelve reactors ranging from 50 MWe to 650 MWe that may be candidates for decommissioning by that time since they will have been operating for 30 to 40 years. Approximately half of these are located on sites with additional reactors and would not likely be immediately dismantled. The Task Force has estimated that the wastes from the remaining reactors could range between 10,000 cubic feet and one million cubic feet depending upon the decommissioning approach selected. This represents a very small percentage of the total radioactive wastes to be disposed of. DOE facility decontamination and decommissioning could produce low level wastes ranging from 10 to 160 million cubic feet (Reference 11). Therefore, improvements in the national waste management program will likely be achieved before the contribution from decommissioning of commercial facilities becomes significant.

Tightening restrictions on the transportation and burial of low level wastes will likely result in increased emphasis on decontamination of components and materials for release as nonradioactive wastes. They will also encourage further application and development of volume reduction techniques that do not appear to be technically or economically attractive at the present time. Centralized volume reduction centers, perhaps located at the national burial sites, would improve the economics of certain volume reduction techniques including meltdown of contaminated materials (Reference 12).

Disposal of nonradioactive wastes also poses some uncertainties for the future. It is difficult to assess the availability of local landfill areas for disposal of wastes 30 to 40 years into the future. It is also difficult to predict the acceptable form of the waste, including the requirements for segregation of waste types. The future value of much of the equipment and material for scrap or salvage is also uncertain. However, it is likely that salvage and scrap value will be significant.

Increased Decommissioning Experience

In addition to the technology advancement that is likely to result from future decommissioning activities at larger power plants, the expanding experience base will also help to more clearly identify the major problem areas and to refine the cost estimates. Current cost estimates are based on limited experience that may not be directly applicable to the current generation nuclear power plants due to their increased complexity and more stringent design requirements. Although extension of this experience is difficult, the work done to date indicates that decommissioning is feasible and the cost represents a reasonably small proportion of the capital cost of the plant. Within the next twenty years, however, it is possible that several larger plants will be decommissioned.

Another important consideration is the possibility of the development of firms specializing in the decommissioning, dismantlement, and demolition of nuclear facilities. As broader experience is obtained, this will result in improved efficiency and more rapid technological advancement.

Regulatory Requirements

As the nuclear industry continues to mature it is likely that regulatory requirements and design criteria will begin to stabilize. Improved stability of the requiements and increased standardization will improve the feasibility of either refurbishment of the plant for continued use, or development of a new plant using existing structures without extensive demolition. Another potential effect of stabilized criteria is an increase in the salvage value of plant components.

CONCLUSIONS

The present level of experience in decommissioning, coupled with the detailed studies of future requirements, leads to the following important conclusions:

1. Decommissioning of current generation nuclear plants is feasible with reasonable levels of personnel exposure and acceptable costs.

2. Decommissioning planning and financial programs for accumulation of funds should be developed considering the wide range of potential future requirements. Flexibility should be maintained and frequent reviews should be conducted to adjust for changing needs.

3. Estimating of decommissioning costs should be performed to a standard basis. Each estimate should be related to a common base to facilitate comparisons. Work should continue to develop a stronger basis for such comparisons.

4. The uncertainties of extrapolated future costs should be recognized.

5. The feasibility of extending the operating life of nuclear plants should be investigated further.

REFERENCES

1. "Some Rather Bizarre Figures for Decommissioning the Hatch Nuclear Plant," Nucleonics Week, March 1, 1979.
2. "Review of Nuclear Power Reactor Decommissiong Alternatives," prepared by Bechtel National, Inc. for the U.S. Department of Energy, October 1978.
3. G. J. Konzek, C. R. Sample, "Decommissioning of Nuclear Facilities - An Annotated Bibliography," Battelle Pacific Northwest Laboratory, NUREG/C12-0131, October 1978.
4. United Power Association Report COO-651-93, "Final Elk River Reactor Program Report", September 1974 (Revised November 1974.)
5. Wayne Meyers, K. Kittinger, Atomics International Division. (Personal communications regarding SRE decommissioning experience).
6. R. I. Smith, G. J. Konzek, and W. E. Kennedy, Jr., "Technology, Safety and Costs of Decommissioning a Reference Pressurized Water Reactor Power Station," Battelle Pacific Northwest Laboratory, NUREG/CR-0130, Vol. 1 and 2, June 1978.
7. R. Bardtenschlager, D. Bottger, A. Gasch, and N. Majohr, "Decommissioning of Light Water Reactor Nuclear Power Plants", Nuclear Engineering and Design, Vol. 45, 1978.
8. R. J. Stouky and E. J. Ricer, "San Onofre Nuclear Generating Station Decommissioning Alternatives", Report 1851, NUS Corporation, February 22, 1977.
9. W. J. Manion and T. S. LaGuardia, "An Engineering Evaluation of Nuclear Power Reactor Decommissioning Alternatives," Atomic Industrial Forum publication AIF/NESP-009, November 1976.
10. "Project 828-1, Interim Report: Literature Review of Dilute Chemical Decontamination Processes for Water-Cooled Nuclear Reactors," Electric Power Research Institute Report NP-1033, March 1979.
11. U.S. Department of Energy Report DOE/ER-004/D, "Report of Task Force for Review of Nuclear Waste Management," February 1978.
12. G. A. Beitel and P.G. Ortiz, "Volume Reduction of Metallic Wastes," Rockwell Hanford Operations Report RHO-SA-91, October 24, 1978.

OPTIMIZATION OF COSTS VERSUS RADIATION EXPOSURES IN DECOMMISSIONING

G. J. Konzek

Pacific Northwest Laboratory*
Richland, Washington 99352

INTRODUCTION

When considering decommissioning activities, it is desirable to achieve a balance between the cost of doing the work and the cost of minimizing the radiation dose to the workers. Ideally, one would like to minimize both of these costs, but, unfortunately, the conditions for achieving these minima rarely coincide in practice. One must make trade-offs to hold both costs and radiation doses to reasonable limits. In this paper I will discuss the opportunities available during a reference light water reactor's (LWR) lifetime to optimize its eventual decomissioning. I will present an optimization methodology that could easily be applied to other decommissioning projects, regardless of size. For each specific phase of the reactor's life cycle, different considerations and alternatives are used to formulate an empirical methodology to optimize costs and reduce occupational radiation exposure. This methodology, when fully developed, could be used as the data base for the Master Decommissioning Plan (MDP) of the power plant, which ultimately will guide the actual decommissioning.

As this evolutionary plan unfolds, items that are most likely to influence either external occupational radiation exposure or major expenditures are continually being identified, refined, and updated for future consideration by an optimization planner. When utilized in this manner, this concept represents a dynamic, ongoing methodology and constant quest for improvement during the entire lifetime of the reference nuclear project, and not just for the time frame conventionally known as "the decommissioning period."

---

*Operated by Battelle Memorial Institute.

OPPORTUNITIES FOR DECOMMISSIONING OPTIMIZATION AT THE DESIGN STAGE

Prevention and/or reduction of radiation exposure of operating personnel is an important consideration in the design of a nuclear power station. Cost-benefit evaluations of design features to reduce radiation exposure imply a dollar value for a unit of radiation dose. If the radiation-related duties at the station can be readily performed within the bounds of existing occupational radiation limits (10 CFR, Section 20.101), the economic saving due to further reducing the radiation exposure is nil. However, if the radiation dose for the same duties exceeds that permitted by regulation so that additional staff will have to be employed, then efforts to reduce radiation exposure may produce economic benefits. This logic is as true for the decommissioning phase as it is for the operational phase of a facility's life.

A strong case can be made that optimization of decommissioning cost is most effective when it is imbedded into the design effort from the beginning. However, this is not usually the case. Often, the pressure to adhere to a tight design schedule results in putting maintenance (and certainly decommissioning) considerations on the back burner while a design is being formulated. If optimization of decommissioning costs takes place early in the design phase, as it must if the maximum benefit is to be derived, an optimization planner may be the first person to consider decommissioning factors in detail. On balance, the optimization planner should recognize that since there is lower potential risk to the public during decommissioning, the designs of nuclear power reactor facilities for safe and efficient operation are more important than concerns for decommissioning.

Using as a basis studies conducted at Pacific Northwest Laboratory[1,2,3] and others reported in the open literature, we have gained insights into plant design characteristics that could expedite and simplify (i.e., optimize) decommissioning. The relative effort that might be applied throughout the life cycle of the reference LWR to anticipate the decommissioning needs is presented in Figure 1. The "level of effort" scale indicates the relative intensity of this effort by the optimization planner in each part of the life cycle. As indicated in Figure 1, the complete optimization cycle for costs and planned reductions of occupational radiation exposures begins at the design stage. It is here that the potential benefits must be gauged against past experience, since the optimum choice may be innovative, involve a process which has not yet been proven in practice, and/or involve the use of relatively new equipment. For example, with the exception of the reactor pressure vessel, which must meet stringent safety requirements requiring unitized fabrication, components such as pumps, steam generators, and heat exchangers could be

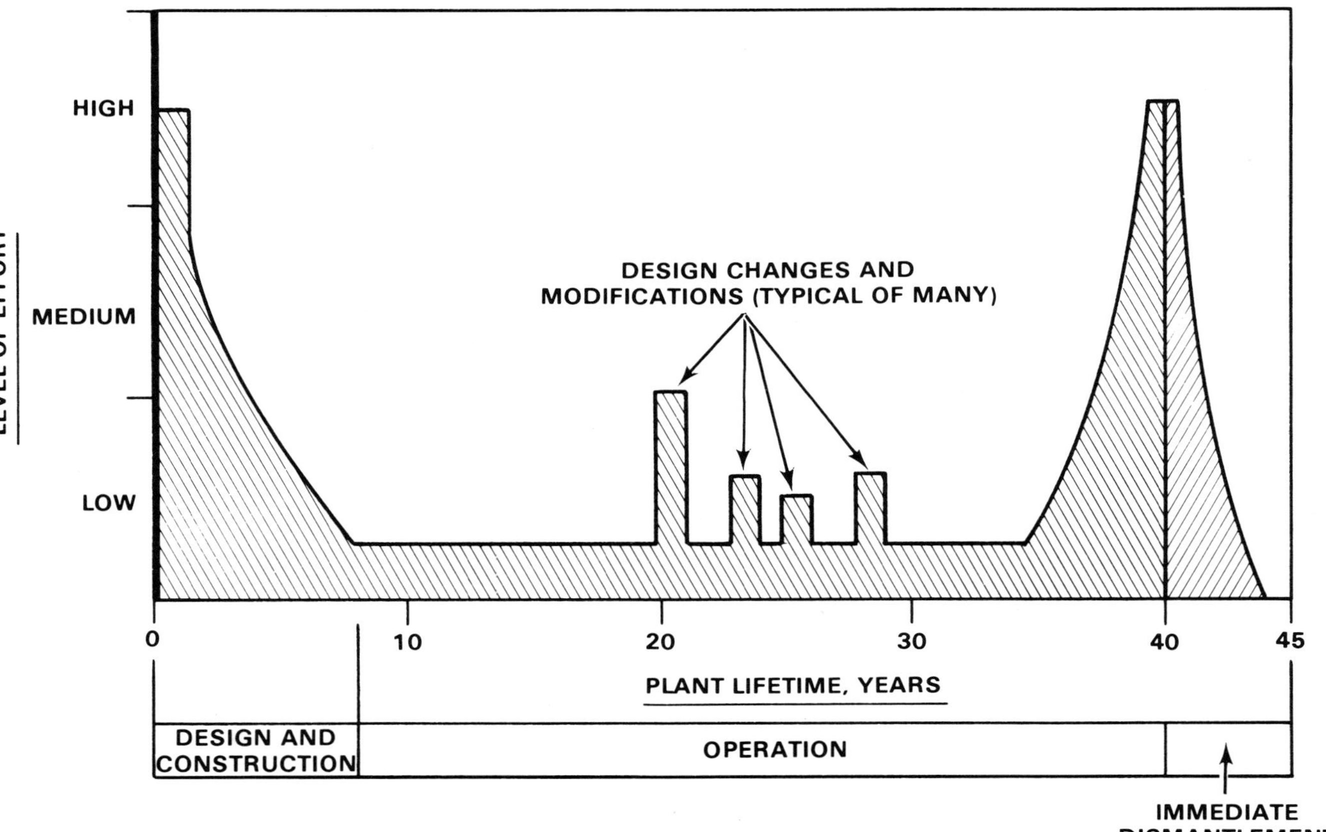

Figure 1. Opportunities and Relative Effort for Optimizing Costs and Reducing Radiation Exposure During the Lifetime of the Reference LWR

built in sections that would facilitate packaging during decommissioning. This feature is typical of the primary conflict which must be addressed by the owner and designer alike. Sectionalizing a liquid container for ease of packaging and shipping will increase gasketing and sealing requirements, result in the formation of crud traps to aggrevate the contamination problems, and provide a higher potential for failure mechanisms during operation. The choice to date has been to minimize operational risks. This procedure aggrevates the decommissioning task.

The need for design features to facilitate the decommissioning of nuclear reactors is widely recognized.[4,5] The proposed changes emphasize in-plant radiation control features, more remotely controlled operations, and increased use of prefabricated parts for concrete structures for later ease of dismantlement. It is stated in a German study on decommissioning[6] that "no fundamental difficulties with the decommissioning are perceptible which could give a reason for changing the basic design of present nuclear power plants." The authors do, however, consider a number of identified difficulties in detail for the purpose of reducing the decommissioning workload, and hence the costs involved. These include the increased use of the abovementioned prefabricated concrete parts, separation of contaminated systems from "clean" systems, and provision of sufficient room for decontamination operations where needed. These approaches, to a large extent, are already practiced at modern nuclear power stations, particularly with regard to the facilitation of repair work. This effort ultimately aids decommissioning.

A 1977 report by the National Research Council/National Academy of Sciences Committee on Radioactive Waste Management[7] indicated that more thought must be given to the design of the basic structure of the buildings in commercial power stations "in order that their useful life can be extended even though their internal operational parts are replaced." By designing so that radioactive parts of a system can be replaced, two important goals can be achieved:

- an extension of the useful lifetime of the power generating station, and
- facilitation of the eventual decommissioning of the station.

In addition, since conceptual design is based in large part on estimated levels of the radiation exposure anticipated during the operating lifetime of the nuclear facility, the level of exposure for that facility is often locked in, or fixed, with the initial design. Since the technology of decommissioning is still evolving, we can be fairly certain that there will be a concomitant evolution of NRC regulatory requirements. For example, if future regulations call for reductions in occupational radiation exposures, the cost of retrofitting could make continued operation of some of the older commercial nuclear power plants prohibitively expensive. With the added impetus of Part 50, Title 10 Code of Federal Regulations, which states: "A design objective shall be to facilitate decontamination and removal of all significant radioactive wastes at the time the facility is permanently decommissioned," an appreciation for the designer's dilemma becomes apparent.

The high costs of past reactor power plant decommissionings relative to nonnuclear facilities are usually attributed to: 1) more massive structures, usually of reinforced concrete; 2) packaging, transporting, and storing or burying radioactive wastes; 3) the need for remote operations, contamination control, and maintenance of radiological surveillance and protection systems; and 4) less cost-efficient manpower use under radiation zone conditions.[8] Therefore, innovative ideas for improvements in the above four areas could offer considerable potential for developing improved, specialized decommissioning technology, directly leading to reduced costs and occupational radiation exposures.

We often observe that the advances achieved to optimize decommissioning activities can sometimes be currently utilized. It can be seen from Figure 2 that optimization planning should include feedback and subsequent evaluation of the information, both from operating plants and from R&D sources. The use of this information can lead to improved design for maintenance. The challenge, however, is to improve maintainability without causing major increases in plant cost and construction schedule.[9] In general, a design feature that reduces exposure to operating personnel can also be expected to reduce radiation exposure to decommissioning personnel. An example of the concept, given in Figure 2, is the recent advance in metal cutting using the arc saw.[10] Widely hailed as a new decommissioning tool, the arc saw obviously has application during the operating years as well. It can be used for various types of maintenance, including equipment removal and replacement activities. Therefore, early consideration given to equipment design, location, accessibility, and shielding requirements also should include consideration for recent advances in decommissioning technology.

The general criteria for selecting design features for consideration should be the beneficial effects such criteria might have in: 1) decreasing decommissioning cost; 2) reducing total decommissioning time; and 3) creating less volume of radioactive waste. An ideal design feature also would reduce or prevent personnel exposure, and assure an adequate margin of occupational or public safety during inspection, maintenance,

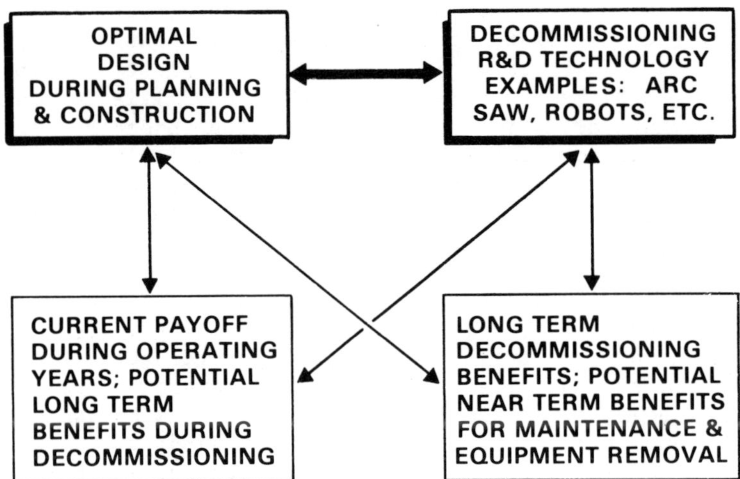

Figure 2. The Interrelationship of Decommissioning Technology and Initial Optimal Design for the Reference LWR

operation, safe storage, and/or dismantlement. Early application of radiation dose reduction designs yield benefits extending over the whole operating life of the plant, whereas the cost to achieve the reduction will be incurred intitially over a relatively short period.

It should be noted that the time-honored methods used to overcome increased radiation levels during reactor plant life, i.e., remote operations and more shielding, usually translate into increased costs. There is no sure-fire formula to guarantee reductions in occupational radiation exposures in every instance. Presently, the shielding groups of A/E firms are actively involved in developing design methods for providing shield systems that 1) meet the radiological criteria set by NRC regulations; and 2) facilitate inspection and maintenance for the purpose of increasing overall plant availability during the plant's operating lifetime. A third function must now be considered: i.e., to facilitate and optimize decommissioning.

OPPORTUNITIES FOR DECOMMISSIONING OPTIMIZATION DURING THE OPERATING YEARS

During the operating years, optimization is occasionally forced by external events. For example, the recent closures of several low-level waste burial grounds in the United States left many nuclear plant operators confronted with a choice: either continue the present radwaste cycle of activities and pay the additional transportation costs for the same volume of waste to more distant, rapidly filling burial grounds, or utilize newer equipment and methods to reduce the radwaste volume. In this instance, industry was prepared with new and improved technology and equipment that allowed the second alternative. The plant operator traded increased capital expenditures for lower operating costs. The use of the newer equipment has also resulted in reduced occupational radiation exposure. This is due, in part, to an increased awareness of this need by the nuclear support industries. The resulting designs and equipment increase the optimization of man-machine interfaces that are an integral part of efficient operation in all nuclear facilities.

Fortunately, the optimization planner should have a somewhat longer time in which to consider the potential problem areas. Thus, he should be better prepared to deal with choices and decisions. Certain facts are clear at the very onset of his task: 1) he can reasonably assess volumes, types, and radionuclide compositions of most of the expected decommissioning radwastes; 2) he can study relevant historical records of decommissioning of other nuclear facilities; and 3) he should establish a data bank relating to his own facility. Thus, he develops a keen awareness of most of the potential pitfalls. Once identified, the pitfalls can be dealt with effectively to eliminate or reduce their contributions to cost and/or radiation exposure, both during the operating years and, ultimately, during decommissioning.

The decommissioning planning field is wide open to new and innovative approaches in the methodology of decommissioning optimization, even though many pieces of such an optimization format would, by necessity, be slanted toward the specific nuclear facility being studied. There are many decommissioning activities common to a broad class of nuclear facilities that are amenable to standardization while still being flexible enough to handle site-specific situations. This flexibility allows the application of innovative thinking (an important human factor) that is so necessary to increase decommissioning productivity.

Accurate record keeping and as-built drawings become increasingly important during this time period, since they provide a major data base for planning the decommissioning of the facility. In addition, modifications and design changes made during the operating years should be examined for both immediate and downstream benefits (i.e., effects on decommissioning). For example, the choice of a simple item such as an epoxy paint might be significant. What is its longevity? Does it penetrate deeply into concrete? A few extra dollars spent now for a paint having specific qualities could mean less radionuclide penetration into the pores of the concrete, making it easier to decontaminate later. It would mean less waste volume of concrete to remove, resulting in lower transportation and burial costs. Considerations like these may not seem very important during the operational period, but they take on increasing importance as decommissioning time approaches.

Each modification and design change represents an opportunity to the optimization planner. He recognizes that these changes have the potential for even greater impact on costs and occupational radiation exposures than the original design. Experience has shown that the following design actions will reduce radiation exposure, both in the operating and in the decommissioning phases. Listed in order of effectiveness, these design actions are:[11]

- stop adding equipment
- eliminate equipment
- simplify equipment
- relocate equipment in lower radiation field
- provide better chemical control and purification
- extend interval between maintenance periods
- arrange for quick removal for shop maintenance
- reduce in-situ maintenance times
- provide more space around equipment
- provide additional shielding.

In the foregoing examples, the considerations discussed are not new. They are not even novel, for most of us already apply this "best option" type selection process in the solutions to most immediate and pressing problems, but such solutions must be disseminated. Decommissioning optimization methodology is advanced only when the most cost-effective equipment and practices are 1) provided to the whole nuclear community for immediate consideration, utilization, and potential improvement; and 2) the applicable parts thereof are incorporated into a decommissioning data bank for future consideration and use. A continuing effort during the operating years should be directed toward ensuring that knowledge already available is expeditiously evaluated and optimally applied.

COUNTDOWN--THE LAST 5 YEARS OF OPERATION

The objective of decommissioning is to place the nuclear power plant in a condition which adequately protects public health and safety and which will eventually result in termination of the licensee's responsbilities to the NRC. This objective may not be met in an optimum manner by a decommissioning plan developed early in the facility's life. In all probability, such a plan would become obsolete. Therefore, the final 5 years of facility operation is an important period for developing practical decommissioning alternatives. By establishing a data bank during the operating years and adding to it as additional experience is gained, the most effective plan from economic, political, and technical viewpoints can be developed by the licensee during this time period using current exposure and contamination limits.

Studies performed to date indicate that radiation doses during decommissioning can be controlled to levels comparable to doses from operating reactors by the use of appropriate work procedures, shielding, and remotely controlled equipment. Therefore, the major effort in this last 5-year time period should be directed toward optimizing costs and reducing radiation exposures through refinements in these general areas. The best utilization of planning time results when an immediate dismantlement mode is the selected decommissioning alternative.

THE DECOMMISSIONING VIA IMMEDIATE DISMANTLEMENT

In establishing initial plans for the actual decommissioning, consideration should be given to cost-benefit analyses that include these important parameters:

- worker radiation exposures
- potential public radiation exposure
- development of realistic decommissioning costs, based on actual experience, when possible
- use of resources
- time and level of surveillance
- the likelihood of various levels of intrusion attempts.

It is preferable that the detailed plans be prepared by those responsible for their implementation. The rigidity required by the operating specifications and procedures during the reactor's life is not necessary in accomplishing the MDP. You will be severely limited if you simply <u>must</u> perform the decommissioning activities within a prescribed framework such as the Critical Path Method. During decommissioning, just as during the preceding planning and preparation period before final shutdown, the plan can and should be flexible. Initial estimates of costs, exposures, and resources should be recognized as just that--engineering estimates to be used for planning the initial work. It is from this base that the optimization planner can set targets and goals for cost and exposure reduction measures. Thus, the optimization planner's professional worth and success will be quite discernible when measured against this yardstick.

In determining and pricing each decommissioning activity, the initial step is to determine possible procedures to follow that will allow decontamination and/or dismantling to proceed safely in a cost-effective manner. This initial procedure and its possible alternatives are then compared and reviewed to arrive at an optimum or "best" procedure for each task. The optimization technical approach for determining the best alternative for a given decommissioning work activity is presented in Figure 3. First, the work activity is clearly defined. Possible alternatives and disposition criteria are considered. Then, plans and techniques are developed to accomplish the task. Cost-benefit and safety analysis are performed. This procedure allows for the comparison of safety, man-rem, and cost from multiple perspectives. The result is the selection of a best procedure for accomplishing each task, based on available information. (One should bear in mind that later difficulties or alternatives could cause this "best" procedure to be upgraded). This best procedure provides the indispensible function of outlining the series of tasks that must be sequenced and priced to come up with a total cost, a job-time estimate, and an occupational radiation exposure estimate.

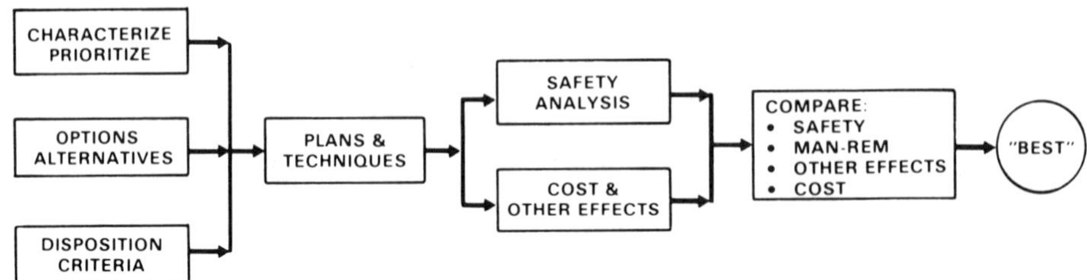

Figure 3. A Postulated Optimization Technical Approach for Individual Decommissioning Activities

A postulated methodology that optimizes the identification, categorization, and prioritization for major and minor classifications of decommissioning activities is shown in Figure 4. Standardized, site-specific criteria are established and then used to categorize each work activity into either the "major" or the "minor" part of the decommissioning program. If a work activity is selected as a major item, concerted efforts to optimize (i.e., further reduce cost and/or radiation exposure from the initial estimate) are made.

Subgroups will emerge in both the major and the minor categories as a result of the selection process illustrated in

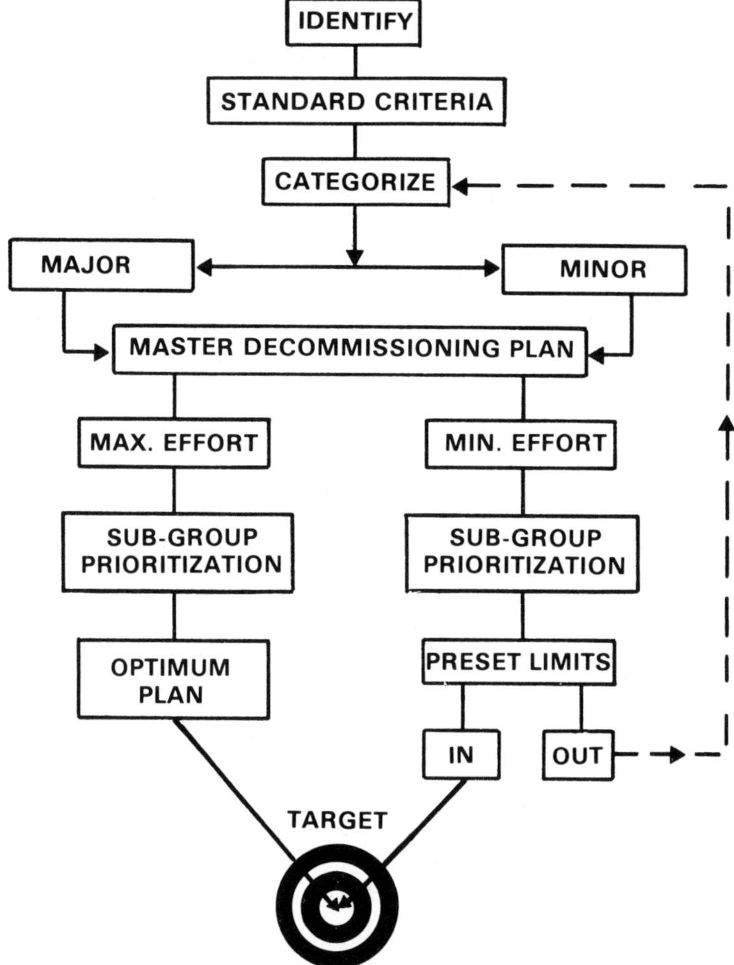

Figure 4. A Postulated Optimization Methodology for the Identification, Categorization, and Prioritization of Major and Minor Decommissioning Activities

Figure 4. Thus, even those decommissioning activities relegated to minor status will be automatically prioritized into the MDP as noncritical exposure or expense work items. Continuing computerized follow-up of the minor work activities, based on regular, pertinent information input, is essential. By flagging costs or exposures that exceed preselected limits, out-of-tolerance items can be placed in a reclassification status. Thus, if the preselected limit for a specific minor task is exceeded at any time, the task could be reinserted into the optimization program and the classification process begun again, based on the new data.

Application of this methodology should obviate any unpleasant surprises in the decommissioning program. It is possible that a highflyer or two within an optimization subgroup might result in a cost overrun or an occupational radiation exposure greater than anticipated. However, the net cumulative result of each optimization subgroup, in terms of both total cost and exposure, should be smaller than those total amounts would have been had this methodology not been used.

CONCLUSIONS AND RECOMMENDATIONS

The estimated worth of decommissioning optimization planning during each phase of the reactor's life cycle is dependent on many variables. The major variables are tabulated and relatively ranked in Table 1. For each phase, optimization qualitative values (i.e., cost, safety, maintainability, ALARA, and decommissioning considerations) are estimated and ranked according to their short-term and long-term potential benefits. These estimates depend on the quality of the input data, interpretation of that data, and engineering judgment. Once identified and ranked, these considerations form an integral part of the information data base from which estimates, decisions, and alternatives are derived. The optimization of costs and the amount of occupational radiation exposure reductions are strongly interrelated during decommissioning. The greatest benefits in either area are directly related to the amount of effort that is applied to each phase of the facility's life cycle by the optimization planner.

Realizing that building the necessary infrastructure for decommissioning will take time is an important first step in any decommissioning plan. It is because of 1) the contemporary sociopolitical realities concerning the nuclear industry; 2) the procurement and logistics realities of the marketplace; and 3) our own professional ethics that the planning and preparation for decommissioning must start early in the life of the nuclear facility. In addition, the following conclusions are established to achieve optimization of costs and reduced occupational radiation exposures:

- the assignment of cost versus man-rem is item-specific and sensitive to the expertise of many interrelated disciplines

- a commitment to long-term decommissioning planning by management will provide the conditions needed to achieve optimization

Table 1. Qualitative Values of Optimization Methodology for All Phases of the Reference LWR Lifetime

Life-Cycle Phase[a]	Qualitative Value (Potential Benefits)		Other Considerations		
	Short-Term	Long-Term	Cost	Characteristics of Radiation Exposure Reduction	Potential Decommissioning Application
• Design and Construction	High	High	Incurred Initially Only[b]	Short-Term and Cumulative	High[c]
• The Operating Years					
Data Bank	Low	High	Nominal	Depends on the Program	High
Modifications and Design Changes	High	Medium to High	Varies	Varies	Varies
Awareness Program[d]	Low to Medium	Low to Medium	Varies	Varies	Medium to High[e]
Exposure Reduction Program	High	High	Varies	Potentially Beneficial to Specific Areas of Operation and Ultimately to Decommissioning	High
• The 5-Year Period Before Final Shutdown	Low to Medium	High	Varies	Varies	High
• The Actual Decommissioning	High	Not Applicable	Varies	Varies	Immediate and High (to others)

(a) Immediate dismantlement is the decommissioning mode selected for the end of the life cycle.
(b) Equipment maintainability _must_ be considered for maximum benefit.
(c) In general, a design feature that reduces external radiation exposure to operations staff also can be expected to reduce radiation exposure to decommissioning workers.
(d) Defined as the expeditious dissemination of current decommissioning technology to line management for the purpose of determining economic and operational applicability for its use.
(e) The results from in-the-field utilization of decommissioning-related equipment provides valuable information to the optimization planner.

- to be most effective, costs and exposure reduction are sensitive to the nearness of the decommissioning operation.

The main attribute to developing the optimization/life cycle methodology is that it can be utilized starting at any time period in the life cycle of the nuclear facility for decommissioning planning. For a new plant, as we have seen, we start at the beginning of the cycle, update continually, consider innovations, and realize full potential and benefits of this concept. For an older plant, the life cycle methodology permits a comprehensive review of the plant history and the formulation of an orderly decommissioning program based on planning, organization, and effort. This program can begin at any time before decommissioning, but the "sooner-the-better" is an essential realization for the optimization/decommissioning planner.

REFERENCES

1. K. J. Schneider and C. E. Jenkins, Technology, Safety and Costs of Decommissioning a Reference Nuclear Fuel Reprocessing Plant, NUREG-0278, U.S. Nuclear Regulatory Commission Report by Pacific Northwest Laboratory, October 1977.

2. R. I. Smith, G. J. Konzek, and W. E. Kennedy, Jr., Technology, Safety and Costs of Decommissioning a Reference Pressurized Water Reactor Power Station, NUREG/CR-0130, U.S. Nuclear Regulatory Commission Report by Pacific Northwest Laboratory, June 1978.

3. C. E. Jenkins, E. S. Murphy, and K. J. Schneider, Technology, Safety and Costs of Decommissioning a Reference Small Mixed Oxide Fuel Fabrication Plant, NUREG/CR-0129, U.S. Nuclear Regulatory Commission Report by Pacific Northwest Laboratory, February 1979.

4. A. Martin, D. T. Read, R. W. Milligan, T. F. Kempe, and D. A. Briaris, A Preliminary Study of the Decommissioning of Nuclear Reactor Installations, ANS Report No. 155, p. 63, Associated Nuclear Services, 123 High Street, Epsom, Surray, KT19 8EB, July 1977.

5. G. F. Stone, "Control of Occupational Radiation Exposures in TVA Nuclear Power Plants Design and Operating Philosophy," ANS-SD-15, Proceedings on the Special Session on Plant and Equipment Design Features for Radiation Protection, CONF-750661, New Orleans, LA, June 8-13, 1975.

6. R. Bardtenschlager, D. Bottger, A. Gasch, and N. Majohr, Decommissioning of Light Water Nuclear Power Plants, Nuclear Engineering and Design, Vol. 45, pp. 1-51, Copyright North-Holland Publishing Company, 1978.

7. The Shallow Land Burial of Low-Level Radioactively Contaminated Solid Waste, National Academy of Sciences, Library of Congress Catalog Card Number 76-56928, Washington, DC, 1976.

8. K. M. Harmon, "Decommissioning Nuclear Facilities," Proceedings of the International Symposium on the Management of Wastes from the LWR Fuel Cycle, Denver, CO, July 11-16, 1976.

9. R. C. Anderson and S. W. W. Shor, "Coordinated Design Reviews To Improve Access and Reduce Radiation Exposure," Nuclear Engineering International, p. 45, March 1978.

10. G. A. Beitel and Max P. Schlienger, Arc Saw Testing, ARH-LD-147, Atlantic Richfield Hanford Company, Richland, WA, and Schlienger, Inc., San Rafael, CA, July 1976.

11. G. G. Legg, Reducing Radiation Exposure in CANDU Power Plants, (Technical paper 5/22 at Nuclex '75, Basle, October 1975), Atomic Energy of Canada Ltd., Chalk River, 1976, report No. 5388.

STATE REGULATORY IMPACT ON REACTOR DECOMMISSIONING:

FINANCING APPROACHES AND THEIR COST

Vincent L. Schwent

California Energy Commission
1111 Howe Avenue
Sacramento, CA 95825

INTRODUCTION

The Nuclear Regulatory Commission (NRC) is moving toward promulgating comprehensive new regulations governing decommissioning. Increasing numbers of state legislatures and utility commissions are becoming more interested in decommissioning activities. In addition, large gatherings such as this are focusing increased attention on various aspects of decommissioning. Given this atmosphere, perhaps it is appropriate to assess how attitudes and policies toward the financial aspects of decommissioning are changing and evolving and the role of state and federal authorities in this process.

If a prediction may be offered as to how the costs and financing of reactor decommissioning will eventually be handled, mine would be this: <u>Specific estimates of the cost of dismantling each reactor will be required and used in turn as the basis for collecting funds from the ratepayers consuming that reactor's power. These monies will be held in tax-exempt funds independent of the control of reactor owners and operators.</u>

Why do I predict this as the future? To see, let us examine the trends evident in the past and present situation in reactor decommissioning.

THE PAST SITUATION

Recent years have seen relatively little activity in the area of decommissioning financing with few state utility regulators considering the subject at all. In 1978, the NRC in conjuction with their resolution of the Public Interest Research Group petition surveyed regulatory agencies in the 38 states with planned, licensed or ordered reactors as to their views on handling decommissioning costs.[1] Twenty-eight of these states did not or could not indicate any formal policy towards financing. The remaining ten indicated a preference for the "negative salvage value" method. The rationale for preferring this approach was generally that allowing utilities to retain decommissioning monies would 1) increase the utilities cash flow, 2) allow them to be invested in the utilities assests which would help offset the need to borrow for new construction and, 3) earn or save ratepayers the utilities' cost of capital on borrowing the equivalent amount of money and which would be greater than the return from investment outside the utility.

At the federal level, prior to the present NRC rulemaking activities, there was very little regulation or guidance available on decommissioning. A number of regulations primarily intended for operating reactors, such as radiation standards, could be construed to also apply to decommissioning activities. Guidance applicable directly to decommissioning, however, appeared to consist basically of Regulatory Guide 1.86. NRC Reg. Guides are not regulations, are not mandatory, and offer only guidance, yet in the absence of any definitive regulation on the subject, Reg. Guide 1.86 assumed a major role in policy formulation at the state level. This document appeared to endorse mothballing, entombment, and complete dismantling as equally complete and satisfactory alternatives for reactor decommissioning.

As a consequence of relying on Reg. Guide 1.86, I believe the few states interested in this issue generally regarded all financing approaches and modes of decommissioning as equally effective, complete and satisfactory. The result of this was that often decisions were made by these state regulatory agencies apparently solely on the basis of minimizing the assumed costs to the ratepayers. In this atmosphere of inadequate federal guidance and cursory financing cost comparisons, decisions were made such as that by the Connecticut PUC that Millstone 1 & 2 and Connecticut Yankee decommissioning financing should be done by negative salvage value and that the funds collected should only be sufficient for mothballing not entombment or dismantling since mothballing was the least expensive alternative.[2]

The costs of financing by methods other than negative salvage value depreciation were examined only superficially if at all. Also, the cost estimates of the alternative modes of decommissioning tended to be either taken directly or simply extrapolated from the thorough but generic AIF study of decommissioning costs,[3] or very roughly estimated by others. The situation prior to last year, therefore, reflected a general lack of state and federal activity in this area, and the decisions which were made were made on the basis of inadequate, insufficient, and misinterpreted data on the requirements and cost of decommissioning and its financing.

THE PRESENT SITUATION

Within the last two years, a number of decisions have been made and reports have been issued which begin to show a change in the direction and certainly the depth of analysis on the question of decommissioning. In some areas trends are developing but in other areas confusion is increasing.

One factor, in particular, necessitates an increasingly larger role for state and federal authorities on the subject of financing: failure to properly decommission a reactor and failure to adequately finance this work will mean health and safety liability and, in turn, financial liability for state and federal governments. Because a single utility may own reactors in a number of states and a single reactor may be owned by utilities based in a number of states, coherent, uniform and consistent governmental policies must be developed regarding the acceptable modes, timing, costs and financing methods for decommissioning.

The modes, timing and costs of decommissioning activities are also necessary precursors to the selection of a financing method and the determination of the costs of financing decommission. In all of these areas, state and federal regulation and law can and will have an impact.

Modes

Recent NRC activities indicate that improved definition will be forthcoming as to what actions constitute complete and terminal decommissioning.[4] Present NRC staff thoughts, as contained in NUREG-0590,[5] would be for commercial reactors to permit only immediate dismantlement or dismantlement preceeded by mothballing with a delay of up to thirty years.

Should this become NRC regulation, a number of existing state financing decisions would have to be restructured to accommodate the higher costs involved in dismantling over simply mothballing or entombment. Some state regulatory authorities are already beginning to make financing decisions based on dismantling, for example, Arkansas, California, and New York, while at least one piece of state legislation specifies dismantlement (see Oregon HB 2509, Table 1).

Timing

The AIF study originally advanced the notion that possible occupational radiation exposure reduction and cost savings might result from delaying dismantling on the order of 100 years. However, delayed dismantling appears to be losing favor both in government and industry. The originally predicted cost savings have been subsequently challenged in both the Battelle generic cost estimate study (NUREG/CR-0130)[6] and the NUS cost study for San Onofre 1.[7] The extent of any radiation exposure reduction gained by delay has also been challenged by the recent recognition that trace amounts of long-lived Ni-59 and Nb-94 in reactor steel will limit the usefulness of long delays as a way of reducing the extent of decommissioning work required and the level of occupational exposure.[5,6] Additionally, state and federal concern over the difficulty of predicting costs over an additional 100 years of inflation, how to guarantee that the financing mechanism chosen will meet these costs, and the social question of leaving this waste disposal problem for future generations, all militate against governmental authorities permitting long delays before terminal decommissioning.

NRC staff opinion, as shown in NUREG-0590, would be to rule out the use of 100 year delays because of the long-lived radioisotope problem. The California PUC is planning for San Onofre 1 to be dismantled as soon as the other two units on the site reach the end of their useful lives.[8]

Cost Estimates

The use of special financing schemes to meet the expense of decommissioning can only be successful if accurate estimates of these eventual costs are provided during the course of reactor operation. Prior to this year, few detailed engineering estimates of the costs of dismantling large commercial reactors were available.[3,6,7] Moreover, two of these, the AIF study and the Battelle study, were generic in nature, presenting representative or average values for decommissioning costs. With little else for utilities to turn to, many state utility agencies, when requesting cost estimates for particular plants, were given estimates from the AIF or Battelle studies or slight adjustments thereto.

This situation, I believe, is changing as state and federal agencies more fully recognize the variability introduced into cost estimates by the specific sites, plants and decommissioning modes under consideration.

The California PUC requires cost estimates for California reactors when financing decommissioning. A detailed decommissioning cost estimate for the San Onofre 1 reactor produced a value two to three times larger than scaling of the AIF or Battelle costs would have produced. An author of the San Onofre estimate has gone on to suggest that a number of site-, facility-, and plan-specific factors can produce variation in cost estimates between otherwise similar facilities of as much as 250%.[9] Factors such as projected duration of decommissioning, local labor costs and productivity, proportion of utility vs. non-utility labor, project extent and complexity, level of decontamination desired and the presence of multiple facilities on one site must be considered if the best possible cost estimates are to be produced. A number of detailed plant-specific cost estimates incorporating such factors are presently underway.

As state regulatory agencies focus more on the financing of decommissioning, it will become increasingly apparent that the success of a financing scheme is limited by the accuracy of the cost estimate. Recent legislation introduced in Maine, New York, and New Hampshire would require authorized state agencies to obtain cost estimates of decommissioning from which to calculate ratepayer contributions to decommissioning financing (see Table 1).

Financing Methods

There appears to be a recognition on the part of some state legislatures, regulatory agencies, the NRC staff and utilities that the selection of a financing method should be based on factors other than solely the bottom line cost to the utility or ratepayers. NRC staff have recently argued in NUREG-0584[10] that in the selection of a financing method, the financial assurance that a given method will provide the requisite funds at decommissioning is a more important selection criterion than the cost of providing this assurance.

NUREG-0584 and NUREG-0590, using this ordering of criteria, endorse financing methods using external funds, either those employing annual contributions or a single large initial deposit, and the use of bonding or insurance, if these later options can be made available.

While several state regulatory agencies continue with the use of negative salvage value depreciation, favor for this method is no longer unanimous. The Connecticut PUC, in apparent recognition of the less than adequate financial assurance that the depreciation method provides, has ruled that the use of depreciation should be accompanied by a bond to assure that the monies collected under the depreciation scheme will be available when necessary.[11] The Pennsylvania PUC has required the owners of Three Mile Island to finance decommissioning through the use of an annuity external to the utility, a method essentially equivalent to the sinking fund method.[12]

Table 1 lists eleven recent pieces of state legislation dealing with reactor decommissioning. These bills range in content from simply requiring a study of costs (Kansas SB 87) to prescribing the use of a sinking fund and specifying how to calculate the required contributions (Indiana HB 1379). Where a financing method is specified, it is either bonding or an external fund, with the use of the sinking fund approach predominating (Maine, New Hampshire, New York, Indiana).

Table 1. Recent State Legislation on Reactor Decommissioning

State Bill No. (Year) Status	Purpose
Kansas SB 87 (1979) Died	Requires Corp. Comm. to study costs; appropriates \$50k for study.
Maine LD 783 (1979) Passed	Est. Committee to study need, procedures, cost and funding.
Arkansas HB 480 (1979) Passed	Creates Ark. DOE and gives it authority to propose decom. recommendations.
Washington HB 599 (1979) Died	Gives Siting Council authority to develop guidelines.
Wisconsin AB 211 (1979) In Committee	Requires PSC to est. methods and publish cost est. and methods.
Oregon HB 2509(1979) Died	Requires Facility Siting Council to adopt rules requiring bonding.
Vermont HB 363 (1979) In Committee	Est. Board to require and oversee bonds or fund to cover accident D&D.
New York AB 5566(1979) SB 3869(1979) In Committee	Sets up sinking fund; PSC to determine methods and costs.
New Hampshire HB 805 (1979) Died in Senate	Sets up sinking fund held and invested by state treas.; est. committee to determine costs, procedures and payments.
Indiana HB 1379(1979) Died	Requires sinking fund; specifies contribution calculation; requires 5 yr. review of fund and costs; est. committee to oversee fund.
Maine LD 1804 (1977) Died in Senate	Requires sinking fund held and invested by state treas.; est. committee to determine costs and procedures.

Although it has been argued that utilities should prefer the negative salvage value depreciation approach because it provides a source of capital to them for present use, utilities also are not unanimous in their preference for the depreciation method over the alternatives. Duquesne Power and Light urged the Pennsylvania PUC to permit them to set up an escrow account for Beaver Valley I decommissioning costs.[13] Jersey Central Power and Light similarly requested of the New Jersey Board of Public Utilities permission to use a "separate independent trust fund" for their Oyster Creek plant.[14] Consolidated Edison endorsed New York's AB 5566, which would require the New York Public Service Commission to set up and use sinking funds to recover decommissioning costs, and Con Edison further suggested that AB 5566 be amended to require that the fund, which would have been held by the utility, should be held by a public agency instead.[15]

The question of how to finance reactor decommissioning is clearly the aspect of decommissioning economics most directly influenced by state and federal activity. The NRC staff have taken the position that assuring that decommissioning will take place is a health and safety question first and a cost question second.[10] This is appropriate given their role in nuclear regulation as compared to that of other federal agencies, such as FERC. The rising interest of state legislators and their preference for financing methods, which provide the highest financial assurance of funds availability, is also significant. With the increasing scope of interest in decommissioning financing in general and the particular interest of parties such as the NRC and state legislatures, whose primary interest is assuring the availability of collected funds, it is likely that this trend towards sinking funds or similar mechanisms, and away from the depreciation method, will continue.

Cost of Financing

Whether cost of financing is used as the primary selection criterion or as a secondary consideration, the cost of utilizing the various alternative financing schemes does need to be considered. The present situation in the comparison of financing costs is one of chaos and confusion. A state regulatory agency, having selected a set of otherwise acceptable financing methods, would, I believe, upon carefully examining the available data on comparative costs, be confused as to which mechanisms would prove least expensive to ratepayers. Such an examination would reveal that, while a number of cost comparisons have been attempted, there is little agreement as to which financing methods are the least costly to use. (See Table 2)

Table 2. Ranking of Decommissioning Financing Methods Based on Least Total Cost

	Expense at Decommissioning	Sinking Fund	Prepaid Fund	Depreciation
Ewers('77)[16]		1		2
Mingst('79)[17]		1	2	3
Smith('78)[6]	1	2	3	
Chapman('79)[18]	1	2		3
Collins('78)[19]		2	3	1
Arkansas Power Light('77)[20]		2	3	1

Moreover, in attempting to compare these comparisons, a reviewer might well be confused by the profusion of names for the mechanisms being contrasted. Methods with very similar names in two different studies may have detailed descriptions indicating that they are actually very different, while mechanisms with quite different labels may, upon close inspection, appear to refer to the same basic scheme (See Table 3). Also, a reviewer would discover that even when the financing mechanisms being compared in two reports appear to be the same, they are quite likely not identical. The authors of these comparisons almost invariably use different assumptions as to the future inflation rate, the potential interest that could be earned by a fund, the utility rate of return, or the appropriate discount rate to apply (Table 4). In addition, the details of how any particular financing mechanism is structured may differ. For example, sinking fund mechanisms in two different studies might vary as to whether contributions to the fund and the interest earned by the fund are taxable, tax deferred or non-taxable; whether the ratepayers pay the taxes when levied or the utility does; or whether any taxes paid by the utility are added to the rate base.

In short, a number of dissimilar assumptions are frequently made regarding details of basically similar mechanisms. Unfortunately a reviewer cannot determine whether these differences are significant to the calculation of relative costs and which, if any, of these differences are the source of the often opposite conclusions reached in the various studies as to which methods are cheapest (See Table 2).

Table 3. A Nomenclature for Decommissioning Financing Methods

Preferred Name	Commonly Used Equivalents
Sinking Fund	Funded Reserve; Annuity; Escrow Account
Prepaid Fund	Prepayment Method; Deposit Method; Prepaid Sinking Fund; Funding at Plant Commissioning; Deposit at Start-Up
Negative Salvage Value Depreciation	Straightline Method; Price Level Adjusted Method; Unfunded Reserve; Funding at Decommissioning; Sinking Fund Method of Depreciation

One method for resolving this present chaotic data base would be some thorough studies analyzing the impact on cost of these myriad variations. As an example, NRC staff have performed one cost comparison which attempted to examine the effect of variations in assumed values for inflation, fund interest earned, and utility discount rate. (See Mingst, Table 2). While this is a start, it would take a massive study, indeed, to work through all the variants that have been introduced in the existing studies.

On the other hand, such a massive study may not be necessary. There is general agreement that the effect of federal tax policy on alternative financing methods is a major source of the cost of financing. Presently, when money is collected for decommissioning during plant operation, by whatever method, utilities are not able to deduct these monies from taxable income as an expense until decommissioning is actually performed. This leads to numerous problems. Basically, utilities argue that they must collect extra money during plant operation from ratepayers to pay these taxes. When decommissioning is actually performed, utilities will be able to expense the work creating a future tax break. This future tax break must, in fairness to previous ratepayers, be normalized or somehow passed back through to the ratepayers paying for the decommissioning work.

Solving the complex problems arising from present tax policies for depreciation-type mechanisms could be done by permitting monies collected to be expensed at the time of collection. This however, will apparently require a federal legislative solution. No such legislation is currently pending in the Congress to solve this problem for depreciation-type approaches. For sinking funds or prepaid funds, however, in addition to the greater financial assurance they provide, several sources have argued that such funds, if properly structured so that utilities can derive no benefit from the monies they contain, could under existing IRS guidelines be possibly granted tax-exempt status. The Pennsylvania PUC's selection of a sinking fund type

Table 4. Range of Some Assumed Parmeters in Financing Cost Comparison Studies

Parameter	Range of Assumed Values
Inflation Rate	2-8%/Yr.
Interest Earned on a Fund	4-11%/Yr.
Utility Rate of Return, Discount Rate	6-15%/Yr.
Facility Life	30-40 Yrs.

financing mechanism[12] was based, at least in part, on the assumption that such favorable tax treatment, and consequently lower financing costs, could be obtained.

NUREG-0584 contains four conditions under which decommissioning deductions from ratepayers would be non-taxible; 1) collected funds are segregated from utility assets, 2) funds are not invested in utility assets, 3) funds are administered by an independent party and 4) excess funds would not be returned to the utility. Several pieces of recent state legislation propose to establish funds in a manner that appears to meet these conditions. (See New York, New Hampshire, Indiana, and Maine, Table 1). In addition, utilities in Pennsylvania and Ohio are in the process of attempting to obtain favorable IRS rulings on proposed sinking fund-type financing schemes.

Regulatory or legislative assistance, then, in structuring some specific sinking fund-type financing proposals that can gain favorable IRS rulings, would have the effect of lowering the financing costs of decommissioning. Sinking funds, as a result, would be more attrative in terms of financing cost as well as for financial assurance reasons.

Premature Shutdown

Previous studies of decommissioning economics have generally failed to consider or attempt to deal with the financial impacts of premature shutdown of a commercial reactor. Decommissioning was assumed to take place on a "normal" reactor with a "normal" amount of contamination and which operated for an "average" lifetime. Events at Three Mile Island have now focused attention on what has been a serious omission from previous decommissioning financing considerations.

Premature reactor shutdown, for whatever reason, will result in inadequate funds being available for even routine decommissioning. Depreciation, sinking funds, and prepaid funds will all be inadequate in such cases.[21] If the cause of premature shutdown is an accident, even a relatively minor one in terms of immediate public consequences, or any other events leading to high in-plant contamination, the costs of decommissioning and the consequent financing shortfall will be even greater than for premature shutdowns of plants with normal levels of contamination. NRC staff were exploring with the nuclear pool insurers, prior to Three Mile Island, the possibility of developing some form of premature shutdown insurance to cover those costs exceeding normal decommissioning. While the events at TMI have sidelined these discussions, there will undoubtedly be a growing recognition of the potential financial liability arising from premature shutdown and the need for state and federal authorities working with the industry to find a means of dealing with this problem.

THE FUTURE SITUATION

Lastly, now that we have examined the past and present, what can be predicted for the future of decommissioning financing? First, dismantling with minimal delay will become the standard mode of decommissioning. Second, because of increased emphasis on the assurance provided by a financing method rather than simply its cost, sinking fund financing will become the norm. Third, costs of financing will be reduced by favorable state and federal tax treatment. Fourth, there will be increased state legislation in this area, especially if the NRC and state regulatory agencies do not develop a uniform and comprehensive set of regulations to insure public health and safety and minimize government's financial liability. And fifth, some form of premature shutdown insurance, provided hopefully by the present nuclear insurers or pools of utilities, will evolve.

REFERENCES

1. U.S. Nuclear Regulatory Commission, Docket PRM 50-22.

2. Nucleonics Week, January 6, 1977, p.5.

3. W.J. Manion and T.S. LaGuardia, An Engineering Evaluation of Nuclear Power Reactor Decommissioning Alternatives, AIF/NESP-009SR, Atomic Industrial Forum (1976).

4. U.S. Nuclear Regulatory Commission, Plan for Reevaluation of NRC Policy on Decommissioning of Nuclear Facilities, NUREG-0436, Rev. 1, (1978).

5. G.D. Calkins, Thoughts on Regulation Changes for Decommissioning, NUREG-0590, U.S. Nuclear Regulatory Commission (1979).

6. R.I. Smith, G.J. Konzek, and W.E. Kennedy, Jr., Technology, Safety, and Costs of Decommissioning a Reference Pressurized Water Reactor Power Station, NUREG/CR-0130, U.S. Nuclear Regulatory Commission (1978).

7. R.J. Stouky and E.J. Ricer, San Onofre Nuclear Generating Station Decommissioning Alternatives, Report 1851, prepared for Southern California Edison by the NUS Corporation (1977).

8. California Public Utilities Commission, Decision No. 89711, December 12, 1978.

9. R.J. Stouky, "Factors Affecting Power Reactor Decommissioning Costs for Complete Removal", presented to the American Nuclear Society Meeting in San Diego, June 19, 1978.

10. R.S. Wood, Assuring the Availability of Funds for Decommissioning Nuclear Facilities, NUREG-0584, U.S. Nuclear Regulatory Commission (1979).

11. Connecticut Public Utilities Control Authority, Docket No.770319, Section IV, I, p.45.

12. _____, "Progress of Regulation: Trends and Topics," Public Utilities Fortnightly, pp. 54-56, April 26, 1979.

13. _____, "Duquesne Light Want Customers to Pay Decommissioning Costs", Nucleonics Week, p.4, October 13, 1977.

14. New Jersey Board of Public Utilities, Docket No. 7610-1021, January 31, 1979.

15. Legislative Memorandum of Ward M. Rockey, Assistant to the Vice President of Consolidated Edison.

16. B.J. Ewers, Jr., "Accounting Today for Future Nuclear Plant Decommissioning Cost", presented to the American Gas Association-Edison Electric Institute Accounting Conference in Dearborn, Mich., May, 1977.

17. B.C. Mingst, DECOST Computer Routine for Decommissioning Cost and Funding Analysis, NUREG-0514, U.S. Nuclear Regulatory Commission (1978).

18. D. Chapman, Nuclear Economics: Taxation, Fuel Cost, and Decommissioning, prepared for the California Energy Commission, (1979)

19. P.A. Collins, "Financing and Accounting Alternatives for Decommissioning Nuclear Plants", presented to the Southeastern Electric Exchange meeting in New Orleans, September 28, 1978.

20. Arkansas Power and Light, Analysis of Decommissioning Arkansas Nuclear One - Unit 1, prepared for the Arkansas Public Service Commission, August 10, 1977.

21. V.L. Schwent, "Costs and Financing of Reactor Decommissioning; Some Considerations", presented to the NARUC Subcommittee of Staff Experts on Accounting Meeting in Seattle, September 12, 1978.

AN ANALYSIS OF DECOMMISSIONING COSTS AND FUNDING

Barry C. Mingst

Low-Level Waste Licensing Branch
Division of Waste Management
U.S. Nuclear Regulatory Commission
Washington, D.C. 20555

INTRODUCTION

This is a study of the effects on the costs of decommissioning nuclear reactors and fuel cycle facilities caused by changes in the economic environment, changes in the accuracy of facility use planning, and the use of various options for funding decommissioning. The results of this study are applicable to the decommissioning of any facility.

A parametric study is performed using the "DECOST" computer routine (Reference 1). This routine was developed to be used in generic or specific case studies for evaluation of decommissioning funding. The cases of a pressurized-water reactor (PWR) and a shallow-land burial, low-level waste site (LLWS) are used as examples. These examples are used to demonstrate the cases of facility decommissioning where the technical costs are, respectively, independent of and dependent on the site lifetime and usage.

The technical costs that this study is based on are taken from References 2 and 3. These costs are important to this study only as examples. The major results are generally independent of the magnitude of the costs involved. Cost figures, where given, are given in 1977 dollars with no contingency additions. A parametric study of the major economic parameters that affect the final cost is done to both demonstrate the range of costs and assurance that can be expected, and to determine which of these factors are the most important in determining the final costs of decommissioning a reactor. The technical costs of decommissioning are assumed to increase at the stated inflation rates.

DIFFERENT MODES OF DECOMMISSIONING

Figure 1 shows the technical costs of decommissioning a PWR by the three basic modes at decommissioning and what funding (in this case a sinking fund) can do to affect the real costs to the customers. The lowering of costs to customers by use of the sinking fund is a result of the interest earned on the fund during plant operation. These curves are based on a pressurized-water reactor and an eight percent tax-free interest rate or an eight percent annual return after taxes. The effects of varying tax rates will be dealt with later. Both families of three curves represent the three basic modes of decommissioning. These three modes are immediate dismantling, indeterminate safe storage or mothballing, and and safe storage with delayed dismantling.

Mothballing or safe storage is often considered to be the least expensive mode of decommissioning due to its very low initial costs. However, Figure 1 shows that if the inflation rate approaches the effective interest rate (within one percent in this case), safe storage is at least as expensive as immediate dismantling. If the inflation rate exceeds the effective interest rate, the total costs of the indeterminate safe storage mode cannot be funded by the users of the facility since maintenance payments will eventually eat away all the capital raised for this purpose. The major conclusion here is that immediate dismantling is the least expensive method of decommissioning when the inflation rate exceeds the effective interest rate.

METHODS OF FUNDING

There are three basic methods of funding decommissioning. The first method is a sinking fund, in which annual payments are made to an interest-bearing account. The second method is a deposit, in which a payment is made to an interest-bearing account at plant startup. The third method is negative-salvage-value depreciation funding, in which the annual payments needed to cover eventual decommissioning are calculated and planned for, but are not set aside from the operator's cash flow. These three types of funding are further broken down into seven separate types of funding. These seven types of funding are illustrated in Figure 2. For illustration the interest rate in Figure 2 is four percent and the inflation rate is two percent.

The deposit method is divided into three methods: one, a deposit that covers 100 percent of the cost of decommissioning at the time of reactor startup (D TRS); two, a deposit that will cover 100 percent of the estimated cost of decommissioning- including inflation- at the end of the expected reactor lifetime (D ERL); and three, a deposit made the same as in method one but annually adjusted to equal 100 percent of the current cost of decommissioning with the difference returned to or paid for by the operator (D RTN).

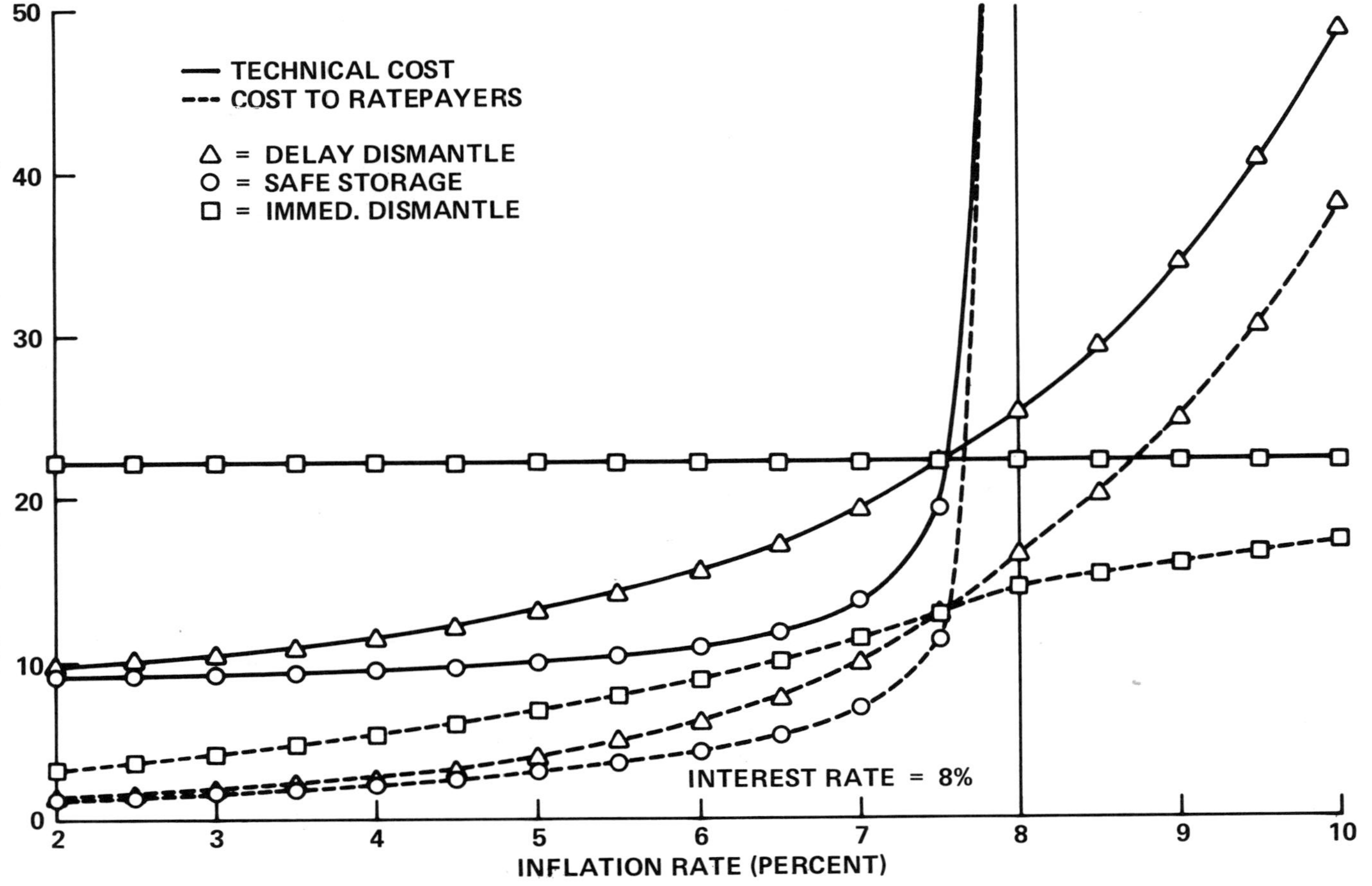

Figure 1. Cost to Decommission a PWR.

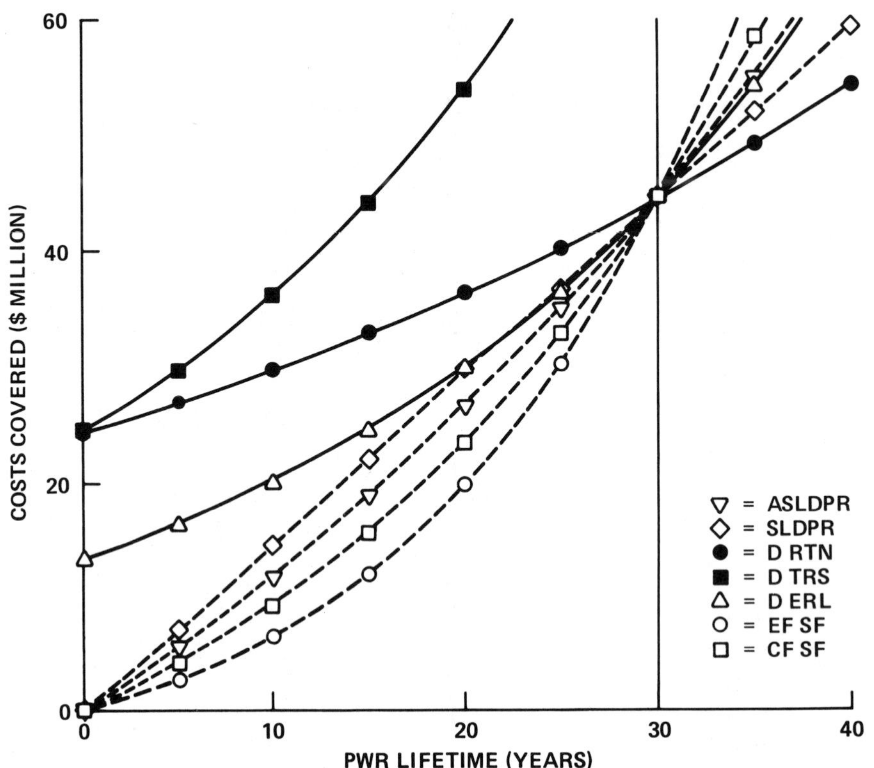

Figure 2. Growth of Funding Methods.

There are two basic methods for making payments into a sinking fund. The first method, called a constant-fee sinking fund (CFSF), allows the annual payments to remain constant in nominal dollars throughout the life of the facility. The second method, called an escalating-fee sinking fund (EFSF), allows the annual payments to escalate at the rate of inflation and the payments to remain the same in constant dollars (dollars adjusted for inflation).

There are two basic methods to account for fund transfers in the negative-salvage-value depreciation funding method. The straight-line depreciation method (SLDPR) projects the inflated cost to decommissioning and divides this amount into equal annual payments. The adjusted, straight-line depreciation method (ASLDPR) reaches the same projected inflated cost by escalating annual payments at the prevailing inflation rate.

Accounting for depreciation funding in this study does not include interest or capitalization, since such inclusion is not always required. Capitalization means the account is credited with the yearly return on capital the utility earns. If one wishes to account for capitalization, the accounting used to determine payments to the fund and the value of the fund becomes identical to that for a sinking fund. The interest rate used in this quasi-sinking fund will be the capitalization rate.

Equitability

One concern in choosing funding methods is the equitability of the funding method chosen. The sinking fund method and the negative-salvage-value depreciation funding method are both easily adjusted to place an equal burden of whatever degree is desired on the users of the facility by adjusting the annual payments. The deposit funding methods can not be adjusted in this manner since payments made on borrowed funds are constant in nominal dollars, and they place a higher burden in constant dollars on earlier users than on later users. Therefore, when comparing the seven basic funding methods, it is useful to remember that there are three rankings of equitability in the seven considered funding methods.

The most equitable methods are (or can be) the adjusted, straight-line depreciation method and the escalating-fee sinking fund. These methods can, in theory, be made completely equitable. The next rank includes the constant-fee sinking fund, the straight-line depreciation method, and the deposit methods except for (in some cases) the deposit with earnings return. These methods are inequitable to the degree of inflation over the period of operation of the facility. The method of deposit with earnings return is a special case. If maintaining the fund requires additional funds every year, the equitability of this method is somewhere between the first two groups. If the fund provides an annual return to the ratepayers, this method of funding is less equitable than either of the first two groups. The result is due both to the later users getting a return on money put into the fund by the earlier users and the later users not being charged as much, even in nominal dollars, as the earlier users.

Costs

The costs, using the different funding methods, for dismantling a PWR are shown in Figure 3. These costs are based on an interest rate of eight percent, inflation rate of six percent, and a varying tax rate as shown. The tax is assessed on all payments made to the fund and all interest earnings by the fund, and a tax deduction is given when the fund is paid to the decommissioning contractor.

This tax deduction is planned for by building a fund of reduced size. In other words, if the tax rate is 48 percent, the fund needed at the time of decommissioning will only be 52 percent of the actual technical cost, since the difference will be made up by the tax deduction. This deduction exactly balances the effect of taxing the payments made into the fund. The tax rate affects only the interest earnings of a funding method. The more the method relies on interest, the greater is the effect of the tax rate.

The tax rate has no effect on the net costs of the negative-salvage-value depreciation funding methods. This is a result of the tax credit exactly balancing the tax assessed on the payments to the operator since no reliance is made on interest in these funding methods.

One interesting result of this study is that the cost of the 100 percent deposit at reactor startup (D TRS) becomes less expensive to the customers as the tax rate increases. This effect is a result of the interest charges on the payback of the loan taken for the deposit (whether obtained within the company or outside) being a large part of the costs of the deposit funding method. As the tax rate increases, an increasing fraction of the interest charges are tax deductible. As noted before, the effective origional deposit is not affected by the tax rate. However, at tax rates that are too high (in this case 25 percent) this deposit method will no longer cover all the costs of decommissioning at planned shutdown since the fund will shrink with time in real terms (the inflation rate exceeds the effective interest rate). The effect of increasing deductions is also the cause of the downturn in the cost curve of the deposit with earnings return method of funding (D RTN).

The tax rate, at least in the range of 0 to 50 percent, does not affect the cost ranking of the funding methods for methods of similar equitability. In the first group of equitability, the

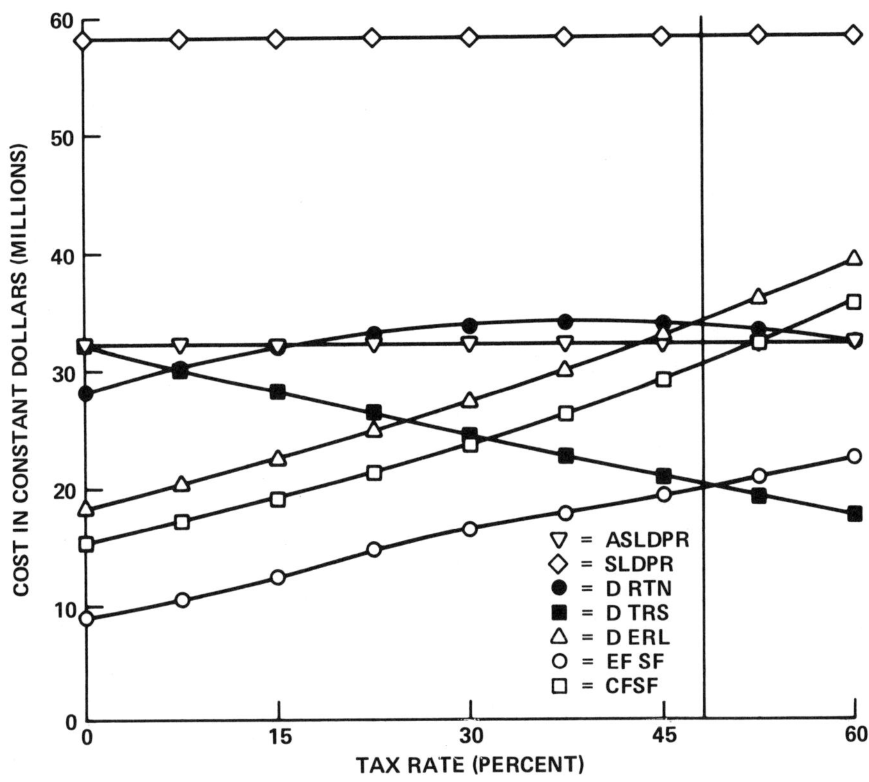

Figure 3. Tax-Rate Dependence of Ratepayer Costs.

adjusted depreciation method is more expensive than the escalating-fee sinking fund. In the second group of equitability; constant-fee sinking funding, deposit funding to reach cost at decommissioning, deposit funding at cost at startup (where the method is effective), and straight-line depreciation funding are, respectively, the least to most expensive. The deposit with earnings return is, in this case, about as expensive as the adjusted depreciation funding method.

Assurance

Figure 4 shows the relative amount of assurance for each of the seven funding variations for the case of a facility, such as a reactor, whose decommissioning costs are independent of the length of operation. These numbers reflect only the "book value" of the funds and do not necessarily reflect the actual level of assurance. For instance, sinking funds and deposits have generally greater assurance than a depreciation method at the same level since they are real funds and not merely accounting proceedures within the operator's finances. In the figure, 100 percent assurance means the fund value will completely cover all the costs of decommissioning. These costs are inflating exponentially with time. The figure shows the growth of each fund during the lifetime of the facility, compared to the inflated costs of decommissioning.

Figure 4 shows that, in general, the deposit funding methods provide the highest level of assurance that decommissioning costs will be paid at either an early or late decommissioning. The negative-salvage-value depreciation method has a higher level of assurance against early decommissioning than does the sinking fund method. This occurs because the sinking fund is expected to accumulate interest, and so there is less money available than in the depreciation method for early decommissioning. The opposite holds true if the facility operates longer than planned (the sinking fund provides more assurance than the depreciation method since the sinking fund is earning interest beyond that which is needed). It should be noted that the latter conclusions are valid only when the depreciation method is not credited with some capitalization or interest. If credit is included, the depreciation method essentially becomes a sinking fund held by the operator; and the method with the higher interest rate will have less assurance against early decommissioning than the other method will have.

Figure 4 (except for numerical values) is valid for any facility whose decommissioning costs (independent of inflation) do not depend on the period of operation of the facility. Figure 5 shows a similar graph for a shallow land, low-level waste burial

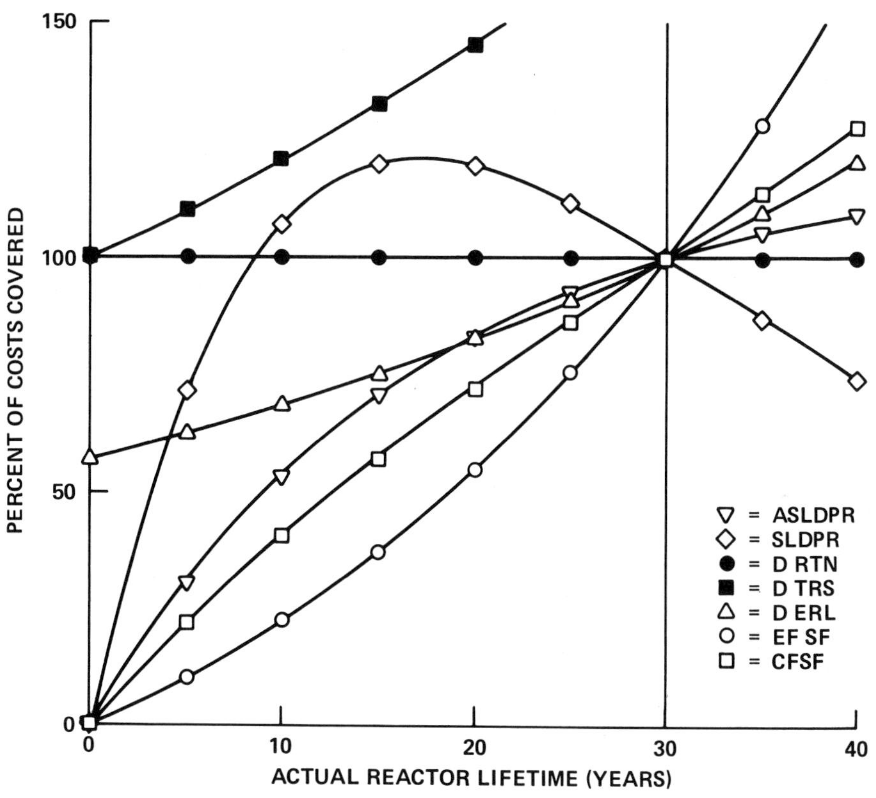

Figure 4. Assurance of Funding Methods.

site. This case is an example of the situation where the decommissioning costs not only depend on the length of time the facility operates, but on the level of use of the facility during operation. Decommissioning, in this instance, includes site stabilization, monitoring, and long-term care for an indeterminate length of time.

It is assumed that the decommissioning fund for the waste is derived from a volume surcharge on waste burial. A burial site could be closed early for administrative reasons and remain unfilled, or it could be filled at a faster rate than was origionally planned. Both these cases for a site planned to operate 20 years, with ongoing active trench reworking, and using a sinking fund funding method are shown in Figure 5.

If the site is filled at the expected rate (FILL 20), the effect of early closure of the site is similar to, but has slightly higher assurance than, the same funding methods would have if they were used for a reactor. This results from the smaller-than-planned area requiring less money to maintain during long-term care.

If the site is filled to capacity earlier than planned (FILL ASL) at the actual site lifetime, an escalating-fee sinking fund will fall short since the expected interest did not have time to accrue. In the case of early filling of the site a constant-fee sinking fund will have built more funds than necessary because the decommissioning charge for waste burial is higher in constant dollars earlier in the site life.

Additional Assurance

Additional assurance may be obtained through the purchase of surety bonds or decommissioning insurance. In both these methods, the facility operator pays an annual fee to another company. The fee is proportional to the shortfall that would occur if the facility shut down that year. The company receiving the fee would then agree to pay the shortfall in the event of early decommissioning. The costs of these modifications to funding methods are greatest during early facility life. These modifications are not analysed in detail in this paper. However, the effect of the use of surety bonds or insurance would be to increase the assurance, increase the cost, and decrease the equitability of the funding methods that they are applied to.

SUMMARY

In summary, one can say that the knowledge of the technical costs of decommissioning is useful when comparing different facilities with similar decommissioning modes. However, comparisons of costs between different modes of decommissioning as well as different methods of funding must take into account the existing economic environment of the facility.

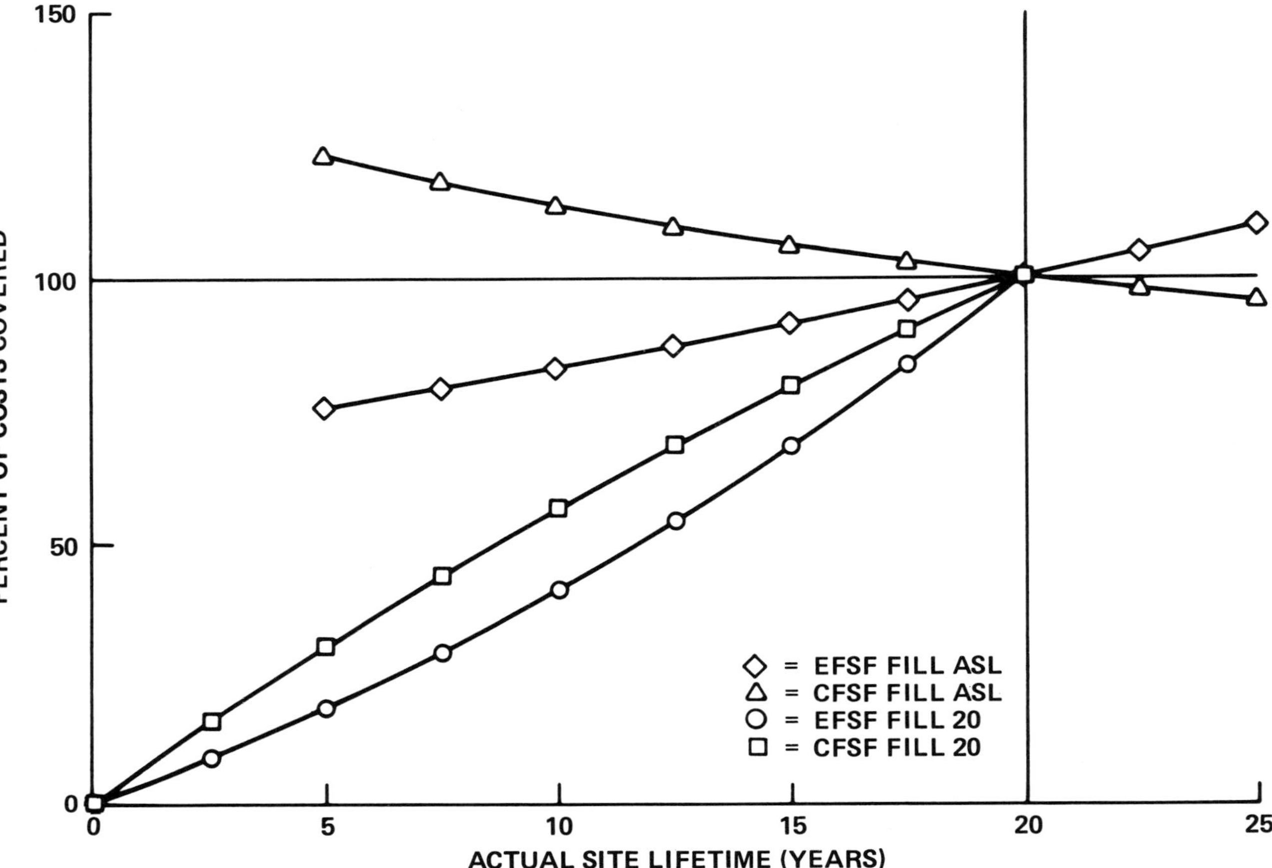

Figure 5. Early Low-Level Waste Site Closure.

REFERENCES

1. B.C. Mingst, NUREG-0514, "DECOST-Computer Routine for Decommissioning Cost and Funding Analysis." Nuclear Regulatory Commission, 1979.

2. NUREG/CR-0130, "Technology, Safety, and Costs of Decommissioning a Reference Pressurized Water Reactor Power Station," Battelle Pacific Northwest Laboratory for U.S. Nuclear Regulatory Commission, June 1978.

3. "Report of the Special Advisory Committee on Nuclear Waste Disposal," No. 142, Kentucky, October 1977.

FINANCIAL ASPECTS OF POWER REACTOR DECOMMISSIONING

John S. Ferguson

Senior Vice President
Middle West Service Company
Dallas, Texas

Annotation

Discussion of financial aspects of decommissioning determined by capital recovery economics consistant with generally accepted depreciation practices and regulatory rules, with emphasis on differences from engineering economics and the significance of inflation and Federal Income Tax regulations.

FINANCIAL ASPECTS OF POWER REACTOR DECOMMISSIONING

As a Consultant to electric utilities, I often become involved in the development of policy for capital recovery and in the determination of the depreciation rates that will carry out the policy. Utility capital recovery economics is controlled by generally accepted depreciation accounting practices and by regulatory commission accounting rules and, as a result, differs significantly from engineering economics. My discussion here will be nuclear decommissioning costs that are required to be handled through utility capital recovery and several related problems that must be faced by utility financial managers and regulators.

PURPOSE OF DEPRECIATION

The Uniform System of Accounts prescribed for electric utilities by the Federal Energy Regulatory Commission and followed by most utilities states that depreciation "as applied to depreciable electric plant, means the loss in service value not restored by current maintenance, incurred in connection with the consumption or prospective retirement of electric plant in the course of service from causes which are known to be in current operation and against which the utility is not protected by insurance. Among the causes to be given consideration are: wear and tear, decay, action of the elements, inadequacy, obsolescence, changes in the art, changes in demand and requirements of public authorities. Service value means the difference between original cost and net salvage value of electric plant."

Depreciation accounting is an allocation process whereby consumption of physical assets is recognized in the income statement of a business enterprise. The purpose of depreciation expense is to provide full recovery of invested capital, and net salvage to be incurred at the time the facilities are decommissioned, over the expected life of the facilities constructed with that capital from those customers receiving benefits from the facilities.

There are three essential aspects to the determination of the depreciation rates: the amount of capital to be recovered; the period of time for recovery; and the pattern of recovery. Of concern here is the determination of the amount to be recovered. Table 1 shows accrual rate calculation formulae for the whole life rate method as well as the remaining life rate method. Decommissioning cost is recognized in the net salvage factor in these formulae. Since nuclear plant cost of removal exceeds gross salvage, net salvage is negative. Net salvage in these formulae is expressed as a percentage of the original construction cost (100) of the facility. Thus, if the average service life is 10 years and salvage 0%, the accrual rate is 10%; whereas, if the salvage is negative 50% the accrual rate is 15%, an increase of 50%.

Perhaps one of the least understood aspects of capital recovery economics is the fact that depreciation accounting practices and regulatory commission rules require that the net salvage to be either received or incurred at the end of life (at the price level at that time) is what must be built into the depreciation accrual rates. It is hard enough to estimate the decommissioning cost today, yet we must attempt to estimate when the cost will be incurred and the price level that will exist at that time. Even though it is difficult to do, expenditure timing and price level must be determined.

REGULATORY CONSTRAINTS

Regulatory bodies are political entities and, whether we like it or not, make political decisions. Regulators have a responsibility to ensure that the utilities are financially viable, as that is the only way adequate service can be provided to customers. However, regulators have a revenue requirement bias. By this I mean that regulators like things that decrease the rates to customers and do not like things that increase rates to customers. Since negative salvage (decommissioning costs) increase rates to customers, by definition, regulators don't like it. Most regulators have not faced up to the need to build sufficient negative net salvage into depreciation rates to adequately provide for decommissioning; and, until recently they have had some fairly logical reasons for doing so. The first really useful decommissioning cost estimates did not start appearing until late 1976 (the AIF study).[1] Therefore, regulators were able to state that they recognize the need, but don't have an adequate cost estimate to use as the basis for setting the depreciation rates. Regulators no longer have this excuse. At this point I must say that the utility industry also had a tendency to ignore the issue.

This situation no longer exists, as now the Battelle studies[2] are being issued. A number of utilities have made detailed analyses as to the applicability of these studies to their facilities.

ENGINEERING ECONOMICS VERSUS CAPITAL RECOVERY

Figure 1 is a comparison of the requirements for engineering economics studies with those of capital recovery. In this example decommissioning consists of three activities. The first, preparation for safe storage, would occur at the end of the life of the plant, and would cost $7 million at the 1976 price level and with 5% annual inflation, $38.6 million at the time it was spent in the year 2011. Annual surveillance would start at that time and would last 25 years, costing $80 thousand per year at the 1976 price level for a total of $2 million at that price level and, with 47.5 years of inflation, would average out to a total cost of $20.3 million in the year 2023.5. Removal would occur at the end of the surveillance period, costing $16 million at the 1976 price level and $298.9 million at the time it would actually be incurred in the year 2036. Using an engineering economics approach, a 9% present worth factor brings the costs to be incurred beyond the year 2011 back to an equivalent 2011 cost. The resulting total is $80.2 million. The annual annuity amount, using 9% interest is $372 thousand per year, and is used to calculate a depreciation rate.

The capital recovery approach could be quite similar, if the $80.2 million were assumed to be invested in the year 2011 and earn 9% interest that is not taxable. The funds thus accumulated would pay the decommissioning costs as they were incurred. With this earnings constraint, the only major difference between engineering economics and capital recovery is the annual amount required to accumulate the $80.2 million. Since this amount would be collected from customers over a 35-year period, the capital recovery requirements would be $2.3 million each year rather than $372 thousand. The $2.3 million is predicated upon untaxed earnings of 9%; highly unrealistic. More realistic would be a 46% tax rate (4.86% after tax earnings) or net earnings equal to inflation. Sensitivity to tax is illustrated when it is realized that $34.70 million at 9% becomes $88.3 million at 5%.

I did not invent this study. It was presented in testimony before a state regulatory body in support of a depreciation rate to use in a feasibility study comparing nuclear and fossil fuel generation sources. Other engineering economics studies comparing the nuclear and fossil fuel alternatives have done such things as to use the same fixed charge rate for both alternatives and to assume that decommissioning costs were expensed as they were incurred.

Utility financial managers tend to get a bit shook up if they find that looking at decommissioning costs from a capital recovery standpoint results in significantly different numbers than looking at it from an engineering economics standpoint. Those involved in the evaluations of nuclear and fossil fuel alternatives in the environmental impact statements now required for licensing should be careful to ensure their studies correctly recognize capital recovery concepts and their income tax ramifications.

DECOMMISSIONING COSTS

In order to more adequately illustrate the impact of inflation than shows up on Figure 1, I have taken decommissioning cost estimates from the AIF report, recognizing contingencies (at 25%), cost savings due to multiple reactors at the same site, and the expenditure timing differences to estimate the actual costs to be incurred. Costs in the AIF report at the 1975 price level were inflated to price levels at the time of expenditure using annual inflation rates of 7.5% through 1985 and 6% per year beyond. The calculations were made for a boiling water reactor plant having two 1178MWe units that would reach the end of their operating life January 1, 2015. The example expenditure pattern illustrated on Figure 2 is for initial mothballing with complete removal after a 104-year cool down period. During 2015, $68.6 million would be spent in the initial mothballing process. Surveillance would begin in 2016, with an expenditure of $5.2 million in that year, and would increase to an annual expenditure of $2.2 billion in 2120. Complete removal would cost nearly $150 billion. The impact of inflation on surveillance and removal cost is shocking, to say the least. The uncertainties surrounding a 104-year mothball period are formidible. These uncertainties aside, the impact of inflation and the difficulties of adequately providing for expenditure to be made far in the future make a good case for prompt removal.

Table 1. Accrual Rate Calculation.

Whole life rates are calculated as follows:

$$\text{Rate} = \frac{100 - S}{ASL}$$

where S is net salvage in per cent, and ASL is average service life.

Remaining life rates are calculated as follows:

$$\text{Rate} = \frac{100 - S - \frac{R \times 100}{B}}{ARL}$$

where B is depreciable plant balance, S is net salvage in per cent, R is book depreciation reserve, and ARL is average remaining life.

Table 2. Fund Required at End of Plant Life.

	Plan A $1000	Plan B $1000	Plan C $1000
AIF Report			
Mothball-Delayed Removal	118,394	1,450,161	473,409
Entombment-Delayed Removal	138,910	1,133,649	370,945
Prompt Removal	365,660	414,962	397,593
Battelle Report			
Mothball-Delayed Removal			
100 year delay	160,253	1,558,619	502,539
50 year delay	213,115	801,877	465,735
30 year delay	293,937	720,922	516,220
Prompt Removal	368,937	418,681	401,157

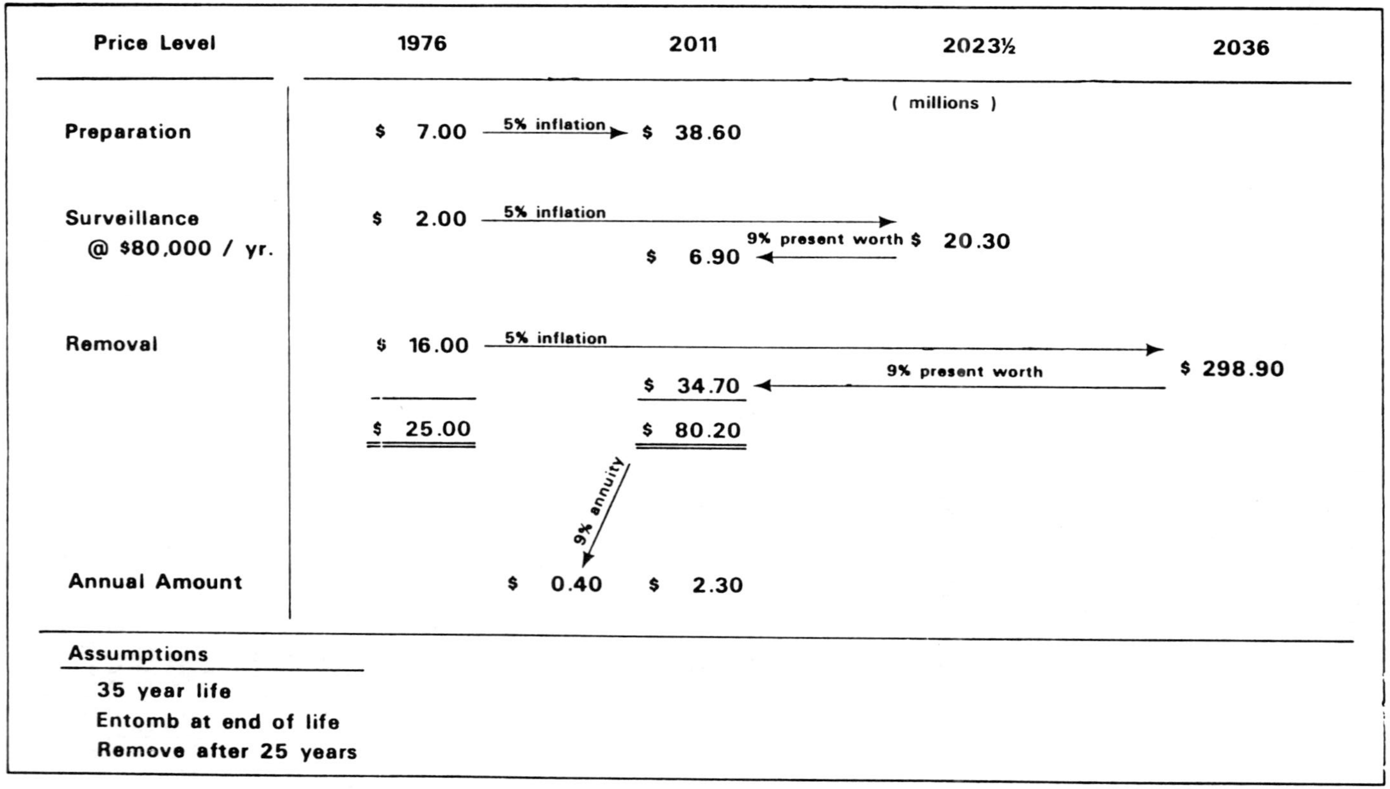

Price Level	1976		2011		2023½	2036
					(millions)	
Preparation	$ 7.00	5% inflation →	$ 38.60			
Surveillance @ $80,000 / yr.	$ 2.00	5% inflation →				
			$ 6.90	← 9% present worth	$ 20.30	
Removal	$ 16.00	5% inflation →				
			$ 34.70	← 9% present worth		$ 298.90
	$ 25.00		$ 80.20			
			9% annuity ↙			
Annual Amount		$ 0.40	$ 2.30			

Assumptions

35 year life
Entomb at end of life
Remove after 25 years

Figure 1. Example of Engineering Approach to Depreciation.

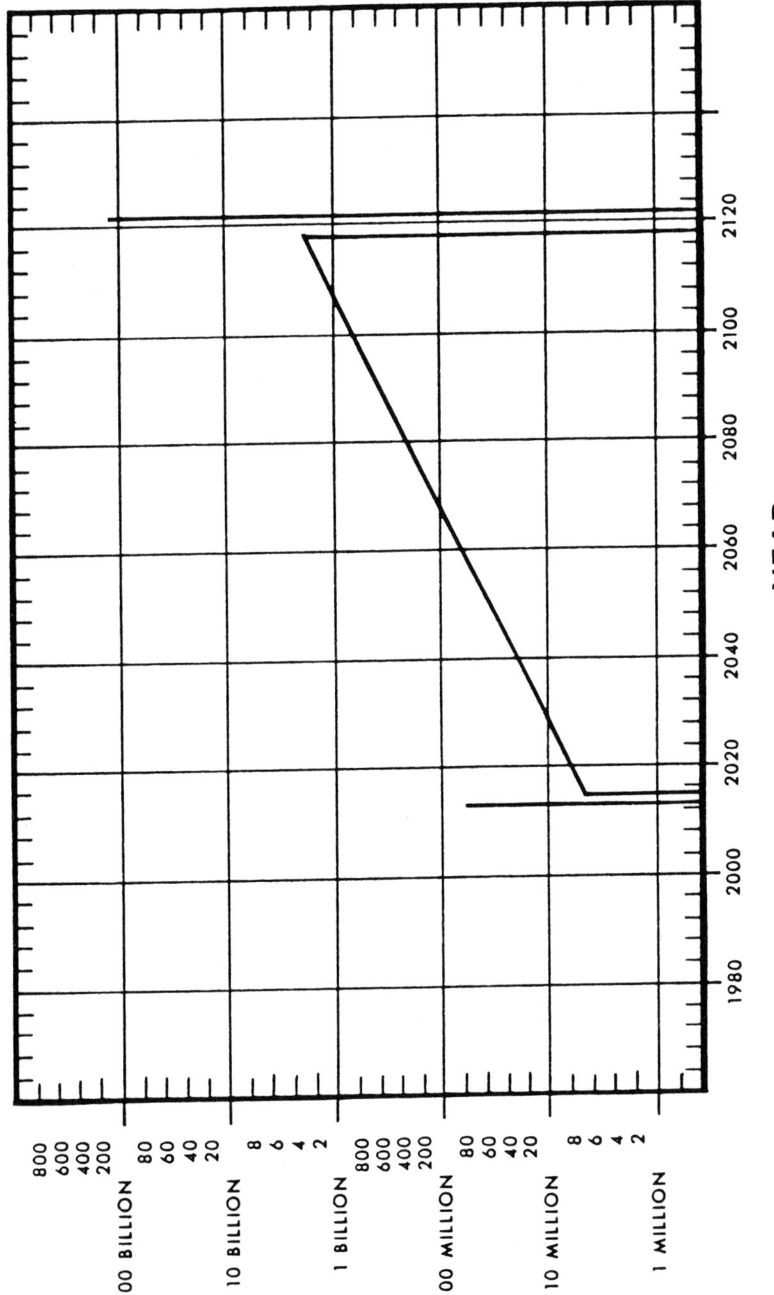

Figure 2. Expenditure for Initial Mothball and Removal After Cooldown Period.

FUND AVAILABILITY

Utility financial managers and regulators are concerned about the availability of the funds required for decommissioning when the decommissioning process must occur. We have previously discussed the impact of inflation on the magnitude of these funds, but there is another hooker in this thing. Current Federal income tax regulations say that removal costs are not a tax deduction until incurred. Therefore, any collections from customers to build up a fund to accomplish the decommissioning are taxable income to the utilities. So for every dollar of decommissioning cost required to be collected, customers have to pay two dollars.

If a delayed removal process is selected for accomplishing the decommissioning, it makes sense to have a fund available at the end of plant life that can be invested to provide earnings that will reduce the apparent amount that must be collected from customers. Investing could also be done as the collections are made, which would further reduce the apparent amount to be collected from customers. Capital recovery funds collected from customers have historically been reinvested by the utilities, reducing the magnitude of their future borrowings. If collections are invested as they are received, use of these funds may be denied, which is a serious deterrent to a utility supporting a funding method. If collections are funded (invested) at the end of plant life, the utility must sell securities to raise the cash for setting up the fund at that time, which may also be a problem.

IMPLICATIONS OF INCOME TAXES

In any case, current Federal income tax law has a significant effect on earnings of a decommissioning fund. Unless these earnings are from tax exempt securities, they will be subject to Federal and perhaps state income tax. Income tax has a significant effect on magnitude of the collections required from the customers. Table 2 shows the funds required at the end of plant life to carry out several different processes of decommissioning a PWR, under three different assumptions concerning the earnings of the fund and their taxability. The figures are from the AIF and Battelle studies, and I have added inflation and the impact of fund earnings. Plan A assumes that the fund is invested in bonds earning 9% not subject to income tax. Plan B assumes that the fund is invested in the same bonds, but that they are subject to a 50% income tax. Plan C assumes that the fund is invested in tax-free municipal bonds with earnings exactly equal to future inflation. You will note that under Plan A, which is not consistent with current tax law, the mothball-delayed removal process is the least expensive of the three decommissioning processes. Under Plan B, which is consistent with current tax law, the mothball process is the most expensive, having increased in cost by a factor of 12 due soley to the impact of income taxes. It is obvious that there are some very serious income tax consequences to decommissioning. Legislation will probably be required to correct them. The figures on Table 2 also make a case from prompt removal.

Conventional wisdom, generated by engineering economics, has it that mothball-delayed removal is the least expensive decommissioning process. The figures on Table 2 illustrate the probable source of this wisdom - failure to recognize capital recovery requirements and their attendant tax consequences.

CONCLUSION

The technical and safety aspects of nuclear generation are enough to make it a controversial and politically sensitive subject. The impact of inflation and federal income tax regulations on the capital recovery requirements for decommissioning further adds to the controversy. The following factors combine to make the capital recovery aspects a significant problem for utility financial managers and regulators:

1. The accounting requirements of capital recovery have evolved over a number of years and are well defined, both in terms of generally accepted accounting practices and regulatory precedent and law.

2. Utility financial managers must provide the cash necessary to accomplish decommissioning, and customers are the source of this cash.

3. Income taxes are a significant component of the revenue requirements to be evaluated by a utility in determining the method and magnitude of capital recovery and by regulators in setting rates.

4. Engineering economic studies often fail to recognize capital recovery requirements, including income tax, and the impact of capital recovery on other components of revenue requirements.

5. Regulators operate under political constraints that make it difficult for them to respond to increased revenue requirements.

The financial aspect of decommissioning nuclear generating plants is an issue whose time is come. The accounting and regulatory rules for handling the issue are in place, and the industry needs to learn how to make the rules operate satisfactorily, recognizing the factors mentioned above. Serious consideration should be given to what is known as the sinking fund method of depreciation and which has seen little use in recent years, due to the complicating impact of income taxes.

REFERENCES

1. An Engineering Evaluation of Nuclear Power Reactor Decommissioning Alternatives, November, 1976, NATIONAL ENVIRONMENTAL STUDIES PROJECT, Atomic Industrial Forum, Inc.
2. Technology, Safety and Costs of Decommissioning a Reference Pressurized Water Reactor Power Station, June 1978, NUREG/CR-0130, BATTELLE PACIFIC NORTHWEST LABORATORY for U.S. Nuclear Regulatory Commission.

FINANCIAL AND ACCOUNTING ALTERNATIVES FOR THE RECOVERY OF NUCLEAR PLANT DECOMMISSIONING COSTS

Preston A. Collins

Gilbert/Commonwealth

Reading, Pennsylvania, U.S.A.

INTRODUCTION

Many views have been presented as to the best method of financing the decommissioning of nuclear power plants. It is the purpose of this presentation to explore these alternative methods and to demonstrate the effect each would have upon the utility's revenue requirement over the life of the plant.

The basic financing alternatives are presented first in their simplest version where income taxes are zero. Then the effects of the present income tax laws will be added and their effect determined.

The basic financing alternatives are:

1. Funding at plant commissioning
2. Funding of the decommissioning reserve
3. Funding at plant decommissioning

Since we are looking at decommissioning costs as an isolated component of the company's operations, the references to the balance sheet and income accounts are in relationship to the net effect upon there accounts, or a sort of superposition theorem of the accounting and revenue requirements.

BASIS OF COMPARISON

Since the method of financing is independent of the decommissioning method, the revenue requirements have been calculated on the basis of each unit of cost as estimated at the time the plant is placed in service. A series of tables has been prepared showing the balance in each affected balance sheet account at the end of each year of operation and the anticipated components of the revenue requirement during each year. A fixed set of assumptions has applied consisting of the following:

Plant Life	35 years
Rate of Return on Capital	10%
Rate of Inflation	8%
Rate of Growth of Fund	8%
Income Tax Rate	46%

The tables are referred to by "Case #". Other information included in the tables is present-worth value of the revenue requirements discounted to age zero using the inflation rate, the price-level cost of decommissioning at the end of each year, and the ratio of the reserve to the price-level cost of decommissioning at the end of each year. The total revenue requirement and the total present-worth revenue requirement are shown at the bottom of their respective columns. It is the data in these two columns upon which the selection of the best method should be based, the criteria being the smallest value of these revenue requirements and an equitable distribution of cost to the accounting periods.

Refer to the tables for Cases #1 through #4 and Figures 1, 2 and 3 on page 7 for a comparison of the depreciation expenses, revenue requirements and present-worth revenue requirements, respectively, over the life span of the plant, as each of the first four cases are discussed.

Case #1: Fund at Plant Commissioning, no Taxes

This alternative establishes a trust fund outside the control of the utility. The funds would be invested in securities of a low risk nature that hopefully would maintain their purchasing power. The amount established in the fund would equal the estimated cost of decommissioning at the time and be amortized over the plant's estimated remaining life. The earnings produced by the fund would be retained therein and be recorded in the asset account for the fund and credited to the reserve for decommissioning. As long as the currently estimated cost of decommissioning--equivalent to the price-level cost in the tables--remains the same as the assets of the fund, the revenue requirement would be as indicated in the table; however, a deviation of the fund below the price-level cost would require a supplementary component to the revenue requirement in order to bring the fund back to its appropriate level. Conversely, earnings producing an excess in the fund would reduce the revenue requirement.

The revenue requirement consists of two components--the amortization expense and the return on capital. Since capital is being

reduced in a straight line, the return on capital reduces proportionately. This is the same pattern experienced by the revenue requirement for investments in depreciable plant. From the table for Case #1, the total revenue requirement is 2.800 and the total present-worth revenue requirement is 1.167.

Case #2: Funding of the Decommissioning Reserve, no Taxes

This method establishes a fund in which revenues equal to the amortization expenses are deposited. The fund earnings are retained and recorded in the company's balance sheet account with the credit going to the reserve for decommissioning. Since there is no initial capital established, amortization expense is computed by dividing the price-level decommissioning cost at the end of the current year less the reserve for decommissioning by the remaining life at the beginning of the year. As in Case #1, the performance of the fund would feed back adjustments to the revenue requirement.

The revenue requirement consists solely of the amortization expense since there is no capital. Note that the present-worth of the revenue requirement is a constant whose total is 1.0 while the total revenue requirement is 5.317.

Case #3: Funding at Decommissioning, no Taxes

There is no fund established with this method and financing is done as part of the company's normal financing procedures. Amortization expense is computed in the same manner as Case #2. Under this method the revenues collected produce a debit balance in the capital account which amounts to a negative rate base. Therefore, a negative return on capital is produced to offset a portion of the amortization expense. The result is a declining revenue requirement until about age 10, then it increases at an ever-increasing rate. A look at the discounted values of the revenue requirement shows that it, too, decreases until about age 18 and then increases at an ever-increasing rate. The result is a total revenue requirement of 6.220 and a total discounted revenue requirement of 0.776.

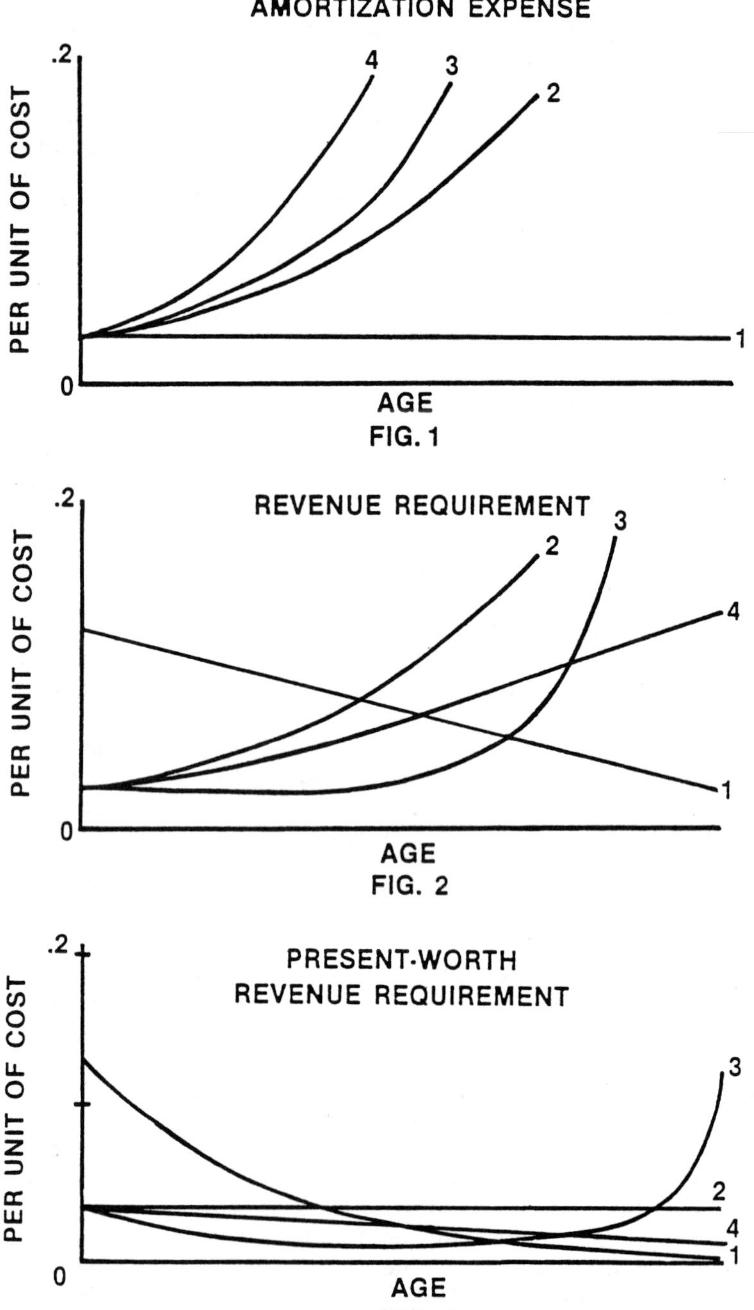

FIG. 1

FIG. 2

FIG. 3

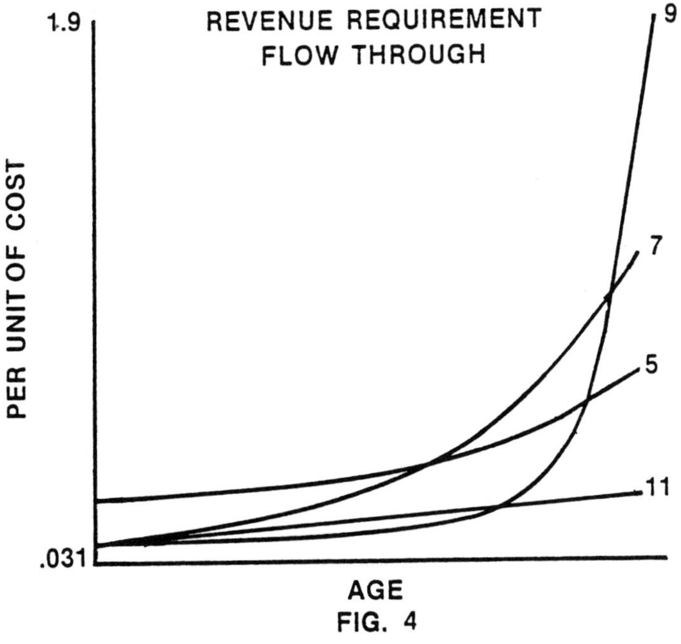
1.9
REVENUE REQUIREMENT
FLOW THROUGH
9
7
5
11
.031
PER UNIT OF COST
AGE

FIG. 4

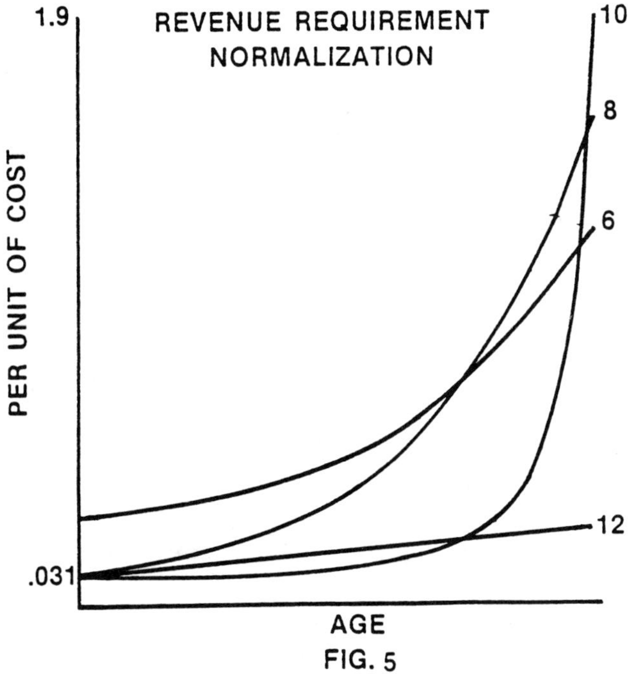
1.9
REVENUE REQUIREMENT
NORMALIZATION
10
8
6
12
.031
PER UNIT OF COST
AGE

FIG. 5

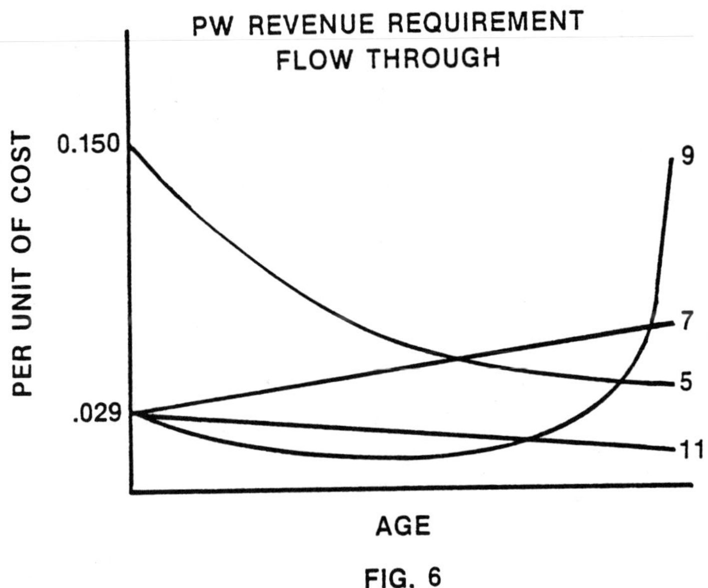
PW REVENUE REQUIREMENT
FLOW THROUGH
PER UNIT OF COST
0.150
.029
9
7
5
11
AGE

FIG. 6

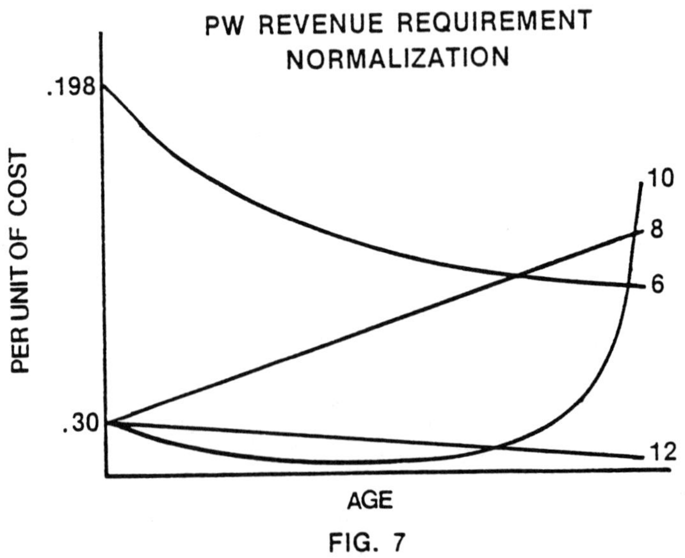
PW REVENUE REQUIREMENT
NORMALIZATION
PER UNIT OF COST
.198
.30
10
8
6
12
AGE

FIG. 7

CASE #1

FINANCING METHOD: FUND AT COMMISSIONING

TAX ACCOUNTING METHOD: NOT APPLICABLE

RATE OF INFLATION: 8.0 %
RATES OF RETURN:
CAPITAL 10.0 %
FUND 8.0 %

TAX RATES:
RETURN ON CAPITAL: 0.0 %
AMORTIZATION 0.0 %
FUND GROWTH 0.0 %

RESERVE COMPOUNDING RATE: 0.0 %

	BALANCE SHEET ACCOUNTS				INCOME ACCOUNTS						
AGE	FUND	CAPITAL	RESERVE	DEF TAXES	AMORT EXPENSE	TAX EXPENSE	RETURN	REVENUE	P W REVENUE	PR LEVEL COST	RATIO
0	1.000	-1.000									
1	1.080	-0.971	-0.109	0.0	0.029	0.0	0.100	-0.129	-0.119	1.080	0.101
2	1.166	-0.943	-0.224	0.0	0.029	0.0	0.097	-0.126	-0.108	1.166	0.192
3	1.260	-0.914	-0.345	0.0	0.029	0.0	0.094	-0.123	-0.098	1.260	0.274
4	1.360	-0.886	-0.475	0.0	0.029	0.0	0.091	-0.120	-0.088	1.360	0.349
5	1.469	-0.857	-0.612	0.0	0.029	0.0	0.089	-0.117	-0.080	1.469	0.417
6	1.587	-0.829	-0.758	0.0	0.029	0.0	0.086	-0.114	-0.072	1.587	0.478
7	1.714	-0.800	-0.914	0.0	0.029	0.0	0.083	-0.111	-0.065	1.714	0.533
8	1.851	-0.771	-1.080	0.0	0.029	0.0	0.080	-0.109	-0.059	1.851	0.583
9	1.999	-0.743	-1.256	0.0	0.029	0.0	0.077	-0.106	-0.053	1.999	0.628
10	2.159	-0.714	-1.445	0.0	0.029	0.0	0.074	-0.103	-0.048	2.159	0.669
11	2.332	-0.686	-1.646	0.0	0.029	0.0	0.071	-0.100	-0.043	2.332	0.706
12	2.518	-0.657	-1.861	0.0	0.029	0.0	0.069	-0.097	-0.039	2.518	0.739
13	2.720	-0.629	-2.091	0.0	0.029	0.0	0.066	-0.094	-0.035	2.720	0.769
14	2.937	-0.600	-2.337	0.0	0.029	0.0	0.063	-0.091	-0.031	2.937	0.796
15	3.172	-0.571	-2.601	0.0	0.029	0.0	0.060	-0.089	-0.028	3.172	0.820
16	3.426	-0.543	-2.883	0.0	0.029	0.0	0.057	-0.086	-0.025	3.426	0.842
17	3.700	-0.514	-3.186	0.0	0.029	0.0	0.054	-0.083	-0.022	3.700	0.861
18	3.996	-0.486	-3.510	0.0	0.029	0.0	0.051	-0.080	-0.020	3.996	0.878
19	4.316	-0.457	-3.859	0.0	0.029	0.0	0.049	-0.077	-0.018	4.316	0.894
20	4.661	-0.429	-4.232	0.0	0.029	0.0	0.046	-0.074	-0.016	4.661	0.908
21	5.034	-0.400	-4.634	0.0	0.029	0.0	0.043	-0.071	-0.014	5.034	0.921
22	5.437	-0.371	-5.065	0.0	0.029	0.0	0.040	-0.069	-0.013	5.437	0.932
23	5.871	-0.343	-5.529	0.0	0.029	0.0	0.037	-0.066	-0.011	5.871	0.942
24	6.341	-0.314	-6.027	0.0	0.029	0.0	0.034	-0.063	-0.010	6.341	0.950
25	6.848	-0.286	-6.563	0.0	0.029	0.0	0.031	-0.060	-0.009	6.848	0.958
26	7.396	-0.257	-7.139	0.0	0.029	0.0	0.029	-0.057	-0.008	7.396	0.965
27	7.988	-0.229	-7.759	0.0	0.029	0.0	0.026	-0.054	-0.007	7.988	0.971
28	8.627	-0.200	-8.427	0.0	0.029	0.0	0.023	-0.051	-0.006	8.627	0.977
29	9.317	-0.171	-9.146	0.0	0.029	0.0	0.020	-0.049	-0.005	9.317	0.982
30	10.063	-0.143	-9.920	0.0	0.029	0.0	0.017	-0.046	-0.005	10.063	0.986
31	10.868	-0.114	-10.753	0.0	0.029	0.0	0.014	-0.043	-0.004	10.868	0.989
32	11.737	-0.086	-11.651	0.0	0.029	0.0	0.011	-0.040	-0.003	11.737	0.993
33	12.676	-0.057	-12.619	0.0	0.029	0.0	0.009	-0.037	-0.003	12.676	0.995
34	13.690	-0.029	-13.662	0.0	0.029	0.0	0.006	-0.034	-0.003	13.690	0.998
35	14.785	-0.000	-14.785	0.0	0.029	0.0	0.003	-0.031	-0.002	14.785	1.000
							TOTAL	-2.800	-1.167		

CASE #2

FINANCING METHOD: FUND THE RESERVE

TAX ACCOUNTING METHOD: NOT APPLICABLE

RATE OF INFLATION: 8.0 %
RATES OF RETURN:
CAPITAL 10.0 %
FUND 8.0 %

TAX RATES:
RETURN ON CAPITAL: 0.0 %
AMORTIZATION 0.0 %
FUND GROWTH 0.0 %

RESERVE COMPOUNDING RATE: 0.0 %

	BALANCE SHEET ACCOUNTS				INCOME ACCOUNTS						
AGE	FUND	CAPITAL	RESERVE	DEF TAXES	AMORT EXPENSE	TAX EXPENSE	RETURN	REVENUE	P W REVENUE	PR LEVEL COST	RATIO
0	0.0	0.0									
1	0.031	0.0	-0.031	0.0	0.031	0.0	0.0	-0.031	-0.029	1.080	0.029
2	0.067	0.0	-0.067	0.0	0.033	0.0	0.0	-0.033	-0.029	1.166	0.057
3	0.108	0.0	-0.108	0.0	0.036	0.0	0.0	-0.036	-0.029	1.260	0.086
4	0.155	0.0	-0.155	0.0	0.039	0.0	0.0	-0.039	-0.029	1.360	0.114
5	0.210	0.0	-0.210	0.0	0.042	0.0	0.0	-0.042	-0.029	1.469	0.143
6	0.272	0.0	-0.272	0.0	0.045	0.0	0.0	-0.045	-0.029	1.587	0.171
7	0.343	0.0	-0.343	0.0	0.049	0.0	0.0	-0.049	-0.029	1.714	0.200
8	0.423	0.0	-0.423	0.0	0.053	0.0	0.0	-0.053	-0.029	1.851	0.229
9	0.514	0.0	-0.514	0.0	0.057	0.0	0.0	-0.057	-0.029	1.999	0.257
10	0.617	0.0	-0.617	0.0	0.062	0.0	0.0	-0.062	-0.029	2.159	0.286
11	0.733	0.0	-0.733	0.0	0.067	0.0	0.0	-0.067	-0.029	2.332	0.314
12	0.863	0.0	-0.863	0.0	0.072	0.0	0.0	-0.072	-0.029	2.518	0.343
13	1.010	0.0	-1.010	0.0	0.078	0.0	0.0	-0.078	-0.029	2.720	0.371
14	1.175	0.0	-1.175	0.0	0.084	0.0	0.0	-0.084	-0.029	2.937	0.400
15	1.360	0.0	-1.360	0.0	0.091	0.0	0.0	-0.091	-0.029	3.172	0.429
16	1.566	0.0	-1.566	0.0	0.098	0.0	0.0	-0.098	-0.029	3.426	0.457
17	1.797	0.0	-1.797	0.0	0.106	0.0	0.0	-0.106	-0.029	3.700	0.486
18	2.055	0.0	-2.055	0.0	0.114	0.0	0.0	-0.114	-0.029	3.996	0.514
19	2.343	0.0	-2.343	0.0	0.123	0.0	0.0	-0.123	-0.029	4.316	0.543
20	2.663	0.0	-2.663	0.0	0.133	0.0	0.0	-0.133	-0.029	4.661	0.571
21	3.020	0.0	-3.020	0.0	0.144	0.0	0.0	-0.144	-0.029	5.034	0.600
22	3.417	0.0	-3.417	0.0	0.155	0.0	0.0	-0.155	-0.029	5.437	0.629
23	3.858	0.0	-3.858	0.0	0.168	0.0	0.0	-0.168	-0.029	5.871	0.657
24	4.348	0.0	-4.348	0.0	0.181	0.0	0.0	-0.181	-0.029	6.341	0.686
25	4.892	0.0	-4.892	0.0	0.196	0.0	0.0	-0.196	-0.029	6.848	0.714
26	5.494	0.0	-5.494	0.0	0.211	0.0	0.0	-0.211	-0.029	7.396	0.743
27	6.162	0.0	-6.162	0.0	0.228	0.0	0.0	-0.228	-0.029	7.988	0.771
28	6.902	0.0	-6.902	0.0	0.246	0.0	0.0	-0.246	-0.029	8.627	0.800
29	7.720	0.0	-7.720	0.0	0.266	0.0	0.0	-0.266	-0.029	9.317	0.829
30	8.625	0.0	-8.625	0.0	0.288	0.0	0.0	-0.288	-0.029	10.063	0.857
31	9.626	0.0	-9.626	0.0	0.311	0.0	0.0	-0.311	-0.029	10.868	0.886
32	10.731	0.0	-10.731	0.0	0.335	0.0	0.0	-0.335	-0.029	11.737	0.914
33	11.952	0.0	-11.952	0.0	0.362	0.0	0.0	-0.362	-0.029	12.676	0.943
34	13.299	0.0	-13.299	0.0	0.391	0.0	0.0	-0.391	-0.029	13.690	0.971
35	14.785	0.0	-14.785	0.0	0.422	0.0	0.0	-0.422	-0.029	14.785	1.000
							TOTAL	-5.317	-1.000		

CASE #3

FINANCING METHOD: FUND AT DECOMMISSIONING

RATE OF INFLATION: 8.0 %
RATES OF RETURN:
CAPITAL 10.0 %
FUND 8.0 %

TAX ACCOUNTING METHOD: NOT APPLICABLE

TAX RATES:
RETURN ON CAPITAL: 0.0 %
AMORTIZATION 0.0 %
FUND GROWTH 0.0 %

RESERVE COMPOUNDING RATE: 0.0 %

	BALANCE SHEET ACCOUNTS				INCOME ACCOUNTS						
AGE	FUND	CAPITAL	RESERVE	DEF TAXES	AMORT EXPENSE	TAX EXPENSE	RETURN	REVENUE	P W REVENUE	PR LEVEL COST	RATIO
0	0.0	0.0									
1	0.0	0.031	-0.031	0.0	0.031	0.0	0.0	-0.031	-0.029	1.080	0.029
2	0.0	0.064	-0.064	0.0	0.033	0.0	-0.003	-0.030	-0.026	1.166	0.055
3	0.0	0.100	-0.100	0.0	0.036	0.0	-0.006	-0.030	-0.024	1.260	0.080
4	0.0	0.140	-0.140	0.0	0.039	0.0	-0.010	-0.029	-0.022	1.360	0.103
5	0.0	0.183	-0.183	0.0	0.043	0.0	-0.014	-0.029	-0.020	1.469	0.124
6	0.0	0.230	-0.230	0.0	0.047	0.0	-0.018	-0.029	-0.018	1.587	0.145
7	0.0	0.281	-0.281	0.0	0.051	0.0	-0.023	-0.028	-0.016	1.714	0.164
8	0.0	0.337	-0.337	0.0	0.056	0.0	-0.028	-0.028	-0.015	1.851	0.182
9	0.0	0.398	-0.398	0.0	0.062	0.0	-0.034	-0.028	-0.014	1.999	0.199
10	0.0	0.466	-0.466	0.0	0.068	0.0	-0.040	-0.028	-0.013	2.159	0.216
11	0.0	0.541	-0.541	0.0	0.075	0.0	-0.047	-0.028	-0.012	2.332	0.232
12	0.0	0.623	-0.623	0.0	0.082	0.0	-0.054	-0.028	-0.011	2.518	0.247
13	0.0	0.714	-0.714	0.0	0.091	0.0	-0.062	-0.029	-0.011	2.720	0.263
14	0.0	0.815	-0.815	0.0	0.101	0.0	-0.071	-0.030	-0.010	2.937	0.278
15	0.0	0.928	-0.928	0.0	0.112	0.0	-0.082	-0.031	-0.010	3.172	0.292
16	0.0	1.052	-1.052	0.0	0.125	0.0	-0.093	-0.032	-0.009	3.426	0.307
17	0.0	1.192	-1.192	0.0	0.139	0.0	-0.105	-0.034	-0.009	3.700	0.322
18	0.0	1.348	-1.348	0.0	0.156	0.0	-0.119	-0.037	-0.009	3.996	0.337
19	0.0	1.522	-1.522	0.0	0.175	0.0	-0.135	-0.040	-0.009	4.316	0.353
20	0.0	1.718	-1.718	0.0	0.196	0.0	-0.152	-0.044	-0.009	4.661	0.369
21	0.0	1.939	-1.939	0.0	0.221	0.0	-0.172	-0.049	-0.010	5.034	0.385
22	0.0	2.189	-2.189	0.0	0.250	0.0	-0.194	-0.056	-0.010	5.437	0.403
23	0.0	2.472	-2.472	0.0	0.283	0.0	-0.219	-0.064	-0.011	5.871	0.421
24	0.0	2.795	-2.795	0.0	0.322	0.0	-0.247	-0.075	-0.012	6.341	0.441
25	0.0	3.163	-3.163	0.0	0.369	0.0	-0.279	-0.089	-0.013	6.848	0.462
26	0.0	3.587	-3.587	0.0	0.423	0.0	-0.316	-0.107	-0.014	7.396	0.485
27	0.0	4.076	-4.076	0.0	0.489	0.0	-0.359	-0.130	-0.016	7.988	0.510
28	0.0	4.645	-4.645	0.0	0.569	0.0	-0.408	-0.161	-0.019	8.627	0.538
29	0.0	5.312	-5.312	0.0	0.668	0.0	-0.464	-0.203	-0.022	9.317	0.570
30	0.0	6.104	-6.104	0.0	0.792	0.0	-0.531	-0.261	-0.026	10.063	0.607
31	0.0	7.057	-7.057	0.0	0.953	0.0	-0.610	-0.342	-0.032	10.868	0.649
32	0.0	8.227	-8.227	0.0	1.170	0.0	-0.706	-0.464	-0.040	11.737	0.701
33	0.0	9.710	-9.710	0.0	1.483	0.0	-0.823	-0.660	-0.052	12.676	0.766
34	0.0	11.700	-11.700	0.0	1.990	0.0	-0.971	-1.019	-0.074	13.690	0.855
35	0.0	14.785	-14.785	0.0	3.085	0.0	-1.170	-1.915	-0.130	14.785	1.000
							TOTAL	-6.220	-0.776		

CASE #4

FINANCING METHOD: FUND AT DECOMMISSIONING

RATE OF INFLATION: 8.0 %
RATES OF RETURN:
CAPITAL 10.0 %
FUND 8.0 %

TAX ACCOUNTING METHOD: NOT APPLICABLE

TAX RATES:
RETURN ON CAPITAL: 0.0 %
AMORTIZATION 0.0 %
FUND GROWTH 0.0 %

RESERVE COMPOUNDING RATE: 8.0 %

	BALANCE SHEET ACCOUNTS				INCOME ACCOUNTS						
AGE	FUND	CAPITAL	RESERVE	DEF TAXES	AMORT EXPENSE	TAX EXPENSE	RETURN	REVENUE	P W REVENUE	PR LEVEL COST	RATIO
0	0.0	0.0									
1	0.0	0.031	-0.031	0.0	0.031	0.0	0.0	-0.031	-0.029	1.080	0.029
2	0.0	0.067	-0.067	0.0	0.036	0.0	-0.003	-0.033	-0.028	1.166	0.057
3	0.0	0.108	-0.108	0.0	0.041	0.0	-0.007	-0.035	-0.028	1.260	0.086
4	0.0	0.155	-0.155	0.0	0.048	0.0	-0.011	-0.037	-0.027	1.360	0.114
5	0.0	0.210	-0.210	0.0	0.054	0.0	-0.016	-0.039	-0.026	1.469	0.143
6	0.0	0.272	-0.272	0.0	0.062	0.0	-0.021	-0.041	-0.026	1.587	0.171
7	0.0	0.343	-0.343	0.0	0.071	0.0	-0.027	-0.044	-0.025	1.714	0.200
8	0.0	0.423	-0.423	0.0	0.080	0.0	-0.034	-0.046	-0.025	1.851	0.229
9	0.0	0.514	-0.514	0.0	0.091	0.0	-0.042	-0.049	-0.024	1.999	0.257
10	0.0	0.617	-0.617	0.0	0.103	0.0	-0.051	-0.051	-0.024	2.159	0.286
11	0.0	0.733	-0.733	0.0	0.116	0.0	-0.062	-0.054	-0.023	2.332	0.314
12	0.0	0.863	-0.863	0.0	0.131	0.0	-0.073	-0.057	-0.023	2.518	0.343
13	0.0	1.010	-1.010	0.0	0.147	0.0	-0.086	-0.060	-0.022	2.720	0.371
14	0.0	1.175	-1.175	0.0	0.165	0.0	-0.101	-0.064	-0.022	2.937	0.400
15	0.0	1.360	-1.360	0.0	0.185	0.0	-0.117	-0.067	-0.021	3.172	0.429
16	0.0	1.566	-1.566	0.0	0.207	0.0	-0.136	-0.071	-0.021	3.426	0.457
17	0.0	1.797	-1.797	0.0	0.231	0.0	-0.157	-0.074	-0.020	3.700	0.486
18	0.0	2.055	-2.055	0.0	0.258	0.0	-0.180	-0.078	-0.020	3.996	0.514
19	0.0	2.343	-2.343	0.0	0.288	0.0	-0.206	-0.082	-0.019	4.316	0.543
20	0.0	2.663	-2.663	0.0	0.321	0.0	-0.234	-0.086	-0.019	4.661	0.571
21	0.0	3.020	-3.020	0.0	0.357	0.0	-0.266	-0.091	-0.018	5.034	0.600
22	0.0	3.417	-3.417	0.0	0.397	0.0	-0.302	-0.095	-0.017	5.437	0.629
23	0.0	3.858	-3.858	0.0	0.441	0.0	-0.342	-0.099	-0.017	5.871	0.657
24	0.0	4.348	-4.348	0.0	0.490	0.0	-0.386	-0.104	-0.016	6.341	0.686
25	0.0	4.892	-4.892	0.0	0.544	0.0	-0.435	-0.109	-0.016	6.848	0.714
26	0.0	5.494	-5.494	0.0	0.603	0.0	-0.489	-0.113	-0.015	7.396	0.743
27	0.0	6.162	-6.162	0.0	0.668	0.0	-0.549	-0.118	-0.015	7.988	0.771
28	0.0	6.902	-6.902	0.0	0.739	0.0	-0.616	-0.123	-0.014	8.627	0.800
29	0.0	7.720	-7.720	0.0	0.818	0.0	-0.690	-0.128	-0.014	9.317	0.829
30	0.0	8.625	-8.625	0.0	0.905	0.0	-0.772	-0.133	-0.013	10.063	0.857
31	0.0	9.626	-9.626	0.0	1.001	0.0	-0.863	-0.138	-0.013	10.868	0.886
32	0.0	10.731	-10.731	0.0	1.105	0.0	-0.963	-0.143	-0.012	11.737	0.914
33	0.0	11.952	-11.952	0.0	1.221	0.0	-1.073	-0.148	-0.012	12.676	0.943
34	0.0	13.299	-13.299	0.0	1.347	0.0	-1.195	-0.152	-0.011	13.690	0.971
35	0.0	14.785	-14.785	0.0	1.486	0.0	-1.330	-0.156	-0.011	14.785	1.000
							TOTAL	-2.950	-0.685		

CASE #5

FINANCING METHOD: FUND AT COMMISSIONING

TAX ACCOUNTING METHOD: FLOW-THROUGH

RATE OF INFLATION: 8.0 %
RATES OF RETURN:
CAPITAL 10.0 %
FUND 8.0 %

TAX RATES:
RETURN ON CAPITAL: 46.0 %
AMORTIZATION 46.0 %
FUND GROWTH 46.0 %

RESERVE COMPOUNDING RATE: 0.0 %

AGE	BALANCE SHEET ACCOUNTS				INCOME ACCOUNTS						
	FUND	CAPITAL	RESERVE	DEF TAXES	AMORT EXPENSE	TAX EXPENSE	RETURN	REVENUE	P W REVENUE	PR LEVEL COST	RATIO
0	0.540	-0.540									
1	0.583	-0.525	-0.059	0.0	0.015	0.096	0.054	-0.165	-0.153	1.080	0.054
2	0.630	-0.509	-0.121	0.0	0.015	0.098	0.052	-0.165	-0.142	1.166	0.103
3	0.680	-0.494	-0.187	0.0	0.015	0.099	0.051	-0.166	-0.132	1.260	0.148
4	0.735	-0.478	-0.256	0.0	0.015	0.102	0.049	-0.166	-0.122	1.360	0.188
5	0.793	-0.463	-0.331	0.0	0.015	0.104	0.048	-0.167	-0.114	1.469	0.225
6	0.857	-0.447	-0.409	0.0	0.015	0.107	0.046	-0.168	-0.106	1.587	0.258
7	0.925	-0.432	-0.493	0.0	0.015	0.110	0.045	-0.170	-0.099	1.714	0.288
8	1.000	-0.417	-0.583	0.0	0.015	0.113	0.043	-0.172	-0.093	1.851	0.315
9	1.079	-0.401	-0.678	0.0	0.015	0.117	0.042	-0.174	-0.087	1.999	0.339
10	1.166	-0.386	-0.780	0.0	0.015	0.121	0.040	-0.176	-0.082	2.159	0.361
11	1.259	-0.370	-0.889	0.0	0.015	0.125	0.039	-0.179	-0.077	2.332	0.381
12	1.360	-0.355	-1.005	0.0	0.015	0.130	0.037	-0.183	-0.073	2.518	0.399
13	1.469	-0.339	-1.129	0.0	0.015	0.136	0.035	-0.187	-0.069	2.720	0.415
14	1.586	-0.324	-1.262	0.0	0.015	0.142	0.034	-0.192	-0.065	2.937	0.430
15	1.713	-0.309	-1.404	0.0	0.015	0.149	0.032	-0.197	-0.062	3.172	0.443
16	1.850	-0.293	-1.557	0.0	0.015	0.156	0.031	-0.202	-0.059	3.426	0.454
17	1.998	-0.278	-1.720	0.0	0.015	0.164	0.029	-0.209	-0.056	3.700	0.465
18	2.158	-0.262	-1.896	0.0	0.015	0.173	0.028	-0.216	-0.054	3.996	0.474
19	2.330	-0.247	-2.084	0.0	0.015	0.183	0.026	-0.224	-0.052	4.316	0.483
20	2.517	-0.231	-2.285	0.0	0.015	0.193	0.025	-0.233	-0.050	4.661	0.490
21	2.718	-0.216	-2.502	0.0	0.015	0.204	0.023	-0.243	-0.048	5.034	0.497
22	2.936	-0.201	-2.735	0.0	0.015	0.217	0.022	-0.254	-0.047	5.437	0.503
23	3.171	-0.185	-2.985	0.0	0.015	0.230	0.020	-0.266	-0.045	5.871	0.508
24	3.424	-0.170	-3.255	0.0	0.015	0.245	0.019	-0.279	-0.044	6.341	0.513
25	3.698	-0.154	-3.544	0.0	0.015	0.261	0.017	-0.293	-0.043	6.848	0.517
26	3.994	-0.139	-3.855	0.0	0.015	0.278	0.015	-0.309	-0.042	7.396	0.521
27	4.314	-0.123	-4.190	0.0	0.015	0.297	0.014	-0.326	-0.041	7.988	0.525
28	4.659	-0.108	-4.551	0.0	0.015	0.318	0.012	-0.345	-0.040	8.627	0.527
29	5.031	-0.093	-4.939	0.0	0.015	0.340	0.011	-0.366	-0.039	9.317	0.530
30	5.434	-0.077	-5.357	0.0	0.015	0.364	0.009	-0.389	-0.039	10.063	0.532
31	5.869	-0.062	-5.807	0.0	0.015	0.390	0.008	-0.413	-0.038	10.868	0.534
32	6.338	-0.046	-6.292	0.0	0.015	0.418	0.006	-0.440	-0.037	11.737	0.536
33	6.845	-0.031	-6.814	0.0	0.015	0.449	0.005	-0.469	-0.037	12.676	0.538
34	7.393	-0.015	-7.377	0.0	0.015	0.482	0.003	-0.501	-0.037	13.690	0.539
35	7.984	-0.000	-7.984	0.0	0.015	0.518	0.002	-0.535	-0.036	14.785	0.540
							TOTAL	-9.141	-2.359		

CASE #6

FINANCING METHOD: FUND AT COMMISSIONING

TAX ACCOUNTING METHOD: NORMALIZATION

RATE OF INFLATION: 8.0 %
RATES OF RETURN:
CAPITAL 10.0 %
FUND 8.0 %

TAX RATES:
RETURN ON CAPITAL: 46.0 %
AMORTIZATION 46.0 %
FUND GROWTH 46.0 %

RESERVE COMPOUNDING RATE: 0.0 %

AGE	BALANCE SHEET ACCOUNTS				INCOME ACCOUNTS						
	FUND	CAPITAL	RESERVE	DEF TAXES	AMORT EXPENSE	TAX EXPENSE	RETURN	REVENUE	P W REVENUE	PR LEVEL COST	RATIO
0	1.000	-1.000									
1	1.080	-1.021	-0.109	0.050	0.029	0.085	0.100	-0.214	-0.198	1.080	0.101
2	1.166	-1.046	-0.224	0.103	0.029	0.087	0.102	-0.218	-0.187	1.166	0.192
3	1.260	-1.073	-0.345	0.159	0.029	0.089	0.105	-0.222	-0.176	1.260	0.274
4	1.360	-1.104	-0.475	0.218	0.029	0.091	0.107	-0.227	-0.167	1.360	0.349
5	1.469	-1.139	-0.612	0.282	0.029	0.094	0.110	-0.233	-0.159	1.469	0.417
6	1.587	-1.177	-0.758	0.349	0.029	0.097	0.114	-0.239	-0.151	1.587	0.478
7	1.714	-1.220	-0.914	0.420	0.029	0.100	0.118	-0.247	-0.144	1.714	0.533
8	1.851	-1.268	-1.080	0.497	0.029	0.104	0.122	-0.255	-0.138	1.851	0.583
9	1.999	-1.321	-1.256	0.578	0.029	0.108	0.127	-0.263	-0.134	1.999	0.628
10	2.159	-1.379	-1.445	0.665	0.029	0.113	0.132	-0.273	-0.127	2.159	0.669
11	2.332	-1.443	-1.646	0.757	0.029	0.117	0.138	-0.284	-0.122	2.332	0.706
12	2.518	-1.513	-1.861	0.856	0.029	0.123	0.144	-0.296	-0.117	2.518	0.739
13	2.720	-1.590	-2.091	0.962	0.029	0.129	0.151	-0.309	-0.114	2.720	0.769
14	2.937	-1.675	-2.337	1.075	0.029	0.135	0.159	-0.323	-0.110	2.937	0.796
15	3.172	-1.768	-2.601	1.196	0.029	0.143	0.168	-0.339	-0.107	3.172	0.820
16	3.426	-1.869	-2.883	1.326	0.029	0.151	0.177	-0.356	-0.104	3.426	0.842
17	3.700	-1.980	-3.186	1.465	0.029	0.159	0.187	-0.375	-0.101	3.700	0.861
18	3.996	-2.100	-3.510	1.615	0.029	0.169	0.198	-0.395	-0.099	3.996	0.878
19	4.316	-2.232	-3.859	1.775	0.029	0.179	0.210	-0.418	-0.097	4.316	0.894
20	4.661	-2.375	-4.232	1.947	0.029	0.190	0.223	-0.442	-0.095	4.661	0.908
21	5.034	-2.532	-4.634	2.132	0.029	0.202	0.238	-0.468	-0.093	5.034	0.921
22	5.437	-2.701	-5.065	2.330	0.029	0.216	0.253	-0.497	-0.091	5.437	0.932
23	5.871	-2.886	-5.529	2.543	0.029	0.230	0.270	-0.529	-0.090	5.871	0.942
24	6.341	-3.087	-6.027	2.772	0.029	0.246	0.289	-0.563	-0.089	6.341	0.950
25	6.848	-3.305	-6.563	3.019	0.029	0.263	0.309	-0.600	-0.088	6.848	0.958
26	7.396	-3.541	-7.139	3.284	0.029	0.282	0.330	-0.641	-0.087	7.396	0.965
27	7.988	-3.798	-7.759	3.569	0.029	0.302	0.354	-0.684	-0.086	7.988	0.971
28	8.627	-4.076	-8.427	3.876	0.029	0.324	0.380	-0.732	-0.085	8.627	0.977
29	9.317	-4.379	-9.146	4.207	0.029	0.347	0.408	-0.783	-0.084	9.317	0.982
30	10.063	-4.706	-9.920	4.563	0.029	0.373	0.438	-0.839	-0.083	10.063	0.986
31	10.868	-5.061	-10.753	4.947	0.029	0.401	0.471	-0.900	-0.083	10.868	0.989
32	11.737	-5.445	-11.651	5.360	0.029	0.431	0.506	-0.966	-0.082	11.737	0.993
33	12.676	-5.862	-12.619	5.805	0.029	0.464	0.545	-1.037	-0.082	12.676	0.995
34	13.690	-6.313	-13.662	6.284	0.029	0.499	0.586	-1.114	-0.081	13.690	0.998
35	14.785	-6.801	-14.785	6.801	0.029	0.538	0.631	-1.198	-0.081	14.785	1.000
							TOTAL	-17.479	-3.927		

CASE #7

FINANCING METHOD: FUND THE RESERVE

TAX ACCOUNTING METHOD: FLOW-THROUGH

RATE OF INFLATION: 8.0 %
RATES OF RETURN:
CAPITAL 10.0 %
FUND 8.0 %

TAX RATES:
RETURN ON CAPITAL: 46.0 %
AMORTIZATION 46.0 %
FUND GROWTH 46.0 %

RESERVE COMPOUNDING RATE: 0.0 %

	BALANCE SHEET ACCOUNTS				INCOME ACCOUNTS						
AGE	FUND	CAPITAL	RESERVE	DEF TAXES	AMORT EXPENSE	TAX EXPENSE	RETURN	REVENUE	P W REVENUE	PR LEVEL COST	RATIO
0	0.0	0.0									
1	0.017	0.0	-0.017	0.0	0.017	0.014	0.0	-0.031	-0.029	1.080	0.015
2	0.036	0.0	-0.036	0.0	0.018	0.016	0.0	-0.034	-0.030	1.166	0.031
3	0.058	0.0	-0.058	0.0	0.019	0.019	0.0	-0.038	-0.031	1.260	0.046
4	0.084	0.0	-0.084	0.0	0.021	0.022	0.0	-0.043	-0.031	1.360	0.062
5	0.113	0.0	-0.113	0.0	0.023	0.025	0.0	-0.048	-0.032	1.469	0.077
6	0.147	0.0	-0.147	0.0	0.024	0.029	0.0	-0.053	-0.033	1.587	0.093
7	0.185	0.0	-0.185	0.0	0.026	0.033	0.0	-0.059	-0.034	1.714	0.108
8	0.228	0.0	-0.228	0.0	0.029	0.037	0.0	-0.065	-0.035	1.851	0.123
9	0.278	0.0	-0.278	0.0	0.031	0.042	0.0	-0.073	-0.036	1.999	0.139
10	0.333	0.0	-0.333	0.0	0.033	0.047	0.0	-0.081	-0.037	2.159	0.154
11	0.396	0.0	-0.396	0.0	0.036	0.053	0.0	-0.089	-0.038	2.332	0.170
12	0.466	0.0	-0.466	0.0	0.039	0.060	0.0	-0.099	-0.039	2.518	0.185
13	0.545	0.0	-0.545	0.0	0.042	0.068	0.0	-0.109	-0.040	2.720	0.201
14	0.634	0.0	-0.634	0.0	0.045	0.076	0.0	-0.121	-0.041	2.937	0.216
15	0.734	0.0	-0.734	0.0	0.049	0.085	0.0	-0.134	-0.042	3.172	0.231
16	0.846	0.0	-0.846	0.0	0.053	0.095	0.0	-0.148	-0.043	3.426	0.247
17	0.970	0.0	-0.970	0.0	0.057	0.106	0.0	-0.163	-0.044	3.700	0.262
18	1.110	0.0	-1.110	0.0	0.062	0.119	0.0	-0.180	-0.045	3.996	0.278
19	1.265	0.0	-1.265	0.0	0.067	0.132	0.0	-0.199	-0.046	4.316	0.293
20	1.438	0.0	-1.438	0.0	0.072	0.147	0.0	-0.219	-0.047	4.661	0.309
21	1.631	0.0	-1.631	0.0	0.078	0.164	0.0	-0.242	-0.048	5.034	0.324
22	1.845	0.0	-1.845	0.0	0.084	0.183	0.0	-0.266	-0.049	5.437	0.339
23	2.084	0.0	-2.084	0.0	0.091	0.203	0.0	-0.294	-0.050	5.871	0.355
24	2.348	0.0	-2.348	0.0	0.098	0.225	0.0	-0.323	-0.051	6.341	0.370
25	2.642	0.0	-2.642	0.0	0.106	0.250	0.0	-0.356	-0.052	6.848	0.386
26	2.967	0.0	-2.967	0.0	0.114	0.277	0.0	-0.391	-0.053	7.396	0.401
27	3.328	0.0	-3.328	0.0	0.123	0.307	0.0	-0.430	-0.054	7.988	0.417
28	3.727	0.0	-3.727	0.0	0.133	0.340	0.0	-0.473	-0.055	8.627	0.432
29	4.169	0.0	-4.169	0.0	0.144	0.376	0.0	-0.520	-0.056	9.317	0.447
30	4.658	0.0	-4.658	0.0	0.155	0.416	0.0	-0.572	-0.057	10.063	0.463
31	5.198	0.0	-5.198	0.0	0.168	0.460	0.0	-0.628	-0.058	10.868	0.478
32	5.795	0.0	-5.795	0.0	0.181	0.508	0.0	-0.690	-0.059	11.737	0.494
33	6.454	0.0	-6.454	0.0	0.196	0.562	0.0	-0.757	-0.060	12.676	0.509
34	7.181	0.0	-7.181	0.0	0.211	0.620	0.0	-0.831	-0.061	13.690	0.525
35	7.984	0.0	-7.984	0.0	0.228	0.684	0.0	-0.912	-0.062	14.785	0.540
							TOTAL	-9.673	-1.579		

CASE #8

FINANCING METHOD: FUND THE RESERVE

TAX ACCOUNTING METHOD: NORMALIZATION

RATE OF INFLATION: 8.0 %
RATES OF RETURN:
CAPITAL 10.0 %
FUND 8.0 %

TAX RATES:
RETURN ON CAPITAL: 46.0 %
AMORTIZATION 46.0 %
FUND GROWTH 46.0 %

RESERVE COMPOUNDING RATE: 0.0 %

	BALANCE SHEET ACCOUNTS				INCOME ACCOUNTS						
AGE	FUND	CAPITAL	RESERVE	DEF TAXES	AMORT EXPENSE	TAX EXPENSE	RETURN	REVENUE	P W REVENUE	PR LEVEL COST	RATIO
0	0.0	0.0									
1	0.031	-0.014	-0.031	0.014	0.031	0.0	0.0	-0.031	-0.029	1.080	0.029
2	0.067	-0.031	-0.067	0.031	0.033	0.001	0.001	-0.036	-0.031	1.166	0.057
3	0.108	-0.050	-0.108	0.050	0.036	0.003	0.003	-0.042	-0.033	1.260	0.086
4	0.155	-0.072	-0.155	0.072	0.039	0.004	0.005	-0.048	-0.035	1.360	0.114
5	0.210	-0.097	-0.210	0.097	0.042	0.006	0.007	-0.055	-0.038	1.469	0.143
6	0.272	-0.125	-0.272	0.125	0.045	0.008	0.010	-0.063	-0.040	1.587	0.171
7	0.343	-0.158	-0.343	0.158	0.049	0.011	0.013	-0.072	-0.042	1.714	0.200
8	0.423	-0.195	-0.423	0.195	0.053	0.013	0.016	-0.082	-0.044	1.851	0.229
9	0.514	-0.236	-0.514	0.236	0.057	0.017	0.019	-0.093	-0.047	1.999	0.257
10	0.617	-0.284	-0.617	0.284	0.062	0.020	0.024	-0.105	-0.049	2.159	0.286
11	0.733	-0.337	-0.733	0.337	0.067	0.024	0.028	-0.119	-0.051	2.332	0.314
12	0.863	-0.397	-0.863	0.397	0.072	0.029	0.034	-0.134	-0.053	2.518	0.343
13	1.010	-0.465	-1.010	0.465	0.078	0.034	0.040	-0.151	-0.056	2.720	0.371
14	1.175	-0.540	-1.175	0.540	0.084	0.040	0.046	-0.170	-0.058	2.937	0.400
15	1.360	-0.625	-1.360	0.625	0.091	0.046	0.054	-0.191	-0.060	3.172	0.429
16	1.566	-0.720	-1.566	0.720	0.098	0.053	0.063	-0.214	-0.062	3.426	0.457
17	1.797	-0.827	-1.797	0.827	0.106	0.061	0.072	-0.239	-0.065	3.700	0.486
18	2.055	-0.945	-2.055	0.945	0.114	0.070	0.083	-0.267	-0.067	3.996	0.514
19	2.343	-1.078	-2.343	1.078	0.123	0.081	0.095	-0.298	-0.069	4.316	0.543
20	2.663	-1.225	-2.663	1.225	0.133	0.092	0.108	-0.333	-0.071	4.661	0.571
21	3.020	-1.389	-3.020	1.389	0.144	0.104	0.123	-0.371	-0.074	5.034	0.600
22	3.417	-1.572	-3.417	1.572	0.155	0.118	0.139	-0.413	-0.076	5.437	0.629
23	3.858	-1.775	-3.858	1.775	0.168	0.134	0.157	-0.459	-0.078	5.871	0.657
24	4.348	-2.000	-4.348	2.000	0.181	0.151	0.177	-0.510	-0.080	6.341	0.686
25	4.892	-2.250	-4.892	2.250	0.196	0.170	0.200	-0.566	-0.083	6.848	0.714
26	5.494	-2.527	-5.494	2.527	0.211	0.192	0.225	-0.628	-0.085	7.396	0.743
27	6.162	-2.835	-6.162	2.835	0.228	0.215	0.253	-0.696	-0.087	7.988	0.771
28	6.902	-3.175	-6.902	3.175	0.246	0.241	0.283	-0.771	-0.089	8.627	0.800
29	7.720	-3.551	-7.720	3.551	0.266	0.270	0.317	-0.854	-0.092	9.317	0.829
30	8.625	-3.968	-8.625	3.968	0.288	0.303	0.355	-0.945	-0.094	10.063	0.857
31	9.626	-4.428	-9.626	4.428	0.311	0.338	0.397	-1.045	-0.096	10.868	0.886
32	10.731	-4.936	-10.731	4.936	0.335	0.377	0.443	-1.155	-0.098	11.737	0.914
33	11.952	-5.498	-11.952	5.498	0.362	0.420	0.494	-1.276	-0.101	12.676	0.943
34	13.299	-6.118	-13.299	6.118	0.391	0.468	0.550	-1.409	-0.103	13.690	0.971
35	14.785	-6.801	-14.785	6.801	0.422	0.521	0.612	-1.555	-0.105	14.785	1.000
							TOTAL	-15.399	-2.341		

CASE #9

FINANCING METHOD: FUND AT DECOMMISSIONING

TAX ACCOUNTING METHOD: FLOW-THROUGH

RATE OF INFLATION: 8.0 %
RATES OF RETURN:
CAPITAL 10.0 %
FUND 8.0 %

TAX RATES:
RETURN ON CAPITAL: 46.0 %
AMORTIZATION 46.0 %
FUND GROWTH 46.0 %

RESERVE COMPOUNDING RATE: 0.0 %

	BALANCE SHEET ACCOUNTS				INCOME ACCOUNTS						
AGE	FUND	CAPITAL	RESERVE	DEF TAXES	AMORT EXPENSE	TAX EXPENSE	RETURN	REVENUE	P W REVENUE	PR LEVEL COST	RATIO
0	0.0	0.0									
1	0.0	0.017	-0.017	0.0	0.017	0.014	0.0	-0.031	-0.029	1.080	0.015
2	0.0	0.035	-0.035	0.0	0.018	0.014	-0.002	-0.030	-0.026	1.166	0.030
3	0.0	0.054	-0.054	0.0	0.020	0.014	-0.003	-0.030	-0.024	1.260	0.043
4	0.0	0.076	-0.076	0.0	0.021	0.013	-0.005	-0.029	-0.022	1.360	0.056
5	0.0	0.099	-0.099	0.0	0.023	0.013	-0.008	-0.029	-0.020	1.469	0.067
6	0.0	0.124	-0.124	0.0	0.025	0.013	-0.010	-0.029	-0.018	1.587	0.078
7	0.0	0.152	-0.152	0.0	0.028	0.013	-0.012	-0.028	-0.016	1.714	0.088
8	0.0	0.182	-0.182	0.0	0.030	0.013	-0.015	-0.028	-0.015	1.851	0.098
9	0.0	0.215	-0.215	0.0	0.033	0.013	-0.018	-0.028	-0.014	1.999	0.108
10	0.0	0.252	-0.252	0.0	0.037	0.013	-0.022	-0.028	-0.013	2.159	0.117
11	0.0	0.292	-0.292	0.0	0.040	0.013	-0.025	-0.028	-0.012	2.332	0.125
12	0.0	0.336	-0.336	0.0	0.044	0.013	-0.029	-0.028	-0.011	2.518	0.134
13	0.0	0.386	-0.386	0.0	0.049	0.013	-0.034	-0.029	-0.011	2.720	0.142
14	0.0	0.440	-0.440	0.0	0.055	0.014	-0.039	-0.030	-0.010	2.937	0.150
15	0.0	0.501	-0.501	0.0	0.061	0.014	-0.044	-0.031	-0.010	3.172	0.158
16	0.0	0.568	-0.568	0.0	0.067	0.015	-0.050	-0.032	-0.009	3.426	0.166
17	0.0	0.644	-0.644	0.0	0.075	0.016	-0.057	-0.034	-0.009	3.700	0.174
18	0.0	0.728	-0.728	0.0	0.084	0.017	-0.064	-0.037	-0.009	3.996	0.182
19	0.0	0.822	-0.822	0.0	0.094	0.018	-0.073	-0.040	-0.009	4.316	0.190
20	0.0	0.928	-0.928	0.0	0.106	0.020	-0.082	-0.044	-0.009	4.661	0.199
21	0.0	1.047	-1.047	0.0	0.119	0.023	-0.093	-0.049	-0.010	5.034	0.208
22	0.0	1.182	-1.182	0.0	0.135	0.026	-0.105	-0.056	-0.010	5.437	0.217
23	0.0	1.335	-1.335	0.0	0.153	0.030	-0.118	-0.064	-0.011	5.871	0.227
24	0.0	1.509	-1.509	0.0	0.174	0.035	-0.134	-0.075	-0.012	6.341	0.238
25	0.0	1.708	-1.708	0.0	0.199	0.041	-0.151	-0.089	-0.013	6.848	0.249
26	0.0	1.937	-1.937	0.0	0.229	0.049	-0.171	-0.107	-0.014	7.396	0.262
27	0.0	2.201	-2.201	0.0	0.264	0.060	-0.194	-0.130	-0.016	7.988	0.276
28	0.0	2.508	-2.508	0.0	0.307	0.074	-0.220	-0.161	-0.019	8.627	0.291
29	0.0	2.869	-2.869	0.0	0.360	0.093	-0.251	-0.203	-0.022	9.317	0.308
30	0.0	3.296	-3.296	0.0	0.428	0.120	-0.287	-0.261	-0.026	10.063	0.328
31	0.0	3.811	-3.811	0.0	0.514	0.157	-0.330	-0.342	-0.032	10.868	0.351
32	0.0	4.442	-4.442	0.0	0.632	0.214	-0.381	-0.464	-0.040	11.737	0.378
33	0.0	5.243	-5.243	0.0	0.801	0.304	-0.444	-0.660	-0.052	12.676	0.414
34	0.0	6.318	-6.318	0.0	1.075	0.469	-0.524	-1.019	-0.074	13.690	0.462
35	0.0	7.984	-7.984	0.0	1.666	0.881	-0.632	-1.915	-0.130	14.785	0.540
							TOTAL	-6.220	-0.776		

CASE #10

FINANCING METHOD: FUND AT DECOMMISSIONING

TAX ACCOUNTING METHOD: NORMALIZATION

RATE OF INFLATION: 8.0 %
RATES OF RETURN:
CAPITAL 10.0 %
FUND 8.0 %

TAX RATES:
RETURN ON CAPITAL: 46.0 %
AMORTIZATION 46.0 %
FUND GROWTH 46.0 %

RESERVE COMPOUNDING RATE: 0.0 %

	BALANCE SHEET ACCOUNTS				INCOME ACCOUNTS						
AGE	FUND	CAPITAL	RESERVE	DEF TAXES	AMORT EXPENSE	TAX EXPENSE	RETURN	REVENUE	P W REVENUE	PR LEVEL COST	RATIO
0	0.0	0.0									
1	0.0	0.017	-0.031	0.014	0.031	0.0	0.0	-0.031	-0.029	1.080	0.029
2	0.0	0.035	-0.064	0.030	0.033	-0.001	-0.002	-0.030	-0.026	1.166	0.055
3	0.0	0.054	-0.100	0.046	0.036	-0.003	-0.003	-0.030	-0.024	1.260	0.080
4	0.0	0.076	-0.140	0.064	0.039	-0.005	-0.005	-0.029	-0.022	1.360	0.103
5	0.0	0.099	-0.183	0.084	0.043	-0.006	-0.008	-0.029	-0.020	1.469	0.124
6	0.0	0.124	-0.230	0.106	0.047	-0.008	-0.010	-0.029	-0.018	1.587	0.145
7	0.0	0.152	-0.281	0.129	0.051	-0.011	-0.012	-0.028	-0.016	1.714	0.164
8	0.0	0.182	-0.337	0.155	0.056	-0.013	-0.015	-0.028	-0.015	1.851	0.182
9	0.0	0.215	-0.398	0.183	0.062	-0.015	-0.018	-0.028	-0.014	1.999	0.199
10	0.0	0.252	-0.466	0.214	0.068	-0.018	-0.022	-0.028	-0.013	2.159	0.216
11	0.0	0.292	-0.541	0.249	0.075	-0.021	-0.025	-0.028	-0.012	2.332	0.232
12	0.0	0.336	-0.623	0.287	0.082	-0.025	-0.029	-0.028	-0.011	2.518	0.247
13	0.0	0.386	-0.714	0.329	0.091	-0.029	-0.034	-0.029	-0.011	2.720	0.263
14	0.0	0.440	-0.815	0.375	0.101	-0.033	-0.039	-0.030	-0.010	2.937	0.278
15	0.0	0.501	-0.928	0.427	0.112	-0.038	-0.044	-0.031	-0.010	3.172	0.292
16	0.0	0.568	-1.052	0.484	0.125	-0.043	-0.050	-0.032	-0.009	3.426	0.307
17	0.0	0.644	-1.192	0.548	0.139	-0.048	-0.057	-0.034	-0.009	3.700	0.322
18	0.0	0.728	-1.348	0.620	0.156	-0.055	-0.064	-0.037	-0.009	3.996	0.337
19	0.0	0.822	-1.522	0.700	0.175	-0.062	-0.073	-0.040	-0.009	4.316	0.353
20	0.0	0.928	-1.718	0.790	0.196	-0.070	-0.082	-0.044	-0.009	4.661	0.369
21	0.0	1.047	-1.939	0.892	0.221	-0.079	-0.093	-0.049	-0.010	5.034	0.385
22	0.0	1.182	-2.189	1.007	0.250	-0.089	-0.105	-0.056	-0.010	5.437	0.403
23	0.0	1.335	-2.472	1.137	0.283	-0.101	-0.118	-0.064	-0.011	5.871	0.421
24	0.0	1.509	-2.795	1.286	0.322	-0.114	-0.134	-0.075	-0.012	6.341	0.441
25	0.0	1.708	-3.163	1.455	0.369	-0.129	-0.151	-0.089	-0.013	6.848	0.462
26	0.0	1.937	-3.587	1.650	0.423	-0.146	-0.171	-0.107	-0.014	7.396	0.485
27	0.0	2.201	-4.076	1.875	0.489	-0.165	-0.194	-0.130	-0.016	7.988	0.510
28	0.0	2.508	-4.645	2.137	0.569	-0.187	-0.220	-0.161	-0.019	8.627	0.538
29	0.0	2.869	-5.312	2.444	0.668	-0.214	-0.251	-0.203	-0.022	9.317	0.570
30	0.0	3.296	-6.104	2.808	0.792	-0.244	-0.287	-0.261	-0.026	10.063	0.607
31	0.0	3.811	-7.057	3.246	0.953	-0.281	-0.330	-0.342	-0.032	10.868	0.649
32	0.0	4.442	-8.227	3.784	1.170	-0.325	-0.381	-0.464	-0.040	11.737	0.701
33	0.0	5.243	-9.710	4.467	1.483	-0.378	-0.444	-0.660	-0.052	12.676	0.766
34	0.0	6.318	-11.700	5.382	1.990	-0.447	-0.524	-1.019	-0.074	13.690	0.855
35	0.0	7.984	-14.785	6.801	3.085	-0.538	-0.632	-1.915	-0.130	14.785	1.000
							TOTAL	-6.220	-0.776		

CASE #11

FINANCING METHOD: FUND AT DECOMMISSIONING

TAX ACCOUNTING METHOD: FLOW-THROUGH

RATE OF INFLATION: 8.0 %
RATES OF RETURN:
CAPITAL 10.0 %
FUND 8.0 %

TAX RATES:
RETURN ON CAPITAL: 46.0 %
AMORTIZATION 46.0 %
FUND GROWTH 46.0 %

RESERVE COMPOUNDING RATE: 8.0 %

	BALANCE SHEET ACCOUNTS				INCOME ACCOUNTS						
AGE	FUND	CAPITAL	RESERVE	DEF TAXES	AMORT EXPENSE	TAX EXPENSE	RETURN	REVENUE	P W REVENUE	PR LEVEL COST	RATIO
0	0.0	0.0									
1	0.0	0.017	-0.017	0.0	0.017	0.014	0.0	-0.031	-0.029	1.080	0.015
2	0.0	0.036	-0.036	0.0	0.019	0.015	-0.002	-0.033	-0.028	1.166	0.031
3	0.0	0.058	-0.058	0.0	0.022	0.016	-0.004	-0.035	-0.028	1.260	0.046
4	0.0	0.084	-0.084	0.0	0.026	0.017	-0.006	-0.037	-0.027	1.360	0.062
5	0.0	0.113	-0.113	0.0	0.029	0.018	-0.008	-0.039	-0.026	1.469	0.077
6	0.0	0.147	-0.147	0.0	0.034	0.019	-0.011	-0.041	-0.026	1.587	0.093
7	0.0	0.185	-0.185	0.0	0.038	0.020	-0.015	-0.044	-0.025	1.714	0.108
8	0.0	0.228	-0.228	0.0	0.043	0.021	-0.019	-0.046	-0.025	1.851	0.123
9	0.0	0.278	-0.278	0.0	0.049	0.022	-0.023	-0.049	-0.024	1.999	0.139
10	0.0	0.333	-0.333	0.0	0.056	0.024	-0.028	-0.051	-0.024	2.159	0.154
11	0.0	0.396	-0.396	0.0	0.063	0.025	-0.033	-0.054	-0.023	2.332	0.170
12	0.0	0.466	-0.466	0.0	0.071	0.026	-0.040	-0.057	-0.023	2.518	0.185
13	0.0	0.545	-0.545	0.0	0.079	0.028	-0.047	-0.060	-0.022	2.720	0.201
14	0.0	0.634	-0.634	0.0	0.089	0.029	-0.055	-0.064	-0.022	2.937	0.216
15	0.0	0.734	-0.734	0.0	0.100	0.031	-0.063	-0.067	-0.021	3.172	0.231
16	0.0	0.846	-0.846	0.0	0.112	0.033	-0.073	-0.071	-0.021	3.426	0.247
17	0.0	0.970	-0.970	0.0	0.125	0.034	-0.085	-0.074	-0.020	3.700	0.262
18	0.0	1.110	-1.110	0.0	0.139	0.036	-0.097	-0.078	-0.020	3.996	0.278
19	0.0	1.265	-1.265	0.0	0.155	0.038	-0.111	-0.082	-0.019	4.316	0.293
20	0.0	1.438	-1.438	0.0	0.173	0.040	-0.127	-0.086	-0.019	4.661	0.309
21	0.0	1.631	-1.631	0.0	0.193	0.042	-0.144	-0.091	-0.018	5.034	0.324
22	0.0	1.845	-1.845	0.0	0.214	0.044	-0.163	-0.095	-0.017	5.437	0.339
23	0.0	2.084	-2.084	0.0	0.238	0.046	-0.185	-0.099	-0.017	5.871	0.355
24	0.0	2.348	-2.348	0.0	0.265	0.048	-0.208	-0.104	-0.016	6.341	0.370
25	0.0	2.642	-2.642	0.0	0.294	0.050	-0.235	-0.109	-0.016	6.848	0.386
26	0.0	2.967	-2.967	0.0	0.325	0.052	-0.264	-0.113	-0.015	7.396	0.401
27	0.0	3.328	-3.328	0.0	0.361	0.054	-0.297	-0.118	-0.015	7.988	0.417
28	0.0	3.727	-3.727	0.0	0.399	0.057	-0.333	-0.123	-0.014	8.627	0.432
29	0.0	4.169	-4.169	0.0	0.442	0.059	-0.373	-0.128	-0.014	9.317	0.447
30	0.0	4.658	-4.658	0.0	0.489	0.061	-0.417	-0.133	-0.013	10.063	0.463
31	0.0	5.198	-5.198	0.0	0.540	0.063	-0.466	-0.138	-0.013	10.868	0.478
32	0.0	5.795	-5.795	0.0	0.597	0.066	-0.520	-0.143	-0.012	11.737	0.494
33	0.0	6.454	-6.454	0.0	0.659	0.068	-0.579	-0.148	-0.012	12.676	0.509
34	0.0	7.181	-7.181	0.0	0.728	0.070	-0.645	-0.152	-0.011	13.690	0.525
35	0.0	7.984	-7.984	0.0	0.803	0.072	-0.718	-0.156	-0.011	14.785	0.540
							TOTAL	-2.950	-0.685		

CASE #12

FINANCING METHOD: FUND AT DECOMMISSIONING

TAX ACCOUNTING METHOD: NORMALIZATION

RATE OF INFLATION: 8.0 %
RATES OF RETURN:
CAPITAL 10.0 %
FUND 8.0 %

TAX RATES:
RETURN ON CAPITAL: 46.0 %
AMORTIZATION 46.0 %
FUND GROWTH 46.0 %

RESERVE COMPOUNDING RATE: 8.0 %

	BALANCE SHEET ACCOUNTS				INCOME ACCOUNTS						
AGE	FUND	CAPITAL	RESERVE	DEF TAXES	AMORT EXPENSE	TAX EXPENSE	RETURN	REVENUE	P W REVENUE	PR LEVEL COST	RATIO
0	0.0	0.0									
1	0.0	0.017	-0.031	0.014	0.031	0.0	0.0	-0.031	-0.029	1.080	0.029
2	0.0	0.036	-0.067	0.031	0.036	-0.001	-0.002	-0.033	-0.028	1.166	0.057
3	0.0	0.058	-0.108	0.050	0.041	-0.003	-0.004	-0.035	-0.028	1.260	0.086
4	0.0	0.084	-0.155	0.072	0.048	-0.005	-0.006	-0.037	-0.027	1.360	0.114
5	0.0	0.113	-0.210	0.097	0.054	-0.007	-0.008	-0.039	-0.026	1.469	0.143
6	0.0	0.147	-0.272	0.125	0.062	-0.010	-0.011	-0.041	-0.026	1.587	0.171
7	0.0	0.185	-0.343	0.158	0.071	-0.013	-0.015	-0.044	-0.025	1.714	0.200
8	0.0	0.228	-0.423	0.195	0.080	-0.016	-0.019	-0.046	-0.025	1.851	0.229
9	0.0	0.278	-0.514	0.236	0.091	-0.019	-0.023	-0.049	-0.024	1.999	0.257
10	0.0	0.333	-0.617	0.284	0.103	-0.024	-0.028	-0.051	-0.024	2.159	0.286
11	0.0	0.396	-0.733	0.337	0.116	-0.028	-0.033	-0.054	-0.023	2.332	0.314
12	0.0	0.466	-0.863	0.397	0.131	-0.034	-0.040	-0.057	-0.023	2.518	0.343
13	0.0	0.545	-1.010	0.465	0.147	-0.040	-0.047	-0.060	-0.022	2.720	0.371
14	0.0	0.634	-1.175	0.540	0.165	-0.046	-0.055	-0.064	-0.022	2.937	0.400
15	0.0	0.734	-1.360	0.625	0.185	-0.054	-0.063	-0.067	-0.021	3.172	0.429
16	0.0	0.846	-1.566	0.720	0.207	-0.063	-0.073	-0.071	-0.021	3.426	0.457
17	0.0	0.970	-1.797	0.827	0.231	-0.072	-0.085	-0.074	-0.020	3.700	0.486
18	0.0	1.110	-2.055	0.945	0.258	-0.083	-0.097	-0.078	-0.020	3.996	0.514
19	0.0	1.265	-2.343	1.078	0.288	-0.095	-0.111	-0.082	-0.019	4.316	0.543
20	0.0	1.438	-2.663	1.225	0.321	-0.108	-0.127	-0.086	-0.019	4.661	0.571
21	0.0	1.631	-3.020	1.389	0.357	-0.123	-0.144	-0.091	-0.018	5.034	0.600
22	0.0	1.845	-3.417	1.572	0.397	-0.139	-0.163	-0.095	-0.017	5.437	0.629
23	0.0	2.084	-3.858	1.775	0.441	-0.157	-0.185	-0.099	-0.017	5.871	0.657
24	0.0	2.348	-4.348	2.000	0.490	-0.177	-0.208	-0.104	-0.016	6.341	0.686
25	0.0	2.642	-4.892	2.250	0.544	-0.200	-0.235	-0.109	-0.016	6.848	0.714
26	0.0	2.967	-5.494	2.527	0.603	-0.225	-0.264	-0.113	-0.015	7.396	0.743
27	0.0	3.328	-6.162	2.835	0.668	-0.253	-0.297	-0.118	-0.015	7.988	0.771
28	0.0	3.727	-6.902	3.175	0.739	-0.283	-0.333	-0.123	-0.014	8.627	0.800
29	0.0	4.169	-7.720	3.551	0.818	-0.317	-0.373	-0.128	-0.014	9.317	0.829
30	0.0	4.658	-8.625	3.968	0.905	-0.355	-0.417	-0.133	-0.013	10.063	0.857
31	0.0	5.198	-9.626	4.428	1.001	-0.397	-0.466	-0.138	-0.013	10.868	0.886
32	0.0	5.795	-10.731	4.936	1.105	-0.443	-0.520	-0.143	-0.012	11.737	0.914
33	0.0	6.454	-11.952	5.498	1.221	-0.494	-0.579	-0.148	-0.012	12.676	0.943
34	0.0	7.181	-13.299	6.118	1.347	-0.550	-0.645	-0.152	-0.011	13.690	0.971
35	0.0	7.984	-14.785	6.801	1.486	-0.612	-0.718	-0.156	-0.011	14.785	1.000
							TOTAL	-2.950	-0.685		

Case #4: Funding at Decommissioning, no Taxes, Compounding the Reserve

In order to overcome the inequitable distribution of costs produced in Case #3, compounding of the reserve for decommissioning was introduced. If inflation did not exist, this would not be a problem. The rationale for compounding is that current customers are major contributers to the inflation and, therefore, should protect the capital contributed by customers of earlier accounting periods. The effect of this is to produce an amortization expense that is increasing at a constant rate, a revenue requirement that increases at a constant rate and a discounted revenue requirement that is decreasing, the latter two totaling 2.950 and 0.685, respectively.

THE EFFECT OF TAXES

There are two methods of accounting for taxes under present law--flow-through and normalization. Normalization is required under the Asset Depreciation Range (ADR) System for the timing differences between accelerated tax depreciation and book depreciation, but the law is silent on normalization of net salvage components, saying only that removal cost is an expense for tax purposes when it is incurred. In effect this makes the amortization expense taxable income.

Flow-through Tax Accounting

This is the method used by most utilities, even under the ADR System. Since removal cost is an expense when incurred, the utility will receive a tax credit amounting to 46% of the removal cost. This means that the reserve for decommissioning and the fund need only accrue to a total of 54% of the decommissioning cost. At decommissioning, the reserve will receive the credit from taxes instead of tax expense. In a strict sense, the method should be called "modified flow-through." For "Funding at Commissioning," Case #5, the initial fund is only 54%, producing an amortization expense of 54% of Case #1. Since amortization expense is treated as income, the revenue requirement to cover the taxes is

$$\frac{\text{amortization expense} \times \text{tax rate}}{1 - \text{tax rate}}$$

In addition to the tax on the amortization expense, there is a tax on the fund growth unless some special provision has been made with the Internal Revenue Service or the funds are invested in tax-free bonds. Since it is otherwise highly unlikely that the fund will be able to earn at a rate sufficient to earn 8% after taxes, it is assumed that these tax expenses will be covered in the tax expense component of the revenue requirement.

Refer to Figures 4 and 5 for comparisons of the revenue requirement and the present-worth revenue requirement for cases covering flow-through.

Normalized Tax Accounting

Normalized tax accounting means that the revenue requirement is computed by assuming that the amortization expense is not taxable. However, based upon the fact that the tax is still payable, it is calculated as a straight percentage of the amortization expense and paid from the company's capital. This in turn produces a revenue requirement for tax expense on the return. The revenue requirement and present-worth revenue requirement for normalized taxes are shown in Figures 6 and 7.

For the "Funding at Commissioning" method this increases the total revenue requirement and total present-worth revenue requirement by 1.91 and 1.66, respectively, over modified flow-through method. For the "Funded Reserve" method these ratios are 1.59 and 1.48, respectively. Interestingly enough, the differences between flow-through and normalized procedures for "Funding at Decommissioning" are zero.

At decommissioning, the tax credit relieves the capital commitment created by the company's paying the taxes on the amortization expense.

SUMMARY

The cases covering the effects of taxes are illustrated in their respective tables numbered Cases 5 through 12, with the odd-numbered cases covering flow-through and the even-numbered cases covering normalization. The total revenue requirements and the total present-worth revenue requirements are summarized as follows.

Case	Financing Method	Taxes Rate	Taxes Accounting	Total Revenue	Present Worth
1	Funding at Commissioning	0		2.800	1.167
2	Funding the Reserve	0		5.317	1.000
3	Funding at Decommissioning	0		6.220	.776
*4	Funding at Decomm., 8% Comp.	0		2.950	.685
5	Funding at Commissioning	46%	Flow-through	9.141	2.359
6	Funding at Commissioning	46%	Normalized	17.479	3.927
7	Funding the Reserve	46%	Flow-through	9.673	1.579
8	Funding the Reserve	46%	Normalized	15.399	2.341
9	Funding at Decommissioning	46%	Flow-through	6.220	.776
10	Funding at Decommissioning	46%	Normalized	6.220	.776
*11	Funding at Decomm., 8% Comp.	46%	Flow-through	2.950	.685
*12	Funding at Decomm., 8% Comp.	46%	Normalized	2.950	.685

* Least expensive on a present-worth basis.

CONCLUSION

From these analyses, the least expensive in terms of total revenue is "Funding at Commissioning" with "Funding at Decommissioning with Compounding" a close second. However, on a present-worth basis, "Funding at Decommissioning with Compounding," Case #4 is the least expensive. A look at Figure 3 shows that only "Funding the Reserve," Case #2, produces a more equitable cost distribution.

It is also interesting to note that the addition of income taxes to Cases #3 and #4 when using either the modified flow-through or normalized tax accounting methods, producing Cases #9, #10, #11, and #12, does not add to the cost of financing.

The selection of a funding method other than "Funding at Decommissioning" is going to cost the customers more, particularly if the tax laws are not modified. In order to further minimize the cost to the customers, a revision to the accounting procedures needs to be implemented wherein the amortization expense is supplemented by compounding the reserve and the amortization expense is calculated each year using a realistic estimate of decommissioning in current dollars.

METHODS OF POWER REACTOR

DECOMMISSIONING COST RECOVERY

N. Barrie McLeod and Y. M. Park

NUS Corporation
4 Research Place
Rockville, MD 20850

Electric utilities have for many years accumulated funds in order to pay for the future decommissioning of equipment and/or facilities. The uniqueness of nuclear plant decommissioning is therefore not in the basic concept, but in the unique combination of time duration, the associated uncertainties, and the cost magnitudes associated with nuclear plants. There are two basic elements which govern the decommissioning revenues to be received from the rate-payers who will benefit from use of the facility's energy output: 1) the actual method and cost of decommissioning and 2) the manner in which the anticipated costs will be collected during the facility's useful life and used on an interim basis prior to being paid out during decommissioning. The first element will normally involve a reactor-specific engineering cost estimate for decommissioning, based on current costs and regulations, and is a necessary input for the second of these two basic elements. Specifically, the purpose of this paper is to describe the various factors related to the second element including rate-regulatory factors, income taxes, and accounting practices; to analyze alternative interim use of accumulated funds; to examine alternative approaches for cost recovery, and to examine one particular alternative that appears to have unique features. Also, parallels are drawn between decommissioning cost recovery and spent fuel disposal cost recovery.

A. BASIC REGULATORY AND FINANCIAL PRINCIPLES

A realistic cost recovery method should of course be based on established regulatory and financial principles, which are generally self-evident but are worth listing for the sake of completeness:

1. Sufficient funds should have been collected over the facility's useful life to pay all costs when they occur. Taxation and all relevant cost changes including inflation should be accommodated.

2. The lowest possible "net cost" to the ratepayer should be achieved, and either the actual or an attributed credit for use-of-funds should be included in the definition of "net cost."

3. Costs should be equitably shared by the beneficiaries of the energy production, whether they be earlier or later beneficiaries/ratepayers.

In addition to the above principles, there is also a good practice which deals well with the generic problem of future cost changes which will originate with the almost-inevitable future changes in regulations, technology, scope and timing affecting decommissioning. The good practice is not to attempt predicting the changes themselves but to employ a method which readily accommodates change once it has occurred. This means that one can consistently use "best current estimates" of cost based on current regulations, technology and scope, coupled with a cost recovery method involving unit cost assessments related to:

$$\frac{\text{Remaining "best current" cost}}{\text{Remaining time or energy}}$$

This "good practice" method basically recovers original costs over the facility life and subsequent cost changes over the remaining life. As will be shown later, cost changes due to inflation can be handled this same way, but there are supplemental methods which may accommodate the special problem of inflation in an even better way.

B. SUMMARY OF INCOME TAX FACTORS

It is generally acknowledged that under current IRS regulations, revenues for future decommissioning costs do not have an offsetting deductible expense and therefore are treated as taxable income when received and actual decommissioning costs are tax deductible when they are actually incurred. What this means is that the IRS receives 46% (as of September 1979) of decommissioning revenues as they are charged, but then pays 46% of the decommissioning costs when they are actually incurred (assuming that the tax rate stays the same, and that sufficient taxable income is available elsewhere to utilize the deduction). When state income taxes are included, effective income tax rates approaching 50% are encountered.

It is also generally acknowledged that income taxes on decommissioning revenues cannot be avoided by turning the revenues over to an external entity who will then invest and manage the funds. If the funds are externally invested in tax-free bonds, the interest income from the bonds is tax-free, but the revenues received must be taxed before the bonds are purchased.

The basic impact of income taxation is therefore that it reduces the funds available for interim use to about one-half of what they would be otherwise. Not surprisingly, taxation thereby increases the net cost of decommissioning by so reducing the interim earning power of accumulated revenues.

C. SUMMARY OF ACCOUNTING PRACTICES

It is conventional practice in the depreciation of electric utility capital investment, to collect the difference between original cost and net salvage value over the useful life of the plant. The accumulated depreciation accounts are subtracted from the original cost accounts to obtain the rate base. This practice is being used to accommodate nuclear plant decommissioning cost recovery by simply treating decommissioning cost as a negative net salvage value. In these circumstances decommissioning cost recovery is accomplished by basing the depreciation rate on, say, 115% or 120% of original cost, of which the 15% or 20% excess represents the depreciation cost expressed relative to original plant cost. Because this approach is both logical and compatible with existing accounting practice, it seems to have been widely adopted.

It is extremely important to recognize, however, that this approach requires careful financial analysis because there is no provision by which the accumulated decommissioning funds can earn and accrue interest. Figure 1 portrays the difference between the current practice (Option 1) and the more typical presumption of financial analysis (Option 2) in which the rate base is not reduced by the amount of the accumulated decommissioning fund and these funds are separately accumulated and used as part of the source of funds which supports the rate base. Under Option 2, the extra earnings associated with the higher rate base are directly attributable to use of the accumulated decommissioning fund and this fund can accrue these earnings. Because of this accrual of earnings, similar to a sinking fund, the direct revenue required for decommissioning can be reduced from that required under Option 1. A standard present value analysis (a single payment by the ratepayers) shows that the two options are identical from a present value viewpoint. However, significant timing differences between the two options can develop, such that there can be marked differences as to whether earlier or later ratepayers bear the majority of the burden. Returning to the statement at the beginning of this paragraph, in doing the financial analyses associated with decommissioning it is important to recognize that Option 1 is the approach being used by most, if not all utilities that have addressed the subject of decommissioning, and who are using the funds internally.

Under Option 1, the accumulated decommissioning fund is used without explicit credit. If inflation were zero, this would not generate difficulties. However, interest rates and rates-of-return have an implicit component which compensates for inflation, in effect maintaining the purchasing power of the funds. Under Option 1, because there is no explicit interest, there is no explicit way to maintain the purchasing power of the funds accumulated for future decommissioning. This indicates that unless some explicit recognition of inflation is made in the cost recovery approach, the later ratepayers will bear not only their share of decommissioning cost, but also the burden of restoring the purchasing power of previously collected funds.

There is an additional accounting-related matter having to do with the impact on decommissioning of normalized vs flow-through accounting practices. Normalized Accounting is the practice of including, as a current cost (and hence as a revenue requirement), those income taxes applicable to the current period, but for which payment has been deferred to a future year by using some form of accelerated tax depreciation for income tax purposes. When used for rate making

FIGURE 1

ACCOUNTING PRACTICE ALTERNATIVES

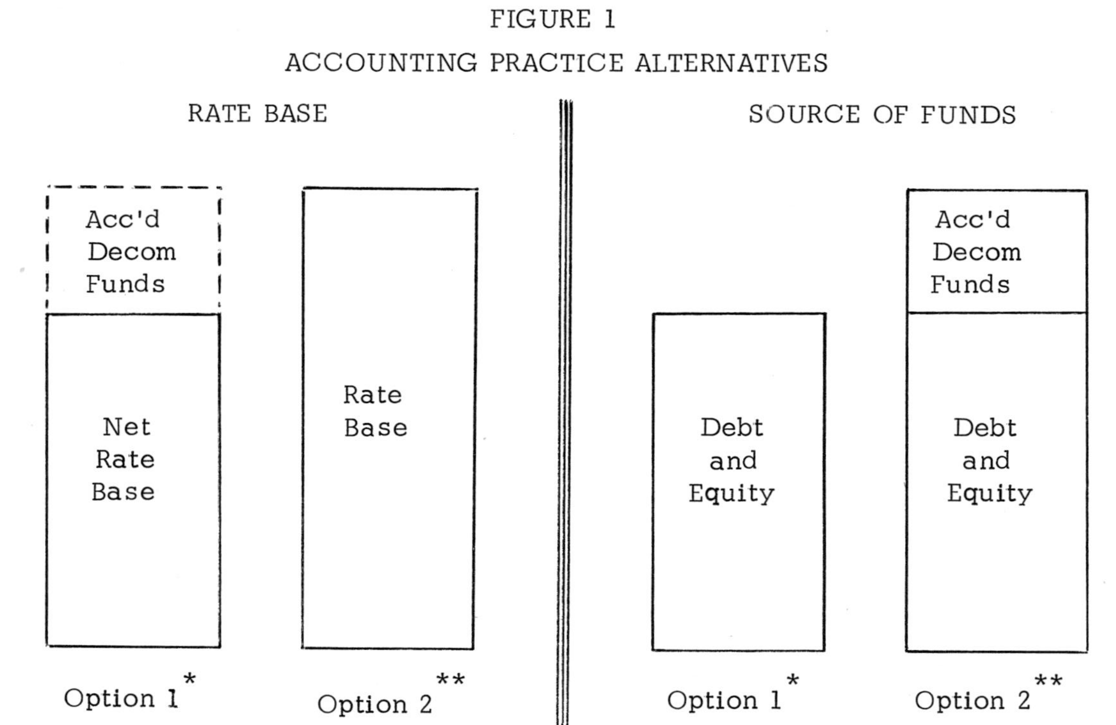

* Option 1, the Negative Rate Base approach is the predominant, if not the only present practice for internal use-of-funds

** Option 2, would include internal sinking fund approaches

purposes, normalizing enables the utility to accumulate these deferred tax funds and use them to reduce the rate base until the tax is ultimately paid. The use of these funds without the requirement of interest payments reduces the utility's revenue requirements and is therefore a benefit to its current customers, in the amount of the interest payments that would otherwise have to be paid. Utilities using normalized accounting would not alter their annual revenue requirements for decommissioning cost as a result of this tax treatment of fund accruals. However, utilities using flow-through accounting would have a revenue requirement almost double the annual decommissioning charges, in order that the accumulated funds would be sufficient to pay for both the decommissioning cost and the flow-through tax benefit when the decommissioning costs become tax-deductible. Basically, as applied to normal depreciation, flow-through accounting is a device to benefit early users of a facility at the expense of later users. As applied to decommissioning cost recovery, it works exactly the opposite, benefiting the later users, or more accurately, the post-shutdown ratepayers, at the expense of the earlier ratepayers.

D. INTERIM USE OF ACCUMULATED FUNDS

A fundamental question which can have a significant cost impact is whether the accumulated decommissioning funds should be utilized internally, or should be accumulated separately and externally. An answer dealing directly with ratepayer interest can be obtained by determining whether internal or external use of funds results in a lower "net cost" to the ratepayer. Net cost must include either a credit for the actual (after-tax) interest received or an attributed credit for use-of-funds, when these funds are accumulated internally and used to reduce the rate base. In the latter case, the credit is a very real savings to the ratepayer and is equal to the accumulated funds times the after-tax rate of return. For analysis purposes it is very properly attributable to the decommissioning reserves. In the following, the term Net Annual Cost refers to the direct annual revenue requirement from the ratepayers, less the actual after-tax interest or the attributed credit for use-of-funds.

The comparison of internal and external use-of-funds can probably be done in a number of ways, but a commonly-understood and meaningful approach is the present value approach. Specifically, one can ask what is the amount of a single ratepayer payment in the present which will result, with proper credits given for interest income or use-of-funds, in the amount required for future decommissioning. It can be shown that this single payment is simply the future decommissioning cost, discounted from the future to the present using a compounded discount rate equal to the after-tax interest received (external) or the after-tax rate of return (internal).

Table 1 gives a comparison of relative present values, using representative current interest rates for external investment in tax-free bonds (Case 1), in taxable corporate bonds (Case 2) and internal use for a utility with a rate base financed with 35% Common Stock, 15% Preferred Stock and 50% Debt (Case 3). Results are presented assuming that decommissioning expenditure occurs 30 years in the future.

Table 1 clearly indicates that unless external investments can produce net interest income approaching the internal net rate, the internal use of funds is decidedly lower cost to the ratepayer. In the examples given, the best external use (tax-free bonds) costs almost double the present value when the funds are used internally. Because the interest rates shown are generally representative of the relationship between after-tax rates of return and the interest rate on tax-free bonds, the internal use of accumulated decommissioning funds is strongly recommended on the basis that it results in significantly lower costs to the ratepayer.

E. COST RECOVERY APPROACHES

One of the principal challenges that must be met by any cost recovery approach is to accommodate inflation, and the principal differences between various cost recovery approaches are due to the different ways in which they accommodate inflation. In general, the approaches can be categorized according to the extent to which they 1) favor early ratepayer/users of the facility or 2) favor later ratepayer/users. In addition, there is an intermediate category. The basic alternatives for decommissioning cost recovery are:

- Alternative 1

The inflated cost of decommissioning is predicted from the current cost, and is collected in uniform annual amounts over the predicted period. Future changes in either base cost or inflation can be accommodated by defining the annual charge as:

TABLE 1

PRESENT VALUE COMPARISON

Decommissioning Revenues Taxable (Normalized)

	Case	Interest Factors	Tax Rate	Discount Rate[1]	$\frac{1}{(1+i)^{30}}$	Relative Present Rate
1.	External, tax free	6.5%	0	6.5%	0.151	1.98
2.	External, taxable	9.0%	48%*	4.68%	0.254	3.32
3.	Internal	0.35 x 15%	-	8.96%	0.0763	1.00 (ref)
		0.15 x 9.1%	-			
		0.50 x 9.0%	48%			

* 48% is the approximate net rate of a composite 4% state income tax rate and a 46% federal rate.

1 Discount rate is the after-tax interest rate or rate-of-return.

$$\text{Annual Charge} = \frac{\text{Remaining Inflated (End Point) Cost}}{\text{Remaining Years}}$$

An obvious feature of this alternative is the requirement for predicting the rate of future inflation. It is also evident that because of collecting inflated dollars in the present, annual charges are going to be relatively large in the early years. In effect, early user/ratepayers bear a substantial portion of the burden of maintaining the purchasing power of the accumulated funds.

- Alternative 2

The current-dollar cost of decommissioning is developed, and the annual amounts collected vary in time so that base cost changes and inflation are ultimately accommodated.

$$\text{Annual Charge} = \frac{\text{Remaining Current-Dollar Cost}}{\text{Remaining Years}}$$

A notable feature of this alternative is that it accommodates inflation by requiring comparatively hugh annual charges in the last few remaining years. Because of this characteristic, later user/ratepayers bear most of the burden of restoring the purchasing power of previously-accumulated funds.

- Alternative 3

The current-dollar cost of decommissioning is used for the pro rata charge, and in addition, a supplemental charge is made for the past-year loss of purchasing power of the previously-accumulated funds, if any. This approach has the conceptual advantage of dealing directly and explicitly with the problem of inflation. Because it deals directly and continuously with inflation, use of this method results in the consistent accumulation of the proper pro rata (in time) proportion of current dollar total decommissioning costs. Future changes in the base cost can be accommodated by defining the annual revenue requirement as:

$$\text{Annual Charge} = (\text{Previous year's inflation rate})\text{x}(\text{Accum. Funds}) + \frac{\text{Remaining Current-Dollar Cost}}{\text{Remaining Years}}$$

In a practical regulatory environment, rate hearings are likely to occur less frequently than annually. The above formulation for maintaining purchasing power can be readily adapted to that circumstance by simply using the (geometric) average inflation rate since the last hearing. The above formula for the annual charge in these circumstances then becomes:

$$\text{Annual Charge} = (\text{Avg. Infla. Rate between hearings}) \times (\text{Accum. Funds}) + \frac{\text{Remaining Current-Dollar Cost}}{\text{Remaining Years}}$$

The important feature of Alternative 3 is that it continuously maintains the purchasing power of the accumulated funds, thereby avoiding gross distortions due to inflation. The highly important distinction between Alternatives 2 and 3 is that in Alternative 2, last year's inflation is recovered over the remaining life whereas in Alternative 3, last year's inflation is recovered this year.

An evaluation of the above alternatives necessarily involves a comparison among the alternatives themselves, and also an evaluation against an appropriate criterion of ratepayer equity. The Net Annual Cost was previously defined as the Annual Charge, less the actual or an attributed credit for the use of the accumulated funds. It is postulated that the most equitable time behavior of the Net Annual Cost is one in which it increases at the general rate of inflation. In these circumstances, the electric ratepayer, whose income generally increases with the general rate of inflation would bear the same percentage-of-income burden whether he is an early or late beneficiary of the output of the facility.

For purposes of comparison, the three alternatives described above can be analyzed and then evaluated as to how they compare with each other, and how they compare with the uniformly-escalating cost which is postulated to be the most equitable to both earlier and later ratepayer/beneficiaries of the energy output.

Table 2 (revenues taxed, normalized basis) makes specific numerical comparisons of revenue and net annual costs for an investor-owned utility with a nuclear unit that will be decommissioned following a 30-year life-time. All costs are relative to a time-zero decommissioning cost of 100, which escalates uniformly at 7.7% per annum. The Net Annual Cost is equal to the annual assessment less a credit for the use of accumulated funds at a net rate of 9%.

Table 2 does not make a comparison with the uniformly inflating Net Annual Cost which is postulated to be the most equitable behavior.

TABLE 2

COMPARISON OF DECOMMISSIONING COST RECOVERY ALTERNATIVES

(Relative Units)

Basis: Revenues for future decommissioning are taxable at 48% (normalized basis), escalation is at 7.7% per annum and the Use-of-Funds credit is at 9.0% per annum.

Year	0	1	10	20	30	30*
Current-Dollar Decom Cost	100	107.7	210.0	440.9	925.7	(After Tax)
ALTERNATIVE 1						
o Annual Charge		30.86	30.86	30.86	30.86	-
Cumulative (Net of Taxes)		16.05	160.47	320.94	481.4	925.7
Use of Funds		-	13.00	27.44	43.43	-
Net Annual Cost		30.86	17.86	3.42	-12.57	-
ALTERNATIVE 2						
o Annual Charge		3.59	7.81	23.88	189.01	-
Cumulative (Net of Taxes)		1.87	27.99	105.08	481.4	925.7
Use of Funds		-	2.15	8.34	34.48	-
Net Annual Cost		3.59	5.66	15.54	154.53	-
ALTERNATIVE 3						
Base Charge		3.59	7.00	14.70	30.86	-
Inflation Charge		-	4.50	19.96	63.97	
o Annual Charge		3.59	11.50	34.66	94.83	-
Cumulative (Net of Taxes)		1.87	36.40	152.84	481.4	925.7
Use of Funds		-	2.74	12.13	38.89	-
Net Annual Cost		3.59	8.76	22.53	55.94	-

*Indicates that if decommissioning were to occur in year 30 the correct after-tax amount has been accrued.

This comparison is made in Table 3, and is done by bringing escalated future dollar amounts to constant dollar amounts by dividing by the escalation factor. When this is done, the most equitable behavior of Net Annual Cost is constant in time, and the alternative which is most nearly constant in current dollars is the most equitable of the alternatives.

It is evident from an inspection of Table 2 that Alternative 3 has smaller extremes than the other alternatives. Alternative 2 appears to unduly penalize early energy users and to dramatically benefit later users. Alternative 2 benefits the early and mid-term users, at the expense of dramatically and inequitably penalizing the later users, particularly in the last three or four years, when the charges escalate dramatically to catch up with inflation. An inspection of the Constant Dollar portion of Table 3 shows the relative behavior more clearly. Alternative 3 exhibits significantly more stable time behavior than the other alternatives, being a factor of eight lower than the early years of Alternative 1, and a factor of amost three lower than the last years of Alternative 2. These are truly significant differences.

Comparison of Alternatives 1 and 2 indicates that these two are at either extreme of favoring later users or earlier users. This implies the need for a method which is intermediate, and Alternative 3 is just such a method. This alternative invokes a cost basis which though new, appears soundly based. The key concept is that it is most equitable to make up for any loss in purchasing power on the accumulated funds as quickly as possible so that future ratepayers are not unfairly burdened. The consequence of this assumption is a method which equitably allocates costs between early and late users of the facility, and which deals directly with inflation.

The reason that Alternative 3 is more equitable than the others is that it directly corrects what would otherwise be a significant defect originating from the typical method of accruing these funds. As previously mentioned they typically accrue in an account which is used to reduce the rate base, which in effect gives the ratepayer use of the funds without paying interest. Although this is an equitable approach, what tends to be forgotten (because it is not explicit) is that the purchasing power of the funds being used must be maintained where those funds are to be used to discharge future expenses. The first term in the above Annual Charge equation for Alternative 3 provides for that necessary maintenance of purchasing power.

TABLE 3

COMPARISON OF DECOMMISSIONING COST RECOVERY ALTERNATIVES ON A CONSTANT DOLLAR BASIS

Relative Units: Basis of 100 in Year 0

Year	0	1	10	20	30
Net Annual Cost, Current Dollars (from Table 2)					
Alternative 1	-	30.86	17.86	3.42	-12.57
Alternative 2	-	3.59	5.66	15.54	154.53
Alternative 3	-	3.59	8.76	22.53	55.94
Escalation Factor	1.000	1.077	2.100	4.409	9.257
Net Annual Cost, Constant (Year 0) Dollars					
Alternative 1	-	28.65	8.50	0.78	-1.36
Alternative 2	-	3.33	2.70	3.52	16.69
Alternative 3	-	3.33	4.17	5.11	6.04

In summary, the Alternative 3 cost recovery approach is strongly recommended on the basis of its advantages:

- It provides the most equitable treatment of both earlier and later ratepayers.
- It results in an annual charge that escalates approximately as the rate of inflation.
- It deals directly and explicitly with inflation.
- It does not require any prediction of future inflation, and relies only on current and past costs.
- It is compatible with existing accounting practice and in fact corrects for an inflation-indiced deficiency in typical current approaches.

F. CONCLUSION

The above analyses of rate-regulatory, tax, accounting and cost recovery factors have led to the following overall conclusions in connection with decommissioning cost recovery.

1. The internal use of accumulated decommissioning funds is strongly recommended because it results in the lowest net ratepayer cost of decommissioning.

2. The most equitable decommissioning cost recovery method is based on current costs and on the prompt and continuous maintenance of the purchasing power of accumulated funds.

Finally, it is noted that all of the above discussion of the tax, accounting, rate-regulatory, and cost recovery factors associated with decommissioning apply almost equally to the problems of spent fuel cost recovery. This indicates that the cost recovery approach recommended for decommissioning would have similar advantages if applied to spent fuel cost recovery as well.

THE CALL FOR RESPONSIBLE REGULATION

Vincent S. Boyer

Philadelphia Electric Co.
2301 Market St.
Philadelphia, PA 19101

INTRODUCTION

One of the most pressing problems the United States faces in the remaining years of this century is the balancing of energy supplies against the increased demand anticipated. Energy is needed to prevent further unemployment, to provide new jobs for the young people coming into the labor market, to move forward in creating a higher average standard of living for Americans and to increase the productivity of workers, thereby offsetting other inflationary impacts.

Conservation and increased efficiency of energy utilization will have a moderating effect on the rate of energy growth, but I believe most analysts would agree that we will be using at least twice as much electrical energy in the year 2000 as we presently use. Now the production of electricity has certain environmental effects, and these have been the subject of much debate in recent years. Another complicating factor is our dependence on foreign energy sources, primarily oil, to meet the nation's present requirements and the effect this has on costs, balance of payments and reliability of supply. I might note that we presently import more than one half of our total oil needs and the recent situation in the Middle East highlights the importance of taking actions to move away from this dependence.

With this critical situation facing the nation, we citizens would think that Congress would take action to improve it. But an affirmative action program has not yet gotten underway. To some extent, this is due to the pressures from no growth, protect the environment regardless of cost groups and to the appointment by the present administration to key positions of people not having the background or clarity of vision of engineers. The reorganization of the Dept. of Energy will put another perturbation in the planning cycle.

I would like to review with you some recent legislation enacted by Congress and the effect these laws will have, not only on your health and pocketbook, but in resolving the critical energy situation. I will also present examples of the misuse of current laws and regulations and the unnecessary costs resulting thereby. No one seems to be concerned over the high costs associated with project delays.

THE NATIONAL ENERGY ACT OF 1978

Thus far, the answer of Congress to the energy situation is the National Energy Act of 1978. This Act is composed of five parts which cover the following areas: 1) Energy Conservation, 2) Power Plant and Industrial Fuel Use, 3) Public Utility Regulation, 4) Natural Gas Price Regulation, and 5) Taxation. You will note that there is no clearly identified topic promoting increased supplies of energy, and I will come back to this point after highlighting the content of the Act.

Section 1

The energy conservation section of the National Energy Act requires, and I quote, "utilities to offer energy audits to their residential customers that would identify appropriate energy conservation and solar energy measures and estimate their likely costs and savings" (end of quote). In addition, the government will subsidize the installation of conservation equipment for low income families, the elderly, schools, and hospitals, and solar installations anywhere will be subsidized through reduced interest loans. Other provisions deal with improving the efficiency of industrial equipment, automobiles and home appliances. Solar energy is promoted by a $100 million package for solar demonstration projects and $98 million for purchases of photovoltaic cells. Recent proposals would add to this amount. Congress apparently feels that throwing money into solar energy will advance the technology, but I might point out that it will not necessarily make it economical.

Another section of the Bill placed great emphasis on the assessment of the conservation potential of bicycles. To quote the law "The Congress recognizes that bicycles are THE most efficient means of transportation..." I might add that bicycles may be efficient, but Larry Christiansen, pitcher for the Phillies, would not agree that they are the safest mode of transportation.

How could the use of bicycles to save energy be implemented? Perhaps a law would be passed requiring all persons living within 10 miles of their place of business to ride bicycles to and from work.

Section 2

Section 2, the Power Plant and Industrial Fuel Use Act of the National Energy Act, prohibits the use of oil and natural gas as fuels in new, large boilers and requires the substitution of coal in boilers presently using these fuels. This Act could require Philadelphia Electric to reconvert one of the boilers at our Cromby Station back to coal, but it's not as simple as it sounds. We would also have to install flue gas cleanup equipment and the total conversion and compliance expenditures would amount to some $40 million, an expenditure equal to the original cost of the plant. This investment would have to be amortized over the 10-12 remaining years of the plant's life, and the power requirements of the cleanup equipment would result in a decrease in the net electrical output of the unit. We certainly do not see this as a cost effective procedure and believe the money could be more effectively utilized in advancing the construction of our nuclear units.

Section 3

Section 3, the Public Utility Regulatory Policies Act, meddles in the traditional province of state regulatory agencies and is viewed as increasing the cost of energy without commensurate benefits. State public utility commissions are required to look at utility rates based, not on the traditional cost to serve, but on social concerns. Rates to be considered are "lifeline," time-of-day, seasonal and interruptible together with load management techniques. The funding of intervenors in rate hearings is also proposed.

Utilities would be required to purchase energy from private power suppliers, which we presently do, although few industries are situated where it is economical for them to generate their own power and at times have more than they require. The current vogue of having a windmill in your backyard is hitting some snags, I understand, due to zoning regulations. Would you like one of these monsters twirling away over your children's play area? Recent visit to Block Island and Cuddy Hunk installations.

Section 4

Section 4 covers the area of deregulation of natural gas prices and is the only provision of the National Energy Act which provides an incentive to increase supplies of any fuel -- namely, natural gas. By 1985, or six years from now, all newly discovered gas would be free from regulation. Industry would be subject to the higher costs, but such costs are eventually passed on to the consumer in the form of higher prices of goods and services.

Section 5

Section 5 provides tax credits for people willing to install energy saving materials and devices or solar energy equipment in the home. Tax credits are also available to industries which change from oil and natural gas to other fuels.

Summary

In summary, the National Energy Act is directed toward reducing energy demands rather than increasing energy supplies. How much better it would have been to modify the burdensome regulations associated with our energy resources, particularly coal and uranium which must be counted on to supply the bulk of our energy needs through the year 2000.

The usual mechanism for keeping supplies of fuels ahead of the demand is the marketplace -- the law of supply and demand. However, since 1954 the federal government has regulated the price of natural gas and, since 1973, the price of oil. Thus, two of our least plentiful sources, which supply about 75% of our annual usage of energy, have been kept at artificially low prices. This has resulted in a reduction of the oil and gas companies' efforts to expand domestic supplies. But even with further exploration, we all realize that oil and gas availability is time limited and we must use these fuels with discretion.

There are only two other major sources of energy available for use now and in the near future -- coal and uranium. While coal is in plentiful supply, current federal regulations stemming from the Clean Air Act Amendments are impeding its expanded use as a fuel. Requiring scrubbers on all coal-fired boilers regardless of the cleanliness of the coal being burned does not provide an incentive to burn coal and especially when the scrubber technology isn't as yet perfected. Uranium use in nuclear reactors has also been hindered by burdensome federal regulation; namely, through the prohibition of recycling of nuclear fuel, and inaction on the disposal program for nuclear wastes.

Decreasing the demand and stressing the conservation of energy is vital if we are to prolong the life of our fuel resources until such time as new ones can be developed; that is, the breeder reactor, solar, and fusion. Of these, the breeder reactor is the closest to being able to supply our nation with large amounts of energy, and its development must certainly be listed as a high national priority. While present generation nuclear plants tap about only 1% of the energy available in uranium fuel, breeder reactors, as you know, utilize about 60% and can satisfy our energy needs for hundreds of years. Delaying the demonstration of breeder reactors will result in increased uranium costs, will limit industrial commitments in the nuclear industry while promoting the leadership of foreign nations, and will create significant pressures on other fuel supplies to meet the needs of the electric utilities.

CLEAN AIR ACT

Next, I would like to review the Clean Air Act and its many amendments, the latest of which were recently issued. This Act, which inhibits the use of coal by industry, led to the Environmental Protection Agency's issuance of proposed performance standards for new generating units. Now note, this is for new, not existing, units. These standards require for all coal fired plants the removal of 85% of the sulfur dioxide in the flue gases on a 24-hour averaging basis. A plant burning a low sulfur coal would be required to achieve the same percentage reduction in sulfur dioxide emissions as a plant burning a high sulfur coal.

The utility industry has recommended that a sliding scale be employed, one which relates sulfur dioxide removal requirements to the sulfur content of the fuel being burned. Compliance would be based on a 30-day averaging period rather than the 24-hour averaging period. The financial benefits of this kind of standard are difficult to estimate accurately, but savings estimates from now to 1990 range between $11 billion and $35 billion.

When the new source performance standards were announced, the proposed regulations completely omitted any reference to the health effects of pollutants. Critical examination of many reports throws doubt on the statistical connection between mortality and the concentration of sulfates in the air. Some scientists feel that the standards are based on guesses and that the regulations could be relaxed without adversely affecting public health. For example, in Britain, elimination of coal smoke has essentially solved their air pollution problem and the present relatively high levels of sulfur dioxide are of much less concern to their experts than are the health effects of tobacco.

A recently released report by the Council on Environmental Quality indicated a general improvement in the air quality of major cities in spite of the fact that we have more people, more cars, and more industry. A comparison with prior years indicated a reduction in the number of days the air was considered to be unhealthy. Yet the Council estimated the nation will spend an additional $645 billion to improve the cleanliness of our air, water and land from 1977 through 1986.

Americans want a clean and healthy environment. They also deserve to have the cost of the clean air regulations bear some relationship to the benefits received.

The 1976 Resource and Conservation Act is another law affecting industry. In 1978, the implementation schedule for the Act's regulations pertaining to hazardous and non-hazardous waste was issued by EPA. This Act deals with criteria for transporting, treating, storing and disposing of all forms of solid waste, including the ash and sludge produced from the burning of fuels. We will be following these proposed regulations closely, as if a reasonable approach incorporating a cost effective analysis is not utilized, our industry can be saddled with a major problem -- growing piles of ash with no reasonable means of disposal. This would be costly to resolve and would place both nuclear and coal fired power plants in the same pickle barrel over the handling of wastes.

REGULATIONS ON THE NUCLEAR INDUSTRY

I would like now to turn to some of our problems in the nuclear regulatory area. One of the ramifications of the President's anti-proliferation, no reprocessing program is the question of what to do with the nuclear fuel elements once they have given up their immediately available energy in a power reactor. Of course, our original intent was to store the spent fuel in water-cooled pits adjacent to the reactor for a period of six to nine months and then to ship the fuel in shielded containers to a reprocessing plant. With this option presently not available to us, the on-site spent fuel pit storate capacity has been increased as an interim measure. This is not without cost, however, and involves an expenditure of several million dollars.

Public Service Electric & Gas, who are responsible for the Salem Station, planned a similar change which they were expediting so that it could be completed before the first discharge of spent fuel into the pit scheduled for the spring of this year. Completing the work before spent fuel is present simplifies the operation and reduces the cost by some $600,000 to $800,000 in addition to avoiding modest levels of radiation exposure to the workmen performing the task. The safety aspects of the use of new racks has been extensively studied and has been approved by the NRC for many other facilities, including Peach Bottom.

A group of intervenors opposed to nuclear energy expressed concerns relating to the increased storage of fuel on site. The intervenor's actions, while raising no new safety issues, delayed approval of the spent fuel rack modifications. The incremental dollars associated with the delay were not tremendous as nuclear plant expenditures go, but they would pay the total home energy bills of everyone in this room for a few years. Our regulatory process must be streamlined to prevent the occurrence of delays of this type.

The Seabrook Nuclear Generating Station in New Hampshire is another case which dramatically illustrates the serious failure of our present governmental practices and procedures to resolve pertinent issues in a timely and coordinated manner. As such, the history to date, is worth reviewing. Most of you present are familiar with parts of this story, and I am sure our registrant from Public Service of New Hampshire could tell it better than I, but the entire story dramatically demonstrates a point, and thus, I am capsulizing it for you.

The Seabrook site is on the New England coastline some 50 miles north of Boston and the plant consists of two 1150 Mw pressurized water reactors. Part of the site was previously being used as the town dump and is near a large area of salt marshes which are considered to play an important part in the ecology of marine life. The plant is designed to leave this important area undisturbed by protecting it with special buffer zones.

One of the major contentions voiced by opponents to nuclear power plants is the effect of the condenser cooling water system on the body of water on which the plant is located. The entrainment of plankton, larvae and fish in the intake water and the effect of the warmer discharge water have all received attention and considerable verbage. As a result, the applicant for a nuclear power plant is required to make extensive studies to adequately prove the acceptability of the design of the proposed cooling system. It is much easier, I might note, to claim the ecology might be harmed than it is to conclusively prove that it will not be adversely affected.

The design of the Seabrook cooling system, after considerable study and review, consisted of two tunnels from the plant out into the ocean. One tunnel is for the intake water to the condensers of the plant's main turbines and the other is for the discharge of the same water, though slightly higher in temperature, back to the ocean. This cooling system was the major source of contention.

Public Service of New Hampshire took the first step toward receiving the necessary approvals and permits for Seabrook in February 1972 when it applied to the State of New Hampshire Site Evaluation Committee and to the Public Utility Commission for site approval. While the State approval was forthcoming in two years, all of the necessary federal approvals were not received until July 1976 when construction began. During this time, several special environmental impact studies were conducted by Battelle Columbus Laboratories for the State, by the EPA, and by the Atomic Energy Commission. All studies eventually indicated the site to be acceptable although this fact seems to have been lost in the tangled regulatory morass which beset the project. The construction permit, granted in July 1976, was suspended in January 1977, reinstated in July 1977, suspended again in July 1978, and reinstated in August 1978. This off-again, on-again approval history, together with the extended period to obtain the permits, has necessitated a delay in the completion date of Unit 1 from 1979 to late 1982.

How did these delays and interruptions come about? The answer lies in the extended and redundant reviews conducted by the various agencies and in the ability of intervenors, demonstrators and lawyers to slow the regulatory review process. Virtually every issue raised during the State reviews was rehashed during the NRC hearings. In many instances, the same groups of people were involved -- people whose motive "concern for the environment" is open to question.

The basic cooling system design was approved by both the State of New Hampshire and by the Region I EPA administrator by mid-1975. In the fall of 1975, however, the EPA required the intake tunnels to be extended 4000 feet further off-shore, more than twice their original length. The intervenors were not satisfied, however, and appealed the ruling. Subsequently, in November 1976, affer construction had begun, the EPA administrator reversed himself and revoked his previous determinations. As a result, the construction permit was lifted. The intervenor's lawyers then flew into the fray requesting that the NRC reopen the subjects of seismic criteria, need for power, fuel cycle, and use of cooling towers. Meanwhile, the EPA was reviewing the once-through cooling system and in June 1977, the EPA head administrator in Washington approved the system as previously planned. This was the third time the EPA had approved this system. Construction resumed, but only for a year. Once again, construction ceased through a court ruling which stated that the intervenors had not had sufficient opportunity to cross examine the experts which the EPA administrator had consulted. Three weeks later, after meeting the Court's requirement, the decision was reaffirmed and construction was resumed in August 1978. Appeals still are lurking in the courts, however, and the end of this Laurel and Hardy episode may not yet have been reached.

What is the cost of all these delays? The increased plant costs are estimated at $500 million and together with the replacement energy charges associated with the two years' delay will result in an estimated 11% higher electric bill for the utility's customers. The average residential customer will pay an additional $1440 for his electricity over the life of the plant. For all customers, this will mean an approximate $2 billion increase in costs expressed in current 1978 dollars.

SUMMARY: RESPONSIBLE REGULATION

The Seabrook situation highlights the need for regulatory reform. The importance of obtaining approval for a site well in advance of its immediate need is apparent. I am not opposed to the public's right to discuss issues in siting a plant, but once reviewed, the matter must be resolved on a timely basis and not reopened for any minor perturbation of the issue. The duplication of reviews by a multiplicity of agencies must be eliminated and a firm timetable established for the resolution of issues. To an engineer this is self-evident, but to a lawyer whose interest lies in promoting the views of his client, this does not sit well.

It is estimated that our total regulatory bill reached $100 billion last year, equivalent to $2000 per year for a wage earner heading a family of four people. We must encourage the public to support our demand for responsible regulation which will eliminate the high costs of delay yet permit the examination of differing views. Together, we must assist our nation in its fight against inflation and in resolving the energy crisis, and at the same time, we must work to improve the quality of regulation.

INDEX